Natural Computing Series

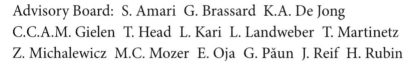

Patrick Siarry · Zbigniew Michalewicz (Eds.)

Advances in Metaheuristics for Hard Optimization

With 167 Figures and 82 Tables

 Springer

Editors

Patrick Siarry
University of Paris 12
Laboratory LiSSi
61 Avenue du Général de Gaulle
94010 Créteil, France
siarry@univ-paris12.fr

Zbigniew Michalewicz
School of Computer Science
University of Adelaide
Adelaide, SA 5005, Australia
zbyszek@cs.adelaide.edu.au

Series Editors

G. Rozenberg (Managing Editor)
rozenber@liacs.nl

Th. Bäck, J.N. Kok, H.P. Spaink
Leiden Center for Natural Computing
Leiden University
Niels Bohrweg 1
2333 CA Leiden, The Netherlands

A.E. Eiben
Vrije Universiteit Amsterdam
The Netherlands

Library of Congress Control Number: 2007929485

ACM Computing Classification (1998): F.2, G.1, I.2, J.6

ISSN 1619-7127
ISBN 978-3-540-72959-4 Springer Berlin Heidelberg New York

Springer is a part of Springer Science+Business Media

springer.com

© Springer-Verlag Berlin Heidelberg 2008

Cover Design: KünkelLopka, Werbeagentur, Heidelberg
Typesetting and Production: LE-TEX Jelonek, Schmidt & Vöckler GbR, Leipzig

Printed on acid-free paper   45/3180/YL   5 4 3 2 1 0

Preface

The community of researchers claiming the relevance of their work to the field of metaheuristics is growing faster and faster, despite the fact that the term itself has not been precisely defined. Numerous books have been published specializing in any one of the most widely known methods, namely, simulated annealing, tabu search, evolutionary algorithms, ant colony algorithms, particle swarm optimization, but attempts to bring metaheuristics closer together are scarce. Yet some common features clearly appear in most metaheuristics, such as the use of diversification to force the exploration of regions of the search space, rarely visited until now, and the use of intensification, to investigate thoroughly some promising regions. Another common feature is the use of memory to archive the best solutions encountered. One common shortcoming of most metaheuristics is the delicate tuning of numerous parameters; the theoretical results available are not sufficient to help the user facing a new, difficult optimization problem.

The goal of this book is to collect state-of-the-art contributions that discuss recent developments in a particular metaheuristic or highlight some general ideas that proved effective in adapting a metaheuristic to a specific problem. Some chapters are overview-oriented while others describe recent advances in one method or its adaptation to a real-world application. The book consists of 21 chapters covering topics from various areas of metaheuristics, including simulated annealing (2 chapters), tabu search (2 chapters), ant colony algorithms (3 chapters), general purpose studies on evolutionary algorithms (5 chapters), applications of evolutionary algorithms (5 chapters), and miscellaneous metaheuristics (4 chapters).

The first chapter on simulated annealing, by Chandra Sekhar Pedamallu and Linet Özdamar, is devoted to a comparison of a simulated annealing (SA) algorithm, an interval partitioning (IP) algorithm, and a hybrid algorithm integrating SA into IP. All three methods, developed for solving the continuous constrained optimization problem, are equipped with a local solver that helps to identify feasible stationary points. The performances are tested on a suite of 32 benchmark problems collected from different sources in the literature.

In the second chapter, Horacio Martínez-Alfaro applies simulated annealing to linkage synthesis of a four-bar mechanism for a given number of desired path points.

Several examples are shown to demonstrate that a path can be better specified, since the user is able to provide more prescribed points than the usual limited number of five allowed by the classical methods.

The following chapter by Ricardo P. Beausoleil deals with nonlinear multiobjective optimization. The chapter introduces a new version of the Multiobjective Scatter Search (MOSS) algorithm, applying a multi-start tabu search and convex combination methods as a diversification generation method. A constraint-handling mechanism is incorporated to deal with constrained problems. The performance of this approach is tested through 16 test problems.

Muhammad A. Tahir and James E. Smith then propose a new approach to improving the performance of the nearest neighbor (NN) classifier. The technique combines multiple NN classifiers, where each classifier uses a different distance function and potentially a different set of features, determined through a combination of tabu search and simple local neighborhood search. Comparison with different well-known classifiers is performed using benchmark data sets available in the literature.

The following chapter by Adem Kalinli and Fatih Sarikoc presents a new parallel ant colony optimization algorithm aimed at solving continuous-type engineering problems. Its performance is evaluated by means of a set of classical test problems, and then it is successfully applied to the training of recurrent neural networks to identify linear and nonlinear dynamic plants.

Alberto V. Donati, Vince Darley, and Bala Ramachandran describe the integration of an ant-based algorithm with a greedy algorithm for optimizing the scheduling of a multi-stage plant in the consumer packaged goods industry. The scheduling must provide both optimal and flexible solutions to respond to fluctuations in demand. "Phase transitions" can be identified in a multidimensional space, where it is possible to vary the number of resources available.

The following chapter by Sung-Soo Kim, Alice E. Smith, and Soon-Jung Hong presents an ant colony approach to optimally load balance code division multiple access micro-cellular mobile communication systems. Load balancing is achieved by assigning each micro-cell to a sector. The cost function considers handoff cost and blocked calls cost, while the sectorization must meet a minimum level of compactness. The problem is formulated as a routing problem where the route of a single ant creates a sector of micro-cells. There is an ant for each sector in the system, multiple ants comprise a colony and multiple colonies operate to find the sectorization with the lowest cost. The method is shown to be effective and highly reliable, and is computationally practical, even for large problems.

Gusz Eiben and Martijn Schut discuss new ways of calibrating evolutionary algorithms, through a suitable control of their parameters on-the-fly. They first review the main options available in the literature and present some statistics on the most popular ones. They then provide three case studies indicating the high potential of uncommon variants. In particular, they recommend focusing on parameters regulating selection and population size, rather than those concerning crossover and mutation.

The chapter by Marc Schoenauer, Pierre Savéant, and Vincent Vidal describes a new sequential hybridization strategy, called "Divide-and-Evolve", that evolutionarily builds a sequential slicing of the problem at hand into several, hopefully easier, sub-problems. The embedded (meta-)heuristic is only asked to solve the "small" problems. Divide-and-Evolve is thus able to globally solve problems that are intractable when fed directly into the heuristic. A prominent advantage of this approach is that it opens up an avenue for multi-objective optimization, even when using a single-objective embedded algorithm.

In their chapter, Carlos García-Martínez and Manuel Lozano are interested in local search based on a genetic algorithm (GA). They propose a binary-coded GA that applies a crowding replacement method in order to keep, within the population, different niches with high-quality solutions. Local search can then be performed by orientating the search in the nearest niches to a solution of interest. The local GA designed consistently outperformed several local search procedures from the literature.

The chapter by Francisco B. Pereira, Jorge M.C. Marques, Tiago Leitão, and Jorge Tavares presents a study on locality in hybrid evolutionary cluster optimization. Since a cluster is defined as an aggregate of between a few and many millions of atoms or molecules, the problem is to find the arrangement of the particles that corresponds to the lowest energy. The authors argue that locality is an important requisite of evolutionary computation to ensure the efficient search for a globally optimal solution.

In the following chapter, Pankaj Kumar, Ankur Gupta, Rajshekhar, Valadi K. Jayaraman, and Bhaskar D. Kulkarni present a genetic algorithm-based learning methodology for classification of benchmark time series problems in medical diagnosis and process fault detection. The results indicate that the constrained window warping method with genetically learned multiple bands could be reliably employed for a variety of classification and clustering problems.

The chapter by Jong-Hwan Kim, Chi-Ho Lee, Kang-Hee Lee, and Naveen S. Kuppuswamy focuses on evolving the personality of an artificial creature by using its computer-coded genomes and evolutionary GA in a simulated environment. The artificial creature, named Rity, is developed in a 3D virtual world to observe the outcome of its reactions, according to its genome (personality) obtained through evolution.

In their chapter, Antonin Ponsich, Catherine Azzaro-Pantel, Serge Domenech, and Luc Pibouleau provide some guidelines for GA implementation in mixed integer nonlinear programming problems. The support for the work is the optimal batch plant design. This study deals with the two main issues for a GA, i.e., the processing of continuous variables by specific encoding and the handling of constraints.

Márcia Marcondes Altimari Samed and Mauro Antonio da Silva Sa Ravagnani present approaches based on GAs to solve the economic dispatch problem. To eliminate the cost of the preliminary tuning of parameters, they have performed a co-evolutionary hybrid GA whose parameters are adjusted in the course of the optimization.

Efrén Mezura-Montes, Edgar A. Portilla-Flores, Carlos A. Coello Coello, Jaime Alvarez-Gallegos, and Carlos A. Cruz-Villar then describe an evolutionary approach to solving a novel mechatronic multiobjective optimization problem, namely that of pinion-rack continuously variable transmission. Both the mechanical structure and the controller performance are concurrently optimized in order to produce mechanical, electronic, and control flexibility for the designed system.

In the last chapter devoted to evolutionary algorithms, Hélcio Vieira Junior implements a GA for an aeronautic military application. The goal is to determine an optimal sequence of flare launch such that the survival probability of an aircraft against a missile is maximized.

The four remaining chapters deal with miscellaneous metaheuristics. First, Jörn Grahl, Stefan Minner, and Peter A.N. Bosman present a condensed overview of the theory and application of the estimation of distribution algorithms (EDAs) in both the discrete and the continuous problem domain. What differentiates EDAs from other evolutionary and non-evolutionary optimizers is that they replace fixed variation operators like crossover and mutation with adaptive variation that comes from a probabilistic model. EDAs have been successfully applied to many problems that are notoriously hard for standard genetic algorithms.

In the following chapter, Zhenyu Yang, Jingsong He, and Xin Yao propose Neighborhood Search (NS) to be embedded with Differential Evolution (DE). The advantages of NS strategy in DE are analyzed theoretically. These analyses focus mainly on the change in search step size and population diversity after using neighborhood search. Experimental results have shown that DE with neighborhood search has significant advantages over other existing algorithms in a broad range of different benchmark functions. The scalability of the new algorithm is also evaluated in a number of benchmark problems, whose dimensions range from 50 to 200.

Sébastien Aupetit, Nicolas Monmarché, and Mohamed Slimane are interested in the training of Hidden Markov Models (HMMs) using population-based metaheuristics. They highlight the use of three methods (GA, ant colony algorithm, and particle swarm optimization) with and without a local optimizer. The study is first performed from a theoretical point of view; the results of experiments on different sets of artificial and real data are then discussed.

The last chapter of the book by Fred Glover deals with inequalities and target objectives that were recently introduced to guide the search in adaptive memory and evolutionary metaheuristics for mixed integer programming. These guidance approaches are useful in intensification and diversification strategies related to fixing subsets of variables at particular values, and in strategies that use linear programming to generate trial solutions whose variables are induced to receive integer values. The author shows how to improve such approaches in the case of 0-1 mixed integer programming.

We do hope you will find the volume interesting and thought provoking. Enjoy!

Adelaide, Australia   *Zbigniew Michalewicz*
Paris, France   *Patrick Siarry*
March 2007

List of Contents

List of Contributors

Jaime Alvarez-Gallegos
Cinvestav-Ipn
Departamento de Ingenieria Eléctrica
México D.F. 07360, Mexico
jalvarez@cinvestav.mx

Sébastien Aupetit
Université de Tours
Laboratoire d'Informatique
37200 Tours, France
sebastien.aupetit@univ-tours.fr

Catherine Azzaro-Pantel
Laboratoire de Génie Chimique de
Toulouse
31106 Toulouse, France
Catherine.Azzaro@ensiacet.fr

Ricardo P. Beausoleil
Instituto de Cibernetica Matematica y
Fisica
Departmento de Optimizacion
Ciudad Habana, Cuba
rbeausol@icmf.inf.cu

Peter A.N. Bosman
Centrum voor Wiskunde en Informatica
1090 GB Amsterdam, The Netherlands
Peter.Bosman@cwi.nl

Carlos A. Coello Coello
Cinvestav-Ipn
Departamento de Computación
México D.F. 07360, Mexico
ccoello@cs.cinvestav.mx

Carlos A. Cruz-Villar
Cinvestav-Ipn
Departamento de Ingenieria Eléctrica
México D.F. 07360, Mexico
cacruz@cinvestav.mx

Vince Darley
Eurobios UK Ltd.
London EC4A 4AB, UK
vince.darley@eurobios.com

Serge Domenech
Laboratoire de Génie Chimique de
Toulouse
31106 Toulouse, France
Serge.Domenech@ensiacet.fr

Alberto V. Donati
Joint Research Center
European Commission
21020 Ispra (VA), Italy
alberto.donati@jrc.it

Gusz Eiben
Department of Computer Science
Vrije Universiteit
NL-1081 HV Amsterdam
The Netherlands
ae.eiben@few.vu.nl

Carlos García-Martínez
Dept. of Computing and Numerical
Analysis
University of Córdoba
14071 Córdoba, Spain
cgarcia@uco.es

Fred Glover
University of Colorado
Boulder, Campus Box 419
Colorado 80309, USA
Fred.Glover@Colorado.EDU

Jörn Grahl
University of Mannheim
Department of Logistics
68131 Mannheim
Germany
joern.grahl@bwl.uni-mannheim.de

Ankur Gupta
Summer Trainee
Chemical Engineering Department
IIT Kharagpur, India
Ankur.Gupta@iitkgp.ac.in

Jingsong He
Nature Inspired Computation and
Applications Laboratory
University of Science and Technology
of China
Hefei, Anhui, China
hjss@ustc.edu.cn

Soon-Jung Hong
SCM Research and Development Team
Korea Integrated Freight Terminal Co.,
Ltd.
Seoul, 100-101, Korea
sjhong75@kift.kumho.co.kr

Valadi K. Jayaraman
Chemical Engineering Division
National Chemical Laboratory
Pune-411008, India
vk.jayaraman@ncl.res.in

Adem Kalinli
Erciyes Universitesi
Kayseri Meslek Yuksekokulu
38039, Kayseri, Turkey
kalinlia@erciyes.edu.tr

Jong-Hwan Kim
Department of Electrical Engineering
and Computer Science
Korea Advanced Institute of Science and
Technology
Daejeon, Korea
johkim@rit.kaist.ac.kr

Sung-Soo Kim
Systems Optimization Lab.
Dept. of Industrial Engineering
Kangwon National University
Chunchon, 200-701, Korea
kimss@kangwon.ac.kr

Bhaskar D. Kulkarni
Chemical Engineering Division
National Chemical Laboratory
Pune-411008, India
bd.kulkarni@ncl.res.in

Pankaj Kumar
Summer Trainee
Chemical Engineering Department
IIT Kharagpur, India

Naveen S. Kuppuswamy
Department of Electrical Engineering
and Computer Science
Korea Advanced Institute of Science and
Technology
Daejeon, Korea
naveen@rit.kaist.ac.kr

Chi-Ho Lee
Department of Electrical Engineering
and Computer Science
Korea Advanced Institute of Science and
Technology
Daejeon, Korea
chiho@rit.kaist.ac.kr

Kang-Hee Lee
Telecommunication R&D Center
Telecommunication Network Business
Samsung Electronics Co., Ltd.
Suwon-si, Gyeonggi-do, Korea
kanghee76.lee@samsung.com

Tiago Leitão
Centro de Informatica e Sistemas
Universidade de Coimbra
3030 Coimbra, Portugal
tleitao@dei.uc.pt

Manuel Lozano
Dept. of Computer Science and
Artificial Intelligence
University of Granada
18071 Granada, Spain
lozano@decsai.ugr.es

Jorge M.C. Marques
Departamento de Quimica
Universidade de Coimbra
3004-535 Coimbra, Portugal
qtmarque@ci.uc.pt

Horacio Martínez-Alfaro
Center for Intelligent Systems
Tecnologico de Monterrey
Monterrey
N.L. 64849, Mexico
hma@itesm.mx

Efrén Mezura-Montes
Laboratorio Nacional de Informática
Avanzada
Rébsamen 80, Xalapa
Veracruz, 91000 Mexico
emezura@lania.mx

Stefan Minner
University of Mannheim
Department of Logistics
68131 Mannheim
Germany
minner@bwl.uni-mannheim.de

Nicolas Monmarché
Université de Tours
Laboratoire d'Informatique
37200 Tours, France
monmarche@univ-tours.fr

Linet Özdamar
Izmir University of Economics
Sakarya Cad. No.156
35330 Balcova
Izmir, Turkey
linetozdamar@lycos.com

Chandra Sekhar Pedamallu
University of Szeged
Institute of Informatics
Szeged, Hungary
pcs_murali@lycos.com

Francisco B. Pereira
Instituto Superior de Engenharia de
Coimbra
3030 Coimbra, Portugal
xico@dei.uc.pt

Luc Pibouleau
Laboratoire de Génie Chimique de
Toulouse
31106 Toulouse, France
Luc.Pibouleau@ensiacet.fr

Antonin Ponsich
Laboratoire de Génie Chimique de
Toulouse
31106 Toulouse, France
Antonin.Ponsich@ensiacet.fr

Edgar A. Portilla-Flores
Autonomous University of Tlaxcala
Engineering and Technology Department
Apizaco Tlax, 90300 Mexico
eportilla@ingenieria.uatx.mx

Rajshekhar
Summer Trainee
Chemical Engineering Department
IIT Kharagpur, India

Bala Ramachandran
IBM T.J. Watson Research Center
Yorktown Heights
10598 NY, US
rbala@us.ibm.com

Márcia Marcondes Altimari Samed
Maringá State University
Computer Science Department
Maringá, Paraná, Brazil
samed@din.uem.br

Mauro Antonio da Silva Sa Ravagnani
Maringá State University
Department of Chemical Engineering
Maringá, Paraná, Brazil
ravag@deq.uem.br

Fatih Sarikoc
Department of Computer Engineering
Erciyes University
Institute of Science and Technology
38039 Kayseri, Turkey
fsarikoc@yahoo.com

Pierre Savéant
Thales Research and Technology
Palaiseau, France
pierre.saveant@thalesgroup.com

Marc Schoenauer
INRIA Futurs
Parc Orsay Université
91893 Orsay, France
Marc.Schoenauer@inria.fr

Martijn C. Schut
Department of Computer Science
Vrije Universiteit
NL-1081 HV Amsterdam
The Netherlands
mc.schut@few.vu.nl

Mohamed Slimane
Université de Tours
Laboratoire d'Informatique
37200 Tours, France
slimane@univ-tours.fr

Alice E. Smith
Industrial and Systems Engineering
Auburn University
AL 36849-5346, USA
smithae@auburn.edu

James E. Smith
School of Computer Science
University of the West of England
United Kingdom
James.Smith@uwe.ac.uk

Muhammad A. Tahir
School of Computer Science
University of the West of England
United Kingdom
Muhammad.Tahir@uwe.ac.uk

Jorge Tavares
Centro de Informatica e Sistemas
Universidade de Coimbra
3030 Coimbra, Portugal
jast@dei.uc.pt

Vincent Vidal
CRIL, IUT de Lens
62307 Lens, France
vidal@cril.univ-artois.fr

Hélcio Vieira Junior
Instituto Technológico de Aeronáutical
São José dos Campos – SP, Brazil
junior_hv@yahoo.com.br

Zhenyu Yang
Nature Inspired Computation and
Applications Laboratory
University of Science and Technology
of China
Hefei, Anhui, China
zhyuyang@mail.ustc.edu.cn

Xin Yao
School of Computer Science
University of Birmingham
United Kingdom
x.yao@cs.bham.ac.uk

Comparison of Simulated Annealing, Interval Partitioning and Hybrid Algorithms in Constrained Global Optimization

Chandra Sekhar Pedamallu[1] and Linet Özdamar[2]

[1] Nanyang Technological University, School of Mechanical and Aerospace Engineering, Singapore.
On overseas attachment to:
University of Szeged, Institute of Informatics, Szeged, Hungary.
pcs.murali@gmail.com

[2] Izmir University of Economics, Sakarya Cad. No.156, 35330, Balçova, Izmir, Turkey.
Fax : 90(232)279 26 26
linetozdamar@lycos.com, lozdamar@gmail.com

Abstract

The continuous Constrained Optimization Problem (COP) often occurs in industrial applications. In this study, we compare three novel algorithms developed for solving the COP. The first approach consists of an Interval Partitioning Algorithm (IPA) that is exhaustive in covering the whole feasible space. IPA has the capability of discarding sub-spaces that are sub-optimal and/or infeasible, similar to available Branch and Bound techniques. The difference of IPA lies in its use of Interval Arithmetic rather than conventional bounding techniques described in the literature. The second approach tested here is the novel dual-sequence Simulated Annealing (SA) algorithm that eliminates the use of penalties for constraint handling. Here, we also introduce a hybrid algorithm that integrates SA in IPA (IPA-SA) and compare its performance with stand-alone SA and IPA algorithms. All three methods have a local COP solver, Feasible Sequential Quadratic Programming (FSQP) incorporated so as to identify feasible stationary points. The performances of these three methods are tested on a suite of COP benchmarks and the results are discussed.

Key words: Constrained Global Optimization, Interval Partitioning Algorithms, Simulated Annealing, Hybrid Algorithms

1 Introduction

Many important real world problems can be expressed in terms of a set of nonlinear constraints that restrict the real domain over which a given performance criterion is optimized, that is, as a Constrained Optimization Problem (COP). The COP is expressed as: *minimize $f(\boldsymbol{x})$*: $\boldsymbol{x} = (x_1, \ldots, x_n)^t \in \xi \subset \mathbb{R}^n$ where ξ is the feasible domain. ξ is defined by k inequality constraints $(g_i(\boldsymbol{x}) \le 0, i = 1, \ldots, k)$, $(m - k)$ equality constraints $(h_i(\boldsymbol{x}) = 0, \ i = k + 1, \ldots, m)$ and domain lower and upper bounds $(LB_x \le \boldsymbol{x} \le UB_x)$. The expressions $g(\boldsymbol{x})$ and $h(\boldsymbol{x})$ may involve nonlinear and linear relations. The objective function, $f(\boldsymbol{x})$, is minimized by an optimum solution vector $\boldsymbol{x}^* = (x_1, \ldots, x_n)^t \in \xi \subset \mathbb{R}^n$ where $f(\boldsymbol{x}^*) \le f(\boldsymbol{x})$ for all $\boldsymbol{x} \in \xi$. In the general COP with a non-convex objective function (and a general non-convex feasible domain), discovering the location of the global optimum is NP-hard. Derivative based solution approaches developed for solving the COP might often be trapped in infeasible and/or sub-optimal sub-spaces if the combined topology of the constraints is too rugged.

Existing global optimization algorithms designed to solve the COP can be categorized as deterministic and stochastic methods. Surveys on global optimization are abundant in the literature (see the recent one [34]). Examples of deterministic approaches are: Lipschitzian methods [37]; branch and bound methods [1]; reformulation techniques [41]; interior point methods [12]; Branch and Reduce (BARON [39] and interval methods [15], [19]. An extensive list of up to date references of stochastic approaches for the COP is maintained by Coello Coello (http://www.cs.cinvestav.mx/~constraint/) where evolutionary approaches, genetic algorithms, ant colony approaches, simulated annealing and many other techniques are cited for continuous and discrete problems.

Here we focus on the Interval Partitioning Algorithms (IPA) and Simulated Annealing (SA) to solve the COP. IPA is a branch and bound technique that uses inclusion functions. Similar to the branch and bound technique, IPA is complete and reliable in the sense that it explores the whole feasible domain and discards sub-spaces in the feasible domain only if they are guaranteed to exclude feasible solutions and/or local stationary points better than the ones already found. On the other hand, SA is a black box stochastic algorithm that generates a sequence of random solutions converging to a global optimum. SA employs a slow annealing process that accepts worse solutions more easily in the beginning stages of the search as compared to later phases [22]. Using this feature, SA escapes from local optima and overcomes the difficulties encountered by derivative based numerical methods. A convergence proof for SA in the real domain is provided in [8]. We now discuss the available literature on SA and IPA designed for the COP.

As Hedar and Fukushima [17] also mention in their report, publications concerning the implementation of SA in the COP are rather scarce. Some successful special case SA applications for constrained engineering problems exist in the literature (e.g., structural optimization problems [2], [25], SA combined with genetic algorithms in economic dispatch [47], in power generator scheduling [48], [49], [50], in thermoelastic scaling behavior [51]). There has also been theoretical work conducted related to the use of SA in the general COP. For instance, Wah and Wang [44], [45] introduce a new penalty method where penalty parameters are also perturbed by SA. Hedar and Fukushima [17] apply multi-start SA from solutions that are preferably pareto optimal according to the infeasibility and optimality criteria. The latter technique aims at achieving a better exploration in both feasible and infeasible regions. Similar to constrained SA applications, interval research on the COP is also relatively scarce when compared with bound constrained optimization. Hansen and Sengupta [16] discuss the inequality COP whereas Ratschek and Rokne [38] describe interval techniques for the COP with both inequality and equality constraints. Numerical results using these techniques are published later in Wolfe [46] and Kearfott [20]. Dallwig et al. [6] propose the software (GLOPT) for solving bound constrained optimization and the COP where a new reduction technique is proposed. More recently, Kearfott [21] presents a software named GlobSol and Markot [26] develops an IPA for the COP with inequalities where new adaptive multi-section rules and a new box selection criterion are presented [27]. Here, we propose a deterministic IPA algorithm having an adaptive tree search management approach that coordinates calls to a local solver, Feasible Sequential Quadratic Programming (FSQP – [52], [24]). In IPA, FSQP is activated within the confinement of each sub-space stored in the list of boxes to be explored. The second approach proposed is the dual-sequence SA, DSA, where the sequences of infeasible and feasible solutions are traced separately. In each SA iteration, a feasible candidate neighbor is compared with the last feasible solution obtained in the feasible sequence, and similarly an infeasible one is compared with the last infeasible solution. Thus, two sequences are constructed in parallel, and, the problems (e.g., the magnitude of penalty parameters) encountered by penalty methods are avoided. This approach is different from other approaches. For instance in Hedar and Fukushima's algorithm [17], each new sequence started from a non-dominated solution is a single sequence. The diversification scheme implemented in DSA is also much simpler and requires minimal memory space. DSA also incorporates FSQP as a local solver, but it has its own invoking policy. In the third approach proposed here, we create a hybrid IPA-DSA algorithm by integrating DSA into IPA where DSA works within the confinement of the specific sub-domain to be explored. This time, FSQP is invoked by DSA. This hybrid approach targets the total coverage of the search domain while enabling a global search in the sub-domains explored. Further, the total search space is reduced by IPA's reliable elimination of infeasible and sub-optimal sub-spaces. All three approaches are compared using a test suite of COP benchmarks. In the next sections we provide the basics of Interval Arithmetic, and brief descriptions of IPA, DSA and IPA-DSA.

2 Basics of Interval Arithmetic

Denote the real numbers by x, y, . . . , the set of compact intervals by $\mathbb{I} := \{[a, b] \mid a \leq b;\ a, b \in \mathbb{R}\}$ and the set of n dimensional intervals (also called simply intervals or boxes) by $\mathbb{I}^n$. Italic letters will be used for intervals. Every interval $x \in \mathbb{I}$ is denoted by $[\underline{x}, \overline{x}]$, where its bounds are defined by $\underline{x} = \inf x$ and $\overline{x} = \sup x$. For every $a \in \mathbb{R}$, the interval point $[a, a]$ is also denoted by a. The width of an interval x is the real number $w(x) = \overline{x} - \underline{x}$. Given two real intervals x and y, x is said to be tighter than y if $w(x) < w(y)$.

Given $(x_1, \ldots, x_n) \in \mathbb{I}$, the corresponding box x is the Cartesian product of intervals, $x = x_1 \times, \ldots, \times x_n$, where $x \in \mathbb{I}^n$. A subset of x, $y \subseteq x$, is a sub-box of x. The notion of width is defined as follows:

$$w(x_1 \times \ldots \times x_n) = \max_{1 \leq i \leq n} w(x_i) \quad \text{and} \quad w(x_i) = \overline{x}_i - \underline{x}_i \qquad (1)$$

Interval Arithmetic operations are set theoretic extensions of the corresponding real operations. Given $x, y \in \mathbb{I}$, and an operation $\lozenge \in \{+, -, \times, \div\}$, we have: $x \lozenge y = \{x \lozenge y | x \in x, y \in y\}$.

Due to properties of monotonicity, these operations can be implemented by real computations over the bounds of intervals. Given two intervals $x = [a, b]$ and $y = [c, d]$, we have:

$$[a, b] + [c, d] = [a + c, b + d],$$
$$[a, b] - [c, d] = [a - d, b - c],$$
$$[a, b] \times [c, d] = [\min\{ac, ad, bc, bd\}, \max\{ac, ad, bc, bd\}],$$
$$[a, b] \div [c, d] = [a, b] \times [1/d, 1/c] \quad \text{if} \quad 0 \notin [c, d].$$

The associative law and the commutative law are preserved over these operations, however, the distributive law does not hold. In general, only a weaker law is verified, called subdistributivity.

Interval arithmetic is particularly appropriate to represent outer approximations of real quantities. The range of a real function f over an interval x is denoted by $f(x)$, and it can be computed by interval extensions.

Definition 1. (*Interval extension*): *An interval extension of a real function* $f : D_f \subset \mathbb{R}^n \to \mathbb{R}$ *is a function* $F : \mathbb{I}^n \to \mathbb{I}$ *such that* $\forall x \in \mathbb{I}^n, x \in D_f \Rightarrow f(x) = \{f(x) \mid x \in x\} \subseteq F(x)$.

Interval extensions are also called *interval forms* or *inclusion functions*. This definition implies the existence of infinitely many interval extensions of a given real function. In a proper implementation of interval extension based inclusion functions the outward rounding must be made to be able to provide a mathematical strength reliability.

The most common extension is known as the natural extension. Natural extensions are obtained from the expressions of real functions, and are inclusion monotonic (this property follows from the monotonicity of interval operations). Hence, given a real function f, whose natural extension is denoted by F, and two intervals x and y such that $x \subseteq y$, the following holds:

$F(x) \subseteq F(y)$. We denote the lower and upper bounds of the function interval range over a given box y as $\underline{F}(y)$ and $\overline{F}(y)$, respectively.

Here, it is assumed that for the studied COP, the natural interval extensions of f, g and h over x are defined in the real domain. Furthermore, (F and similarly, G and H) is α-convergent over x, that is, for all $y \subseteq x$, $w(F(y)) - w(f(y)) \leq cw(y)^\alpha$ where c and α are positive constants.

An *interval constraint* is built from an interval function and a relation symbol, which is extended to intervals. A constraint being defined by its expression (atomic formula and relation symbol), its variables, and their domains, we will consider that an interval constraint has interval variables (variables that take interval values), and that each associated domain is an interval.

The main feature of interval constraints is that if its solution set is empty, i.e., it has no solution over a given box y, then it follows that the solution set of the COP is also empty and the box y can be reliably discarded. In a similar manner, if the upper bound of the objective function range, $\overline{F}(y)$, over a given box y is less than or equal to the objective function value of a known feasible solution (the Current Lower Bound, CLB), then y can be reliably discarded since it cannot contain a better solution than the CLB.

Below we formally provide the conditions where a given box y can be discarded reliably based on the ranges of interval constraints and the objective function.

In a partitioning algorithm, each box y is assessed for its optimality and feasibility status by calculating the ranges for F, G, and H over the domain of y.

Definition 2. *(Cut-off test based on optimality:) If $\underline{F}(y) < CLB$, then box y is called a <u>sub-optimal</u> box.*

Definition 3. *(Cut-off test based on feasibility:) If $\underline{G}_i(y) > 0$, or $0 \notin H_i(y)$ for any i, then box y is called an infeasible box.*

Definition 4. *If $\underline{F}(y) \leq CLB$, and $\overline{F}(y) > CLB$, then y is called an indeterminate box with regard to optimality. Such a box holds the potential of containing x^* if it is not an infeasible box.*

Definition 5. *If $(\underline{G}_i(y) < 0$, and $\overline{G}_i(y) > 0)$, or $(0 \in H_i(y) \neq 0)$ for some i, and other constraints are consistent over y, then y is called an indeterminate box with regard to feasibility and it holds the potential of containing x^* if it is not a sub-optimal box.*

The IPA described in the following section uses the feasibility and optimality cut-off tests in discarding boxes reliably and sub-divides indeterminate boxes repetitively until either they are discarded or they are small enough (these boxes have a potential of holding x^* and finally the local solver identifies it). However, at certain points in this process, available indeterminate boxes are occasionally (given that some conditions hold) subjected to local search before they reach the tolerance size. The latter is undertaken to speed up convergence.

3 The Interval Partitioning Algorithm (IPA)

Under reasonable assumptions, IPA is a reliable convergent algorithm that sub-divides indeterminate boxes to reduce the uncertainties related to feasibility and optimality by nested partitioning. In terms of subdivision direction selection (the choice of the variables to partition in a given indeterminate box), convergence depends on whether the direction selection rule is balanced [5]. The contraction and the α-convergence properties enable this. Here, the rule that selects the variable with the widest variable domain is utilized (Rule A) for this purpose. The reduction in the uncertainty levels of boxes finally lead to their elimination due to sub-optimality or infeasibility while helping IPA in ranking remaining indeterminate boxes in a better fashion.

A box that becomes feasible after nested partitioning still has uncertainty with regard to optimality unless it is proven that it is sub-optimal. The convergence rate of IPA might be very slow if we require nested partitioning to reduce a box to a point interval that is the global optimum. Hence, we need to use the local search procedure FSQP that might also identify stationary points in larger boxes. FSQP calls are coordinated by a special adaptive tree search procedure that is developed in Pedamallu et al. [36].

The adaptive tree management system maintains a stage-wise branching scheme that is conceptually similar to the iterative deepening approach [23]. The iterative deepening approach explores all nodes generated at a given tree level (stage) before it starts assessing the nodes at the next stage. Exploration of boxes at the same stage can be done in any order, the sweep may start from best-first box or the one on the most right or most left of that stage (depth-first). On the other hand, in the proposed adaptive tree management system, a node (parent box) at the current stage is permitted to grow a sub-tree forming partial succeeding tree levels and nodes in this sub-tree are explored before exhausting the nodes at the current stage. In the proposed IPA, if a feasible solution (CLB) is not identified yet, boxes in the sub-tree are ranked according to descending total constraint infeasibilty, otherwise they are ranked in ascending order of $\underline{F}(y)$. A box is selected among the children of the same parent according to either box selection criterion, and the child box is partitioned again continuing to build the same sub-tree. This sub-tree grows until the Total Area Deleted (TAD) by discarding boxes fails to improve in two consecutive partitioning iterations in this sub-tree. Such failure triggers a call to FSQP for all boxes that have not been previously subjected to local search. The boxes that have undergone local search are placed back in the list of pending boxes and exploration is resumed among the nodes at the current stage. Feasible and improving solutions found by FSQP are stored (that is, if a feasible solution with a better objective function value is found, CLB is updated and the solution is stored).

The above adaptive tree management scheme is achieved by maintaining two lists of boxes, B_s and B_{s+1} that are the lists of boxes to be explored at the current stage s and at the next stage $s + 1$, respectively. Initially, the set of indeterminate or feasible boxes in the pending list B_s consists only of x and B_{s+1} is empty. As child boxes are added to a selected parent box, they are ordered according to the current ranking criterion. Boxes in the sub-tree stemming from the selected parent at the current stage are explored and partitioned until there is no improvement in TAD in two consecutive partitioning iterations. At that point, partitioning of the selected parent box is stopped and all boxes that have not been processed by local search are sent to FSQP module and processed to identify feasible and improving point solutions if FSQP is successful in doing so.[1] From that moment onwards, child boxes generated from any other selected parent in B_s are stored in B_{s+1} irrespective of further calls to FSQP in the current stage. When all boxes in B_s have been assessed (discarded or partitioned), the search moves to the next stage, $s + 1$, starting to explore the boxes stored in B_{s+1}.

The tree continues to grow in this manner taking up the list of boxes of the next stage after the current stage's list of boxes is exhausted. The algorithm stops either when there are no boxes remaining in B_s and B_{s+1} or when there is no improvement in CLB as compared with the previous stage. The proposed IPA algorithm is described below.

IP with adaptive tree management

Step 0. Set tree stage, $s = 1$. Set future stage, $r = 1$. Set non-improvement counter for TAD: $nc = 0$. Set B_s, the list of pending boxes at stage s equal to x, $B_s = \{x\}$, and $B_{s+1} = \varnothing$.

Step 1. If the number of function evaluations or CPU time reaches a given limit, or, if both $B_s = \varnothing$ and $B_{s+1} = \varnothing$, then STOP.

Else, if $B_s = \varnothing$ and $B_{s+1} \neq \varnothing$, then set $s \leftarrow s + 1$, set $r \leftarrow s$, and continue. Pick the first box y in B_s and continue.

1.1 If y is infeasible or suboptimal, discard y, and go to Step 1.

1.2 Else if y is sufficiently small, evaluate m, its mid-point, and if it is a feasible improving solution, update CLB, reset $nc \leftarrow 0$, and store m. Remove y from B_s and go to Step 1.

Step 2. Select variable(s) to partition (sort variables according to descending width and select first v variables whose widths exceed the average width of candidate variables).

Step 3. Partition y into 2^v non-overlapping child boxes. Check TAD, if it improves, then reset $nc \leftarrow 0$, else set $nc \leftarrow nc + 1$.

Step 4. Remove y from B_s, add 2^v boxes to B_r.

4.1. If $nc > 2$, apply FSQP to all (previously unprocessed by FSQP) boxes in B_s and B_{s+1}, reset $nc \leftarrow 0$. If FSQP is called for the first time in stage s, then set $r \leftarrow s + 1$. Go to Step 1.

4.2. Else, go to Step 1.

---

[1] It should be noted that, whether or not FSQP fails to find an improving solution, IPA will continue to partition indeterminate boxes as long as they pass both cutoff tests. Finally, the algorithm encloses potential improving solutions in sufficiently small boxes where FSQP can identify them.

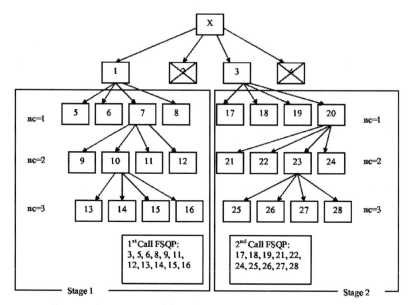

Fig. 1 Implementation of the adaptive iterative deepening procedure

The adaptive tree management system in IP is illustrated in Fig. 1 on a small tree where node labels indicate the order of nodes visited.

4 The Dual-sequence SA Algorithm: DSA

In contrast to the deterministic exhaustive IPA, DSA is a stochastic global search algorithm that does not necessarily cover the whole search space within a finite number of iterations. DSA is shown to be superior to single sequence SA applications where constraint handling is carried out via penalty functions [35]. DSA also utilizes FSQP to identify stationary points nearby good solutions. Before we describe DSA algorithm, we provide notation related to infeasibility degrees of constraints.

We denote the degree of infeasibility of constraint i at a given solution vector x by $INF_i(x)$ and define it below for infeasible equality and inequality constraints.

$$INF_i(x) = \begin{cases} 0 & \text{if} \quad h_i(x) = 0 \quad \text{or} \quad g_i(x) \le 0; \\ |h_i(x)| & \text{or} \quad g_i(x) \quad \text{otherwise.} \end{cases} \tag{2}$$

Total infeasibility degree of a solution x is given as: $TIF(x) = \Sigma_{i=1}^{m} INF_i(x)$.

In Fig. 2 we provide the pseudocode of DSA and the relevant notation. DSA starts with a random initial sample, so called "seed", within the hypercube defined by $[LB_x, UB_x]$. The iteration counter of DSA is denoted as q. In each iteration, a candidate solution, x^q, that is a neighbor to x^{q-1} is generated by perturbing the value of a selected coordinate of x^{q-1} within the coordinate lower and upper bounds. This is

Notation and Functions used in DSA:

q: iteration counter

q_max: maximum allowed number of iterations for DSA

Gq: global counter for consecutive number of non-improving feasible solutions

Gq_max: maximum allowable value of Gq before new sequence is started

Lq: local counter for consecutive number of non-improving feasible or infeasible solutions

Lq_max: maximum allowable value of Lq before accepting a worse solution probabilistically

x^q: solution vector obtained by perturbing x^{q-1}.

temp: annealing temperature.

t_f: minimum temperature allowed.

feas_obj^q: *f(x)* of the last feasible solution encountered until iteration q.

f^*: best feasible $f(x)$ found until iteration q.

Perturb(x^q): Function that selects a coordinate j of the solution vector x^{q-1} and increases or decreases x_j randomly in the interval $[LB_j, UB_j]$.

Paccept(x^q): Function that calculates acceptance probability for x^q. (if a randomly generated float number in [0,1] is less than acceptance probability, it reports that x^q is acceptable.)

Re-anneal: Set *temp* to one.

Procedure DSA

Generate a starting solution for a new sequence;

Do{

   If $(Gq > Gq\_max)$ {   /* *Diversification I* */

    Generate a starting solution for a new sequence;

    x^{q-1}←new seed;

    If (x^{q-1} is feasible) *feas_obj*$^{q-1}$ ← $f(x^{q-1})$; TIF^{q-1} ← ∞;

    Else TIF^{q-1} ← $TIF(x^{q-1})$; *feas_obj*$^{q-1}$ ← ∞;

    Reset Gq, Lq ←0; *Re-anneal*;

    }

   x^q ← *Perturb*(x^{q-1});

   If (x^q is feasible){

    Set Fq←Fq+1;

    If $(Fq > Fq\_max)$ Reset Fq ←0; $TIF^{\,q}$← ∞;   /* *Diversification II* */

    Else $TIF^{\,q}$← $TIF^{\,q-1}$;   /* *Update inf. sequence information* */

    feas_obj^q ← $f(x^q)$;   /* *Update feas. sequence information* */

    If $(f(x^q) < f^*)$ Update f^*; Reset Gq ← 0;   /* *continue feasible sequence* */

    Else Gq ← Gq +1;   /* *increase counter for Diversification I* */

    If $(f(x^q) <$ *feas_obj*$^{q-1}$) Invoke probabilistic FSQP call;

    } /**endif**/

    Else {

      $TIF^{\,q}$←$TIF(x^q)$;   /* *Update inf. sequence information* */

      feas_obj^q ← *feas_obj*$^{q-1}$;   /* *Update feas. sequence information* */

      }

    If $(TIF^{\,q} < TIF^{\,q-1})$ OR (*feas_obj*$^q <$ *feas_obj*$^{q-1}$) Reset Lq ← 0; /**better sol.**/

    Else {   /**worse sol.**/

    Lq ← Lq+1;   /**increase intensification counter**/

    Invoke *Paccept*(x^q);   /**calculate prob. of acceptance**/

    If $(Lq > Lq\_max)$ AND (x^q is accepted) { /**accept. of worse sol. enabled**/

    Reset Lq ← 0;   /**reset intensification counter**/

    Reduce *temp*;

    If $(temp < t_f)$ *Re-anneal*;

    If $(q ≈ q\_max)$ Invoke FSQP call;   /**FSQP call near worse solution**/

    } /**endif**/

    Else x^q ← x^{q-1};   /**preserve old solution**/

    }/**endelse**/

   q ← q +1;

  } while $(q ≤ q\_max)$;

Fig. 2 Pseudocode of DSA

achieved by procedure Perturb. In DSA, every solution is classified so as to belong to the feasible or infeasible sequence.

In each iteration, the information regarding both sequences is updated according to the neighbor's feasibility status. If x^q is feasible, then, $feas\_obj^q$ becomes equal to $f(x^q)$ and TIF^q takes its old value TIF^{q-1} since there is no change in the infeasible sequence. Otherwise, if x^q is infeasible, TIF^q is set equal to $TIF(x^q)$ and $feas\_obj^q$ takes its old value. These are indicated in Fig. 2 by *update sequence information* comments.

Diversification in DSA: In DSA, we introduce two diversification schemes. These are indicated as comments in Fig. 2. The first scheme (Diversification I) is controlled by a counter, G_q that stops a sequence of solutions once it stagnates. Stagnation occurs when the best feasible solution found so far, f^*, does not improve during a long sequence of moves. Thus, G_q records the number of consecutive non-improving feasible solutions and it is reset to zero whenever a feasible objective value lower than f^*, is encountered. When G_q exceeds the limit G_{q_max}, $(G_{q\_max} = 0.1 \cdot q\_max)$, the procedure generates a new seed, which is not a neighbor to the previous solution, x^{q-1}. In other words, a new sequence of solutions is started, replaces x^{q-1}, and all parameters are re-initialized. The temperature *temp* is re-annealed, G_q is set to 0. According to the feasibility status of the new seed, $feas\_obj^{q-1}$ and TIF^{q-1} are re-set.

The second diversification scheme (Diversification II) in DSA is designed for avoiding traps set by recurrent feasible solutions. We define a counter F_q that counts the number of feasible solutions obtained so far. When $F_q > F_{q\_max}$, we re-set TIF^q to a large number so that new and worse infeasible solutions are accepted and the search moves away from the vicinity of a trapping feasible solution. F_q is reset to zero after it exceeds F_{q_max} $(F_{q\_max} = 0.01 \cdot q\_max)$.

Intensification in DSA: In DSA, we do not immediately allow probabilistic acceptance of a worse solution. A number of consecutive hill-climbing iterations are applied before enabling its acceptance. We carry out this intensification process by a control counter, L_q that records the consecutive number of feasible or infeasible non-improving solutions. Each non-improving solution is directly rejected and not given any probabilistic chance of acceptance until L_q reaches its limit, L_{q_max}. Here, we set $L_{q\_max} = n$, that is, we set it to the number of dimensions in the COP, because we would like to have a chance to find a better neighbor in each coordinate before we may accept a worse solution. This approach has an intensification effect on the search.

When L_q exceeds L_{q_max}, it is re-set to zero, and DSA is permitted to accept a worse candidate solution x_q with the annealing probability defined in Equation (3). After a worse solution is accepted, the next cycle of hill-climbing starts. This intensification approach reduces the number of moves that stray from a good sequence.

$$\text{Prob(accept)} = \begin{cases} \exp\left(-\dfrac{f(x^q) - feas\_obj^{q-1}}{f(x^q)temp}\right) & \text{if } x^q \text{ is feasible} \\ \exp\left(-\dfrac{TIF(x^q) - TIF^{q-1}}{TIF(x^q)temp}\right) & \text{otherwise} \end{cases} \quad (3)$$

According to the equation above, if x^q turns out to be feasible, $f(x^q)$ is assessed against $feas\_obj^{q-1}$ and if infeasible, assessment is carried out according to the comparison between $TIF(x^q)$ and TIF^{q-1}. This enables the procedure to compare a feasible candidate solution x^q (generated after an infeasible solution) with the last feasible solution in the feasible sequence, rather than with x^{q-1}. Hence, x^q might not be immediately accepted if it is not better than $feas\_obj^{q-1}$. The latter eliminates trapping in feasible areas. In any case, if a new feasible solution is good enough, its neighborhood is scanned by FSQP. The situation is similar when an infeasible candidate solution is generated. That particular infeasible candidate is accepted or rejected according to the last solution's TIF in the infeasible sequence. Thus, an infeasible solution arriving just after a feasible solution can still be accepted if it improves TIF. In this manner, DSA enhances the exploration power of SA in infeasible regions and reduces wasted function calls in feasible regions.

The cooling temperature $temp$ found in Equation (3) is managed as follows. Since the probability of acceptance is based on normalized deterioration, initially $temp$ is set to 1.0. Whenever a worse solution is accepted, the annealing temperature $temp$ is reduced geometrically ($temp \longleftarrow temp/(1 + \theta)$, here, $\theta = 0.005$). This approach is different from the standard SA algorithm that waits for an epoch length to reduce $temp$. Here, the intensification duration becomes an adaptive epoch length. Ozdamar and Demirhan [33] test this approach against several SA algorithms from the literature and find that this is the most effective method in bound constrained optimization. In order to increase Prob(accept) in later stages of the search and give more freedom to DSA, the temperature $temp$ is re-annealed when it falls below its minimum allowable value, t_f.

DSA stops when the maximum number of moves, q_max, is reached. The best feasible solution, f^*, encountered during the search is reported.

FSQP calls: Throughout the search, DSA interacts with FSQP by invoking a probabilistic call to local search whenever a *better* feasible solution is found. Hence, FSQP contributes to SA by providing exploration ability around the feasible solutions identified by DSA. The relation between FSQP and DSA is such that DSA provides the current feasible solution x^q to FSQP as a starting point. FSQP seeks for a local or global stationary point around x^q and simply updates f^* if it finds a better solution. The sequence of solutions generated by DSA is not affected by invoking FSQP.

In order to reduce the number of FSQP calls, we introduce an annealing type of activation probability that depends on the annealing temperature $temp$ as well as on how much the new feasible solution x^q improves $f(x)$ compared to the last feasible solution, $feas\_obj^{q-1}$. The probability of calling FSQP is calculated as follows.

$$\text{Prob}(\text{call\_FSQP}) = \exp\left(\frac{-1}{(feas\_obj^{q-1} - f(x^q)) \cdot \sqrt{temp}}\right) \tag{4}$$

FSQP is also allowed to explore around feasible or infeasible worse solutions accepted near the end of the search, that is, when the iteration index q exceeds, for instance, $0.9 \cdot q\_max$. The latter FSQP calls do not require the calculation of Prob(call_FSQP).

5 The Hybrid IPA-DSA Algorithm

The hybrid IPA-DSA algorithm is basically the IPA algorithm described in Sect. 3 with the exception that rather than invoking FSQP in Step 4.1, the DSA algorithm is invoked. In the hybrid algorithm DSA works as described in Sect. 4, however, the search space is confined by the boundaries of box y. DSA invokes FSQP using its own probabilistic calling scheme. Our goal in developing this hybrid is to use the advantage of activating global search in larger boxes within an exhaustive algorithm framework. We make the following parameter adjustments in the hybrid algorithm. In each box y that DSA is activated, the number of iterations allowed is $q\_max =$ $500 \cdot \text{sizeof}(y)/\text{sizeof}(x)$, and similarly, the number of iterations the solver FSQP is allowed in a given box y is equal to $150 \cdot \text{sizeof}(y)/\text{sizeof}(x)$. Thus, these parameters are based on the box size.

6 Numerical Experiments and Comparisons with Other Deterministic and Stochastic Search Methods

6.1 Test Problems

We test the performance of all methods on 32 COPs collected from different sources in the literature. These are listed in Table 1 with their characteristics (number of non-linear/linear equalities and inequalities as well as expression types), optimal or best results reported, and their source references. A few test problems are tested in their revised versions. Most of the problems involve polynomial constraints and objective functions with nine exceptions where trigonometric expressions are found (problems P3, P22, P23, P24, P25, P26, P27, P30, P31). Some problems have a single variable in their objective function (P4, P6, P7, P8, P10, P11, P12, P17), and perturbations made by DSA in such problems do not affect the objective function directly. This might lead to some difficulties in finding the direction of descent.

6.2 Comparison with Single Sequence Penalty-based SA Algorithms

DSA and the hybrid IPA-DSA are compared with five SA algorithms that handle constraint feasibility by using an objective function augmented with constraint infeasibility degrees. All these algorithms are single sequence techniques, that is, solutions are not differentiated according to feasibility status. Similar to DSA, they all call FSQP in the same manner to identify feasible stationary solutions.

We call penalty based SA algorithms SAP. In SAP, the augmented objective $f'(x)$ of the COP includes a penalty function, $p(x)$ that is non-decreasing in $TIF(x)$. $f'(x)$ is defined as

$$\min f'(x) = \begin{cases} f(x) + p(x) & \text{if} \quad TIF(x) > 0 \\ f(x) & \text{otherwise} \end{cases} \tag{5}$$

Table 1 List of Test problems

Second column: N: dimension, NE: Nonlinear equations, LE: linear equations, NI: Nonlinear
inequalities, LI: Linear inequalities
*: indicates that result is obtained in this chapter

Prob. Id.	N,# NE, # LE,# NI, # LI	Notes	$f(x^*)$	Source
P1	13,0,0,0,9	Quadratic $f(x)$	-15	[29] (G1)
P2	8,0,0,3,3	Linear $f(x)$, 2nd degree NI	3514.81*	[29] (G10)
P3	20,0,0,1,1	Trigonometric $f(x)$, 20th degree NI	0.8036	[29] (G2)
P4	21,0,0,1,10	Linear single variable $f(x)$, 2nd degree concave NI	-8695.012	[11] (Chapter 2, Test problem 7)
P5	11,0,0,1,5	Linear single variable $f(x)$, 2nd degree concave NI	-39	[11] (Chapter 2, Test problem 6)
P6	6,0,0,7,0	Linear single variable $f(x)$, 2nd degree NI	-30665.53	[11] (Chapter 2, Test problem 2)
P7	3,0,0,1,0	Linear single variable $f(x)$, 2nd degree NI	-3	[10] (e_1.def)
P8	5,0,0,1,6	Linear single variable $f(x)$, 2nd degree NI	-13	[43] (Example 1)
P9	2,0,0,2,2	Linear single variable $f(x)$, 2nd degree NI, convex and non-convex	-2.8284	[43] (Example 4)
P10	3,0,0,2,1	Linear single variable $f(x)$, 2nd degree NI	0.741	[42] (Example 1)
P11	3,0,0,2,0	Linear single variable $f(x)$, 2nd degree NI	-0.5	[42] (Example 2)
P12	8,0,0,3,3	Linear single variable $f(x)$, 2nd degree NI	100	[14]
P13	5,0,0,6,0	Nonlinear $f(x)$, 2nd degree NI	-30665.53	[14]
P14	5,0,0,6,0	Nonlinear $f(x)$, 2nd degree NI	-31535.37	[7]
P15	5,0,0,6,0	Nonlinear $f(x)$, 2nd degree NI	-31022	[32]
P16	10,0,1,0,0	2nd degree $f(x)$ with many , interactive terms	-0.375	[11] (Problem 2.10)
P17	21,0,0,1,10	Linear single variable $f(x)$, 2nd degree NI	-4150.410	Revised P4
P18	7,0,0,12,2	Nonlinear $f(x)$, Geometric fractional NI, 2nd degree NI	1227.183	[7]
P19	13,0,0,12,1	Linear $f(x)$, Geometric fractional NI, 2nd degree NI	97.59	[7]
P20	14,6,1,0,0	2nd degree $f(x)$, 2nd degree NE	-1.765	[4] (Alkyl.gms)
P21	10,0,7,0,0	2nd degree $f(x)$	0.814*	[4] (Genhs28.gms)
P22	14,0,0,2,0	Quadratic $f(x)$, trigonometric NI	0.00*	[4] (revised robot.gms)
P23	6,0,0,4,3	2nd $f(x)$, trigonometric NI	-2355.39*	[4] (revised mathopt3.gms)
P24	14,2,0,0,0	Quadratic $f(x)$, trigonometric NE	0.00	[4] (robot.gms)
P25	6,4,3,0,0	Quadratic $f(x)$, trigonometric NE	-1071.6	[4] (mathopt3.gms)
P26	11,0,0,4,2	Linear $f(x)$, trigonometric NI	0*	[4] (revised hs087.gms)
P27	9,0,0,8,2	Linear $f(x)$, trigonometric NI, 2nd degree NI	5319.58*	[4] (revised hs109.gms)
P28	17,0,0,7,4	Linear $f(x)$, 4th degree NI, 3rd degree NI, 2nd degree NI	-1000*	[4] (revised rk23.gms)
P29	4,0,0,0,6	2nd $f(x)$	-333.333*	[4] (revised hs044.gms)
P30	4,3,0,0,0	2nd $f(x)$, trigonometric NE	5126.49	[40] (G5)
P31	2,0,0,2,0	Trigonometric $f(x)$, 2nd degree NI	-0.1	[40] (G8)
P32	10,0,0,5,3	2nd $f(x)$, 2nd degree NI	24.306	[40] (G7)

In SAP, the augmented $f'(x)$ is used to assess solutions rather than $f(x)$. The annealing probability of acceptance is expressed below.

$$\text{Prob(accept)} = \exp\left(\frac{-(f'(x^q) - f'(x^{q-1}))}{f'(x^q)temp}\right) \tag{6}$$

SAP contains the intensification scheme in DSA and the first diversification scheme. The second diversification scheme in DSA is omitted since the feasible solution sequence is not traced separately in SAP. In other words, all features of SAP and DSA are the same with this exception and this performance comparison only involves the maintenance of a single sequence or a dual sequence.

We now list the types of penalty functions $p(x)$ considered in this comparison.

Penalty functions. We utilize five different expressions for $p(x)$. These are adapted from the GA literature. The first two penalty functions are static in the sense that they are not affected by the status of the search, i.e., on the number of function evaluations already spent until the current solution is obtained. The third one is dynamic, and the last two are classified as annealing penalty functions. These are listed below.

1. MQ [30]. MQ is a static penalty function that depends only on the number of infeasible constraints rather than on $TIF(x)$. The penalty function is expressed below.

$$p(x) = B - \frac{Bs}{m} \tag{7}$$

Here, s is the number of feasible constraints (equalities plus inequalities), m is the total number of constraints, and B is a large positive number (10^9 is the suggested value by the developers of the function) that guarantees that an infeasible solution has a larger objective function value than a feasible solution.

2. QP (Quadratic Penalty function). QP is the static classical quadratic penalty function expressed below.

$$p(x) = B \sum_{i=1}^{m} INF_i^2(x) \tag{8}$$

Again, B is a sufficiently big positive number (e.g., 10^6) such that a solution having a larger $TIF(x)$ is guaranteed to have a worse objective function value than another with a smaller $TIF(x)$.

3. JH [18]. JH is a dynamic penalty function that depends both on the progress of the search and on $TIF(x)$. The idea in JH is to allow exploration in earlier stages of the search and become strict in later stages. Here, for representing the progress of the search we take the number of SAP iterations, q, made until the current solution is reached. The penalty function is adapted as follows:

$$p(x) = (qC)^\alpha \sum_{i=1}^{m} INF_i^\beta(x) \tag{9}$$

The value of the parameter C is suggested as 0.5, and for α, β, the values of 1 or 2 are suggested. Here, we prefer the former value because q becomes large in later stages of the search.

The advantage of JH and the following two annealing penalties over static ones is that the infeasibility (and its related $p(x^q)$) of a candidate solution, x^q, can be traded off with the reduction in $f(x^q)$. That is, $(f(x^q) + p(x^q)) < f(x^{q-1})$ where x^{q-1} is the feasible predecessor solution. This increases the exploration capability of SAP.

4. *MA* [28]. MA is an annealing penalty function where the quadratic penalty weight is divided by the annealing temperature, *temp*, that is reduced as the search runtime increases. Similar to JH, the penalty becomes quite high near the end of the search, and infeasible solutions tend to be rejected. The function $p(x)$ is adapted as follows.

$$p(x) = \frac{0.5 \sum_{i=1}^{m} INF_i^2(x)}{temp} \tag{10}$$

Here, we let *temp* take the value of the geometric cooling temperature that the SAP algorithm uses in that iteration. Hence, the rejection criterion of non-improving solutions in SAP is aligned with $p(x)$.

5. *CSBA* [3]. CSBA is an annealing penalty function of multiplicative type where $f(x)$ is multiplied by an exponential function whose arguments are $TIF(x)$ and *temp*. As *temp* goes to zero, the penalty function goes to one leading to increased $f(x)$. The penalty function is adapted as follows.

$$p(x) = f(x)e^{\frac{-\sqrt{temp}}{TIF(x)}} \tag{11}$$

6.3 Comparison with IPA

The IPA described here utilizes an adaptive tree management. For the purpose of the comparison, here, we also include two more IPA versions: IPA with best-first tree management scheme and IPA with depth-first scheme. These two versions are similar to IPA-adaptive scheme. However, in the best-first approach, the one with least $TIF(y)$ among indeterminate boxes is selected for re-partitioning until a feasible solution is identified, and, then, once a feasible solution is identified, the box with the lowest $\underline{F}(y)$ is selected. On the other hand, in the depth-first approach, the box on the left branch of the last partitioned box is always selected for re-partitioning. Calls to FSQP are managed in the same way (according to the improvement in TAD) in all three IPA versions.

Furthermore, we also provide the hybrid IPA-DSA algorithm in three versions: with adaptive, best-first and depth-first tree management schemes.

6.4 Comparison with Other Deterministic Approaches

To complete the comparison portfolio, we also solve the test problems with five well-known solvers that are linked to the commercial software GAMS (www.gams.com) and also with stand-alone FSQP [52], [24] whose code has been provided by AEM (www.aemdesign.com/FSQPmanyobj.htm). The solvers used in this comparison are BARON 7.0 [39], Conopt 3.0 [9], MINOS 5.5 [31], Snopt 5.3.4 [13] and FSQP. Among these solvers, BARON is an exhaustive global solver and all others are local solvers. We allow every solver to complete its run without imposing additional stopping criteria except the maximum CPU time.

6.5 Results

In the experiments, SAP, DSA, and the hybrid IPA-DSA are re-run 100 times for every test problem using a 1.7 GHz PC with 256 MB RAM in Windows operating system. In SAP and DSA, every problem is run with a maximum iteration counter of $q\_max = 2000(n + m)$ or maximum CPU time of 900 CPUseconds. In IPA-DSA hybrid, the number of function evaluations, excluding those made by DSA, is limited to $1000(n + m)$ and DSA is allowed a number of function calls that depends on the relative size of the box as described in Sect. 5. One run of IPA-DSA takes a longer time than stand-alone DSA and therefore, the maximum number of function evaluations is halved. Similarly, the stand-alone IPA methods are allowed $1000(n + m)$ function evaluations. All methods in the comparison have a limited run time of 900 CPUseconds.

In Tables 2 and 3, we illustrate the results obtained by the hybrid stochastic and stochastic methods. In Table 2, we summarize the results obtained by the three tree management schemes in the hybrid IPA-DSA method. In Table 3, we provide the results obtained by stand-alone DSA and five SAP methods.

We provide the results in the following format. For all SA methods and hybrid IPA-DSA, we provide the average absolute deviations from the global optima (and standard deviations) of three results: the average deviation of the worst solution obtained in 100 runs (1st results column in Tables 2 and 3) over 32 problems, the average deviation of the best solution in 100 runs (2nd column in Tables 2 and 3) over

Table 2 Hybrid IPA-DSA results

Hybrid IPA-DSA	Abs. Dev. (Worst)	Abs. Dev. (Best)	Abs. Dev. (Average)	Ratio of Unsolved Probs. Over 100 runs	CPU Secs
IPA-DSA **(Adaptive Tree)**					
Average	327.40	96.73	144.58	0.000	38.37
Std. Dev.	800.53	327.87	356.42	0.000	136.98
# of Optimal Solutions	17	23			
# of Unsolved Probs.	0	0			
IPA-DSA **(Best-first Tree)**					
Average	237.990	95.80	152.44	0.001	44.55
Std. Dev.	653.53	328.02	372.54	0.004	160.23
# of optimal Solutions	15	21			
# of Unsolved Probs.	1	0			
IPA-DSA **(Depth-first Tree)**					
Average	256.17	100.50	137.41	0.014	12.48
Std. Dev.	634.01	327.27	305.58	0.073	45.53
# of Optimal Solutions	16	20			
# of Unsolved Probs.	2	0			

Table 3 Hybrid IPA-DSA results

SA Methods	Abs. Dev. (Worst)	Abs.Dev. (Best)	Abs. Dev. (Average)	Ratio of Unsolved Probs. Over 100 runs	CPU (Secs)
without Penalty					
DSA					
Average	218.62	109.54	136.29	0.004	1.842
Std. Dev.	571.53	332.49	357.25	0.02	2.325
# of Optimal Solutions	15	21			
# of Unsolved Probs.	1	0			
with Penalty (SAP)					
Morales-Quezada					
Average	543.87	111.82	298.11	0.02	1.55
Std. Dev.	1279.56	330.91	666.60	0.08	2.74
# of Optimal Solutions	14	20			
# of Unsolved Probs.	4	0			
Static Quadratic					
Average	340.39	107.46	156.85	0.01	1.98
Std. Dev.	730.49	331.39	362.95	0.03	4.19
# of Optimal Solutions	13	22			
# of Unsolved Probs.	4	0			
Joines and Houck					
Average	385.82	107.26	157.45	0.01	1.73
Std. Dev.	841.84	331.44	366.32	0.05	3.55
# of Optimal Solutions	13	20			
# of Unsolved Probs.	4	0			
Michalewicz and Attia					
Average	622.50	116.22	334.69	0.01	0.75
Std. Dev.	1381.33	334.51	657.46	0.04	0.81
# of Optimal Solutions	9	20			
# of Unsolved Probs.	6	0			
Carlson et al.					
Average	647.40	116.22	334.69	0.03	0.54
Std. Dev.	1403.84	334.51	657.46	0.10	0.63
# of Optimal Solutions	8	20			
# of Unsolved Probs.	7	0			

32 problems and the average of 100 runs' average deviation (3rd column in Tables 2 and 3) over 32 problems. Finally, the average ratio of unsolved problems in 100 runs where no feasible solution was found (4th column) and the average CPU times per run are reported (5th column). We also report the number of optimal solutions found among worst and best solutions (3rd row of every method), and the number of problems where no feasible solution was found at all in 100 runs, i.e., total failure of the procedure (4th row of every method).

In Table 4, we provide the results for deterministic methods. These are illustrated in terms of the average absolute deviation from the global optimum, standard deviation, number of optimal solutions obtained and number of problems for which no feasible solution was found.

Let us first compare the results in Tables 2 and 3. We observe that the average results obtained by DSA and the hybrid IPA-DSA (adaptive tree) methods are not

Table 4 Hybrid IPA-DSA results

Deterministic Methods	Abs. Dev.	CPU Secs	Deterministic Methods	Abs. Dev.	CPU Secs
FSQP			**MINOS**		
Average	300.75	0.04	Average	393.90	0.27
Std. Dev.	939.19	0.07	Std. Dev.	955.30	0.16
# of Optimal Solutions	16		# of Optimal Solutions	11	
# of Unsolved Probs.	0		# of Unsolved Probs.	3	
Baron (Exhaustive)			**Conopt**		
Average	108.99	0.13	Average	516.90	0.29
Std. Dev.	374.99	0.12	Std. Dev.	1215.03	0.33
# of Optimal Solutions	13		# of Optimal Solutions	9	
# of Unsolved Probs.	0		# of Unsolved Probs.	7	
Snopt					
Average	442.84	0.38			
Std. Dev.	960.95	0.58			
# of Optimal Solutions	10				
# of Unsolved Probs.	5				

			IPA		
IPA **Adaptive Tree**	Abs. Dev.	CPU Secs	IPA **Best-first Tree**	Abs. Dev.	CPU Secs
Average	114.48	39.27	Average	117.70	282.00
Std. Dev.	330.59	158.54	Std. Dev.	342.01	352.53
# of Optimal Solutions	17		# of Optimal Solutions	17	
# of Unsolved Probs.	0		# of Unsolved Probs.	3	
IPA **(Depth-first Tree)**					
Average	380.74	4.73			
Std. Dev.	980.11	10.12			
# of Optimal Solutions	17				
# of Unsolved Probs	2				

significantly different. However, in terms of the number of optimal solutions found and in terms of the criterion of achieving feasible solutions in every run, the hybrid method is better than DSA. If we consider the CPU times taken by both methods, the hybrid method is seen to be computationally very expensive. In general, the best-first tree management approach takes longer CPU times due to its memory and sorting requirements and the depth-first approach is fastest among the hybrids. Yet, in terms of achieving a feasible solution in every run, the depth-first approach is inferior to the other two approaches (a feasible solution is not identified in about 1.4 runs over 100). In terms of the worst solutions obtained, DSA's performance is superior to those of SAP, and the best-first and depth-first approaches in the hybrid method perform as well as DSA in this respect. These results show that using DSA as a stand-alone solver is a reasonable choice for the COP because CPU times are small and average case and worst case quality of solutions are acceptable. On the other hand, if one wishes to improve the best solution found at the expense of longer computation times, the hybrid IPA-DSA method can be used.

We now take a look at Table 4 and compare deterministic methods with SA-based approaches. As expected, the best method among GAMS solvers is the exhaustive approach BARON. The computation time taken by this solver is very small, however, the number of optimal solutions found is much lower than the best solutions found by all SA-based methods. Again, this advantage comes with more computational effort. For DSA, we can obtain best results in 100 runs while BARON carries out only one run. However, one should not forget that BARON would not be able to find a better solution within the allowed CPU time while it is always possible to obtain a better result by running a stochastic method more than once. Using the stand-alone IPA-adaptive tree might not be a good option under these circumstances. Though the number of optimal solutions obtained is higher than BARON and problem solving capability is as good as that of BARON, the computation times are higher.

We can summarize our findings as follows. Considering the deterministic and stochastic COP methods tested in these experiments, SA-based methods seem to be the best option to use in terms of CPU times and solution quality. If a user is willing to afford longer computation times, then he/she can resort to the hybrid IPA-DSA solver to have a higher chance of identifying the optimum solution to the COP at hand. The IPA-DSA adaptive tree approach described here uses its exhaustive partitioning and exploration tools that result in the highest number of optimal solutions and best capability in identifying feasible solutions in every run.

7 Conclusions

We have described different SA-based approaches to solve the COP. Among the SA methods described here, the performance of the novel dual-sequence SA (DSA) and those of SA approaches with different penalty functions (SAP) are illustrated. Furthermore, a new stand-alone exhaustive interval partitioning algorithm (IPA) with an adaptive tree management scheme is described. This method is also tested with different tree management schemes. IPA provides us with an exhaustive exploration framework that guides DSA in conducting the search. By combining IPA with DSA, we obtain a hybrid algorithm (IPA-DSA) that uses the advantage of exploring the whole search space by systematic partitioning while discarding sub-spaces that are guaranteed to exclude feasible solutions or better stationary points reliably. Finally, we have included the well-known deterministic commercial solvers in the experiments. Our empirical results indicate that it is desirable to use DSA to solve the COP if one wishes to obtain a high quality solution within small computation times. However, the solution quality can be even better if the hybrid IPA-DSA is used (in terms of the number of optimal solutions found and identifying feasible solutions). The performance of the exhaustive deterministic method BARON is also quite good and fast, but it lacks the flexibility of finding even better solutions by multiple re-runs.

Acknowledgement. We wish to thank Professor Andre Tits (Electrical Engineering and the Institute for Systems Research, University of Maryland, USA) for providing the source code of CFSQP.

References

1. Al-Khayyal F A, Sherali H D (2000) SIAM Journal on Optimization, 10:1049–1057
2. Bennage W A, Dhingra A K (1995) International Journal of Numerical Methods in Engineering, 38:2753–2773
3. Carlson S, Shonkwiler R, Babar S, Aral M (1998) Annealing a genetic algorithm over constraints. SMC 98 Conference, available from http://www.math.gatech.edu/shenk/body.html
4. Coconut Test Problems on Constrained Optimization Problems – Library2. http://www.mat.univie.ac.at/~neum/glopt/coconut/
5. Csendes T, Ratz D (1997) SIAM Journal of Numerical Analysis 34:922–938
6. Dallwig S, Neumaier A, Schichl H (1997) GLOPT – A Program for Constrained Global Optimization. In: Bomze I M, Csendes T, Horst R, Pardalos P M (eds) Developments in Global Optimization. Kluwer, Dordrecht
7. Dembo R S (1976) Mathematical Programming 10:192–213
8. Dekkers A, Aarts E (1991) Mathematical Programming 50:367–393
9. Drud A S (1996) CONOPT: A System for Large Scale Nonlinear Optimization. Reference Manual for CONOPT Subroutine Library, ARKI Consulting and Development A/S, Bagsvaerd Denmark
10. Epperly T G (1995) Global optimization of nonconvex nonlinear programs using parallel branch and bound. Ph. D dissertation, University of Wisconsin-Madison, Madison
11. Floudas C A, Pardalos P M (1990) A collection of Test Problems for Constrained Global Optimization Algorithms. Volume 455 of Lecture Notes in Computer Science. Springer-Verlag, Berlin Heidelberg New York
12. Forsgren A, Gill P E, Wright M H (2002) SIAM Review 44:525–597
13. Gill P E, Murray W, Saunders M A (1997) SNOPT: An SQP algorithm for large-scale constrained optimization. Numerical Analysis Report 97-2, Department of Mathematics, University of California, San Diego, La Jolla, CA
14. Hansen P, Jaumard B, Lu S-H (1991) Mathematical Programming 52:227–254

15. Hansen E R(1992) Global Optimization Using Interval Analysis. Marcel Dekker, New York
16. Hansen E, Sengupta S (1980) Global constrained optimization using interval analysis. In: Nickel K L (eds), Interval Mathematics
17. Hedar A-R, Fukushima M (2006) Derivative-free filter simulated annealing method for constrained continuous global optimization. Journal of Global Optimization (to appear)
18. Joines J, Houck C (1994) On the use of non-stationary penalty functions to solve non-linear constrained optimization problems with GAs. Proceedings of the First IEEE International Conference on Evolutionary Computation, IEEE Press. 579–584
19. Kearfott R B (1996a) Rigorous Global Search: Continuous Problems. Kluwer, Dordrecht, Netherlands
20. Kearfott R B (1996b) A Review of Techniques in the Verified Solution of Constrained Global Optimization Problems. In: Kearfott R B, Kreinovich V (eds) Applications of Interval Computations. Kluwer, Dordrecht, Netherlands
21. Kearfott R B(2003) An overview of the GlobSol Package for Verified Global Optimization. Talk given for the Department of Computing and Software, McMaster University, Cannada
22. Kirkpatrick A, Gelatt Jr C D, Vechi M P (1983) Science 220:671–680
23. Korf R E (1985) Artificial Intelligence 27:97–109
24. Lawrence C T, Zhou J L, Tits A L (1997) User's Guide for CFSQP version 2.5: A Code for Solving (Large Scale) Constrained Nonlinear (minimax) Optimization Problems, Generating Iterates Satisfying All Inequality Constraints. Institute for Systems Research, University of Maryland, College Park, MD
25. Leite J P B, Topping B H V (1999) Computers and Structures 73:545–564
26. Markot M C (2003) Reliable Global Optimization Methods for Constrained Problems and Their Application for Solving Circle Packing Problems. PhD dissertation, University of Szeged, Hungary
27. Markot M C, Fernandez J, Casado L G, Csendes T (2006) New interval methods for constrained global optimization. Mathematical Programming 106:287–318
28. Michalewicz Z, Attia N (1994) Evolutionary optimization of constrained problems. Proceedings of the Third Annual Conference on Evolutionary Programming, World Scientific
29. Michalewicz Z (1995) Genetic algorithms, numerical optimization, and constraints. Proceedings of the Sixth International Conference on Genetic Algorithms, Morgan Kaufmann
30. Morales K A, Quezada C C (1998) A universal eclectic genetic algorithm for constrained optimization. Proceedings 6th European Congress on Intelligent Techniques and Soft Computing, EUFIT'98
31. Murtagh B A, Saunders M A (1987) MINOS 5.0 User's Guide. Report SOL 83-20, Department of Operations Research, Stanford University, USA
32. Myung H, Kim J-H, Fogel D B (1995) Evolutionary Programming 449–463
33. Özdamar L, Demirhan M (2000) Computers and Operations Research, 27:841–865
34. Pardalos P M, Romeijn H E (2002) Handbook of Global Optimization Volume 2. Springer, Boston Dordrecht London
35. Pedamallu C S, Özdamar L (2006) Investigating a hybrid simulated annealing and local search algorithm for constrained optimization. Inpress EJOR (http://dx.doi.org/10.1016/j.ejor.2006.06.050)
36. Pedamallu C S, Özdamar L, Csendes T (2006) An interval partitioning approach for continuous constrained optimization. Accepted for publication in Models and Algorithms in Global Optimization. Springer, Berlin Heidelberg New York

37. Pinter J D (1997) LGO – A program system for continuous and Lipschitz global optimization. In: Bomze I M, Csendes T, Horst R, Pardalos P M (eds) Developments in Global Optimization. Kluwer Academic Publishers, Dordrecht Boston London
38. Ratschek H, Rokne J (1988) New Computer Methods for Global Optimization. Ellis Horwood, Chichester
39. Sahinidis N V (2003) Global Optimization and Constraint Satisfaction: The Branch-and-Reduce Approach. In: Bliek C, Jermann C, Neumaier A (eds) COCOS 2002, LNCS. Springer, Boston Dordrecht London
40. Schoenauer M, Michalewicz Z (1999) Evolutionary Computation 7:19–44
41. Smith E M B, Pantelides C. C (1999) Computers and Chemical Engineering 23:457–478
42. Swaney R E (1990) Global Solution of algebraic nonlinear programs. AIChE Annual Meeting, Chicago, IL
43. Visweswaran V, Floudas C A (1990) Computers and Chemical Engineering 14:1419–1434
44. Wah B W, Wang T (2000) International Journal on Artificial Intelligence Tools 9:3–25
45. Wah B W, Chen Y X (2000) Optimal anytime constrained simulated annealing for constrained global optimization. In: Dechter R (ed.) LNCS 1894. Springer, Berlin Heidelberg New York
46. Wolfe M A (1994) Journal of Computational and Applied Mathematics 50:605–612
47. Wong K P, Fung C C (1993) IEE Proceedings: Part C 140:509–515
48. Wong K P, Wong Y W (1995) IEE Proceedings: Generation Transmission and Distribution 142:372–380
49. Wong K P, Wong Y W (1996) IEEE Transactions on Power Systems 11:112–118
50. Wong K P, Wong Y W (1997) IEEE Transactions on Power Systems 12:776–784
51. Wong Y C, Leung K S, Wong C K (2000) IEEE Transactions on Systems, Man, and Cybernetics, Part C 30: 506–516
52. Zhou J L, Tits A L (1996) SIAM Journal on Optimization 6:461–487

Four-bar Mechanism Synthesis for *n* Desired Path Points Using Simulated Annealing

Horacio Martínez-Alfaro

Center for Intelligent Systems, Tecnológico de Monterrey, Monterrey, N.L. 64849, México.
+52. 81. 8328.4381
hma@itesm.mx

Abstract

This chapter presents an alternative method of linkage synthesis (finding link lengths and its initial position) using a computational intelligence technique: Simulated Annealing. The technique allows one to define *n* desired path points to be followed by a four-bar linkage (path generation problem). The synthesis problem is transformed into an optimization problem in order to use the Simulated Annealing algorithm. With this approach, a path can be better specified since the user will be able to provide more "samples" than the usual limited number of five allowed by the classical methods. Several examples are shown to demonstrate the advantages of this alternative synthesis technique.

Key words: Mechanism, Synthesis, Simulated Annealing, Optimization

1 Introduction

Linkage synthesis or finding link lengths and its initial position is well known and studied in many text books [7, 21, 26]. Two methods are used: graphical and analytical. However, their main limitation is the number of prescribed points [7, 23]. Some authors [24, 25, 28] have worked successfully on solving the problem when more than four points are prescribed. However, an elimination process has to be undertaken in order to find the satisfied constraint solutions, i.e. when a linkage solution does not follow the prescribed points in the specified order, or when the link lengths are very large or small.

The main disadvantage is the number of prescribed points: there has not been any previous publication in which any of the above methods, graphical and analytical, guarantees what happens between any two prescribed precision points. The larger the number of prescribed precision points, the greater the effort needed to solve the problem. This is what has been the "problem".

An alternative way of linkage synthesis is presented which is capable of dealing with *n* (> 5) prescribed points. The technique requires a desired trajectory in order

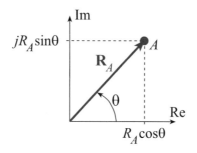

Fig. 1 Location of a point in the complex plane

to obtain "samples" from it. The synthesis problem is then transformed into an optimization problem and solved using the Simulated Annealing (SA) algorithm [19]. Our study is focused on synthesizing a four-bar linkage which consists in finding its link lengths and its initial position with additional constraints such as limiting link lengths, displacement angles, and/or linkage type.

2 Position Analysis of a Four-bar Linkage

The position of a point in a complex plane may be determined by a position vector as shown in Fig. 1 where $\mathbf{R}_A = R_A \cos\theta + jR_A \sin\theta$ with R_A being the magnitude of the vector. A two-dimensional vector has two attributes that can be expressed in either polar or Cartesian coordinates. In Cartesian form, the x (real part) and y (imaginary part) components are specified and in polar form, a magnitude and an angle are

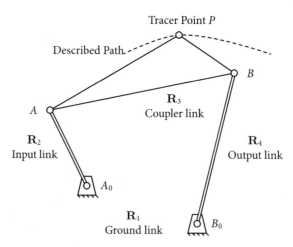

Fig. 2 Four-bar mechanism notation

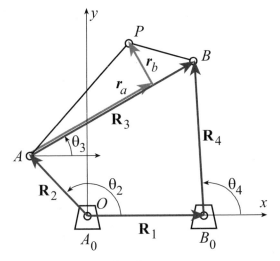

Fig. 3 Close-loop notation

provided. Vectors are commonly used for the analysis of mechanisms. For the analysis of the four-bar mechanism shown in Fig. 2, links can be expressed as position vectors which are arranged in a closed-loop form. The sum of these vectors must be zero in order to have a closed trajectory. For this study, the lengths and angles for each link are defined as shown in Fig. 3. For simplicity, x (real) axis is aligned with the ground link, $\mathbf{R}_1$ (see Fig. 2).

3 Problem formulation

There are two classical methods for four-bar linkage synthesis: graphical and analytical, also called "the complex number" approach [7, 21]. The complex number approach has been the most widely used method for computer-based applications. As mentioned earlier, the effort needed to solve a synthesis problem increases with the number of prescribed points.

The problem statement is to design a four-bar linkage that moves its coupler link such that a point on that link, a tracer point, which in this case is determined by two vectors $(\mathbf{r}_a, \mathbf{r}_b)$, will be at the first point of the desired trajectory, and after the input link has rotated an angle ϕ_1, the tracer point will be at the second point of the trajectory, and so forth until the nth point is reached. Since the problem is transformed into optimization one, an objective or cost function is needed. This objective function involves the tracer point coordinates, which are functions of the link positions and lengths, and the n prescribed points of the desired trajectory.

The technique used to minimize the error between the trajectory described by the four-bar mechanism, and the one given by the user was the Simulated Annealing (SA) algorithm.

Our input data for the vector loop equation are the four link lengths, link angles θ_1 and θ_2, and ground pivot coordinates A_0 or $\mathbf{R}_{A_0}$.

The simplicity of this linkage synthesis approach is based on the fact that the objective function to be minimized is the squared error/difference between the desired trajectory and the one generated by the synthesized four-bar linkage tracer point. In this way we can obtain the following expression:

$$C(\mathbf{P}, \hat{\mathbf{P}}) = \sum_{i=1}^{n} (\mathbf{p}_i - \hat{\mathbf{p}}_i)^2 \tag{1}$$

where $\mathbf{P}$ is an array with $\mathbf{p}_i$ as the location of the ith prescribed point and $\hat{\mathbf{P}}$ an array with $\hat{\mathbf{p}}_i$ as the ith location of the synthesized linkage tracer point.

Our final goal is to reduce the error between the desired trajectory and the actual trajectory described by the synthesized linkage. This can be seen also as a fitting curve problem: *the goal is to fit a curve that is generated by a four-bar linkage of unknown link lengths and initial position.*

Our original synthesis problem is now transformed into a combinatorial optimization problem [16–19]. This type of problem requires a substantial amount of CPU time but with the evolving computer technology, this has become a non-important factor.

Simulated annealing has proven to be a powerful tool in solving large combinatorial optimization problems [1–3, 9–13, 15, 27]. Although the solution obtained with this technique does not guarantee passing exactly through all the prescribed points, the global behavior of the trajectory can be better controlled.

For this case it should be noticed that beside vectors $\mathbf{R}_1$, $\mathbf{R}_2$, $\mathbf{R}_3$, and $\mathbf{R}_4$ and pin points A, B, and B_0, there are two vectors, $\mathbf{r}_a$ and $\mathbf{r}_b$. $\mathbf{r}_a$ has the same direction (angle) as link 3, $\mathbf{R}_3$, and ends where a perpendicular vector $\mathbf{r}_b$ (to the linkage tracer point) starts. The tracer point is located at the tip of the $\mathbf{r}_b$ vector (as shown in Fig. 4).

With direction chosen as shown on Fig. 4, we have the following closed-loop equation:

$$\mathbf{R}_2 + \mathbf{R}_3 - \mathbf{R}_4 - \mathbf{R}_1 = 0 \tag{2}$$

The user determines the number of precision points, n, and then gives their initial coordinates. These coordinates are used to determine if the initial linkage satisfies the Grashof crank-rocker linkage type condition.

$$L_{\max} + L_{\min} \leq L_a + L_b \tag{3}$$

where $L_{\max}$ is the length of the longest link, $L_{\min}$ is the length of the shortest link, L_a and L_b are the lengths of the remaining two links.

Our input data for the vector loop equation are the four link lengths, and angles θ_1 and θ_2. θ_2 is the independent variable and is the one to be varied during the analysis. θ_1 can be considered also as an input since it is modified by the simulated annealing algorithm for each design attempt. Using the closed-loop vector equation in the

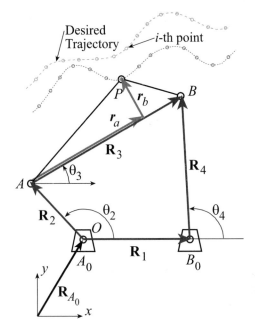

Fig. 4 Reference path and described path of a four-bar linkage

Cartesian form, we can solve for the two unknown variables θ_3 and θ_4 with constant link length inputs, and θ_2 as independent variable:

$$0 = R_2 \left(\cos \theta_2 + j \sin \theta_2 \right) + R_3 \left(\cos \theta_3 + j \sin \theta_3 \right) - \tag{4}$$
$$R_4 \left(\cos \theta_4 + j \sin \theta_4 \right) - R_1 \left(\cos \theta_1 + j \sin \theta_1 \right)$$

This equation is separated into its real part (x component):

$$R_2 \cos \theta_2 + R_3 \cos \theta_3 - R_4 \cos \theta_4 - R_1 \cos \theta_1 = 0 \tag{5}$$

and imaginary part (y component):

$$R_2 \sin \theta_2 + R_3 \sin \theta_3 - R_4 \sin \theta_4 - R_1 \sin \theta_1 = 0 \tag{6}$$

These equations are solved simultaneously [21] for θ_3 and θ_4 to obtain:

$$\theta_3 = 2 \arctan \left(\frac{-B_1 \pm \sqrt{A_1^2 + B_1^2 - C_1^2}}{C_2 - A_1} \right) \tag{7}$$

$$\theta_4 = 2 \arctan \left(\frac{-B_1 \pm \sqrt{A_1^2 + B_1^2 - C_2^2}}{C_1 - A_1} \right) \tag{8}$$

where:

$$A_1 = R_1 \cos \theta_1 - R_2 \cos \theta_2, \qquad B_1 = R_1 \sin \theta_1 - R_2 \sin \theta_2,$$
$$C_1 = R_4^2 + A_1^2 + B_1^2 - R_3^2/2\, R_4, \qquad C_2 = R_4^2 - A_1^2 - B_1^2 - R_3^2/2\, R_3 \tag{9}$$

The next step is to calculate the coordinates of the pin joints between each pair of links in order to place the mechanism in a global coordinate system. Between links R_2 and R_3:

$$A_x = R_{A_0 x} + R_2 \cos \theta_2 \tag{10}$$

$$A_y = R_{A_0 y} + R_2 \sin \theta_3 \tag{11}$$

Between links R_3 and R_4:

$$B_x = R_x + R_2 \cos \theta_2 + R_3 \cos \theta_3 \tag{12}$$

$$B_y = R_y + R_2 \sin \theta_2 + R_3 \sin \theta_3 \tag{13}$$

Between links R_4 and R_1:

$$B_{0x} = B_x - R_4 \cos \theta_4 \tag{14}$$

$$B_{0y} = B_y - R_4 \sin \theta_4 \tag{15}$$

In order to obtain the cost of the proposed solution we obtain the coordinates of the tracer point:

$$t_x = R_x + R_2 \cos \theta_2 + r_a \cos \theta_3 - r_b \sin \theta_3 \tag{16}$$
$$t_y = R_y + R_2 \sin \theta_2 + r_a \sin \theta_3 + r_b \cos \theta_3 \tag{17}$$

4 Simulated Annealing Algorithm

Simulated annealing is basically an iterative improvement strategy augmented by a criterion for occasionally accepting higher cost configurations [15,22]. Given a cost function $C(\mathbf{z})$ (analog to energy) and an initial solution (state) $\mathbf{z}_0$, the iterative improvement approach seeks to improve the current solution by randomly perturbing $\mathbf{z}_0$. The Metropolis algorithm [15] was used for acceptance/rejection of the new state $\mathbf{z}'$ at a given temperature T, i.e.,

- randomly perturb $\mathbf{z}$, the current state, to obtain a neighbor $\mathbf{z}'$, and calculate the corresponding change in cost $\delta C = \mathbf{z}' - \mathbf{z}$
- if $\delta C < 0$, accept the state
- otherwise if $\delta C > 0$, accept the state with probability

$$P(\delta C) = \exp\left(-\delta C/T\right) \tag{18}$$

This represents the *acceptance–rejection loop* of the SA algorithm. The acceptance criterion is implemented by generating a random number, $\rho \in [0, 1]$ and comparing it to $P(\delta C)$; if $\rho < P(\delta C)$, then the new state is accepted. The *outer loop* of the algorithm is referred to as the cooling schedule, and specifies the equation by which the temperature is decreased. The algorithm terminates when the cost function remains approximately unchanged, i.e., for n_{no} consecutive outer loop iterations.

Any implementation of simulated annealing generally requires four components:

1. **Problem configuration** (domain over which the solution will be sought).

2. **Neighborhood definition** (which governs the nature and magnitude of allowable perturbations).

3. **Cost function**.

4. **Cooling schedule** (which controls both the rate of temperature decrement and the number of inner loop iterations).

The domain for our problem is the real plane for link lengths and θ_1. The cost function was described in detail in Sec. 3. The neighborhood function used here is the same as used by Martínez-Alfaro and Flugrad [17], which is modeled as an ϵ ball around each link length and θ_1. To determine a perturbation for any given link length, two random numbers are generated. One of them is used to specify the magnitude of the perturbation and the other one to determine the sign of the perturbation. The allowable perturbations are reduced by the following limiting function:

$$\epsilon = \epsilon_{max} \frac{\log(T - T_f)}{\log(T_0 - T_f)} \tag{19}$$

where ϵ_{max} is an input parameter and specifies the maximum link length perturbation, and T, T_0, T_f are the current, initial and final temperatures, respectively.

The cooling schedule in this chapter is the same *hybrid* one introduced by Martínez-Alfaro and Flugrad [16] in which both the temperature and the inner loop criterion vary continuously through the annealing process [4]. The outer loop behaves nominally as a constant decrement factor,

$$T_{i+1} = \alpha\, T_i \tag{20}$$

where $\alpha = 0.9$ for this research. The temperature throughout the inner loop is allowed to vary proportionally with the current optimal value of the cost function. So, denoting the inner loop index as j, the temperature is modified when a state is accepted, i.e.,

$$T_j = \frac{C_j}{C_{last}} T_{last} \tag{21}$$

where C_{last} and T_{last} are the cost and temperature associated with the last accepted state. Note that at high temperatures, a high percentage of states are accepted, so the temperature can fluctuate by a substantial magnitude within the inner loop.

The following function was used to determine the number of acceptance–rejection loop iterations:

$$N_{in} = N_{dof} \left[2 + 8 \left(1 - \frac{\log(T - T_f)}{\log(T_0 - T_f)} \right) \right] \tag{22}$$

where N_{dof} is the number of degrees of freedom of the system.

The initial temperature must be chosen such that the system has sufficient energy to visit the entire solution space. The system is sufficiently melted if a large percentage, i.e. 80%, of state transitions are accepted. If the initial guess for the temperature yields less than this percentage, T_0 can be scaled linearly and the process repeated. The algorithm will proceed to a reasonable solution when there is excessive energy; it is simply less computationally efficient. Besides the stopping criterion mentioned above, which indicates convergence to a global minimum, the algorithm is also terminated by setting a final temperature given by

$$T_f = \alpha^{N_{out}} T_0 \tag{23}$$

where N_{out} is the number of outer loop iterations and is given as data to our problem. Figure 5 shows a flowchart of the implementation.

5 Results

The following results were obtained by a computer program implemented in C programming language. Figure 6 shows a path with five prescribed precision points and the synthesized four-bar mechanism. It also shows both paths, desired and real, and the mechanism in its first position.

The following example, Fig. 7, shows a desired path with 20 precision points. Due to the characteristics of the prescribed trajectory, the resulting mechanism generates a close trajectory to the prescribed one.

Figure 8 shows the following example which is a path with $n = 32$ precision points. In this example we can observe how a larger number of prescribed points can describe a smoother path and the global path looks acceptable.

Figure 9 shows the following test consisting of $n = 20$ prescribed points of the desired trajectory with a space constraint given by a polygon with vertices at $(0, -2.5)$, $(7.5, 0)$, $(10, 0)$, $(10, 7.5)$, $(2.5, 10)$, $(-2.5, 5)$. The additional constraint helps to obtain a smoother trajectory by limiting link lengths. Figure 9 shows the prescribed points and the generated ones with the resulting mechanism changed.

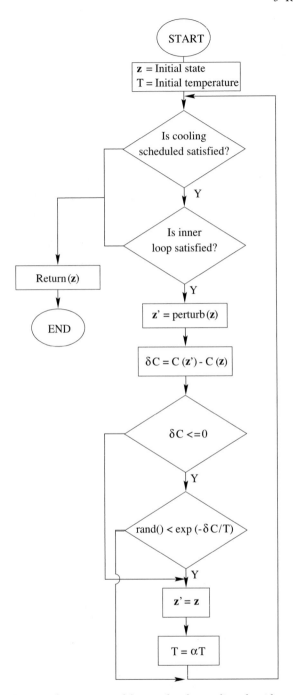

Fig. 5 Implementation of the simulated annealing algorithm

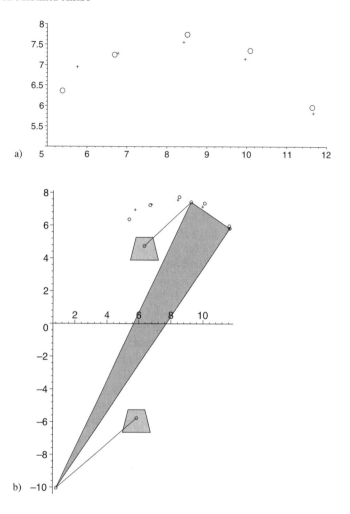

Fig. 6 Desired and synthesized paths for $n = 5$ prescribed points

Many other tests were performed with different trajectories and different seeds for the random number generator and the numerical results were very similar for each set of prescribed points or trajectory. However, the resulting mechanism changed in initial position and link lengths, even when the cost was almost the same (an average difference of only 0.0001 units). Thus, small differences in cost may generate very different mechanisms.

6 Conclusions and Future Research

Four-bar mechanism synthesis is possible when more than five prescribed precision points are provided. The simulated annealing algorithm could be an alternative syn-

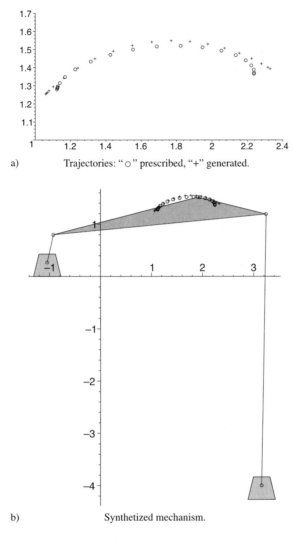

a)   Trajectories: " o " prescribed, "+" generated.

b)   Synthetized mechanism.

Fig. 7 Trajectory with 20 prescribed points

thesis procedure when more than the classical number of points are provided. It is possible to apply the simulated algorithm even when just a few points are provided. Some advantages can be mentioned:

- The synthesis problem is transformed into an optimization one: this transformation is based on a simple four-bar mechanism position analysis.

- More constraints can be included in addition to the available space or link lengths constraints.

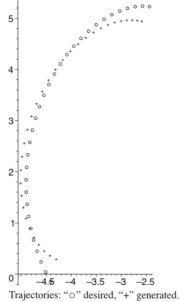

a)   Trajectories: "○" desired, "+" generated.

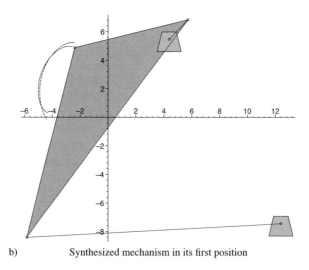

b)   Synthesized mechanism in its first position

Fig. 8 Desired and synthesized paths for $n = 32$ prescribed points

Although the mechanisms obtained with this technique do not pass exactly through all specified points, we have good control of the global trajectory of the mechanism.

The simulated annealing algorithm can be applied to more synthesis types: prescribed timing, function generator, etc., and some other additional constraints could

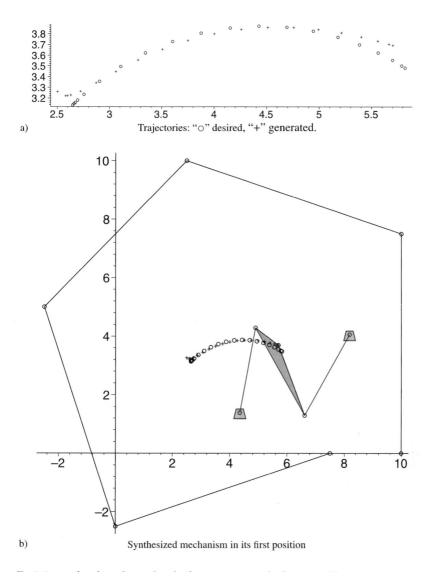

a) Trajectories: "o" desired, "+" generated.

b) Synthesized mechanism in its first position

Fig. 9 Desired and synthesized paths for $n = 20$ prescribed points with space constraints

be included, specially when the mechanism is part of a machine and must fit inside the dimensions of the machine.

Another future research area could be spatial mechanisms synthesis. As noted above, application of the simulated annealing algorithm converts the synthesis problem into an analysis problem with a specific link dimension and position search criteria.

References

1. I. Bohachevsky, M.E. Johnson, and M.L. Stein. Generalized simulated annealing for function optimization. *Technometrics*, 28:209–217, 1986
2. R.E. Burkard and F. Rendl. A thermodynamically motivated simulation procedure for combinatorial optimization problems. *European Journal of Operational Research*, 17:169–174, 1984
3. A. Casotto, F. Romeo, and A.L. Sangiovanni-Vincentelli. A parallel simulated annealing algorithm for the placement of macro-cells. In *Proceedings IEEE International Conference on Computer-Aided Design*, pp. 30–33, Santa Clara, November 1986
4. T. Elperin. Monte carlo structural optimization in discrete variables with annealing algorithm. *Internatrional Journal for Numerical Methods in Engineering*, 26:815–821, 1988
5. A.G. Erdman. Three and four precision point kinematic synthesis of planar synthesis. *Journal of Mechanism and Machine Theory*, 16:227–245, 1981
6. A.G. Erdman. Improved centerpoint curve generation technique for four-precision position synthesis using the complex number approach. *Journal of Mechanisms, Transmisions, and Automation in Design*, 107:370–376, 1985
7. A.G. Erdman and W.E. Sandor. *Mechanism Design: Analysis and Synthesis*. Prentice-Hall, New Jersey, 1987
8. U. Faigle and W. Kern. Note on the convergence of simulated annealing algorithms. *SIAM Journal on Control and Optimization*, 29:153–9, January 1991
9. T.L. Friesz, H.-J. Cho, and N.J. Mehta. A simulated annealing approach to the network design problem with variational inequality constraints. *Transportation Science*, 26:18–26, February 1992
10. S. B. Gelfand and S. K. Mitter. Analysis of simulated annealing for optimization. In *Proceedings of 24th Conference on Decision and Control*, pp. 779–786, Ft Lauderdale, December 1985
11. G.G.E. Gielen, Herman C.C. Walscharts, and W.M.C. Sansen. Analog circuit design optimization based on symbolic simulation and simulated annealing. *IEEE Journal of Solid-State Circuits*, 25:707–713, June 1990
12. S. Kirkpatrick, C.D. Gelatt, and M.P. Vecchi. Optimization by simulated annealing. *Science*, 220(4598):671–680, 1983
13. S. Kirkpatrick and G. Toulouse. Configuration space analysis of travelling salesman problems. *Journal Physique*, 46:1277–1292, 1985
14. P. J. M. van Laarhoven and E. H. L. Aarts. *Simulated Annealing: Theory and Applications*. D. Reidel, Dordrecht, Holland, 1987
15. A. Malhotra, J. H. Oliver, and W. Tu. Synthesis of spatially and intrinsically constrained curves using simulated annealing. *ASME Advances in Design Automation*, DE-32:145–155, 1991
16. H. Martínez-Alfaro and D.R. Flugrad. Collision-free path planning of an object using B-splines and simulated annealing. In *IEEE Internactional Conference on Systems, Man, & Cybernetics*, 1994
17. H. Martínez-Alfaro and D.R. Flugrad. Collision-free path planning of robots and/or AGVs using B-splines and simulated annealing. In *ASME DTEC/Mechanisms Conference*, 1994
18. H. Martínez-Alfaro and A. Ulloa-Pérez. Computing near optimal paths in C-space using simulated annealing. In *ASME Design Engineering Technical Conference/Mechanisms Conference*, Irvine, CA, 1996
19. H. Martínez-Alfaro, H. Valdez, and J. Ortega. Linkage synthesis of a four-bar mechanism for *n* precision points using simulated annealing. In *ASME Design Engineering Technical Conferences/Mechanisms Conference*, 1998

20. N. Metropolis, A. Rosenbluth, M. Rosenbluth, and A. Teller. Equations of state calculations by fast computing machines. *Journal of Chemical Physics*, 21:1087–1091, 1953

21. R.L. Norton. *Design of Machinery. An Introduction to the Synthesis and Analysis of Mechanisms.* McGraw-Hill, New York, 1992

22. R.A. Rutenbar. Simulated annealing algorithms: An overview. *IEEE Circuits and Devices*, Vol. 5, January:19–26, 1989

23. G.N. Sandor, D. Kohli, C.F. Reinholtz, and A. Ghosal. Closed-form analytic synthesis of a five-link spatial mmotion generator. *Journal of Mechanism and Machine Theory*, (1):97–105, 1984

24. T. Subbian. *Use of continuation methods for kinematic synthesis and analysis.* PhD thesis, Iowa State University, Ames, IA, 1990

25. T. Subbian and D.R. Flugrad. Four-bar path generation synthesis by a continuation method. In *ASME Advances in Design Automation*, ASME DE-19-3:425–432, 1989

26. C. H. Suh and C. W. Radcliffe. *Kinematics and Mechanism Design.* John Wiley & Sons, Inc., New York, 1978

27. D. Vanderbilt and S.G. Louie. A Monte Carlo simulated annealing approach to optimization over continuous variables. *Journal of Computational Physics*, 36:259–271, 1984

28. C. W. Wampler, A. P. Morgan, and A. J. Sommese. Complete solution of nine-point path synthesis problem for four-bar linkage. In *ASME Advances in Design Automation*, ASME DE-25:361–368, 1990

"MOSS-II" Tabu/Scatter Search for Nonlinear Multiobjective Optimization

Ricardo P. Beausoleil

Instituto de Cibernética Matemática y Física, Departmento de Optimización,
Ciudad Habana, Cuba.
rbeausol@icmf.inf.cu

Abstract

This chapter introduces a new version of our multiobjective scatter search to deal with nonlinear continuous vector optimization problems, applying a multi-star tabu search (TS) and convex combination methods as a diversification generation method. A new mechanism of forbidden solutions to create different families of subsets of solutions to be combined, is applied in this version. A constrain-handling mechanism is incorporated to deal with constrained problems. The performance of our approach is compared with our MOSS implementation [1] for some bi-objective unconstrained test problems, 3-objective test problems were compared with points of the Pareto-optimal front, and our constraint-handling mechanism is illustrated.

Key words: Multiple Objectives, Metaheuristics, Tabu Search, Scatter Search, Nonlinear Optimization

1 Introduction

The great majority of exact methods to deal with complex multiobjective optimization problems, that is, problems with nonlinearities in the objective function and/or constraints, and with a large number of variables have been considered impractical. To deal with these difficulties, different metaheuristic approaches have been developed.

Evolutionary algorithms have been widely applied to multiobjective optimization problems. Here, we present an approach based on tabu search and scatter search as a hybrid method to deal with complex multiobjective optimization problems.

This approach is a new version of our multiobjective scatter search (MOSS) [1], where new techniques are introduced in order to improve it, and to deal with constrained problems.

MOSS is also a hybrid method that, in the tabu search phase, creates restrictions to prevent moves toward solutions that are "too close" to previously visited solutions. A sequential fan candidate list strategy is used to explore solution neighborhoods.

A weighted linear function is used to aggregate the objective function values. The weights used are modified in such a way that a sufficient variety of points can be generated. These solutions are later used as reference points for the scatter phase.

In order to guide the search more quickly to better regions, in MOSS-II we embed, in the tabu phase, convex combinations by joining the new solutions with point of current set of efficient points.

The proposed algorithm incorporates a new strategy to diversify the search using a new mechanism of forbidden solutions to create different families of subsets to be combined. Also, an intensification strategy in the combination method is introduced to improve the search toward the Pareto-optimal front.

MOSS-II incorporates a mechanism to manipulate constraints. While most MOEAs consider combinations of solutions with efficient and non efficient solutions, feasible and infeasible solutions, MOSS-II considers its combinations using only feasible efficient solutions.

The organization of the chapter is as follows: In Sect. 2 the problem is stated. The general scheme of our strategy is presented in Sect. 3. Section 4 presents the tabu search phase. In Sect. 5 we present the scatter search phase. Experimental computation is presented in Sect. 6. Section 7 contains the conclusions.

2 Statement of the Problem

We are interested in finding $x = (x_1, x_2, \ldots, x_n)$ such that it optimizes $F(x) = (f_1(x), f_2(x), \ldots, f_r(x))$ under the following constraints: $\Im = \{x \in \bar{X} : g_j \leq 0, j \in \{1, 2, \ldots, m\}\}$, where $\bar{X} = \Pi_{i=1}^n X_i$, $x_i \in X_i \subset \mathbf{R}$, $(i \in \{1, \ldots, n\})$. Here, $\bar{X}$ is the search space and $\Im$ is the feasible space.

3 General Scheme of the Strategy

We use a multi-start TS as a generator of diverse solutions. This approach can be seen as a sequence of TSs where each TS has its own starting point, recency memory, and aspiration threshold; they all share the frequency memory to bias the search to unvisited or less visited regions. Initially, the starting solutions are systematically generated then, each TS in sequence initiates the search from one starting point. The searches continue while new solutions are incorporated to the set of trial solutions S, when one TS finishes its work, the convex combinations can be applied between solutions of the actual set of reference solutions R and elements of the set S, the new generated solutions are introduced into S, the generated solutions are filtered in the case of constrained problems, and the current nondominated points are subtracted and designated to be reference solutions. When all TS finish their work, new solutions are created consisting of linear combinations of elements of subsets of the reference set. The new potentially Pareto solutions are created by joining the new solutions with those of the reference set, and extracting a collection of efficient solutions. The reference set is diversified by re-starting from the TS approach taking as starting points solutions of the current potentially Pareto set.

We use the following notation in the development of the text:

S	=	a set of trial solutions, from which all other sets derive.
P	=	an approximate Pareto set, containing all non-inferior solutions of S.
$R1$	=	a set of high-quality non-inferior solutions subset of P.
$R2$	=	a set of elements belonging to the difference of P and $R1$.
R	=	a set of current reference solutions belonging to $R1$ or $R2$.
T_D	=	a set of tabu solutions, composing a subset of R excluded from consideration to be combined during t scatter iterations.
$T1_D$	=	a set of tabu solutions, composing a subset of R excluded from consideration to be combined during $t1$ scatter iterations.
C	=	a critical set, consisting of duplicated solutions.
D	=	a diverse subset of the set R.
$\Omega(D)$	=	the set of combined solutions, created from a given set D.
C^K		is the Kramer choice function and
$C^K(P)$	=	a set of selected elements of P.
b	=	the larger size of the trial solutions set.
$b1$	=	the larger size of the reference set.
$b2$	=	the larger size of the approximate Pareto set.
s^*	=	the maximum number of starting points.

Algorithm MultiObjectiveScatterSearch

(*MaxIter* = maximum number of global iterations.
NewParetoSolution = a Boolean variable that takes the value
TRUE if a new non-dominated solution entered to the reference set
FALSE in other case,
CutOffLimit = a limit number of scatter iterations,
tau = the generational distance between two consecutive efficient fronts,
τ = an approximation parameter.)

    While (*MaxIter* is not reached or *tau* $\geq \tau$)
        Create seed solutions used to initiate the approach
        Use of a memory-based strategy to generate
        a set of trial solutions from the starting points
        While (*NewParetoSolution* or *CutOffLimit* is not reached)
            Filter to preserve the feasibility
            Create a set of reference points taking
            the current potentially Pareto solutions
            Separate in two subsets the reference set using a choice function
            Generate diverse subsets of the reference set
            using tabu sets and a measure of dissimilarity
            Apply line search approach to obtain new solutions
        End *while*
        Rebuild
    End *while*

4 Tabu Search Phase

We use a multi-start TS as a generator of diverse solutions. In the new implementation the parameter *NumComb* defines the number of tabu iterations necessary in order to apply the convex combinations method with a set of parameters $\{wc_i : i = 1,\ldots,10\}$, *startpoints* is the number of initial points.

MultiStart_Tabu Phase
    Repeat
        icomb = 1
        tabu_iteration = *tabu_iteration* +1
        Set the tabu list empty
        Generate the new starting point x
        Set the reference point equal to $F(x)$
        Apply our Taboo search method
        If *tabu_iteration* = *icomb* · *NumComb* then
           Apply the convex combination method
           icomb = *icomb* +1
        If *constrained* then
           Filter to preserve the feasibility
        Apply the Pareto relation to $R \cup S$
        Update the set R
    Until *tabu_iteration* = *startpoints*
End of the Phase

This Taboo search method chooses a new solution vector taking into account the aspiration level function, the additive function value and a penalty function that modifies the value of the additive function. Also, tabu restrictions are used to guide the search.

A good balance between the diversification and intensification phase is very important to obtain a good initial set of solutions in a reasonable time. *NonNewParetoSolution*, *Level* and *MaxLevel* are parameters to obtain this balance.

4.1 Move

The range of variables is split into subranges, frequency memory is used to control the random selection of the subranges where the variables take values.

Next we identify our candidate list strategy. We use a simplified version of a sequential fan strategy as a candidate list strategy. The sequential fan generates p best alternative moves at a given step, and then creates a fan of solution streams, one for each alternative. The best available moves for each stream are again examined, and only the p best moves overall provide the p new streams at the next step. In our case, taking $p = 1$, we have in each step one stream and a fan of 60 points to consider.

Now we explain how to transit to a new solution. Let E be the set of efficient moves and D the set of deficient moves, where a deficient move is a move that does not satisfy the aspiration level, otherwise the move is efficient. Then, we define the best move as

$[m \in E(x) : c^*(x') = \max\{c(x'), x' = m(x)\}]$ if $E(x) \neq \emptyset$, in the case where $E(x) = \emptyset$ and $D(x) \neq \emptyset$ then, we select $[m \in D(x) : c^*(x') = \max\{c(x'), x' = m(x)\}]$. In our algorithm, we use the Boolean variable *NewParetoSolution* in order to identify if the set E is nonempty.

MaxLevel is an indicator of the maximum number of times that a non-efficient move is accepted, and *Level* is the number of times that a non-efficient move has occurred. When *Level* is greater than *MaxLevel* then, we take a random solution of S.

Our implementation uses as move attribute, variables that change their values as a result of the move. The change is represented by a difference of values $f_k(x') - z_k^* \; \forall k = 1,\ldots,r, x' \in \bar{X}$ where x' was generated from x by a recent move, x is a current solution and Z^* is a reference solution, $Z^* = (z_1^*, \ldots, z_r^*)$.

4.2 Aspiration Level and Transitions

An aspiration threshold is used to obtain an initial set of solutions as follows: without lost of generality, let us assume that every criterion is maximized. Notationally, let $\Delta f(x') = (\Delta f_1(x'), \ldots, \Delta f_r(x'))$ where $\Delta f_k(x') = f_k(x') - z_k^*, k \in \{1, \ldots, r\}$.

A goal is satisfied, permitting x' to be accepted and introduced in S if $(\exists \Delta f_k(x') \geq 0) or (\forall k \in \{1, \ldots, r\}[\Delta f_k(x') = 0])$, otherwise is rejected.

The point Z^* is updated by $z_k^* = \max f_k(x') \forall k \in \{1, \ldots, r\}, x' \in S$.

In order to measure the quality of the solution we propose to use in our tabu search approach an Additive Function Value *AFV* with weighting coefficients λ_k ($\lambda_k \geq 0$), representing the relative importance of the objectives. We want to set the weights λ_k ($k = 1, \ldots, r$) so that the solution selected is closest to the new aspiration threshold. Therefore each component in the weight vector is set according to the objective function values. We give more importance to those objectives that have greater differences between the quality of the trial solution and the quality of the reference solution. The influence is given by an exponential function $\exp(-s_k)$, where s_k is obtained as $s_k = (f_k(x') - z_k^*)/z_k^*$, $\lambda_k = 2 - \exp(-s_k)$ $(k \in \{1, \ldots, r\})$, then $AFV(x') = \sum_{k=1,r} \lambda_k (f_k(x') - z_k^*)$.

A movement to diversify is executed when $residence[i, j]$ is greater than T_i, where T_i is the threshold that determines the number of times that one sub-range can be visited without penalizing and $residence[i, j]$ is the number of time that the sub-range j has been visited by the variable with index i ($i \in \{1, \ldots, n\}$). We modify the value of $AFV(x')$ as follows: $AFV(x') = (1 - \frac{freq}{FreqTotal}) \cdot AFV(x')$, where *freq* is an addition of the entries of type *residence* associated to the selected variables and subranges that hold the condition $residence[i, j] > T_i$, and $FreqTotal = \sum_i \sum_{j=1}^{subrange} residence[i, j]$ $(\forall i \in I' = \{i : i \text{ was choosen}\})$, we would have for each variable a threshold T_i equal to the maximum between 1 and $Round(\sum_{j=1}^{subrange} residence[i, j]/subrange)$, where *Round* is the closest integer.

4.3 Tabu Restrictions

Tabu restrictions are imposed to prevent moves that bring the values of variables "too close" to values they held previously [6].

The implementation of this rule is as follows: the variable x' is excluded from falling inside the line interval bounded by $x - w(x' - x)$ and $x + w(x' - x)$, where $1 \geq w > 0$, when a move from x to x' is executed.

We have the following escape mechanism: when the forbidden moves grow so much that all movements become tabu and none satisfies the aspiration level, a reduction mechanism is activated and the tabu distance in each list is reduced, then the number of forbidden moves is reduced.

4.4 Frequency Memory

Our approach uses a frequency-based memory denoted by residence, this is a record that has two entries, $residence[i, j]$ explained above, and $residence\_x[i]$ containing the number of times that the variable $x[i]$ has been visited, $i \in \{1, 2, \ldots, n\}$, $j \in \{1, 2, \ldots, subrange\}$.

4.5 Duplicated Points

Avoiding the duplicate points already generated can be a significant factor in producing an effective overall procedure. The generation of a duplicate point is called a critical event. Our algorithm is based on a "critical event design" that monitors the current solutions in R and in the trial solutions set S. The elements considered in the critical event design are the values of the objectives and the decision variables. We consider that a critical event takes place if one trial solution is too close to another solution belonging to the trial solution set or to the reference set.

Let $B(p, \xi)$ be a set of points within distance ξ from p, $\xi > 0$. We call $B(p, \xi)$ a ball with center p and radius ξ. Then, for any point $p \in F[S \cup R]$ we define a ball $B(p, \rho)$ with $0 < \rho \leq 1$, and for any point $p' \in S \cup R$ a ball $B(p', \delta)$ with $0 < \delta \leq 1$.

A critical event takes place if a trial solution satisfies a full "critical condition". A "critical condition" is "full" if it is satisfied and the trial solution belongs to $B(p', \delta)$ defined on $S \cup R$. The "critical condition" is satisfied if the image of the trial solution pertains to $B(p, \rho)$ defined on $F[S \cup R]$.

The radius of the ball defined on the decision space is fixed and the radius of the ball defined on the objective space is an adaptive parameter in the interval $[\min r, \max r]$; initially $\rho = \max r$ then, if the average number of the current solutions per ball defined on $F[S \cup R]$, is greater than the average number of previous solutions per ball defined on previous $F[S \cup R]$ then, the current radius is halved, otherwise the value of the radius is doubled (inside of the permissible range.) The distance from p to any point on $S \cup R$ is defined by the Euclidean distance, and on $F[S \cup R]$ by the Usual distance (that is the absolute value.)

Taboo search method
    $NonNewParetoSolution = 0$
    While $NonNewParetoSolution < b_3$
        $NewParetoSolution = $ FALSE
        Update the reference point

```
      Generate new solutions to include into the set S
      Update the tabu list
      Choose the new solution x
      If NewParetoSolution then
         NonNewParetoSolution = 0
         Intensification = TRUE
         Level = 0
      else
         NonNewParetoSolution = NonNewParetoSolution + 1
         Intensification = FALSE
         If Level < MaxLevel then
            Level = Level + 1
   Endwhile
End
```

5 Scatter Search Phase

Scatter Search (SS) is designed to generate a dispersed set of points from a chosen set of reference points [6]. Its foundation derives from strategies for combining decision rules and constraints, with the goal of enabling a solution procedure based on the combined elements to yield better solutions than one based only on the original elements. SS operates on a set of solutions non-randomly generated called reference set. New solutions, called trial solutions, are created by making combination of subsets of the current set of reference solutions. The new reference set is selected from the current set of reference solutions, and the new trial solutions created.

5.1 Pseudo-code SS

Scatter_Search Phase

```
   scatter_iter = 0
   While CutOffLimit > scatter_iter do
       Apply Kramer choice function to separate
       the Potentially Pareto set
       Apply the Combination Subset method
       Update S
       If constrained then
          Filter to preserve the feasibility
       Apply Pareto relation to R ∪ S
       Update the Reference Point R
       scatter_iter = scatter_iter + 1
   endwhile
   Rebuild
End of the Phase
```

5.2 Kramer Choice Function

In order to split the current Pareto set into two subsets, an optimality principle is used: "Selection by a number of dominant criteria" [13]. For all x, $y \in P$, let $q(x, y)$ be the number of criteria for which the decision variable y improves the decision variable x, then $Q_P(x) = \max_{y \in P} q(x, y)$, $x \in P$ can be seen as a discordance index if x is assumed to be preferred to y. Then the Kramer choice function is defined as follows: $C^K(P) = \{x' \in P | Q_P(x') = \min_{x \in P} Q_P(x)\}$.

5.3 Choosing Diverse Subsets

As a basis for creating combined solutions we generate subsets $D \subseteq R$. Our approach is organized to generate three different collections of diverse subsets, which we refer to as $D1$, $D2$, and $D3$.

Suppose $R1 \neq \emptyset$ and $R2 \neq \emptyset$ then, the type of subsets we consider are as follows:

3-element subsets $D1$, where the first element is in $R1 - T_{D1}$, the second element pertains to $R1 - T1_{D1}$ and it is the most dissimilar to the first, and the third element belongs to $R1 - T1_{D1}$ selected to be the most dissimilar to the former two.

3-element subsets $D2$, where the first element is in $R1 - T_{D2}$, the second element pertains to $R2 - T1_{D2}$ and it is the most dissimilar to the first, and the third element belongs to $R2 - T1_{D2}$ selected to be the most dissimilar to the former two.

3-element subsets $D3$, where the first element is in $R2 - T_{D3}$, the second element pertains to $R1 - T1_{D3}$ and it is the most dissimilar to the first, and a third element that belongs to $R1 - T1_{D3}$ selected to be the most dissimilar to the selected elements. If $R1 - Z$ or $R2 - Z$, for $Z = T_{D \in \{D1, D2, D3\}}$ and $Z = T1_{D \in \{D1, D2, D3\}}$, are empty then, we take a random solution of $R1$ or $R2$, respectively.

If the cardinality of $R1$ or $R2$ is 1, then, the third element is created by a linear combination of the two previously selected elements.

The most dissimilar solution is measured with the max-min criterion, that is, we define the distance between a point $p \in R$ and a non-empty subset $A \subseteq R$ by

$$d(p, A) = \min\{d(p, a) : a \in A\}$$

where d is the Euclidean measure, then the max-min criterion is defined as

$$d^*(p, A) = \max\{d(p, A) : p \in R\}$$

We use two different tabu sets in order to avoid duplicating the diverse subsets to be combined and to obtain more diversity in the search. Having selected the first solution to be considered in the diverse subset, it is included in the tabu set $T_{D \in \{D1, D2, D3\}}$. The second element to be considered in the diverse subset is included in the tabu set $T1_{D \in \{D1, D2, D3\}}$. A simple dynamic rule to create a tabu tenure t and $t1$ of the set $T_{D \in \{D1, D2, D3\}}$ is used. We choose t to vary randomly between $t_{\min} = 1$ and $t_{\max} = 6$, and $t1$ as a progressive tabu restriction, its value being increased by the number of times that this solution has been selected as second element in D. The Tabu restrictions are overridden to create one 3-element subset, if one of the selected solution is a new solution, that is, a solution created in the last scatter iteration.

Another difference with respect to MOSS is the introduction of one intensification strategy that consists in combining solutions with more similar structures. To do this, we define new subsets D4 of two elements where the first element is in $R1 - T_{D4}$, the second element pertains to $R1 - T1_{D4}$ selected to be closest to the first.

5.4 Linear Combinations

Our strategy consists in creating $\Omega(D) = x + @(y - x)$, for $@ = 1/2, 1/3, 2/3, -1/3, -2/3, 4/3, 5/3$, and $x, y \in D$ as follows:

1. Generate new trial points on lines between x and y.
2. Generate new trial points on lines between x and z.
3. Generate new trial points on lines between y and z.
4. Generate one solution by applying $y + \frac{x-y}{2}$.
5. Generate one solution by applying $z + \frac{x-z}{2}$.
6. Generate one solution by applying $z + \frac{y-z}{2}$.

where x, y and z are elements of $D \in \{D1, D2, D3\}$.

The infeasible variables are set equal to the closest bounds to them. The set S is the union of all generated solutions, $S = \cup \Omega(D)$ for all D.

Note that the number of subsets generated depends on the number of solutions not forbidden in the current set R, and on the larger size of the trial solutions set.

5.5 The Reference Set

The reference set R is a subset of the trial solution set S that consists of an approximation to the Pareto-optimal set.

Let $P \subseteq \Omega^P \cup R$, where $\Omega^P \subseteq S \backslash C$, consists of a subset of current non-dominated solutions. If $|P| > b1$, then we use the max-min criterion and a parameter ϵ as measure of the closeness, to obtain a diversified collection of solutions P', that is, set one first element of P into P', then let $x \in P$ maximizes the minimum distance $d(F(x), F(y(i)))$ for $i \leq |P'|$ and hold the condition $d(F(x), F(y(i))) \geq \epsilon$, $y(i)$ pertaining to the non-empty set P' then, when $P' = b1$ set $P = P'$.

5.6 Rebuilding the Reference Set

In order to rebuild the reference set, we take *startpoints* starting points from R using the max-min criterion and we resume the process with tabu search to construct a set S of new trial solutions.

$$\text{Set, } startpoints = \min\{20, |R|\}$$

6 Experimental Computation

In this section we investigate the performance of our MOSS-II described above on a number of standard test functions having two, three and four objectives. The results obtained for 3-objective and 4-objective test problems are compared to the Pareto-optimal front. We also compare our approach with the results obtained by our MOSS [1] for 2-objective test problems, some constrained test problems, taken from the book *Evolutionary Algorithms for Solving Multiobjective-Problems* (Coello Coello C.A. and co-authors [3]) illustrate our constraint-handling mechanism, and a quantitative study with difficulty constrained test problems shows the performance of MOSS-II on constrained problems.

The experiments were done on a personal computer at 800 MHz. The code was written in DELFI-6. The CPU times were measured with the system routine.

6.1 Effect of the ϵ Parameter

Now we show how changes in the value of the ϵ parameter determine changes in the distribution of the solutions.

Unconstrained Problems
Murata test problem:

$$\text{minimize } f_1(x, y) = 2\sqrt{x}$$
$$\text{minimize } f_2(x, y) = x(1 - y) + 5$$
$$1 \leq x \leq 4, \quad 1 \leq y \leq 2$$

Rendon test problem:

$$\text{minimize } f_1(x, y) = \frac{1}{x^2 + y^2 + 1}$$
$$\text{minimize } f_2(x, y) = x^2 + 3y^2 + 1$$
$$-3 \leq x, \quad y \leq 3$$

Constrained Problems
Binh(2) test problem:

$$\text{minimize } f_1(x, y) = 4x^2 + 4y^2$$
$$\text{minimize } f_2(x, y) = (x - 5)^2 + (y - 5)^2$$
$$s.t.$$
$$0 \geq (x - 5)^2 + (3.1)y^2 - 25$$
$$0 \geq -(x - 8)^2 + (y + 3)^2 + 7.7$$
$$x \in [0, 5], \quad y \in [0, 3]$$

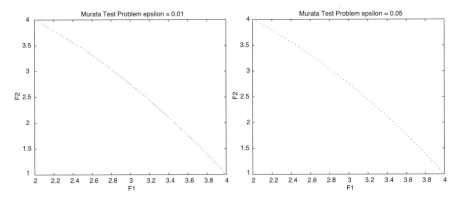

Fig. 1 Pareto front achieved by MOSS-II on Murata problems. On the left $\epsilon = 0.01$ and on the right $\epsilon = 0.05$

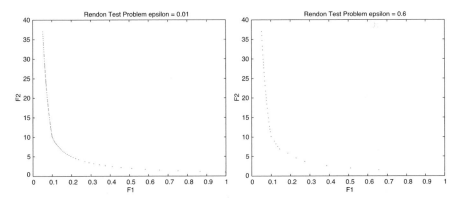

Fig. 2 Pareto front achieved by MOSS-II on Rendon problems. On the left $\epsilon = 0.01$ and on the right $\epsilon = 0.6$

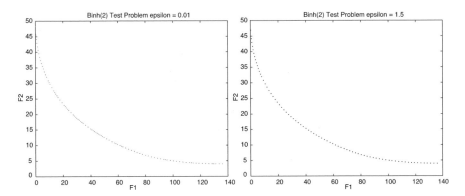

Fig. 3 Pareto front achieved by MOSS-II on Binh(2) problems. On the left $\epsilon = 0.01$ and on the right $\epsilon = 1.5$

Jimenez test problem:

$$\text{maximize } f_1(x, y) = 5x + 3y$$
$$\text{maximize } f_2(x, y) = 2x + 8y$$
$$s.t.$$
$$0 \geq x + 4y - 100$$
$$0 \geq 3x + 2y - 150$$
$$0 \geq -5x - 3y + 200$$
$$0 \geq -2x - 8y + 75$$
$$xy \geq 0$$

6.2 An Illustration of the Approximation to Pareto-optimal Front

Here we illustrate the approximation to Pareto-optimal front with several test problems using one hundred variables. In these the first approximation corresponds with the output of the tabu phase and the last is the achieved front.

Kursawe test problem: Pareto-optimal set and Pareto-optimal front disconnected concave-convex front and isolated point.

$$\text{minimize } f_1(x) = \sum_{i=1}^{n-1} \left(-10 \exp\left(-0.2\sqrt{x_i^2 + x_{i+1}^2} \right) \right)$$

$$\text{minimize } f_2(x) = \sum_{i=1}^{n} \left(|x_i|^{0.8} + 5 \sin(x_i)^3 + 3.5828 \right)$$

$$x_i \in \left[-10^3, 10^3 \right]$$

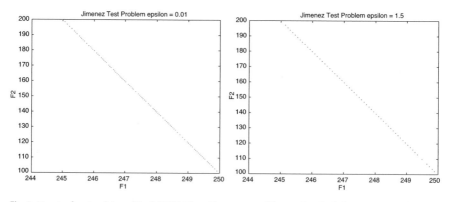

Fig. 4 Pareto front achieved by MOSS-II on Jimenez problems. On the left $\epsilon = 0.01$ and on the right $\epsilon = 1.5$

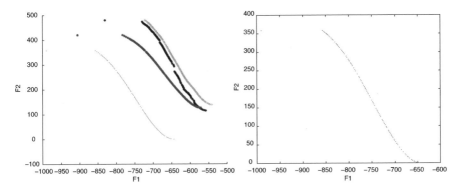

Fig. 5 A sequence of approximation and the last approximation to Pareto-optimal front on the Kursawe test problem

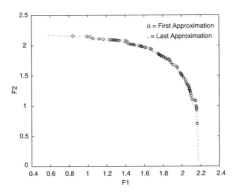

Fig. 6 First (o) and final (.) approximations to Pareto-optimal front on Quagliarella test problem

Quagliarella test problem: Pareto-optimal set disconnected and a concave Pareto-optimal front with a diminishing density of solutions towards the extreme points.

$$\text{minimize } f_1(x) = \left(\frac{1}{n} \sum_{i=1}^{n} \left(x_i^2 - 10\cos(2\pi(x_i) + 10) \right) \right)^{1/4}$$

$$\text{minimize } f_2(x) = \left(\frac{1}{n} \sum_{i=1}^{n} (x_i - 1.5)^2 - 10\cos(2\pi(x_i - 1.5)) + 10 \right)^{1/4}$$

$$x_i \in \left[-5, 5 \right]$$

Minimize ZDT Test Problems [4] [9]:

ZDT1 test problem: The ZDT1 problem has a convex Pareto-optimal front.
ZDT2 test problem: The ZDT2 problem has a non-convex Pareto-optimal front.

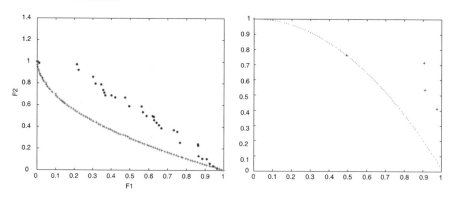

Fig. 7 First (+) and final (.) approximations to Pareto-optimal front on ZDT1, ZDT2

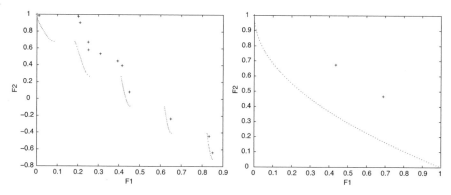

Fig. 8 First (+) and final (.) approximations to Pareto-optimal front on ZDT3, ZDT4

ZDT3 test problem: This problem provides some difficulties by introducing discontinuities in the Pareto-optimal front.

ZDT4 test problem: This problem has several local Pareto-optimal fronts, providing difficulties converging to the global Pareto-optimal front.

6.3 A Comparative Study MOSS-II vs MOSS (2-Objective, 100 variables)

The performance of our MOSS-II was compared with the results obtained by our first variant MOSS (MOSS showed a good performance versus SPEA2, NSGA-II and PESA; see [1]). The parameters were taken as follows: $\rho \in [0.0078125, 1]$ and $\delta = 0.0001$, $\epsilon = 0.01$, $\tau = 0.01$, *MaxIter* = 6, *CutoffLimit* = 30, *NumComb* = *startpoints*/2, *MaxLevel* = 3, $b = 240$, $b1 = 100$, $s^* = 100$ in order to obtain 100 solutions. Initially, *startpoints* = s^*.

A statistical test was carried out on the 2-objective test problems described above. The following box-plots describe the performance on the test problems Kursawe, Quagliarella, ZDT1, ZDT2, ZDT3 and ZDT4, these box-plots show the coverage set, the $Q_{RS}(A)$ defining the relative number of nondominated solutions in the approxi-

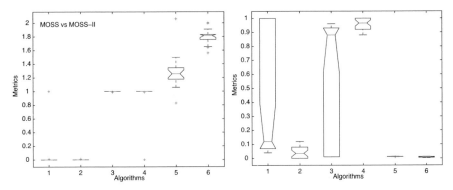

Fig. 9 Box plot representing the distribution of the coverage set, $Q_{RS}(A)$ and spacing metrics on Kursawe and Quagliarella problems (MOSS = 1, MOSS-II = 2, MOSS = 3, MOSS-II = 4, MOSS = 5 and MOSS-II = 6)

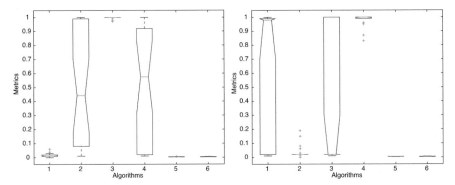

Fig. 10 Box plot representing the distribution of the coverage set, $Q_{RS}(A)$ and spacing metrics on ZDT1 and ZDT2 problems (MOSS = 1, MOSS-II = 2, MOSS = 3, MOSS-II = 4, MOSS = 5 and MOSS-II = 6)

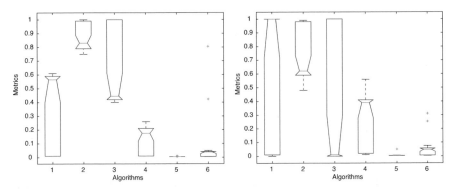

Fig. 11 Box plot representing the distribution of the coverage set, and spacing metrics on the ZDT3 problem at left, and the coverage set, $Q_{RS}(A)$ and spacing metrics on the ZDT4 problem at right (MOSS = 1, MOSS-II = 2, MOSS = 3, MOSS-II = 4, MOSS = 5 and MOSS-II = 6)

Table 1 MOSS performance for 2-objective test problems

Problems	CS(MOSS-II,MOSS)			CS(MOSS,MOSS-II)		
	Maximum	Minimum	Median	Maximum	Minimum	Median
Kursawe	1.000	0.000	0.000	0.010	0.000	0.000
Quagliarella	1.000	0.040	0.120	0.120	0.000	0.035
ZDT1	0.060	0.000	0.010	1.000	0.010	0.440
ZDT2	1.000	0.010	0.980	0.190	0.000	0.020
ZDT3	0.610	0.010	0.565	1.000	0.750	0.830
ZDT4	1.000	0.000	1.000	0.990	0.480	0.620
	$Q_{MOSS-II}(MOSS)$			$Q_{MOSS}(MOSS-II)$		
	Maximum	Minimum	Median	Maximum	Minimum	Median
Kursawe	1.000	0.990	1.000	1.000	0.000	1.000
Quagliarella	0.960	0.010	0.880	1.00	0.880	0.965
ZDT1	1.000	0.970	1.000	1.000	0.010	0.575
ZDT2	1.000	0.010	0.020	1.00	0.830	1.000
ZDT3	1.000	0.400	0.445	0.260	0.010	0.175
ZDT4	1.000	0.000	0.010	0.560	0.010	0.390
	Spacing(MOSS)			Spacing(MOSS-II)		
Kursawe	2.060	0.830	1.260	2.000	1.560	1.810
Quagliarella	0.013	0.008	0.012	0.012	0.003	0.010
ZDT1	0.007	0.003	0.003	0.006	0.003	0.003
ZDT2	0.005	0.003	0.005	0.006	0.003	0.003
ZDT3	0.011	0.004	0.005	0.806	0.005	0.037
ZDT4	0.050	0.003	0.004	0.312	0.005	0.047

mated set of solutions A, and the spacing. The first two columns represent the coverage set measure, the following columns represent the $Q_{RS}(A)$ metric, and the last two columns the spacing metric.

Table 1 provides evidence that, for problems ZDT1, ZDT2, ZDT3, ZDT4, Kursawe, and Quagliarella, MOSS-II performs as well as MOSS in terms of convergence and distribution.

The results shown in the box plots and tables, indicate that the algorithm was able to find well-distributed solutions on all Pareto-achieved fronts. To obtain 100 solutions for the above test functions the following average computational times were needed: ZDT1 41 seconds, ZDT2 43 seconds, ZDT3 34.9 seconds, ZDT4 51.2 seconds, Kursawe 64.3 seconds, and Quagliarella 31.2 seconds.

6.4 A Comparative Study of 3-objective and 4-objective Test Problems

In this section we examine the performance of our algorithm on four test problems taken from the literature [10]. The k-paramter for the 3-objective and 4-objective test problems was equal to 18 and 10, respectively (see Deb et al. for details about the k-paramter).

Problem DTLZ1. The following problem is a 3-objective problem with a linear Pareto-optimal front and $n = 20$ variables. The objective function values lie on the linear hyper-plane satisfying $f_1 + f_2 + f_3 = 0.5$ in the range $f_1, f_2, f_3 \in [0, 0.5]$. Here we use the following parameters: $MaxIter = 25$, $CutoffLimit = 30$, $NumComb = startpoints/2$, $MaxLevel = 3$, $b = 240$, $b1 = 100$, $s^* = 100$. To obtain 100 solutions with average time of 236 seconds, we set $\rho \in [0.0078125, 1]$, $\delta = 0.0001$, $\epsilon = 0.01$, $\tau = 0.01$.

Problems DTLZ2, DTLZ3, DTLZ4 and DTLZ2(4). The following test problems prove the ability of our approach to converge on the global Pareto-optimal front, sat-

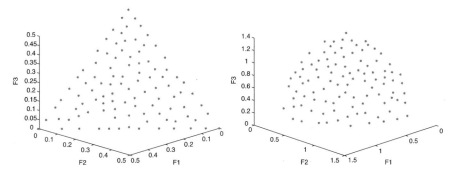

Fig. 12 Pareto front achieved by MOSS-II on DTLZ1 and DTLZ2 problems

isfying $f_1^2 + f_2^2 + f_3^2 = 1$ in the range f_1, f_2, $f_3 \in [0, 1]$. We considered 20 variables for the DTLZ2, DTLZ3, DTLZ4 problems and for the 4-objective DTLZ2(4) problem, 12 variables were considered. Parameters was taken as: *MaxIter* = 15 for the 3-objective problems, and for the 4-objective problem *MaxIter* = 30, the rest as in DTLZ1. For DTLZ2, DTLZ3, and DTLZ4 an average time of 144 seconds was needed to obtain 100 solutions, and for DTLZ2(4) 304 seconds.

DTLZ4 problems have variable density in the search space. Observe that our algorithm achieved a distributed set of solutions.

All the above figures correspond with the median associated to the spacing measure.

Statistics

Here, we present a quantitative study of the case illustrated above. The test problems have known Pareto-optimal fronts, and the results of the four algorithms were compared to it.

In all cases, the median is close to the value zero, indicating that the solutions are found to lie close to the Pareto-optimal fronts and with well-distributed solutions.

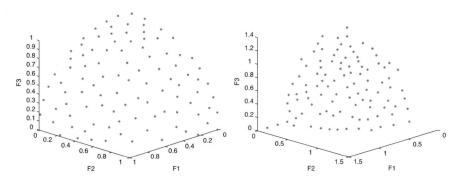

Fig. 13 Pareto front achieved by MOSS-II on DTLZ3 and DTLZ4 problems

Table 2 MOSS performance for 3-objective and 4-objective test problems

Problems	Generational Distance			Spacing		
	Maximum	Minimum	Median	Maximum	Minimum	Median
DTLZ1	0.327	0.000	0.002	0.367	0.009	0.011
DTLZ2	0.034	0.032	0.033	0.031	0.021	0.026
DTLZ3	0.544	0.030	0.032	0.444	0.016	0.025
DTLZ4	0.106	0.031	0.036	0.131	0.023	0.035
DTLZ2(4)	0.074	0.052	0.062	0.116	0.062	0.079

6.5 An Illustration with Constrained Test Problems

This section examines the performance of our algorithm on several constrained test problems taken from the literature [3], [11], [16], [12].

Many researchers have explored the problem of how to deal with constraints in a multiobjective metaheuristic. Many techniques have been developed, for example, static penalties, dynamic penalties, adaptive penalties, rejection infeasible solutions, multiobjective optimization, repair algorithms, co-evolutionary models, and others (see [14], [15], [8]). We examine 25 different side-constraint problems with different

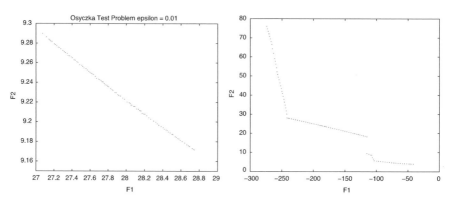

Fig. 14 Pareto front achieved by MOSS-II on Osyczka and Osyczka2 problems

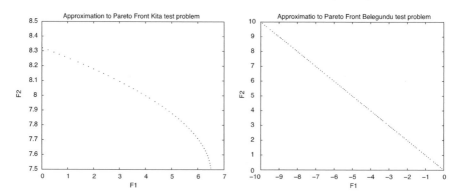

Fig. 15 Pareto front achieved by MOSS-II on Kita and Belegundu problems

levels of difficulty, using rejection infeasible solutions in each iteration. The Pareto
fronts achieved are shown in Figs. 14–20.

Belegundu test problem:

$$\text{minimize } f_1(x, y) = -2x + y$$
$$\text{minimize } f_2(x, y) = 2x + y$$
$$s.t.$$
$$0 \geq -x + y - 1$$
$$0 \geq x + y - 7$$
$$x \in [0, 5], \quad y \in [0, 3]$$

Kita test problem:

$$\text{minimize } f_1(x, y) = -x^2 + y$$
$$\text{minimize } f_2(x, y) = (1/2)x + y + 1$$
$$s.t.$$
$$0 \geq (1/6)x + y - 13/2$$
$$0 \geq (1/2)x + y - 15/2$$
$$0 \geq 5x + y - 30$$
$$xy \geq 0$$

Osyczka test problem:

$$\text{minimize } f_1(x, y) = x + y^2$$
$$\text{minimize } f_2(x, y) = x^2 + y$$
$$s.t.$$
$$0 \geq -x - y + 12$$
$$0 \geq x^2 + 10x - y^2 + 16y - 80$$
$$x \in [2, 7], \quad y \in [5, 10]$$

Osyczka2 test problem:

$$\text{minimize } f_1(x) = -(25(x_1 - 2)^2 + (x_2 - 2)^2 + (x_3 - 1)^2 + (x_4 - 4)^2 + (x_5 - 1)^2$$
$$\text{minimize } f_2(x) = x_1^2 + x_2^2 + x_3^2 + x_4^2 + x_5^2 + x_6^2$$
$$s.t.$$
$$0 \geq x_1 + x_2 - 2$$
$$0 \geq 6 - x_1 - x_2$$
$$0 \geq -2 - x_2 + x_1$$
$$0 \geq -2 - x_1 + 3x_2$$
$$0 \geq -4 - (x_3 - 3)^2 - x_4$$

$$0 \ge (x_5 - 3)^2 - x_6 - 4$$
$$x_1, x_2, x_6 \in [0,10], \quad x_3, x_5 \in [1,5], \quad x_4 \in [0,6]$$

Tanaka test problems:

$$\text{minimize } f_1(x, y) = x$$
$$\text{minimize } f_2(x, y) = y$$
$$s.t.$$
$$0 \le -x^2 - y^2 + 1 + \left(a \cos(b \arctan(x/y)) \right)$$
$$0 \ge (x - 0.5)^2 + (y - 0.5)^2 - 0.5$$
$$0 \ge x, y \le \pi$$

Tanaka1 with $a = 0.1$ and $b = 16$
Tanaka2 with $a = 0.1$ and $b = 32$
Tanaka(abs) with $a = 0.1$ and $b = 16$
Tanaka(abs) with $a = 0.1$ and $b = 32$
Deeper Tanaka with $a = 0.1(x^2 + y^2 + 5xy)$ and $b = 32$
Non-periodic Tanaka with $a = 0.1(x^2 + y^2 + 5xy)$ and $b = 8(x^2 + y^2)$

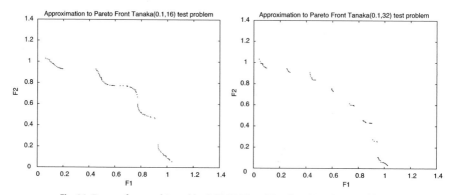

Fig. 16 Pareto front achieved by MOSS-II on Tanaka1, Tanaka2 problems

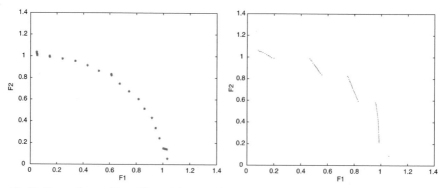

Fig. 17 Pareto front achieved by MOSS-II on deeper Tanaka, non-periodic Tanaka problems

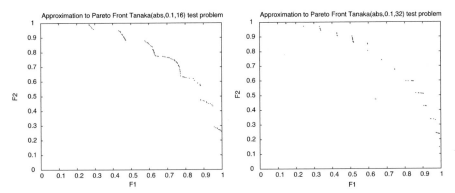

Fig. 18 Pareto front achieved by MOSS-II on Tanaka abs. $a = 0.1$ $b = 16$ and Tanaka abs $a = 0.1$ $b = 32$ problems

TBU test problem:

$$\text{minimize } f_1(x, y) = 4x^2 + 4y^2$$
$$\text{minimize } f_2(x, y) = (x - 5)^2 + (y - 5)^2$$
$$s.t.$$
$$(x - 1)^2 + y^2 - 25 \leq 0$$
$$-(x - 8)^2 - (y + 3)^2 + 7.7 \leq 0$$
$$-15 \leq x, \quad y \leq 30$$

Deb-Gupta test problem:

$$\text{minimize } f_1(x) = x_1^2$$
$$\text{minimize } f_2(x) = h(x_1) + G(x)S(x_1)$$
$$s.t.$$
$$g_j(x) \geq 0, \quad j = 1, \ldots, m$$

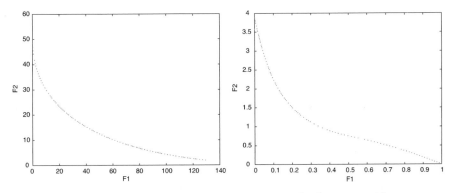

Fig. 19 Pareto front achieved by MOSS-II on TBU and Deb-Gupta problems

where,

$$h(x_1) = 1 - x_1^2 \quad , \quad G(x) = \sum_{i=2}^{5} 50x_i^2 \quad , \quad S(x_1) = \frac{1}{0.2 + x_1} + x_1^2$$

$$0 \le x_1 \le 1, \ -1 \le x_i \le 1$$

Test. $g_1(x) = 0.2x_1 + x_2 - 0.1$

Viennet test problem:

$$\text{minimize } f_1(x, y) = \frac{(x-2)^2}{2} + \frac{(y+1)^2}{13} + 3$$

$$\text{minimize } f_2(x, y) = \frac{(x+y-3)^2}{175} + \frac{(2y-x)^2}{17} - 13$$

$$\text{minimize } f_3(x, y) = \frac{(3x-2y+4)^2}{8} + \frac{(x-y+1)^2}{27} + 15$$

$$s.t.$$

$$y < -4x + 4$$

$$x > -1$$

$$y > x - 2$$

$$-4 \le x, y \le 4$$

Tamaki test problem:

$$\text{maximize } f_1(x, y, z) = x$$

$$\text{maximize } f_2(x, y, z) = y$$

$$\text{maximize } f_2(x, y, z) = z$$

$$s.t.$$

$$x^2 + y^2 + z^2 \le 1$$

$$0 \le x, y, z \le 4$$

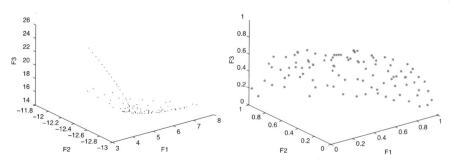

Approximation to Pareto Front Viennet4 test problem

Fig. 20 Pareto front achieved by MOSS-II on Viennet and Tamaki problems

6.6 Quantitative Study of Constrained Problems

Now we concentrate our study on several constrained test problems proposed by K. Deb in [11]. We study the performance of our approach on test functions from CTP1 to CTP7, these problems present difficulties near the Pareto-optimal front and also in the entire search space. In all problems, five decision variables are used.

In this experiment the parameter *maxiter* was set equal to 25, the other parameters were taken as above. Twenty simulation runs were performed for each problem.

Test Problems:

Problem CTP1:

$$\text{minimize } f_1(x) = x_1,$$
$$\text{minimize } f_2(x) = g(x) \exp(-f_1/g(x)),$$
$$s.t.$$
$$c_j = f_2 - a_j \exp(-b_j f_1(x)) \geq 0, \quad j = 1, 2, 3, 4$$
$$x_1 \in [0, 1], \quad -5 \leq x_i \leq 5, \quad (i = 2, 3, 4, 5)$$

where

$$a_1 = 0.909, \quad a_2 = 0.823, \quad a_3 = 0.760, \quad a_4 = 0.719$$
$$b_1 = 0.525, \quad b_2 = 0.276, \quad b_4 = 0.144, \quad b_4 = 0.074$$

In this problem one-fifth of the solutions lie on the boundary of the constrained Pareto-optimal set, the rest on the constraints.

Problems CTP2 until CTP7:

$$\text{minimize } f_1(x) = x_1,$$
$$\text{minimize } f_2(x) = g(x)(1 - f_1(x)/g(x)).$$
$$s.t.$$
$$c(x) = \cos(\theta)(f_2(x) - e) - \sin(\theta)f_1(x) \geq$$
$$a|\sin(b\pi(\sin(\theta)(f_2(x) - e) + \cos(\theta)f_1(x))^c)|^d$$
$$x_1 \in [0, 1], \quad -5 \leq x_i \leq 5, \quad (i = 2, 3, 4, 5)$$

The following parameters are used to define the problems:

CTP2: $\theta = -0.2\pi$, $a = 0.2$, $b = 10$, $c = 1$, $d = 0.5$, $e = 1$.
CTP3: $a = 0.1$, $d = 6$, the others as above.
CTP4: $a = 0.75$, the others as above.
CTP5: $c = 2$, the others as above.

For the above problems, the Pareto-optimal solutions lie on the straight line $(f_2(x) - e) \cos(\theta) = f_1(x) \sin(\theta)$.

CTP6: $\theta = 0.1\pi$, $a = 40$, $b = 5$, $c = 1$, $d = 2$, $e = -2$.

In this test problem, the Pareto-optimal region corresponds to all solutions satisfying the following inequalities $1 \le ((f_2 - e) \sin(\theta) + f_1(x) \cos(\theta)) \le 2$.

CTP7: $\theta = -0.05\pi$, $a = 40$, $b = 5$, $c = 1$, $d = 6$, $e = 0$.

The Pareto-optimal region is a disconnected set of the unconstrained Pareto-optimal feasible region. For more details about the above test functions see Deb [11].
The function $g(x) = 41 + \sum_{i=1}^{5}(x_i^2 - 10 \cos(2\pi x_i))$ was used in our experiment.
A knowledge of the relations that satisfy the solutions of the Pareto-optimal set permits one to obtain a measure of the convergence property of our approximations in the following way: for each simulation run, using the above relations we can calculate how far each solution lies from the Pareto-optimal set (a deviation). Hence it is easy to calculate the mean of the deviations for each simulation run as a convergence measure, and finally we obtain the maximum, minimum, and the median of these measures as an overall convergence measure.

Figures 21–24 illustrate our approximations of the above test problems. The Pareto fronts achieved correspond with the median of our simulation runs for the convergence measure.

Figure 21 shows the approximation to the Pareto-optimal front for CTP1 and CTP2. MOSS-II is able to find solutions in all fronts for the CTP1 problem, and for the CTP2 it found solutions in all disconnected regions, showing good convergence and distributed solutions.

MOSS-II is able to find a very close approximation to the true Pareto-optimal front in each region of the CTP3 (Fig. 22) problem. Problem CTP4 did not cause difficulty for MOSS-II, which obtained close convergence to the true Pareto-optimal front and well-distributed solutions.

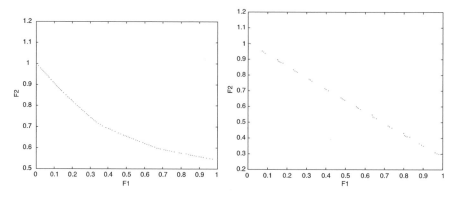

Fig. 21 Pareto front achieved by MOSS-II for CTP1, CTP2

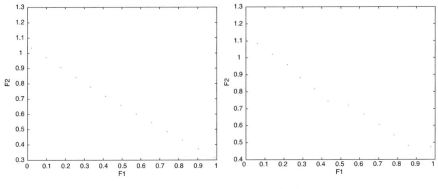

Fig. 22 Pareto front achieved by MOSS-II for CTP3, CTP4

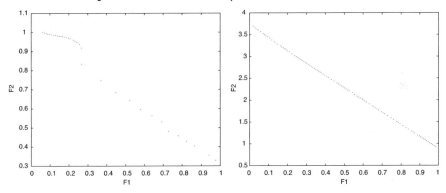

Fig. 23 Pareto front achieved by MOSS-II for CTP5, CTP6

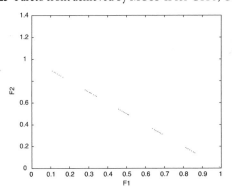

Fig. 24 Pareto front achieved by MOSS-II on CTP7

Figure 23 shows the approximations for CTP5 and CTP6. The non-uniformity in spacing of the Pareto-optimal solutions for CTP5 presents no great difficulty to MOSS-II; it was able to find a good approximation. In the CTP6 problem, observe that MOSS-II was able to converge to the right feasible patch and close to the true Pareto-optimal front.

Perpendicular infeasible patches caused no difficulty for MOSS-II (Fig. 24). Tables 3 and 4 summarize the results explained above and illustrated in the figures. Overall, the median of the convergence measure and the spacing are close to the value zero, indicating that the solutions found lie close to the Pareto-optimal fronts and with well- distributed solutions. The maximum extension front shows evidence that in general, MOSS-II was able to find solutions in all fronts. Also, a reasonable number of solutions was obtained to each problem in a very good computation time.

The box-plots in Fig. 25 illustrate the results given in the above tables for the convergence measure and spacing measure.

Figure 25 shows that the median of the convergence measure and the spacing measure in each problem are close to zero, and the dispersion around the median in all cases is small, showing that MOSS-II performance was consistent.

Table 3 MOSS performance for constrained problems

Problems	Convergence			Spacing		
	Maximum	Minimum	Median	Maximum	Minimum	Median
CTP1	0.097	0.066	0.084	0.006	0.002	0.003
CTP2	0.650	0.002	0.003	0.016	0.002	0.003
CTP3	0.116	0.017	0.028	0.063	0.003	0.043
CTP4	0.764	0.075	0.098	0.060	0.000	0.004
CTP5	0.145	0.046	0.052	0.040	0.014	0.038
CTP6	0.346	0.227	0.306	0.548	0.007	0.010
CTP7	4.933	0.001	0.001	3.947	0.950	1.022

Table 4 MOSS performance for constrained problems

Problems	Maximum extension front			Cardinality		
	Maximum	Minimum	Median	Maximum	Minimum	Median
CTP1	0.581	0.561	0.568	100	63	68
CTP2	1.000	0.828	0.841	100	44	49
CTP3	0.900	0.627	0.892	19	14	15
CTP4	1.017	0.763	0.893	100	9	14
CTP5	0.983	0.667	0.819	93	22	38
CTP6	2.799	1.541	1.685	100	77	100
CTP7	3.947	0.950	1.022	100	40	43

	Time		
	Maximum	Minimum	Median
CTP1	16	7	10
CTP2	36	2	14
CTP3	31	8	22
CTP4	15	9	14
CTP5	30	10	9
CTP6	14	6	9
CTP7	10	3	5

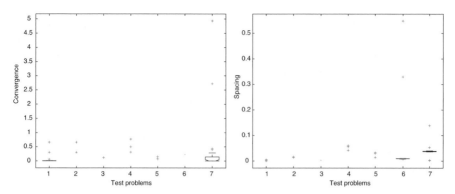

Fig. 25 Box plot representing the distribution of the convergence measure, and spacing measure on CTP problems, from left to right CTP1 = 1, CTP2 = 2, CTP3 = 3, CTP4 = 4, CTP5 = 5, CTP6 = 6, and CTP7 = 7

6.7 Results

Our approach has been executed for different types of problems with different complexities, convex and non-convex search spaces, Pareto fronts with a set of discrete points, disconnected Pareto curves, linear and non-linear constraints. The aim was to improve the first approach to solving problems with more variables, more objective functions, and with constrained problems. We believe that MOSS-II achieved this.

MOSS-II has incorporated new features to obtain different resolutions in the distribution of solutions, and also in the stop condition.

It is important to highlight that MOSS-II can solve all the above problems efficiently, setting the parameter *maxiter* equal to 25, and the rest as a set of invariant parameters, as shown above.

The results illustrated show that MOSS-II achieved a good approximation to feasible Pareto-optimal fronts in all test problems with a widely distributed set of solutions.

7 Conclusions

In these experiment the first TS phase was shown to be useful in generating an initial good Pareto frontier and introduce diversity in the search process. The embedded convex combinations in the tabu phase between points of the trial solution set S and points of the reference set R, improve the tabu phase of MOSS. Convex combinations in continuous problems seem to be a good mechanism to generate new Pareto points. A good approximation to the Pareto frontier was obtained in only a few iterations. A Kramer Choice Function seems a good optimality principle to separate the reference set of current Pareto solutions. Different metrics and different forbidden mechanisms are used to construct diverse subsets of points to obtain a balance between intensification and diversification. Our reference set of points, in which all points are nondominated solutions, permits the use of a simple constraint-handling mechanism with good performance and reasonable computation time.

Sixty different test problems were solved using the proposed approach, and are incorporated into a library, available with an executable code for our approach at rbeausol@icmf.inf.cu.

The proposed approach seems to be a viable strategy to solve unconstrained and constrained multiobjective nonlinear problems.

References

1. Beausoleil R (2006) "MOSS" Multiobjective Scatter Search Applied to Nonlinear Multiple Criteria Optimization, European Journal of Operational Research, Special Issue on Scatter Search, Rafael Martí (ed.), Elsevier Science, 169:426–449
2. Beausoleil R (2004) Bounded Variables Nonlinear Multiple Criteria Optimization using Scatter Search, Revista de Matemática: Teoría y Aplicaciones, CIMPA, 11(1):17–40
3. Coello Coello C A, Van Veldhuizen D A, Lamont G B (2000) Evolutionary Algorithms for Solving Multiobjective Problems, Kluwer Academic Publishers, pp. 455–476
4. Zitzler E, Deb K, Thiele L (2000) Comparison of Multiobjective Evolutionary Algorithms: Empirical Result, Evolutionary Computation Journal, 8(2):125–148
5. Glover F, Laguna M (1993) Modern Heuristic Techniques for Combinatorial Problems, In: Modern Heuristic Techniques for Combinatorial Problems. Halsted Press, John Wiley & Sons, Inc., Chapter 3, pp. 70–147
6. Glover F (1994) Tabu Search for Nonlinear and Parametric Optimization (with links to genetic algorithms), Discrete Applied Mathematics 40:231–255
7. Jaszkiewicz A (1999) Genetic Local Search for Multiple Objective Combinatorial Optimization. Technical Report RA-014/98, Institute of Computing Science, Poznan University of Technology
8. Jiménez F, Verdegay J L (1998) Constrained Multiobjective Optimization by Evolutionary Algorithms. In: Proceedings of the International ICSC Symposium on Engineering of Intelligent Systems (EIS'98). University of La Laguna, Tenerife, Spain
9. Deb K (1999) Multiobjective Genetic Algorithms: Problem Difficulties and Construction of Test Problems, Evolutionary Computation Journal, 7(3):205–230
10. Deb K, Thiele L, Laumanns M, Zitzler E (2002) Scalable Multi-Objective Optimization Test Problems. In: Congress on Evolutionary Computation (CEC'2002) Piscataway, New Jersey, IEEE Service Center, Volume 1, pp. 825–830
11. Deb K, Pratap A, Mayaravian T (2001) Constrained Test Problems for Multi-objective Evolutionary Optimization In: E. Zitzler, Kalyanmoy D, Thiele L, Coello Coello C, Corne D, (eds) First International Conference On Evolutionary Multi-criterion Optimization, Springer-Verlag. Lecture Notes in Computer Science No. 1993:284–298
12. Deb K, Gupta V (2005) A Constraint Handling Strategy for Robust Multi-Criteria Optimization. KanGAL Report Number 2005001
13. Makarov I M, Vinogradskaia T M, Rubinski A A, Sokolov V B (1982) Choice Theory and Decision Making, Nauka, Moscow, pp. 228–234
14. Michalewickz Z, Schoenauer M. (1996) Evolutionary Algorithms for Constrained Parameter Optimization Problems. Evolutionary Computation, 4(1):1–32

15. Richardson J T, Palmer M R, Liepins G, Hilliard M (1989) Some Guidelines for Genetic Algorithms with Penalty Functions. In: Schaffer J. D. (eds), Proceedings of the Third International Conference on Genetic Algorithms, George Mason University. Morgan Kaufmann Publishers, pp. 191–197
16. Binh T T, Korn U (1997) MOBES: A Multiobjective Evolution Strategy for Constrained Optimization Problems, The Third International Conference on Genetic Algorithms (Mendel 97), Brno, Czech Republic, pp. 176–182

Feature Selection for Heterogeneous Ensembles of Nearest-neighbour Classifiers Using Hybrid Tabu Search

Muhammad A. Tahir and James E. Smith

School of Computer Science, University of the West of England, UK
Muhammad.Tahir@uwe.ac.uk James.Smith@uwe.ac.uk

Abstract

The nearest-neighbour (NN) classifier has long been used in pattern recognition, exploratory data analysis, and data mining problems. A vital consideration in obtaining good results with this technique is the choice of distance function, and correspondingly which features to consider when computing distances between samples. In this chapter, a new ensemble technique is proposed to improve the performance of NN classifiers. The proposed approach combines multiple NN classifiers, where each classifier uses a different distance function and potentially a different set of features (feature vector). These feature vectors are determined for each distance metric using a Simple Voting Scheme incorporated in Tabu Search (TS). The proposed ensemble classifier with different distance metrics and different feature vectors (TS–DF/NN) is evaluated using various benchmark data sets from the UCI Machine Learning Repository. Results have indicated a significant increase in the performance when compared with various well-known classifiers. The proposed ensemble method is also compared with an ensemble classifier using different distance metrics but with the same feature vector (with or without Feature Selection (FS)).

Key words: Nearest Neighbour, Tabu Search, Ensemble Classifier, Feature Selection

1 Introduction

The nearest-neighbour (NN) classifier has long been used in pattern recognition, exploratory data analysis, and data mining problems. Typically, the k nearest neighbours of an unknown sample in the training set are computed using a predefined distance metric to measure the similarity between two samples. The class label of the unknown sample is then predicted to be the most frequent one occurring in the k nearest-neighbours. The NN classifier is well explored in the literature and has been proved to have good classification performance on a wide range of real-world data sets [1–3, 27].

The idea of using multiple classifiers instead of a single best classifier has aroused significant interest during the last few years. In general, it is well known that an ensemble of classifiers can provide higher accuracy than a single best classifier if the member classifiers are diverse and accurate. If the classifiers make identical errors, these errors will propagate and hence no accuracy gain can be achieved in combining classifiers. In addition to diversity, accuracy of individual classifiers is also important, since too many poor classifiers can overwhelm the correct predictions of good classifiers [37]. In order to make individual classifiers diverse, three principle approaches can be identified:

- Each member of the ensemble is the same type of classifier, but has a different training set. This is often done in an iterative fashion, by changing the probability distribution from which the training set is resampled. Well-known examples are bagging [24] and boosting [25].
- Training multiple classifiers with different inductive biases to create diverse classifiers, e.g. "stacking" approach [38].
- Using the same training data set and base classifiers, but employing feature selection so that each classifier works with a specific feature set and therefore sees a different snapshot of the data. The premise is that different feature subsets lead to diverse individual classifiers, with uncorrelated errors.

Specific examples of these three different approach can be found in the literature relating to NN techniques. Bao et al. [10] followed the second route, and proposed an ensemble technique where each classifier used a different distance function. However, although this approach does use different distance metrics, it uses the same set of features, so it is possible that some errors will be common, arising from features containing noise, which have high values in certain samples. An alternative approach is proposed by Bay [15] following the third route: each member of the ensemble uses the same distance metric but sees a different randomly selected subset of the features.

Here we propose and evaluate a method which combines features of the second and third approaches, with the aim of taking some initial steps towards the automatic creation and adaptation of classifiers tuned to a specific data set. Building on [10,15], we explore the hypothesis that the overall ensemble accuracy can be improved if the choices of subsets arise from

- iterative heuristics such as tabu search [17] rather than random sampling
- different distance metrics rather than single distance metric.

Furthermore we hypothesise that these choices are best co-adapted, rather than learnt separately, as co-adaptation may permit implicit tackling of the problem of achieving ensemble diversity. In order to do this, and to distinguish the effects of different sources of benefits, a novel ensemble classifier is proposed that consists of multiple NN classifiers, each using a different distance metric and a feature subset derived using tabu search. To increase the diversity, a simple voting scheme is introduced in the cost function of Tabu Search. The proposed ensemble NN classifier (DF–TS–1NN) is then compared with various well-known classifiers.

The rest of this chapter is organized as follows. Section 2 provides review on Feature Selection Algorithms. Section 3 describes a proposed multiple distance function ensemble classifier, followed by experiments in Section 4. Sect. 5 concludes the paper.

2 Feature Selection Algorithms (a Review)

The term feature selection refers to the use of algorithms that attempt to select the best subset of the input feature set. It has been shown to be a useful technique for improving the classification accuracy of NN classifiers [7, 8]. It produces savings in the measuring features (since some of the features are discarded) and the selected features retain their original physical interpretation [9]. Feature selection is used in the design of pattern classifiers with three goals [9, 11]:

1. to reduce the cost of extracting features
2. to improve the classification accuracy
3. to improve the reliability of the estimation of performance.

The feature selection problem can be viewed as a multiobjective optimization problem since it involves minimizing the feature subset and maximizing classification accuracy. Mathematically, the feature selection problem can be formulated as follows. Suppose X is an original feature vector with cardinality n and $\bar{X}$ is the new feature vector with cardinality $\bar{n}$, $\bar{X} \subseteq X$, $J(\bar{X})$ is the selection criterion function for the new feature vector $\bar{X}$. The goal is to optimize $J(\ )$. The problem is NP-hard [29, 30]. Therefore, the optimal solution can only be achieved by performing an exhaustive search in the solution space [1]. However, an exhaustive search is feasible only for small n. A number of heuristic algorithms have been proposed for feature selection to obtain near-optimal solutions [9, 11, 12, 31–34].

The choice of an algorithm for selecting the features from an initial set depends on n. The feature selection problem is said to be of small scale, medium scale, or large scale for n belonging to the intervals [0,19], [20,49], or [50,∞], respectively [11, 12]. Sequential Forward Selection (SFS) [35] is the simplest greedy sequential search algorithm. Other sequential algorithms such as Sequential Forward Floating Search (SFFS) and Sequential Backward Floating Search (SBFS) are more efficient than SFS and usually find fairly good solutions for small and medium scale problems [32]. However, these algorithms suffer from the deficiency of converging to local optimal solutions for large scale problems when $n > 100$ [11, 12]. Recent iterative heuristics such as tabu search and genetic algorithms have proved to be effective in tackling this category of problems, which are characterized by having an exponential and noisy search space with numerous local optima [12, 17, 33, 36].

Tabu search (TS) has been applied to the problem of feature selection by Zhang and Sun [12]. In their work, TS performs the feature selection in combination with an objective function based on the Mahalanobis distance. This objective function is used to evaluate the classification performance of each subset of the features selected by the TS. The feature selection vector in TS is represented by a binary string where a 1 or 0 in the position for a given feature indicates the presence or absence of that feature in the solution. Their experimental results on *synthetic data* have shown that TS not only has a high probability of obtaining an optimal or near-optimal solution, but also requires less computational effort than other suboptimal and genetic algorithm based methods. TS has also been successfully applied in other feature selection problems [8, 13, 14].

3 Proposed Ensemble Multiple Distance Function Classifier (DF–TS–1NN)

In this section, we discuss the proposed ensemble multiple distance function TS/1NN classifier (DF–TS–1NN). The use of n classifiers, each with a different distance function and potentially different set of features is intended to increase the likelihood that the errors of individual classifiers are not correlated. In order to achieve this it is necessary to find appropriate feature sets *within the context of the ensemble as a whole*. However with F features the search space is of size $2^{F \cdot n}$. Initial experiments showed that in order to make the search more tractable it is advantageous to hybridize the global nature of TS in the whole search space, with local search acting only within the sub-space of the features of each classifier. Figure 1 shows the training phase of the proposed classifier.

During each iteration, N random neighbours with *Hamming Distance* 1 from the current feature set FV_i are generated for each classifier $i \in \{1, \dots, n\}$ and evaluated using the NN error rate for the appropriate distance metric Di. From the set of N neighbours, the M best are selected for each classifier.[1] All M^n possible combinations

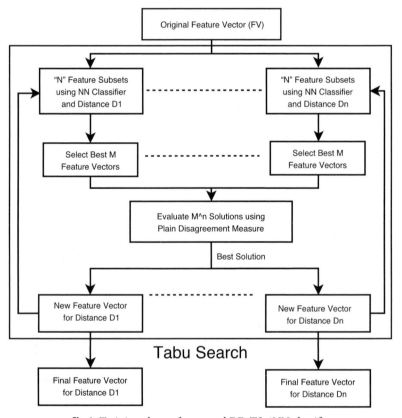

Fig. 1 Training phase of proposed DF–TS–1NN classifier

---

[1] In this study $M = 2$, $n = 5$, and $N = \sqrt{F}$, where F = Total Number of Features

are then evaluated using a simple voting scheme (SVS) and the best is selected to go forward to the next iteration. Thus, the feedback from the SVS allows TS to search iteratively for combinations of feature vectors that improve the classification accuracy. Implicitly it seeks feature vectors for the different distance measures whereby the errors are not correlated, and so provides diversity. By using n distance functions, n feature vectors are obtained using TS in the training phase. In the testing phase, the n NN classifiers with their different feature vectors are combined as shown in Fig. 2.

In the following subsections, feature selection using TS and the various distance metrics described in this paper are discussed as they are at the heart of the proposed algorithm.

3.1 Distance Metrics

The following five distance metrics are used for NN classifiers. All metrics are widely used in the literature.

- Squared Euclidean Distance: $E = \sum_{i=1}^{m}(x_i - y_i)^2$

- Manhattan Distance: $M = \sum_{i=1}^{m}(x_i - y_i)$

- Canberra Distance: $C = \sum_{i=1}^{m}(x_i - y_i)/(x_i + y_i)$

- Squared chord distance: $S_c = \sum_{i=1}^{m}(\sqrt{x_i} - \sqrt{y_i})^2$

- Squared Chi-squared distance: $C_s = \sum_{i=1}^{m}(x_i - y_i)^2/(x_i + y_i)$

  where x and y are the two input vectors and m is the number of features.

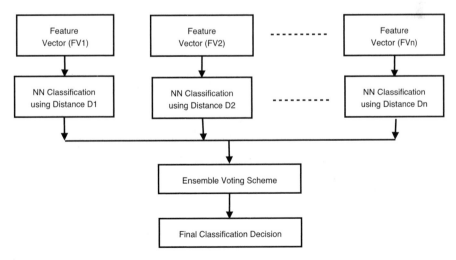

Fig. 2 Testing Phase

3.2 Feature Selection and Diversity using Tabu Search

TS was introduced by Glover [5, 6] as a general iterative metaheuristic for solving combinatorial optimization problems. TS is conceptually simple and elegant. It is a form of local neighbourhood search which starts from an initial solution, and then examines feasible neighbouring solutions. It moves from a solution to its best admissible neighbour, even if this causes the objective function to deteriorate. To avoid cycling, solutions that were recently explored are declared forbidden or tabu for a number of iterations. The tabu list stores a characterization of the moves that led to those solutions. The tabu status of a solution is overridden when certain criteria (aspiration criteria) are satisfied. Sometimes intensification and diversification strategies are used to improve the search. In the first case, the search is accentuated in promising regions of the feasible domain. In the second case, an attempt is made to consider solutions over a broader area of the search space and so provide it with a global nature. The flow chart of the TS algorithm is given in Table 1.

Table 1 Algorithm Tabu Search (TS)

Ω	: **Set of feasible solutions**
S	: Current Solution
S^*	: Best admissible solution
$Cost$	: Objective function
$N(S)$	: Neighbourhood of solution S
V^*	: Sample of neighbourhood solutions
T	: Tabu list
AL	: Aspiration Level

Begin
1.   Start with an initial feasible solution $S \in \Omega$.
2.   Initialize tabu list and aspiration level.
3.   For fixed number of iterations Do
4.   Generate neighbour solutions $V^* \subset N(S)$.
5.   Find best $S^* \in V^*$.
6.   If move S to S^* is not in T Then
7.   Accept move and update best solution.
8.   Update tabu list and aspiration level.
9.   Increment iteration number.
10.   Else
11.   If $Cost(S^*) < AL$ Then
12.   Accept move and update best solution.
13.   Update tabu list and aspiration level.
14.   Increment iteration number.
15.   End If
16.   End If
17.   End For
    End

The size of the tabu list can be determined by experimental runs, watching for the occurrence of cycling when the size is too small, and the deterioration of solution quality when the size is too large [16]. Suggested values of tabu list include $Y, \sqrt{Y}$ (where Y is related to problem size, e.g. number of modules to be assigned in the quadratic assignment problem (QAP), or the number of cities to be visited in the travelling salesman problem (TSP), and so on) [17].

Objective Function
A simple voting scheme is used in each instance of n classifiers. The objective function is the number of instances incorrectly classified using a simple voting scheme. The objective is to minimize

$$Cost = \sum_{i=1}^{S} C_i \qquad (1)$$

where S is the number of samples, $C_i = 1$ if instance is classified incorrectly after simple voting in n classifiers, else $C_i = 0$.

Initial Solution
The feature selection vector is represented by a 0/1 bit string where 0 indicates that the feature is not included in the solution while 1 indicates that it is. All features are included in the initial solution.

Neighbourhood Solutions
During each iteration, N random neighbours with *Hamming Distance* 1 (HD1) are generated for the feature set for each classifier and evaluated using the NN error rate with the appropriate distance metric as the cost function. Neighbours are generated by randomly adding or deleting a feature from the feature vector of size F. Among the neighbours, M best solutions are selected, yielding M possible classifiers for each of the n distance metrics. The M^n resulting ensembles are then evaluated using Equation (1) and the one with the best cost (i.e. the solution which results in the minimum value of Equation (1)) is selected and considered as a new current solution for the next iteration. Note that these ensembles may be quickly evaluated since we pre-computed the decision of each of the $M \times n$ classifiers during the local search phase. Figure 3 shows an example showing neighbourhood solutions during one iteration. Let us assume that the cost of the three different feature subsets in the solution are 50, 48, and 47 using distance metrics 1, 2, and 3, respectively. $N = 4$ neighbours are then randomly generated for each distance metric using $HD1$. $M = 2$ best solutions are selected and $M^n = 2^3 = 8$ solutions are evaluated using the ensemble cost function. The best solution is then selected for the next iteration.

Tabu Moves
A tabu list is maintained to avoid returning to previously visited solutions. In our approach, if an ensemble solution (move) is selected at iteration i, then selecting the same ensemble solution (move) for T subsequent iterations (tabu list size) is tabu.

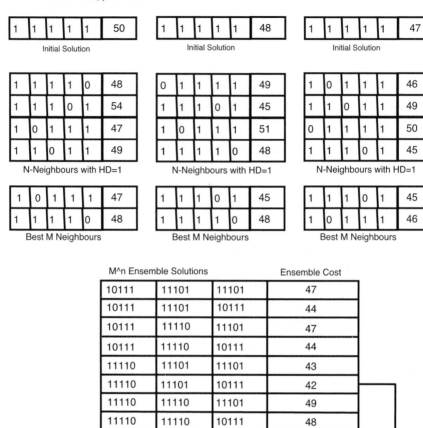

Fig. 3 An example showing neighbourhood solutions during one iteration of the proposed TS method. $n = 3$, $N = 4$, and $M = 2$

Aspiration Criterion

The aspiration criterion is a mechanism used to override the tabu status of moves. It temporarily overrides the tabu status if the move is sufficiently good. In our approach, if an ensemble solution is selected at iteration i and this move results in a best cost for all previous iterations, then that solution is selected even if that feature is in the tabu list.

Termination Rule

The most commonly used stopping criteria in TS are

- after a fixed number of iterations
- after some number of iterations when there has been no increase in the objective function value
- when the objective function reaches a pre-specified value.

In this work, the termination condition is a fixed number of iterations.

4 Experiments

To evaluate the effectiveness of our method, extensive experiments were carried out, and comparisons with several methods performed.

4.1 Methods

The proposed (DF–TS–1NN) algorithm is compared with the following methods. All methods are implemented using the WEKA library [26].

- Decision Tree Method (C4.5): A classifier in the form of a tree structure, where each node is either a leaf node or a decision node [3, 20].
- Decision Table (DT): It uses a simple decision table majority classifier [21].
- Random Forest (RF): Ensemble Classifier using a forest of random trees [22].
- Naive Bayes Algorithm (NBayes): The Naive Bayes Classifier technique is based on Bayes' theorem. Despite its simplicity, Naive Bayes can often outperform numerous sophisticated classification methods [23].
- Bagging: A method for generating multiple versions of a predictor and using these to get an aggregated predictor (ensemble) [24]. C4.5 is used as base classifier.
- AdaBoost1: A meta-algorithm for constructing ensembles which can be used in conjunction with many other learning algorithms to improve their performance [25]. C4.5 is used as base classifier.

In addition, we compare the following variations of the proposed ensemble algorithms:

1. DF–1NN: Ensemble Classifier using NN classifiers with each classifier having different distance metrics (DF) and without FS.
2. DF–TS1–1NN: Ensemble Classifier using NN classifiers, each using a different distance metric. FS using TS is applied independently for each data set.
3. DF–TS2–1NN: Ensemble Classifier as above but with a single common feature set selected by TS. Subsets for various distance metrics are derived using TS.
4. DF–TS3–1NN: Proposed Ensemble Classifier. Different feature subsets for each classifier derived simultaneously using TS.

4.2 Data Sets Descriptions and Experimental Setup

We have performed a number of experiments and comparisons with several benchmarks from the UCI [4] in order to demonstrate the performance of the proposed classification system. A short description of the benchmarks used, along with TS runtime parameters are given in Table 2.

The tabu list size and number of neighbourhood solutions are determined using the following equation:

$$T = N = ceil\left(\sqrt{F}\right) \tag{2}$$

where T is the tabu list size, N is the number of neighbourhood solutions and F is the number of features.

Table 2 Data sets description. P = Prototypes, F = Features, C = Classes, T = Tabu list size, N = Number of neighbourhood solutions

Name	P	F	C	T	N
Statlog Diabetes	768	8	2	3	3
Statlog Heart	270	13	2	4	4
Statlog Australian	690	14	2	4	4
Statlog Vehicle	846	18	4	5	5
Statlog German	1000	20	2	5	5
Breast Cancer	569	32	2	6	6
Ionosphere	351	34	2	6	6
Sonar	208	60	2	8	8
Musk	476	166	2	13	13

In all data sets, *B*-fold cross-validation has been used to estimate error rates [18]. For *B*-fold CV, each data set is divided into *B* blocks using *B*–1 blocks as a training set and the remaining block as a test set. Therefore, each block is used exactly once as a test set. Each experiment was run 100 times using different random 10-CV partitions and the results were averaged over the 100 runs [19].

The number of iterations for FS using TS is 200 for all data sets, which was chosen after preliminary experimentation.

In order to offset any bias due to the different range of values for the original features in the NN classifier, the input feature values are normalized over the range [1,10] using Equation (3) [7]. Normalizing the data is important to ensure that the distance measure allocates equal weight to each variable. Without normalization, the variable with the largest scale will dominate the measure.

$$x'_{i,j} = \left(\frac{x_{i,j} - \min_{k=1...n} x_{(k,j)}}{\max_{k=1...n} x_{(k,j)} - \min_{k=1...n} x_{(k,j)}} * 10 \right) \tag{3}$$

where $x_{i,j}$ is the *j*th feature of the *i*-th pattern, $x'_{i,j}$ is the corresponding normalized feature, and *n* is the total number of patterns.

4.3 Comparison of Different ways of Creating Feature Sets

Table 3 shows the classification accuracy using various distance functions within single classifiers, and for the ensemble technique without feature selection. As can be seen, on some data sets there is a wide discrepancy between the accuracy obtained with different distance metrics. With the simple voting scheme used here the votes of the less accurate classifiers can dominate, so that the ensemble performs worse than the best single classifier on those datasets.

Table 4 shows the classification accuracy using various distance functions and with FS and compared with the various variations of the proposed method. Comparing the results for individual classifiers with feature selection ($\{E, M, C, C_s, S_c\}$) to those without (Table 3) it can be seen that the accuracy is increased in every case – a nice example of the value of performing feature selection.

Table 3 Classification accuracy (%) using individual classifiers and various variations of the proposed classifier. M = Manhattan, E = Euclidean, C = Canberra, C_s = Chi-squared, S_c = Squared-chord

Data Set	E	M	C	C_s	S_c	DF–1NN
Australian	82.1	82.0	**85.7**	82.3	82.4	84.0
Breast Cancer	95.3	95.2	95.2	95.4	95.4	**95.6**
Diabetes	**70.5**	69.7	66.0	69.4	69.6	70.2
German	70.9	71.1	70.2	70.5	70.0	**71.8**
Heart	78.1	79.6	**80.8**	79.0	78.3	79.0
Ionosphere	87.0	90.7	**92.2**	89.1	89.0	90.3
Musk	85.4	83.3	84.0	**86.1**	86.0	86.0
Sonar	82.5	84.6	**86.6**	86.0	86.4	85.4
Vehicle	69.6	69.5	69.6	70.4	70.4	**70.7**

Turning to the use of feature selection to derive a common subset for all classifiers (DF–TS2–1NN), not only do we see improved performance compared to the same algorithm without feature selection (DF–1NN in Table 3), but now the mean accuracy is higher than the best individual classifier on most data sets. This is a good example, which indicates that in order for ensembles to work well, the member classifiers should be accurate.

The other condition for ensembles to work well is diversity, and the performance improves further when feature selection is done independently for each classifier (DF–TS1–1NN), as they can now use potentially different feature sets. However, this approach only implicitly (at best) tackles the diversity issue, and the performance is further increased when different feature subsets co-adapt, so that each feature set is optimized in the context of the ensemble as whole (DF–TS3–1NN). In all but two cases our proposed method (DF–TS3–1NN) outperforms the others and the means differ by more than the combined standard deviations, indicating a high probability that these are truly significantly different results. In the two cases where DF–TS1–1NN

Table 4 Mean and standard deviation of classification accuracy (%) using individual classifiers and variations of the proposed classifier. M = Manhattan, E = Euclidean, C = Canberra, C_s = Chi-squared, S_c = Squared-chord

Data Set	E	M	C	C_s	S_c	DF–TS1–1NN	DF–TS2–1NN	DF–TS3–1NN
Australian	86.5	88.1	86.4	85.9	86.8	89.0(0.61)	85.1(0.45)	**90.5**(0.48)
Breast Cancer	97.4	97.8	97.5	97.4	97.5	97.9(0.22)	97.6(0.32)	**98.0**(0.25)
Diabetes	71.7	70.8	71.1	70.1	70.3	**75.5**(0.71)	72.5(0.82)	74.5(0.85)
German	72.3	73.8	74.1	74.5	73.4	76.5(0.63)	74.2(0.71)	**79.8**(0.62)
Heart	83.2	82.6	82.2	84.0	83.0	85.0(1.11)	83.8(1.45)	**86.3**(0.90)
Ionosphere	93.3	95.4	96.2	91.1	94.3	95.3(0.41)	95.1(0.37)	**96.3**(0.52)
Musk	91.2	89.9	89.8	92.3	91.8	91.6(0.67)	92.3(0.82)	**94.5**(0.72)
Sonar	91.0	90.9	93.1	91.5	93.0	93.5(0.82)	93.4(1.00)	**94.7**(1.09)
Vehicle	73.9	75.1	74.2	74.9	74.2	**77.2**(0.60)	74.5(0.59)	76.9(0.61)

has a higher observed mean than DF–TS3–1NN, the differences are less than the standard deviation of either set of results, so they are almost certainly not significant.

Table 5 shows the number of features used by the proposed classifier for various data sets. Different features have been used by the individual classifiers that are part of the whole ensemble classifier, thus increasing diversity and producing an overall increase in the classification accuracy. F_{Common} represents those features that are common for ensemble classifier, i.e. that are used by each classifier. As can be seen on most data sets there are few, if any, features that are used by every classifier. This is a cause of diversity among the decisions of the different classifiers, and the fact that these feature sets are learnt rather than simply assigned at random is responsible for the different classifiers all remaining accurate – the other pre-requisite for successful formation of an ensemble.

4.4 Comparison with other Algorithms

Table 6 shows results of a comparison of classification accuracy (in %) between the proposed DF–TS–1NN classifier and others for different data sets. The proposed algorithm achieved higher accuracy on all data sets except Diabetes.

Table 5 Total number of features used by proposed classifier. F_T = Total available features, F_M = Feature using Manhattan distance, F_E = Features using Euclidean distance, F_C = Features using Canberra distance, F_{C_s} = Features using chi-squared distance, F_{S_c} = Feature using squared-chord distance.

Data Set	F_T	F_E	F_M	F_C	F_{C_s}	F_{S_c}	F_{Common}	$F_{Ensemble}$
Australian	14	5	9	9	7	5	1	14
Breast Cancer	32	19	13	15	21	13	3	28
Diabetes	8	3	5	1	3	5	0	8
German	20	9	13	13	13	15	3	19
Heart	13	10	8	10	6	8	2	13
Ionosphere	34	11	13	15	11	11	2	26
Musk	166	84	74	76	86	90	0	124
Sonar	60	31	33	27	35	33	0	58
Vehicle	18	9	11	13	7	13	0	17

Table 6 Average classification accuracy (%) using different classifiers. DT = Decision table. RF = Random forest

Data Set	C4.5	DT	RF	NBayes	Bagging	AdaBoost	1NN	DF–TS3–1NN
Australian	84.3	84.7	86.1	77.1	86.0	85.0	79.6	**90.5**
Breast Cancer	93.6	93.3	95.9	93.3	95.35	96.1	95.4	**98.0**
Diabetes	74.3	74.1	74.7	75.6	**76.0**	72.4	70.3	74.5
German	71.6	72.5	74.7	74.5	74.6	72.50	70.9	**79.8**
Heart	78.2	82.3	80.2	84.0	80.5	79.2	75.7	**86.3**
Ionosphere	89.8	94.2	95.4	92.8	92.2	90.3	87.5	**96.3**
Musk	82.7	80.8	87.8	73.9	88.2	90.0	85.6	**94.5**
Sonar	73.0	72.6	80.3	67.9	78.5	80.1	86.5	**94.7**
Vehicle	72.7	66.4	74.7	45.4	74.5	76.4	69.7	**76.9**

- For Australian, German and Ionosphere data sets there is improvement of 1.98%, 5.06% and 0.4% respectively when compared with the best of the other methods (Random Forest Classifier).
- For Heart, there is an improvement of 3.3% when compared with the best of the other methods (Naive Bayes Classifier).
- For Vehicle, Breast Cancer and Musk data sets, there is an improvement of 0.5%, 0.76%, and 4.55% respectively when compared with the best of the other methods (AdaBoost).
- For Sonar, there is an improvement of 7.8% when compared with the best of the other methods (1NN).
- Since Diabetes has only eight features, the proposed algorithm is unable to combine the benefits of feature selection and ensemble classifiers using different distance metrics.

As can be seen, the proposed method performs consistently well and outperforms other methods on all but one data set. Moreover, for the other methods there is considerable variation in performance according to how well the indicative bias of each method suits each data set. It is worth noting that the two methods of producing ensembles always improve the performance compared to the base C4.5 classifiers, apart from Ada-Boost on the Diabetes data set.

Figure 4 shows the standard deviation obtained over 100 runs of random 10-fold cross-validation of each data set for different algorithms. From the graph, it is clear that the standard deviation of the proposed classifier compares favorably with other

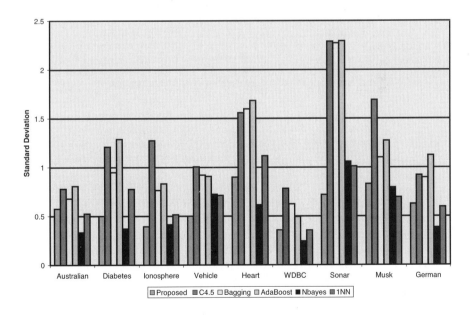

Fig. 4 Standard deviation for different algorithms on various data sets

algorithms, and is usually less than the observed difference in mean accuracies, suggesting that these are significant. In particular it is always less than that of the two other boosting algorithms. Thus if we think in terms of the Bias-Variance decomposition of classifier errors, it might initially appear that both the bias and the variance terms are reduced for this method, but this must be studied in more detail.

4.5 Analysis of Learning

Figures 5–7 show the classification accuracy (%) versus number of iterations for Australian, Ionosphere and German data sets using one run of the solution search space using TS. The figure clearly indicates that TS focuses on a good solution space. The proposed TS algorithm progressively zooms towards a better solution subspace as time elapses; a desirable characteristics of approximation iterative heuristics.

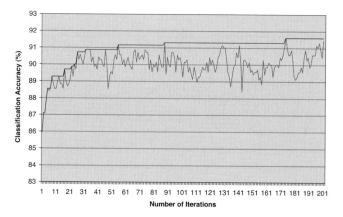

Fig. 5 Error rate vs number of iterations for Australian data set

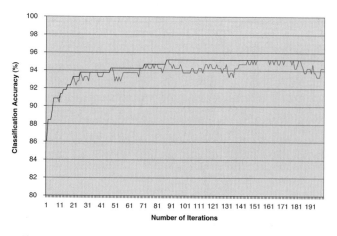

Fig. 6 Error rate vs number of iterations for Ionosphere data set

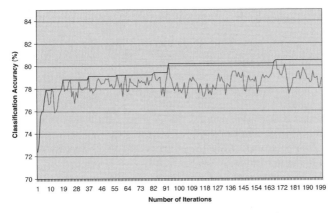

Fig. 7 Error rate vs number of iterations for German data set

5 Conclusions

A new ensemble technique is proposed in this paper to improve the performance of NN classifiers. The proposed approach combines multiple NN classifiers, where each classifier uses a different distance function and potentially a different set of features (feature vector). These feature vectors are determined using a combination of Tabu Search (at the level of the ensemble) and simple local neighbourhood search (at the level of the individual classifiers).

We show that rather than optimizing the feature set independently for each distance metric, it is preferable to co-adapt them, so that each feature set is optimized in the context of the ensemble as whole. This approach also implicitly deals with the problem tackled by many authors, namely of how to find an appropriate measure for the diversity of an ensemble so that it can be optimized. Our solution is to simply do this explicitly by letting TS operate, using the ensemble error rate as its cost function.

The proposed ensemble DF–TS–1NN classifier is evaluated using various benchmark data sets from the UCI Machine Learning Repository. Results indicate a significant increase in performance compared with other different well-known classifiers.

This work is intended as a step towards the automatic creation of classifiers tuned to specific data sets. Having done our initial "proof of concept", the next stages of this research programme will be concerned with automating the choice of distance metric and k for each of our $k - NN$ classifiers. We will also consider ways of automatically selecting subsets of the training examples to use for classification, as a way of tackling the well-known scalability problems of NN as the number of training examples increases.

Acknowledgement

This work is supported by the European Commission (Project No. STRP016429, acronym DynaVis). This publication reflects only the authors' views.

References

1. T.M. Cover, and P.E. Hart (1967). *Nearest Neighbor Pattern Classification.* IEEE Transactions on Information Theory. *13(1)*, 21–27
2. C. Domeniconi, J. Peng, and D. Gunopulos (2002). *Locally Adaptive Metric Nearest-Neighbor Classification.* IEEE Transactions on Pattern Analysis and Machine Intelligence. *24(9)*, 1281–1285
3. D. Michie, D.J. Spiegelhalter and C.C. Taylor (1994). *Machine Learning, Neural and Statistical Classification.* Ellis Horwood
4. C. Blake, E. Keogh, and C. J. Merz. UCI Repository of machine learning databases, University of California, Irvine
5. F. Glover (1989). *Tabu Search I.* ORSA Journal on Computing, *1(3)*, 190–206
6. F. Glover (1990). *Tabu Search II.* ORSA Journal on Computing, *2(1)*, 4–32
7. M.L. Raymer et al (2000). *Dimensionality Reduction using Genetic Algorithms.* IEEE Transactions on Evolutionary Computation, *4(2)*, 164–171
8. M.A. Tahir et al (2006). *Novel Round-Robin Tabu Search Algorithm for Prostate Cancer Classification and Diagnosis using Multispectral Imagery.* IEEE Transactions on Information Technology in Biomedicine, *10(4)*, 782–793
9. A.K. Jain, and R.P.W. Duin, and J. Mao (2000). *Statistical Pattern Recognition: A Review.* IEEE Transactions on Pattern Analysis and Machine Intelligence, *22(1)*, 4–37
10. Y. Bao and N. Ishii and X. Du (2004). *Combining Multiple k-Nearest Neighbor Classifiers Using Different Distance Functions.* Lecture Notes in Computer Science (LNCS 3177), 5th International Conference on Intelligent Data Engineering and Automated Learning (IDEAL 2004), Exeter, UK.
11. M. Kudo and J. Sklansky (2000). *Comparison of Algorithms that Select Features for Pattern Classifiers.* Pattern Recognition, *33*, 25–41
12. H. Zhang and G. Sun (2002). *Feature Selection using Tabu Search Method.* Pattern Recognition,, *35*, 701–711
13. M.A. Tahir, A. Bouridane, and F. Kurugollu (2007). *Simultaneous Feature Selection and Feature Weighting using Hybrid Tabu Search/K-Nearest Neighbor Classifier.* Pattern Recognition Letters, 28, 2007
14. D. Korycinski, M. Crawford, J. W Barnes, and J. Ghosh (2003). *Adaptive Feature Selection for Hyperspectral Data Analysis using a Binary Hierarchical Classifier and Tabu Search.* Proceedings of the IEEE International Geoscience and Remote Sensing Symposium, IGARSS
15. S.D. Bay (1998). *Combining Nearest Neighbor Classifiers Through Multiple Feature Subsets* Proceedings of the Fifteenth International Conference on Machine Learning, 37–45
16. F. Glover, E. Taillard, and D. de Werra (1993). *A User's Guide to Tabu Search.* Annals of Operations Research, *41*, 3–28
17. S.M. Sait and H. Youssef (1999). *General Iterative Algorithms for Combinatorial Optimization.* IEEE Computer Society
18. S. Raudys and A. Jain (1991). *Small Sample Effects in Statistical Pattern Recognition: Recommendations for Practitioners* IEEE Transactions on Pattern Analysis and Machine Intelligence, *13(3)*, 252–264
19. R. Paredes and E. Vidal (2006). *Learning Weighted Metrics to Minimize Nearest-Neighbor Classification Error* IEEE Transactions on Pattern Analysis and Machine Intelligence, *28(7)*, 1100–1110
20. J.R. Quinlan (1993). *C4.5: Programs for Machine Learning.* Morgan Kaufmann
21. R. Kohavi (1995). *The Power of Decision Tables.* Proceedings of the 8th European Conference on Machine Learning

22. L. Breiman (2001). *Random Forests.* Machine Learning, *45(1)*, 5–32
23. R. Duda and P. Hart (1973). *Pattern Classification and Scene Analysis* Wiley, New York.
24. L. Breiman (1996). *Bagging Predictors.* Machine Learning, *24(2)*, 123–140
25. Y. Freund and R.E. Schapire (1996). *Experiments with a New Boosting Algorithm.* Proceedings of International Conference on Machine Learning, 148–156
26. I.H. Witten and E. Frank (2005). *Data Mining: Practical Machine Learning Tools and Techniques*, 2nd edn, Morgan Kaufmann, San Francisco
27. J. Wang, P. Neskovic, and L. Cooper (2007). *Improving Nearest Neighbor Rule with a Simple Adaptive Distance Measure* Pattern Recognition Letters, *28*, 207–213
28. M.A. Tahir and J. Smith (2006). *Improving Nearest Neighbor Classifier using Tabu Search and Ensemble Distance Metrics.* Proceedings of the IEEE International Conference on Data Mining (ICDM)
29. E. Amaldi, and V. Kann (1998). *On the Approximability of Minimizing Nonzero Variables or Unsatisfied Relations in Linear Systems.* Theoretical Computer Science, *209*, 237–260
30. S. Davies, and S. Russell (1994). *NP-completeness of Searches for Smallest Possible Feature Sets.* In Proceedings of the AAAI Fall Symposium on Relevance, AAAI Press, 37–39
31. A.K. Jain and D. Zongker (1997). *Feature Selection: Evaluation, Application, and Small Sample Performance.* IEEE Transactions on Pattern Analysis and Machine Intelligence, *19(2)*, 153–158
32. P. Pudil, J. Novovicova, and J. Kittler (1994). *Floating Search Methods in Feature Selection.* Pattern Recognition Letters, *15*, 1119–1125
33. W. Siedlecki and J. Sklansy (1989). *A Note on Genetic Algorithms for Large-scale Feature Selection.* Pattern Recognition Letters, *10(11)*, 335–347
34. S.B. Serpico, and L. Bruzzone (2001). *A New Search Algorithm for Feature Selection in Hyperspectral Remote Sensing Images.* IEEE Transactions on Geoscience and Remote Sensing, *39(7)*, 1360–1367
35. A.W. Whitney (1971). *A Direct Method of Nonparametric Measurement Selection.* IEEE Transactions on Computers, *20(9)*, 1100–1103
36. S. Yu, S.D. Backer, and P. Scheunders (2002). *Genetic Feature Selection Combined with Composite Fuzzy Nearest Neighbor Classifiers for Hyperspectral Satellite Imagery.* Pattern Recognition Letters, *23*, 183–190
37. O. Okun and H. Proosalut (2005). *Multiple Views in Ensembles of Nearest Neighbor Classifiers.* In Proceedings of the ICML Workshop on Learning with Multiple Views, Bonn, Germany, 51–58
38. D.H. Wolpert (1992). *Stacked Generalization,* Neural Networks, *5*, 241–259

A Parallel Ant Colony Optimization Algorithm Based on Crossover Operation

Adem Kalinli[1] and Fatih Sarikoc[2]

[1] Department of Computer Technologies, Erciyes University,
Kayseri Vocational High School, Kayseri, Turkey.
kalinlia@erciyes.edu.tr

[2] Department of Computer Engineering, Erciyes University,
Institute of Science and Technology, Kayseri, Turkey.
fsarikoc@yahoo.com

Abstract

In this work, we introduce a new parallel ant colony optimization algorithm based on an ant metaphor and the crossover operator from genetic algorithms. The performance of the proposed model is evaluated using well-known numerical test problems and then it is applied to train recurrent neural networks to identify linear and non-linear dynamic plants. The simulation results are compared with results using other algorithms.

Key words: Parallel Ant Colony Optimization, Hybrid Algorithms, Continuous Optimization, Recurrent Neural Network, System Identification

1 Introduction

There are many combinatorial optimization problems of the NP-hard type, and they cannot be solved by deterministic methods within a reasonable amount of time. The great difficulty of optimization problems encountered in practical areas such as production, control, communication and transportation has motivated researchers to develop new powerful algorithms. Therefore, several heuristics have been employed to find acceptable solutions for difficult real-world problems. The most popular of these new algorithms include genetic algorithms (GAs), simulated annealing (SA), ant colony optimization (ACO), tabu search (TS), artificial immune system (AIS), and artificial neural networks (ANNs) [1–4]. Although all of these algorithms convergence to a global optimum, they cannot always guarantee optimum solutions to the problem. Therefore, they are called approximate or heuristic algorithms.

The ACO algorithm is an artificial version of the natural optimization process carried out by real ant colonies. The first ACO algorithm was proposed by Dorigo et al. in 1991, and was called an ant system (AS) [5, 6]. Real ants communicate with each other by leaving a pheromone substance in their path, and this chemical substance leads other ants. Thus, stimergy is provided and swarm intelligence emerges in the colony behaviour. The main features of the algorithm are distributed computation, positive feedback and constructive greedy search. Since 1991, several studies have been carried out on new models of the ACO algorithm and their application to difficult optimization problems. Some of these algorithms are known as AS with elitist strategy (AS_{elit}), rank based version of AS (AS_{rank}), MAX-MIN AS and ant colony system (ACS) [7–10]. In most application areas, these algorithms are mainly used for optimization in discrete space [9–13]. In addition, different kinds of ant algorithms, such as continuous ant colony optimization (CACO), API, continuous interacting ant colony (CIAC) and touring ant colony optimization (TACO), have been introduced for optimization in the continuous field [14–17].

It is known that there is a premature convergence (stagnation) problem in the nature of ant algorithms [6]. Therefore, as the problem size grows, the ability of the algorithm to discover the optimum solution becomes weaker. On the other hand, when the problem size and number of parameters increase, parallel implementation of the algorithm could give more successful results [18,19]. Furthermore, ant colony optimization approaches are population based and they are naturally suited to parallel implementation [18–22]. So, these advantages lead us to consider a parallel version of the ant algorithm.

In this work a parallel ant colony optimization (PACO) algorithm based on the ant metaphor and the crossover operator of GAs is described. Our aim is to avoid premature convergence behaviour of the ant algorithm and to benefit from the advantages of a parallel structure. The performance of the proposed PACO algorithm is compared to that of the basic TS, parallel TS (PTS), GA and TACO algorithms for several well-known numerical test problems. Then, it is employed to train a recurrent neural network to identify linear and nonlinear dynamic plants. The second section of the chapter presents information about parallel ant colony algorithms in the literature. In the third section, the basic principles of TACO algorithms are introduced and the proposed model is described. Simulation results obtained from the test functions optimization and an application of PACO to training recurrent neural network are given in the fourth section. The work is concluded in the fifth section.

2 Parallel Ant Colony Algorithms

There are a few parallel implementations of ant algorithms in the literature. The first of these studies is that of Bolondi and Bondanza. They used fine-grained parallelism and assigned each ant to a single processor. Due to the high overhead for communication, this approach did not increase performance with an increased number of processors. Better results have been obtained with a more course-grained model [20,23].

Bullnheimer et al. propose two parallelization strategies, synchronous and partially asynchronous implementations of the ant system [18]. In simulations made on some TSP instances, it is shown that the synchronization and communication overhead slows down the performance. For this reason, the asynchronous parallel version outperforms the synchronous version as it is reduces the communication frequency.

Stützle, using some TSP instances, empirically tests the simple strategy of executing parallel independent short runs of a MAX-MIN ant system [19]. He compares the solution quality of these short runs with the solution quality of the execution of one long run whose running time equals the sum of the running times of the short runs. He shows that using parallel independent runs with different initial solutions is very effective in comparison with a single long run.

Talbi et al. implemented a synchronous master–worker model for parallel ant colonies to solve the quadratic assignment problem [22]. At each iteration, the master broadcasts the pheromone matrix to all the workers. Each worker receives the pheromone matrix, constructs a complete solution by running an ant process, applies a tabu search for this solution as a local optimization method and sends the solution found to the master. According to all solutions, the master updates the pheromone matrix and the best solution found, and then the process is iterated.

Michel et al. propose an island model approach inspired by GAs [24]. In this approach every processor holds a colony of ants and in a fixed number of generations each colony sends its best solution to another colony. If the received new solution is better, then it becomes the new solution for the colony and pheromone updating is done locally depending on this new solution. Thus, the pheromone matrices of colonies may differ from each other.

Delisle et al. presented a shared memory parallel implementation of an ant colony optimization for an industrial scheduling problem in an OpenMP environment [25].

In another implementation, Krüger et al. indicate that it is better to exchange only best solutions found so far than to exchange the whole pheromone matrix [26].

Middendorf et al. show that information exchanges between colonies in small quantities decrease the run time of the algorithm and improve the quality of the solutions in multi-colony ant systems. They also conclude that it is better to exchange local best solutions only with a neighbour in a directed ring and not too often, instead of exchanging the local best solution very often and between all colonies [20].

3 Touring Ant Colony Optimization and Proposed Parallel Model

3.1 Pheromone Based Feedback in Ant System Metaphor

Real ants are capable of finding the shortest path from their nest to a food source, back or around an object. Also, they have the ability to adapt to changes in the environment. Another interesting point is that ants are almost blind, in other words they cannot see well enough to select directions to follow. Studies on ants show that their ability to find the shortest path is the result of chemical communication among them. They use a chemical substance called pheromone to communicate with each other. This type of indirect interaction through modification of the environment, which is called stimergy, is the main idea of ACO algorithms.

Ants deposit a certain amount of pheromone on their path while walking and each ant probabilistically chooses a direction to follow. The probability degree of being the chosen direction depends on the pheromone amount deposited on that direction. If the pheromone amount of all directions is equal, then all directions have the same probability of being preferred by ants. Since it is assumed that the speed of all ants is the same and, therefore, all ants deposit the same amount of pheromone on their paths, shorter paths will receive more pheromone per time unit. Consequently, large numbers of ants will rapidly choose the shorter paths. This positive feedback strategy is also called an auto-catalytic process. Furthermore, the quantity of pheromone on each path decreases over time because of evaporation. Therefore, longer paths lose their pheromone intensity and become less attractive as time passes. This is called a pheromone-based negative feedback strategy.

If there are only a few ants, the auto-catalytic process usually produces a bad-optimal path very quickly rather than an optimal one. Since there are many ants searching simultaneously for the optimum path, the interaction of these auto-catalytic processes causes the search to converge to the optimum path very quickly and to finally find the shortest path between the nest and the food without getting stuck in a sub-optimal path. The behaviour of real ant colonies when finding the shortest path represents a natural adaptive optimization process.

The ant colony optimization algorithm is an artificial version of the natural optimization process carried out by real ant colonies as described above. A simple schematic algorithm modeling the behaviour of real ant colonies can be summarized as below:

BEGIN
Initialize
REPEAT
  Generate the artificial paths for all ants
  Compute the length of all artificial paths
  Update the amount of pheromone on the artificial paths
  Keep the shortest artificial path found up to now
UNTIL (iteration = maxiteration or a criterion is satisfied)
END.

3.2 Touring Ant Colony Optimization Algorithm

In this algorithm, a solution is a vector of design parameters which are coded as a binary bit string. Therefore, artificial ants search for the value of each bit in the string. The concept of the TACO algorithm is shown in Fig. 1.

At the decision stage for the value of a bit, ants use only the pheromone information. Once an ant completes the decision process for the values of all bits in the string,

Fig. 1 An artifical path (solution) found by an ant

it means that it has produced a solution to the problem. This solution is evaluated in the problem and a numerical value showing its quality is assigned to the solution using a function, often called the fitness function. With respect to this value, an artificial pheromone amount is attached to the links, forming the artificial way, between the chosen bits. An ant on the nth bit position chooses the value of 0 or 1 for the bit on the $(n + 1)$th position depending on the probability defined by the following equation:

$$p_{ij}(t) = \frac{[\tau_{ij}]^\alpha}{\sum\limits_{j=1}^{2} [\tau_{ij}]^\alpha} \tag{1}$$

where $p_{ij}(t)$ is the probability associated with the link between bit i and j, $\tau_{ij}(t)$ is the artificial pheromone of the link, α is a weight parameter. Artificial pheromone is computed by the following formula:

$$\Delta\tau_{ij}^k(t, t + 1) = \begin{cases} \frac{Q}{F_k} & \text{if the ant } k \text{ passes the link } (i, j) \\ 0 & \text{otherwise} \end{cases} \tag{2}$$

where $\Delta\tau_{ij}^k$ is the pheromone quantity attached to the link (i, j) by the artificial ant k, Q is a positive constant and F_k is the objective function value calculated using the solution found by the ant k.

After M ants complete the search process and produce their paths, the pheromone amount to be attached to the sub-path $(0 \rightarrow 1)$ between time t and $(t+1)$ is computed as

$$\Delta\tau_{ij}(t, t + 1) = \sum_{k=1}^{M} \Delta\tau_{ij}^k(t, t + 1) \tag{3}$$

The amount of pheromone on the sub-path (i, j) at the time $(t + 1)$ is calculated using the following equation:

$$\tau_{ij}(t + 1) = \rho\tau_{ij}(t) + \Delta\tau_{ij}(t, t + 1) \tag{4}$$

where ρ is a coefficient called the evaporation parameter.

3.3 Parallel Ant Colony Optimization Algorithm

In this work, we introduce a hybrid algorithm model to avoid premature convergence of ant behaviour and to obtain a robust algorithm. Generally, hybrid models utilize the benefits of different algorithms. The proposed parallel ant colony optimization (PACO) algorithm is based on the data structure of the TACO and the crossover operator of GAs. We combine the convergence capability of the ant metaphor and the global search capability of the genetic algorithm.

In the PACO algorithm, each solution is represented by a binary vector of design parameters and artificial ants search for the value of each bit in the string as TACO. For this reason, the proposed algorithm can search a sampled finite subset of continuous space.

The flowchart and pseudocode of the proposed model is given in Fig. 2 and Fig. 3, respectively. In the model, different independent ant colonies are executed in parallel. Each colony has a copy of the same search space with the same initial pheromone quantities. However, it is possible to use different control parameter values for each colony. As the algorithm runs, the pheromone quantities of each copy may be different. A colony does not change the pheromone quantities of another colony. However, they have the ability to exchange information implicitly. The information exchange process between the ant colonies is based on the crossover operation.

Execution of the colonies is stopped after a given number of iterations (*NumAntCycle*). There is no specific rule to determine this number; it may be defined experimentally. *NumAntCycle* is normally chosen to be sufficiently large to allow the search to complete local searching. When all ants complete their paths in a colony, the quality of the path produced by each ant is evaluated and then the best one found is reserved as the local best of the colony. In every fixed number of *NumAntCycle* iterations, local best solutions of each colony are added to the solution population. Later, this population is altered by a crossover procedure to produce a new population. This new population is formed by implementing the crossover operation among the solutions belonging to the previous solution population. After crossover, the best part of the population survives and the solutions of this part are used to update the pheromone quantities of the best paths in each colony. Thus, one epoch of the algorithm is completed. In successive epochs, the search continues, depending on the pheromone quantities updated in the previous epoch. This process is repeated until a predefined number of epochs (*NumOfEpoch*) is completed. *NumOfEpoch* may change according to the problem, so the value of this parameter is experimentally defined.

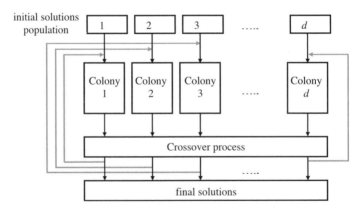

Fig. 2 Flowchart of the PACO algorithm

Step 1) /* initialization */

    For d:=1 to *NumberOfColony* do   /* pheromone initialization */

      For j:=1 to *LenghtOfBitString* do

          $\tau_d(0, j) = c$

          $\tau_d(1, j) = c$

    End

    End

    For d:=1 to *NumberOfColony* do   /* initial population of solutions */

      $BestSolBitStr_d^{old} = FuncRandFeasible()$

Step 2) /* movement of ants */

    For d:=1 to *NumberOfColony* do

      For j:=1 to *LenghtOfBitString* do

        For k:=1 to *NumberOfAnts* do

          Select a bit value 0 or 1 according to equation (5) and (6)

          Assign chosen value to $tabu_d^k(j)$

          chosen bit value $i = tabu_d^k(j)$

          $\tau_d^{new}(i, j) = (1 - \rho)\tau_d^{old}(i, j) + \rho.\tau_0$ /* local update */

        End

      End

    End

Step 3) /* Evaluate ant solutions */

    For d:=1 to *NumberOfColony* do

      For k:=1 to *NumberOfAnts* do

      $CostOfSol_d^k = FuncCost\left(tabu_d^k(j)\right)$ /* j=1 to *LenghtOfBitString* */

      Reserve best solution bit string of each colony as $BestSolBitStr_d^{new}$

      End

    End

Step 4) /* Empty all tabu list */

Step 5) /* Crossover operation */

    If (*NumAntCycle* is completed) Then do   /*crossover condition is provided*/

      Mate $BestSolBitStr_d^{old}$, $BestSolBitStr_d^{new}$

      Produce offspring $BestSolBitStr_d^{offspring}$

      Evaluate cost of offspring solutions

      Survive best part of the population

      Keep the best solutions as $BestSolBitStr_d^{old}$ and the costs as $CostOfBestSol_d$

      For d:=1 to *NumberOfColony* do   /* crossover update */

        For j:=1 to *LenghtOfBitString* do

          $i = BestSolBitStr_d^{old}(j)$

          $\tau_d^{new}(i, j) = (1 - \rho)\tau_d^{old}(i, j) + 1/\left(a + \left|CostOfBestSol_d\right|\right)$

        End

      End

    End-if

Step 6) If (*NumOfEpoch* is completed) Then Stop   /* Stopping criteria is satisfied */

    Else Goto Step 2

Fig. 3 Pseudocode of the proposed algorithm

Different independent ant colonies are sequentially executed in a single processor, for this reason, implementation of the algorithm is virtually parallel. Communication between the colonies is carried out at predetermined moments; therefore the parallelism used in this work is synchronous.

Crossover Procedure

The crossover operator employed by GAs is used to create two new solutions (children) from two existing solutions (parents) in the population. Depending on the method of problem representation in string form, a proper crossover operator must be chosen. When the problem is represented in binary string form, the simplest crossover operation can be applied as follows: two solutions are randomly selected as parent solutions from the population at two randomly selected points. The parts between the points are swapped and two new solutions are produced. A crossover operation can thus yield better solutions by combining the good features of parent solutions. An example of this simple crossover operator is given below:

PresentSolution1   **101 1101 01110**
PresentSolution2   110 0011 11011

NewSolution1   **101 0011 01110**
NewSolution2   110 **1101** 11011

Movements of Ants

In the proposed model, the data representation structure is defined as discrete elements in a matrix form. Rows of this matrix are indicated by the values of i and columns are indicated by the values of j. Since binary data representation is used, i can take the values 0 or 1, but the maximum value of j depends on the parameters of the problem. The element (i, j) of a predefined matrix format addresses a point in the data structure on which artificial ants move as depicted in Fig. 1.

At the beginning of the search, some initial pheromone quantity (c) is allocated to each path of the binary coded search space. In addition, an initial population of random solutions in the feasible region is formed for genetic crossover operations on the succeeding population.

While ants move from one point to another, they search the value of each bit in the string, in other words, they try to decide whether the value of the next bit to be chosen is 0 or 1 according to the state transition rule given in Equation (5).

$$
\text{tabu}_d^k(j) = \begin{cases} \text{argmax}_{i \in \{0,1\}} \{\tau_d(i,j)\} & \text{with the probability of } q_0 \text{ (exploit)} \\ S_d(j) & \text{with the probability of } (1-q_0) \text{ (explore)} \end{cases} \tag{5}
$$

where $\text{tabu}_d^k(j)$ means the jth element of the tabu list for the kth ant in the dth colony. In other words, it represents the selected value for the next bit. $\tau_d(i,j)$ is accumulated pheromone substance on the path (i, j) belonging to the dth colony and $S_d(j)$ is a stochastically found new value of the jth element for the dth colony ($S_d(j) \in \{0, 1\}$). In this equation, ants take a deterministic decision with the probability q_0, which is an initial coefficient balancing deterministic search versus stochastic search. Deterministic search exhausts accumulated pheromone knowledge to find new solutions near to the best one found so far. On the other hand, stochastic search explores possible new paths to enhance the searching area. $S_d(j)$ is defined according to the probability distribution formula

$$P_d(i, j) = \frac{\tau_d(i, j)}{\tau_d(j, 0) + \tau_d(j, 1)}, \quad i \in \{0, 1\} \tag{6}$$

where $P_d(i, j)$ represents probability as an indication of accumulated pheromone attraction of the point (i, j), which is to be selected in the search space of the dth colony. To select a bit value $S_d(j)$ for the jth element of the tabu list, a stochastic decision mechanism can be implemented over the probability distribution formula $P_d(i, j)$. With this formula a roulette wheel mechanism could be used as a decision mechanism in order to define the next value of $S_d(j)$.

After choosing a path each ant updates the pheromone level of the path. This operation is called the *local updating rule* and the formula for the rule is

$$\tau_d^{new}(i, j) = (1 - \rho).\tau_d^{old}(i, j) + \rho.\tau_0 \tag{7}$$

This formula is implemented in order to make the way previously chosen less attractive and to direct the following ants to other paths. ρ is an initial coefficient representing evaporation rate and τ_0 is another coefficient showing the minimum level of pheromone instances in each path.

After a predefined number of cycles (*NumAntCycle*), each colony reserves its best solution found so far and this solution is added to the solution population for the crossover operation. The crossover operation eliminates the worse part of the population and provides a global reinforcement mechanism. Each solution of the surviving part is assigned to one of the colonies, thus, information exchange between colonies is implicitly provided. After this process, pheromone values of each colony are updated depending on the returned solutions according to the formula

$$\tau_d^{new}(i, j) = (1 - \rho).\tau_d^{old}(i, j) + \frac{1}{(a + |cost\_d|)} \tag{8}$$

where cost_d is a value calculated by the cost function related to the assigned solution to the dth colony. a is a constant employed to avoid overflow and scale the pheromone effect of the cost value. Determining a proper value for a is highly dependent on the problem. So, some preliminary experience and knowledge about the range of the cost function values is necessary. This may be considered as a weakness of the algorithm. Furthermore, by employing a crossover procedure, the proposed algorithm moves far from a realistic simulation of ants in order to increase the search capability of the global optimum and to yield better performance.

It is known that there are reports in the literature based on the concepts TACO and Island Model. The proposed algorithm uses a binary data representation, which was earlier used in the TACO algorithm [17]. However, PACO differs from TACO in some features. First, in PACO, selection of the next point depends on a *state transition rule* instead of using only a probability distribution formula. By employing a state transition rule, PACO is able to balance exploitation versus exploration with a certain probability as happened in ACS. Second, after selection of each new point, a *local updating formula* is implemented in order to lead other ants to unselected paths. Moreover, crossover procedure is employed at predetermined intervals to provide information exchange between colonies. Thus, PACO uses synchronous and parallel information exchange structures established in a multi-colony ant system. Island Model is another hybrid algorithm that combines both ACO and GA. However, this model has a different data representation structure and runs more than one processor [24]. Since each processor holds a colony, implementation of the algorithm is more complicated than PACO and TACO.

4 Simulation Results

The simulation work consists of two parts: numeric function optimization and training an Elman network to identify dynamical linear and non-linear systems.

4.1 Continuous Function Optimization

Seven well-known minimization test functions were employed to determine the performance of the proposed PACO algorithm. These test functions are given in Table 1.

Table 1 Numerical test functions used in the simulations

Notation	Name	Function
F1	Sphere	$f_1 = \sum_{i=1}^{4} x_i^2$
F2	Rosenbrock	$f_2 = 100(x_1^2 - x_2)^2 + (1 - x_1)^2$
F3	Step	$f_3 = \sum_{i=1}^{5} [x_i]$, where $[x_i]$ represents the greatest integer less than or equal to x_i
F4	Foxholes	$f_4 = [0.002 + \sum_{j=1}^{25} (j + \sum_{i=1}^{2} (x_i - a_{ij})^6)^{-1}]^{-1}$ $\{(a_{1j}, a_{2j})\}_{j=1}^{25} = (-32,-32), (-16,-32), (0,-32), (16,-32), (32,-32),$ $(-32,-16), (-16,-16), (0,-16), (16,-16), (32,-16),...,$ $(-32,32), (-16,32), (0,32), (16,32), (32,32)$
F5		$f_5 = (x_1^2 + x_2^2)/2 - \cos(20\pi x_1)\cos(20\pi x_2) + 2$
F6	Griewangk	$f_6 = 1 + \sum_{i=1}^{10} \left(\frac{x_i^2}{4000}\right) - \prod_{i=1}^{10} \left(\cos\left(\frac{x_i}{\sqrt{i}}\right)\right)$
F7	Rastrigin	$f_7 = 20A + \sum_{i=1}^{20} \left(x_i^2 - 10\cos(2\pi x_i)\right), A = 10$

The first four test functions were proposed by De Jong [27]. All test functions reflect different degrees of complexity.

Sphere (F1) is smooth, unimodal, strongly convex, and symmetric.

Rosenbrock (F2) is considered to be difficult, because it has a very narrow ridge. The tip of the ridge is very sharp, and it runs around a parabola.

Step (F3) is a representative of the problems of flat surfaces. Flat surfaces are obstacles for optimization algorithms because they do not give any information as to which direction is favourable. The background idea of the step function is to make the search more difficult by introducing small plateaus to the topology of an underlying continuous function.

Foxholes (F4) is an example of a function with many local optima. Many standard optimization algorithms get stuck in the first peak they find.

Function F5 has 40000 local minimum points in the region when x_1 and x_2 are within $[-10, 10]$.

Griewangk (F6) is also a non-linear and multi-modal function. The terms of the summation produce a parabola, while the local optima are above parabola level. The dimensions of the search range increase on the basis of the product.

Rastrigin's function (F7) is a fairly difficult problem due to the large search space and large number of local minima. This function contains millions of local optima in the interval considered.

The solutions for the functions, parameter bounds, resolutions and the length of each solution for each test function are given in Table 2.

In the first stage, simulation results were obtained for the proposed model. The proposed PACO was executed 30 times with different initial solutions. The number of ant colonies running in parallel was 4 (*NumOfCol*) and the number of ants was 30 (*NumOfAnts*). Each colony at any epoch was run for 20 (*NumAntCycle*) iterations for the first five functions and 50 for the others. The total number of cycles made at any epoch was 80 (*NumOfCol · NumAntCycle*) for the first five functions and 200 for the other two functions. This process was repeated through 16800 evaluations for the first five functions, 50000 evaluations for the other two functions in order to compare the performance of the proposed method with the results obtained using GA, TS, PTS,

Table 2 Number of parameters, solutions, parameter bounds and length of solution for the test functions

Function	Number of parameters	Solutions x_i	Solutions $f(x)$	Parameter bounds Lower	Parameter bounds Upper	Length of a solution
F1	4	0.0	0.0	−5.12	5.12	40
F2	2	1.0	0.0	−2.048	2.048	32
F3	5	−5.12	−30.0	−5.12	5.12	50
F4	2	−32.0	1.0	−65536	65536	40
F5	2	0.0	1.0	−10	10	36
F6	10	0.0	0.0	−600	600	200
F7	20	0.0	0.0	−5.12	5.12	400

and TACO algorithms taken from [4] and after that the search was stopped. Other parameters of the PACO were chosen as $c = 0.1$, $\rho = 0.1$, $\tau_0 = 0.1$, $q_0 = 0.9$, $a = 0.001$.

To show the robustness of the proposed model, frequency histograms of the results obtained using GA, TACO, basic TS, PTS and PACO algorithms are given in Figs. 4–10 for the test functions 1–7, respectively.

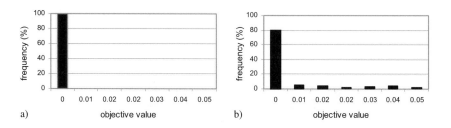

Fig. 4 Histograms drawn from the results obtained for the function F1 by (**a**) TS, PTS, TACO and PACO algorithms, (**b**) GA

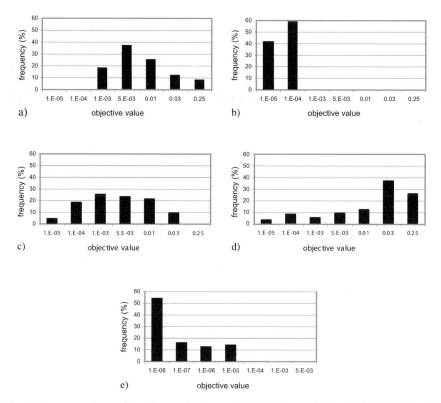

Fig. 5 Histograms drawn from the results obtained for the function F2 by (**a**) TS, (**b**) PTS, (**c**) GA, (**d**) TACO, and (**e**) PACO

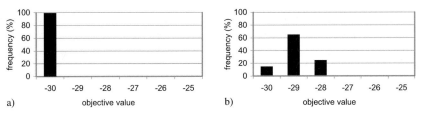

Fig. 6 Histograms drawn from the results obtained for the function F3 by (**a**) TS, PTS, GA, and PACO, (**b**) TACO

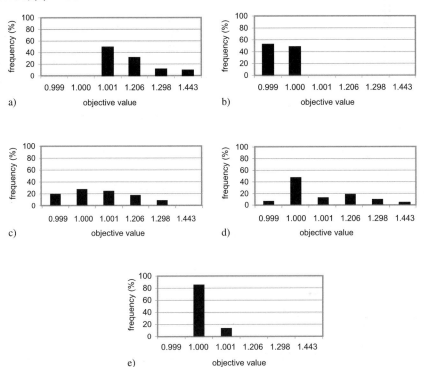

Fig. 7 Histograms drawn from the results obtained for the function F4 by (**a**) TS, (**b**) PTS, (**c**) GA, (**d**) TACO, and (**e**) PACO

4.2 Training Recurrent Neural Network by Using the PACO Algorithm

The use of artificial neural networks (ANNs) to identify or model dynamic inputs is a topic of much research interest. The advantage of neural networks for these types of applications is to learn the behaviour of a plant without much *a priori* knowledge about it. From a structural point of view, there are two main types of neural networks: feedforward neural networks (FNNs) and recurrent neural networks (RNNs) [28]. Connections that allow information to loop back to the same processing element are called recursive and NNs having these types of connections are named RNNs.

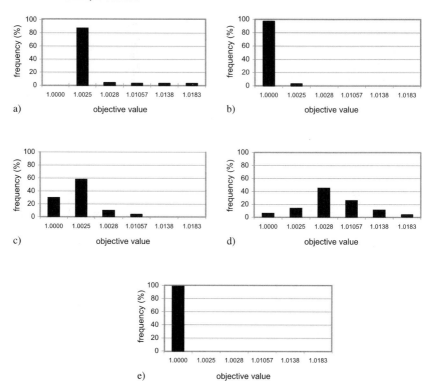

Fig. 8 Histograms drawn from the results obtained for the function F5 by (**a**) TS, (**b**) PTS, (**c**) GA, (**d**) TACO, and (**e**) PACO

RNNs are more suitable than FNNs for representing a dynamic system since they have a dynamic mapping between their output(s) and input(s). RNNs generally require less neurons in the neural structure and less computation time. Moreover they have a low probability of being affected by external noise. Because of these features, RNNs have attracted the attention of researchers in the field of dynamic system identification.

Although gradient based search techniques such as back-propagation (BP) are currently the most widely used optimization techniques for training neural networks, it has been shown that these techniques are severely limited in their ability to find global solutions. Global search techniques such as GA, SA, TS and ACO have been identified as a potential solution to this problem. Although the use of GAs for ANN training has mainly focused on FNNs [29–31], there are several works on training RNNs using GAs in the literature [32–34]. SA and TS have some applications for the training of ANNs [35–37]. Although GA, SA and TS algorithms have been used for training some kinds of neural networks, there are few reports of use of the ACO algorithm to train neural networks [38–40].

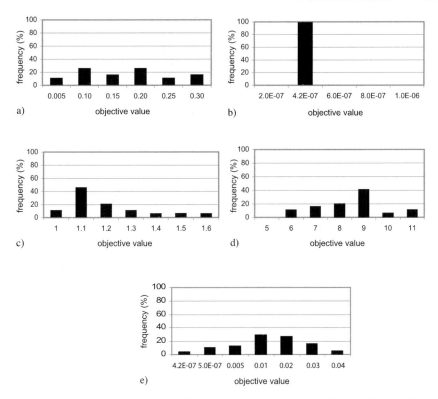

Fig. 9 Histograms drawn from the results obtained for the function F6 by (**a**) TS, (**b**) PTS, (**c**) GA, (**d**) TACO, and (**e**) PACO

A special type of RNN is the Elman network [41]. Elman network and its modified models have been used in applications of system identification. Figure 11 depicts the original Elman network with three layers of neurons. The first layer of this network consists of two different groups of neurons. These are the group of external input neurons and the group of internal input neurons also called context units. Context units are also known as memory units as they store the previous output of the hidden neurons. Elman networks introduced feedback from the hidden layer to the context portion of the input layer. Thus, the Elman network has feedforward and feedback connections. However, so that it can be trained essentially as feedforward networks by means of the simple BP algorithm, the feedback connection weights have to be kept constant. For the training to converge, it is important to select the correct values for the feedback connection weights. However, finding these values manually can be a lengthy trial-and-error process.

In this part of the work, the performance of the proposed PACO algorithm is tested for training the Elman network to identify dynamic plants. The use of the PACO algorithm to train the Elman network to identify a dynamic plant is illustrated in Fig. 12. Here, $y_m(k)$ and $y_p(k)$ are the outputs of the network and plant, at time k, respectively. Training of the network can be considered as a minimization problem defined by

$$\min_{\mathbf{w} \in W} J(\mathbf{w}) \tag{9}$$

where $\mathbf{w} = \begin{bmatrix} w_1 w_2 w_3 ... w_v \end{bmatrix}^T$ is the weight vector of the network. The time-averaged cost function $J(\mathbf{w})$ to be minimized by adaptively adjusting $\mathbf{w}$ can be expressed as

$$\min J(\mathbf{w}) = \left(\frac{1}{N} \sum_{k=1}^{N} \left(y_\mathrm{p}(k) - y_\mathrm{m}(k) \right)^2 \right)^{1/2} \tag{10}$$

where N is the number of samples used for calculation of the cost function.

A solution to the problem is a string of trainable connection weights representing a possible network (Fig. 13). The PACO algorithm searches for the best weight set by means of cost function values calculated for solutions in string form.

The structure of the network employed in this work is selected as in [4] to compare the results. Since the plants to be identified are single-input single-output (SISO), the number of external input neurons and output neurons is equal to one. The number of neurons at the hidden layer is equal to 6. Therefore, the total number of connections is 54, of which 6 are feedback connections. In the case of only

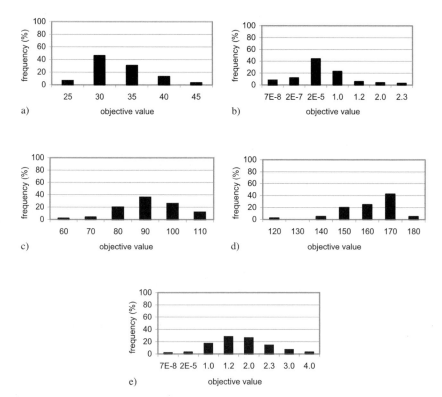

Fig. 10 Histograms drawn from the results obtained for the function F7 by (**a**) TS, (**b**) PTS, (**c**) GA, (**d**) TACO, and (**e**) PACO

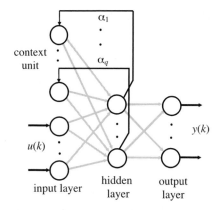

Fig. 11 Structure of the Elman network

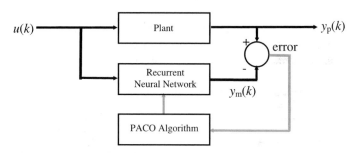

Fig. 12 Scheme for training a network to identify a plant using the PACO algorithm

w_1	w_2	w_3		w_p	α_1	α_2		α_q
1	2	3	$\cdots$	p	$p+1$	$p+2$	$\cdots$	$p+v$

Fig. 13 Representation of the trainable weights of a network in string form

feedforward connections being trainable, a solution is represented as a string of 48 weights. When all connections have trainable weights, then the string consists of 54 weights, of which 6 are feedback connection weights. In both cases, each weight is represented with 16 binary bits. The feedback connections have weight values ranging from 0.0 to 1.0 while feedforward can have positive or negative weights between 1.0 and −1.0. Note that from the point of view of the PACO algorithm, there is no difference between feedback and feedforward connections, and training one type of connections is carried out identically to training the other, unlike in the case of the commonly used BP training algorithm.

In the training stage, first a sequence of input signals $u(k)$, $(k = 0, 1, \dots)$ is fed to both the plant and the recurrent network designed with weights obtained from a solution of the PACO algorithm. Second the rms error value between the plant and recurrent network outputs is computed by means of Equation (10). Next, the rms error values computed for the solutions are used to select the highest evaluation weight set. The weight set with which the minimum rms error was obtained is selected as the highest evaluation weight set. From the point of view of the optimization, this is again a minimization problem. Simulations were conducted to study the ability of RNN trained by PACO to model a linear and non-linear plant. A sampling period of 0.01 s was assumed in all cases.

Linear plant: This is a third-order linear system described with the following discrete-time equation,

$$
\begin{aligned}
y(k) = A_1 y(k-1) + A_2 y(k-2) + A_3 y(k-3) \\
+ B_1 u(k-1) + B_2 u(k-2) + B_3 u(k-3)
\end{aligned} \tag{11}
$$

where $A_1 = 2.627771$, $A_2 = -2.333261$, $A_3 = 0.697676$, $B_1 = 0.017203$, $B_2 = -0.030862$, $B_3 = 0.014086$.

The Elman network with all linear neurons was tested. Training input signal, $u(k)$, $k = 0, 1, \ldots, 199$, was randomly produced and varied between -2.0 and 2.0. First the results were obtained by assuming that only the feedforward connection weights are trainable. Second the results were obtained by considering all connection weights of the Elman network trainable. For each case, experiments were repeated six times for different initial solutions. The results obtained using the BP and the PACO algorithms are given in Fig. 14. As an example, the responses of the plant and the network designed by the PACO are presented in Fig. 15. The average rms error values and the improvement percentages for a linear plant obtained using BP and PACO algorithms are presented in Table 3.

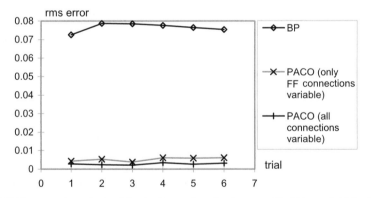

Fig. 14 RMS error values obtained for the linear plant for six runs with different initial solutions

Table 3 Comparison of results for the linear plant

Model	Average rms error	Improvement(%)
Back Propagation (BP)	7.67536×10^{-02}	-
PACO ($\alpha = 1$)	5.26118×10^{-03}	93.15
PACO (all weights trainable)	2.76346×10^{-03}	96.40

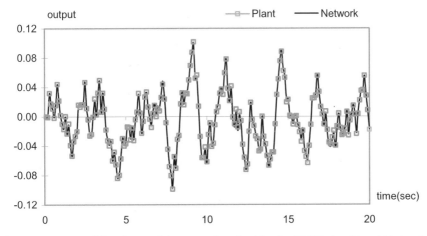

output   —□— Plant   —— Network

Fig. 15 Responses of the plant and the network trained by the PACO algorithm (third order linear plant, rms error = 2.137717×10^{-03})

Non-linear plant: The second plant model adopted for the simulations was that of a simple pendulum swinging through small angles [42]. The discrete-time description of the plant is:

$$y(k) = \left(2 - \frac{\lambda T}{ML^2}\right)y(k-1) + \left(\frac{\lambda T}{ML^2} - 1 - \frac{gT^2}{L}\right)y(k-2)$$
$$+ \frac{gT^2}{6L}y^3(k-2) - \frac{T^2}{ML^2}u(k-2) \qquad (12)$$

where M stands for the mass of the pendulum, L the length, g the acceleration due to gravity, λ the friction coefficient, y the angle of deviation from the vertical position, and u the external force exerted on the pendulum. The parameters used in this model were as follows:

$T = 0.2$ s, $g = 9.8$ m/s^2, $\lambda = 1.2$ kgm^2/s, $M = 1.0$ kg, $L = 0.5$ m.

Replacing the parameters with their values in Equation (12) gives:

$$y(k) = A_1 y(k-1) + A_2 y(k-2) + A_3 y^3(k-2) + B_1 u(k-2) \qquad (13)$$

where $A_1 = 1.04$, $A_2 = -0.824$, $A_3 = 0.130667$, $B_1 = -0.16$.

The Elman network with non-linear neurons in the hidden layer was employed. The hyperbolic tangent function was adopted as the activation function of non-linear neurons. The neural networks were trained using the same sequence of random input signals as mentioned above. As in the case of the linear plant, the results were obtained for six different runs with different initial solutions. The rms error values obtained by the BP algorithm and the PACO are presented in Fig. 16. As an example, the responses of the non-linear plant and the recurrent network with the

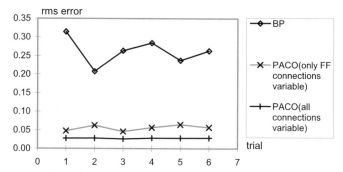

Fig. 16 RMS error values obtained for the non-linear plant for six different runs with different initial solutions

weights obtained by the PACO are shown in Fig. 17. The average rms error values and the improvement percentages for the non-linear plant obtained using BP and PACO algorithms are presented in Table 4.

5 Discussion

From the histograms obtained for the numerical test functions 1 and 3, it is seen that the basic TS, PTS and PACO algorithms are able to find the optimum solution

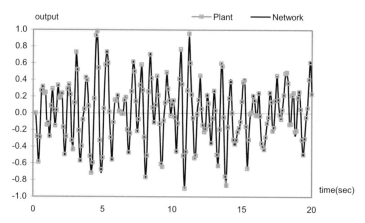

Fig. 17 Responses of the plant and the network trained by the PACO algorithm (non-linear plant, rms error = 2.611416×10^{-02})

Table 4 Comparison of results for the non-linear plant

Model	Average rms error	Improvement(%)
Back Propagation (BP)	0.26182	-
PACO ($\alpha = 1$)	5.59883×10^{-2}	78.62
PACO (all weights trainable)	2.78438×10^{-2}	89.37

for all runs (see Figs. 4 and 6). The reason is that the first problem is a convex and continuous function, and the third is a convex and discrete function. For the rest of the numerical test problems, the basic TS, GA and TACO cannot reach the optimal solution for all runs. However, PTS and the proposed PACO can find the optimum solutions or solutions very near to the optimum for each run. From the histograms presented in Figs. 5, 7 and 8 it can easily be concluded that the proposed PACO is able to find better solutions for F2, F4 and F5 than the basic TS, PTS, GA and TACO. Although the PACO can provide better solutions for the F6 and F7 than the basic TS, GA and TACO algorithms, its performance is not as good as the PTS algorithm. However, the results obtained by the PACO for these functions are also acceptable.

The original Elman network could identify the third-order linear plant successfully. Note that an original Elman network with an identical structure to that adopted for the original Elman network employed in this work and trained using the standard BP algorithm failed to identify even a second-order linear plant. Moreover, when the original Elman network had been trained by the basic GA, the third-order plant could not be identified although the second-order plant had been identified successfully [43]. It can be seen that, apparently for both plants, the training was significantly more successful when all connection weights of the network were trainable than when only feedforward connection weights could be changed. Thus, by using the PACO algorithm not only was it possible and simple to train the feedback connection weights, but the training time required was lower than for the feedforward connection weights alone. It is clearly seen from Figs. 14 and 16 and Tables 3 and 4 that, for both network structures (with all connection weights variable and with only feedforward connection weights trainable), the proposed PACO trained the networks better than the BP algorithm.

In this work, the data representation structure of the TACO and the search strategy of ACS were employed in the proposed model. However, the proposed model is a general structure for parallelization and it is also possible to use other ant based search strategies such as MAX-MIN ant system. Moreover, performance of the proposed model was tested on continuous problems; however this model can be implemented for combinatorial type problems. Using different ant based search strategies and performance examination of combinatorial problems are considered as future studies for the proposed model.

6 Conclusions

In this study a parallel ant colony optimization algorithm was proposed. The performance of the proposed algorithm was compared with that of basic TS, PTS, GA and TACO algorithms for numerical test problems. It was also applied to training a recurrent neural network to identify linear and non-linear plants, and the results obtained were compared with those produced by the BP algorithm. From the simulation results it was concluded that the proposed algorithm can be used to search multi-modal spaces successfully, and can be efficiently applied to train recurrent neural networks to identify dynamic plants accurately. It can be finally concluded that the PACO algorithm might be an efficient tool for solving continuous optimization problems.

References

1. Reeves CR (Ed.) (1995) Modern Heuristic Techniques for Combinatorial Optimization. McGraw-Hill: UK.
2. Corne D, Dorigo M, Glover F (Eds) (1999) New Ideas in Optimization, McGraw-Hill: UK.
3. Farmer JD, Packard NH, Perelson AS (1986) The Immune System, Adaptation, and Machine Learning. Physica, 22D:187–204
4. Kalinli A, Karaboga D (2004) Training recurrent neural networks by using parallel tabu search algorithm based on crossover operation. Engineering Applications of Artificial Inteligence, 17(5):529–542
5. Dorigo M, Maniezzo V, Colorni A (1991) Positive feedback as a search strategy. Technical Report No:91–016 Politecnico di Milano
6. Dorigo M, Maniezzo V, Colorni A (1996) The ant system: Optimization by a colony of cooperating agents. IEEE Trans. on Systems, Man and Cybernetics – Part B, 26(1):1–13
7. Christopher FH et al. (2001) Swarm intelligence: an application of social insect optimization techniques to the traveling salesman problem. Artificial Intelligence I
8. Bullnheimer B, Hartl RF, and Strauss C (1999) A new rank based version of the ant system, a computational study. Central European J for Operations Research and Economics, 7(1):25–38
9. Stützle T, Hoos HH (1997) The MAX-MIN ant system and local search for the traveling salesman problem. In Baeck T, Michalewicz Z, Yao X, (Eds), Proc. of the IEEE Int. Conf. on Evolutionary Computation (ICEC'97):309–314
10. Gambardella LM, Dorigo M (1996) Solving symmetric and asymmetric TSPs by ant colonies. Proc. of IEEE Int. Conf. on Evolutionary Computation, IEEE-EC 96, Nagoya, Japan:622–627
11. Di Caro G, Dorigo M (1998) Mobile agents for adaptive routing. Proc. of 31st Hawaii Conf. on Systems Sciences (HICSS-31):74–83
12. Stützle T, Dorigo M (1999) ACO algorithms for quadratic assignment problem. in: Corne D, Dorigo M, Glover F (Eds), New Ideas in Optimization, McGraw-Hill:33–50
13. Gambardella LM, Taillard E, Agazzi G (1999) MACS-VRPTW: A multiple ant colony system for vehicle routing problems with time windows. Technical Report, IDSIA-06: Switzerland
14. Bilchev G, Parmee IC (1995) The ant colony metaphor for searching continuous design spaces. Lecture Notes in Computer Science, Springer-Verlag, LNCS 993:25–39
15. Monmarché N, Venturini G, Slimane M (2000) On how Pachycondyla apicalis ants suggest a new search algorithm. Future Generation Systems Computer 16(8):937–946
16. Dreo J, Siarry P (2004) Continuous ant colony algorithm based on dense heterarchy. Future Generation Computer Systems, 20(5):841–856
17. Hiroyasu T, Miki M, Ono Y, Minami Y (2000) Ant colony for continuous functions, The Science and Engineering, Doshisha University
18. Bullnheimer B, Kotsis G, Strauss C (1998) Parallelization strategies for the ant system. in: De Leone R, Murli A, Pardalos P, Toraldo G (Eds), High Performance Algorithms and Software in Nonlinear Optimization. Kluwer Series of Applied Optimization, Kluwer Academic Publishers, Dordrecht, The Netherlands, 24:87–100
19. Stützle T (1998) Parallelization strategies for ant colony optimization, in: Eiben AE, Back T, Schoenauer M, Schwefel HP (Eds), Fifth Int. Conf. on Parallel Problem Solving from Nature, Springer-Verlag: 1498:722–731

20. Middendorf M, Reischle F, Schmeck H (2000) Information exchange in multicolony algorithms. in: Rolim J, Chiola G, Conte G, Mansini LV, Ibarra OH., Nakano H. (Eds), Parallel and Distributed Processing: 15 IPDPSP Workshops Mexico, Lecture Notes in Computer Science, Springer-Verlag, Heidelberg, Germany, 1800:645–652

21. Dorigo M (1993) Parallel ant system: An experimental study. Unpublished manuscript, (Downloadable from http://iridia.ulb.ac.be/~mdorigo/ACO/ACO.html)

22. Talbi EG, Roux O, Fonlupt C, Robillard D (1999) Parallel ant colonies for combinatorial optimization problems. in: Rolim J. et al. (Eds) Parallel and Distributed Processing, 11 IPPS/SPDP'99 Workshops, Lecture Notes in Computer Science, Springer-Verlag, London, UK 1586:239–247

23. Bolondi M, Bondanza M (1993) Parallelizzazione di un algoritmo per la risoluzione del problema del commesso viaggiatore. Master's Thesis, Dipartimento di Elettronica e Informazione, Politecnico di Milano: Italy

24. Michel R, Middendorf M (1998) An island model based ant system with lookahead for the shortest supersquence problem. in: Eiben AE, Back T, Schoenauer H, Schwefel P (Eds), Parallel Problem Solving from the Nature, Lecture Notes in Computer Science, Springer-Verlag, Heidelberg, Germany, 1498:692–701

25. Delisle P, Krajecki M, Gravel M, Gagné C (2001) Parallel implementation of an ant colony optimization metaheuristic with openmp. Int. Conf. on Parallel Architectures and Compilation Techniques, Proceedings of the 3rd European Workshop on OpenMP (EWOMP'01), Barcelona, Spain

26. Krüger F, Merkle D, Middendorf M (1998) Studies on a parallel ant system for the BSP model, unpublished manuscript. (Downloadable from http://citeseer.ist.psu.edu/239263.html)

27. De Jong KA (1975) An Analysis of The Behaviour of a Class of Genetic Adaptive Systems. PhD thesis, University of Michigan

28. Pham DT, Liu X (1999) Neural Networks for Identification. Prediction and Control, 4th edn, Springer-Verlag

29. Arifovic J, Gencay R (2001) Using genetic algorithms to select architecture of a feedforward artificial neural network. Physica A, 289:574–594

30. Sexton RS, Gupta JND (2000) Comparative evaluation of genetic algorithm and backpropagation for training neural networks. Information Sciences, 129:45–59

31. Castillo PA, Merelo JJ, Prieto A, Rivas V, Romero G (2000) G-Prop: Global optimization of multilayer percetptrons using Gas. Neurocomputing, 35:149–163

32. Ku KW, Mak MW, Siu WC (1999) Adding learning to cellular genetic algorithms for training recurrent neural networks. IEEE Trans. on Neural Networks, 10(2):239-252

33. Blanco A, Delgado M, Pegalajar MC (2000) A genetic algorithm to obtain the optimal recurrent neural network. Int. J. Approximate Reasoning, 23:67–83

34. Blanco A, Delgado M, Pegalajar MC (2001) A real-coded genetic algorithm for training recurrent neural networks. Neural Networks, 14:93–105

35. Castillo PA, Gonzalez J, Merelo JJ, Prieto A, Rivas V, Romero G (1999) SA-Prop: Optimization of multilayer perceptron parameters using simulated annealing. Lecture Notes in Computer Science, Springer, 606:661-670

36. Sexton RS, Alidaee B, Dorsey RE, Johnson JD (1998) Global optimization for artificial neural networks: A tabu search application. European J of Operational Research, 106:570–584

37. Battiti R, Tecchiolli G (1995) Training neural nets with the reactive tabu search. IEEE Trans. on Neural Networks, 6(5):1185–1200

38. Zhang S-B, Liu Z-M (2001) Neural network training using ant algorithm in ATM traffic control. IEEE Int. Symp. on Circuits and Systems (ISCAS 2001) 2:157–160

39. Blum C, Socha K (2005) Training feed-forward neural networks with ant colony optimization: An application to pattern classification. Fifth Int. Conf. on Hybrid Intelligent Systems

40. Li J-B, Chung Y-K (2005) A novel back-propagation neural network training algorithm designed by an ant colony optimization. Transmission and Distribution Conference and Exhibition: Asia and Pacific:1–5

41. Elman JL (1990) Finding structure in time. Cognitive Science, 14:179–211

42. Liu X (1993) Modelling and Prediction Using Neural Networks. PhD Thesis, University of Wales College of Cardiff, Cardiff, UK.

43. Pham DT, Karaboga D (1999) Training Elman and Jordan networks for system identification using genetic algorithms. J. of Artificial Intelligence in Engineering 13:107–117

An Ant-bidding Algorithm for Multistage Flowshop Scheduling Problem: Optimization and Phase Transitions

Alberto V. Donati[1], Vince Darley[2], and Bala Ramachandran[3]

[1] Joint Research Center, European Commission, Via E. Fermi 1, TP 267 21020 Ispra (VA), Italy.
Corresponding author:
alberto.donati@jrc.it

[2] Eurobios UK Ltd., 26 Farringdon Street, London EC4A 4AB, UK.
vince.darley@eurobios.com

[3] IBM T.J. Watson Research Center, Route 134, Yorktown Heights, 10598 NY, US.
rbala@us.ibm.com

Abstract

In this chapter we present the integration of an ant-based algorithm with a greedy algorithm for optimizing the scheduling of a multistage plant in the consumer packaged goods industry. The multistage manufacturing plant is comprised of different stages: mixing, storage, packing and finished goods storage, and is an extension of the classic Flowshop Scheduling Problem (FSP). We propose a new algorithm for the Multistage Flowshop Scheduling Problem (MSFSP) for finding optimized solutions. The scheduling must provide both optimal and flexible solutions to respond to fluctuations in the demand and operations of the plants while minimizing costs and times of operation. Optimization of each stage in the plant is an increasingly complex task when considering limited capacity and connectivity of the stages, and the constraints they mutually impose on each other.

We discuss how our approach may be useful not just for dynamic scheduling, but also for analyzing the design of the plant in order for it to cope optimally with changes in the demand from one cycle of production to the next. Phase transitions can be identified in a multidimensional space, where it is possible to vary the number of resources available. Lastly we discuss how one can use this approach to understand the global constraints of the problem, and the nature of phase transitions in the difficulty of finding viable solutions.

Key words: Scheduling, Optimization, Ant Colony System, Phase Transitions

This work was co-funded in a Consortium Agreement between Bios Group Inc. and Unilever during 1998–1999. The authors would also like to thank D. Gregg, S. Kauffman and R. MacDonald for useful discussions and support of this work.

1 Introduction and Problem Description

We focus in this chapter on the scheduling of operations of a multistage manufacturing plant in the consumer packaged goods industry, the core of a multistage supply chain. The production plant considered here has three main parts: making (mixers), intermediate storage (tanks), and packing lines; each stage has inputs and/or outputs, with limited connectivity with the adjacent stage. Each finished product or SKU (Stock Keeping Unit) – the final output of the production cycle – differs in terms of variant and pack size.

Given the demand for the period considered (usually one week) with the type and quantity of each SKU required for the period, the optimization consists in finding the most efficient schedule of the resources of each stage, to minimize the latest completion time on the packing lines (also called the *makespan*). In this way, the approach proposed here will globally optimize the schedule of the factory. The plant details are the following.

Making. Raw materials are subject to a number of parallel chemical processes to obtain different variants, each characterized by a number of attributes, such as color, base formulation, and dilution level. The variants produced in a mixer in batches are temporarily stored in tank facilities, and then directed to packing lines. The Making section in the plant is characterized by the number of mixers. Each mixer is then characterized by the variants it can make, the processing duration for every variant, the batch size, the cleaning/changeover times between different variants (which depend on the exact variant sequence). Finally mixers are characterized by the connectivity with the tanks, that is the maximum number of simultaneous connections with tanks for the transfer of the variant made, and their respective flow rate into the tanks.

Intermediate Storage. Storage facilities are temporary storage/caching tanks or silos, connecting the mixers to the packing lines. This stage is characterized by the number of tanks, and for each tank the variants that can be stored, its capacity with respect to the variant, the maximum number of simultaneous connections they can receive from the mixers, the maximum number of simultaneous connections for the transfer of the variant stored to the packing lines, and the setup/changeover times in changing the variant stored.

Packing Lines. The packing lines are the stage where the packing process is completed, which results in the finished product. The plant is characterized by the number of packing lines active (briefly PL). Each PL is characterized by which SKU can be packed and the relative packing rate, by the setup times due to changes in the pack size, and by the setup times due to changes in variant attributes. They are also characterized by the number, time and duration of maintenance operations, and finally by a "preferred attribute sequence" (for example, a color sequence), that might be preferred or required to follow under a specified policy. The maximum number of connections they can receive from the tanks are also specified.

Finished Goods Storage. Forecasted demand and actual demand can sometimes present differences which cannot be covered during the production span. This can be due in part to forecast errors, but mostly due to short notice changes to customer orders, or manufacturing problems may lead to time/quantity problems with the availability of certain products in sufficient quantity. This stage allows for the storage of a small buffer of goods to handle those situations. Nevertheless finished goods storage is wasteful in terms of excess inventory and physical requirements, but allows one to relax the constraints on the rest of the manufacturing plant. Hence, manufacturing operations continuously endeavor to keep the finished goods inventory at a minimum, while satisfying the supply chain and end customer requirements.

Supply Chain. In principle the optimization process could be continued outside the factory to consider interactions between different factories (which may be producing the same or different products), transport of finished goods and raw ingredients. Addressing these interactions would require the analysis of the corresponding supply chain networks. We have limited ourselves here to the scheduling aspects of a given manufacturing plant in the supply chain.

For the plant type considered here, the connectivity between the three stages is limited. Taking into account all the problem details and constraints makes Multistage Factory Scheduling an increasingly complex task, besides its nature as an NP-complete problem.

Flowshop/jobshop scheduling has been widely studied in recent years and there is an abundant literature about all the possible variants/constraints of such problems.

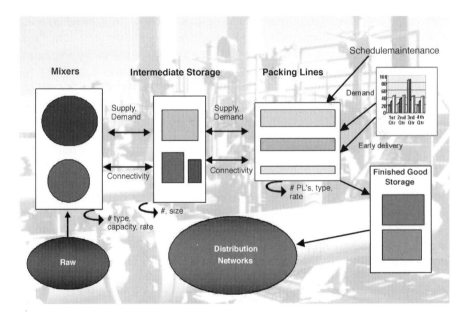

Fig. 1 Supply chain and plant representation. From left to right: mixers, intermediate storage (tanks), and packing lines. Each stage is characterized by number of resources (#), their type/capacity/rate and connectivity with the adjacent stage. The weekly demand profile of the SKUs to be done, is the input of the problem

In some cases a two-objective hierarchical minimization is carried out, the first with respect to the maximum makespan (the time from start to completion of the last job), the other minimizing the total usage time.

Mathematical programming approaches have been studied for model based planning and scheduling in the process industries. The flexible flowshop scheduling problem with more than two stages was first studied by Lee and Vairaktarakis in [1] presenting a worst-case heuristic method. More recently [2] theoretical aspects of a 3-stages flexible flowshop to complete finished products are considered and solved with an agent based model to deal with unforeseen changes in the system. In [3], mathematical programming models and solution methods are presented and strategies for implementing model-based integration of process operations are reviewed to address scheduling of multiproduct plants, integration of planning and scheduling across plant sites and design of multipurpose batch plants under uncertainty.

In general, there are two classes of process scheduling formulations – one based on discretizing time into uniform or non-uniform intervals and another based on treating time in a continuous manner. The advantages and disadvantages of these approaches are reviewed in [4]. Various heuristic/metaheuristics methods have been considered as well: a simulated annealing algorithm for a multistage manufacturing system is proposed in [5], a fast tabu search in [6], simulated annealing and genetic algorithm [7], and various other heuristics like the NEH (polynomial heuristic) presented in [8] and in the survey in [9] of machine scheduling problems. This list is limited, given the abundance of variety of methods and model/problems for these types of scheduling problems and their possible variants.

In this chapter, we present a new version of the permutation flowshop scheduling problem with limited temporary storage and with the presence of a making stage that is intimately related to the schedule. The model has been formulated to suit the process industries with liquid goods, so while on one side there is a possible loss of generality in the problem considered, on the other it constitutes an application to a real case.

We will also analyze the robustness of schedules and related phase transitions. Phase transitions in optimization problems have received considerable research attention recently in that they constitute an essential aspect when dealing with changes that might increase utilization of the system resources, such as spikes in the demand profiles from one period to another. Phase transition in problem difficulty in the general NK search problem has been studied in [10]. Phase transitions have been studied for the jobshop problem [11] and multiprocessor scheduling [12]. This work differs from the earlier phase transition research in that our focus is not on the phase transition with respect to problem difficulty, but the phase transition related to operations management in a production facility.

The chapter is organized as follows. In Sect. 2, we introduce the use of ant-based algorithms for the multistage scheduling flowshop problem. We provide the algorithm details in Sect. 3 and present computation results in Sect. 4. In the remainder of the chapter, we study the design and phase transition aspects of the problem.

2 Ants and Multistage Scheduling Flowshop Problem (MSFSP)

Recently a variety of new distributed heuristics have been developed to solve hard, NP-complete optimization problems. An ant-based algorithm uses a colony of ants, or cooperative agents, to find and reinforce optimal solutions. A variety of Ant Colony Optimization (ACO) algorithms have been proposed for discrete optimization, as discussed in [14], and have been successfully applied to the traveling salesman problem, symmetric and asymmetric ([15] and [16]), the quadratic assignment problem [17], graph-coloring problem [18], sequential ordering problem [19], flowshop [20], jobshop [21], and vehicle routing problem [24]. In particular in [20] a comparison is made with a previous ACO algorithm, the min-max ant system (MMAS, presented in [22]) for the scheduling of a single-stage flowshop problem, and an improved version is also discussed, for the standard Taillard's problems presented in [23].

The use of ants-based algorithms rather than other heuristics or metaheuristics methods, such as Genetic Algorithms, Simulated Annealing or Taboo Search, is motivated by several facts. First, it has been shown that ants algorithms can very effectively explore the search space, thus finding very good solutions in short times for many NP complete problems when combined with local search heuristics. A comparison of several strategies and metaheuristics and theoretical aspects is discussed in [25]; the effects of local search procedures have been discussed in [19] to show that the combination of an ACO and local search methods gives better results with respect to other heuristics combined with the same procedures; in particular for the sequential ordering problem (SOP), which is an asymmetric TSP with precedence constraints, it is reported that the Ant Colony System (ACS) is better than a genetic algorithm in combination with the same local search. This is due to the fact that ACS produces starting solutions that are easily improved by the local search while starting solutions produced by the genetic algorithm quickly bring the local search to a local minimum.

Besides this, ants-based algorithms can deal with a high number and variety of system constraints and it can easily be adapted to new situations where additional constraints are added to the problem. Finally, the fact that they can deal with changes in the system in a very effective way, so that they can be applied in a dynamic situation such as plant scheduling, where changes (e.g. glitches, breakdowns, additional delays, etc.) are likely to happen, so a quick reschedule is needed.

Ant-based optimization is a parallel distributed optimization process. A population of ants collectively discover and combine good pieces of different solutions. The mechanism is to encode the information not in a population of individual solutions, but in a modification of the environment: a *shared* space of pheromones. Artificial ants, searching for solutions, construct a solution step by step, and examine at each step the pheromones in the neighborhood. What step to do next will be determined by weighting the immediate reward of that move with the pheromone distribution, left by other ants in situations similar to the one they find themselves in. Once an ant has completed a solution,it calculates the fitness of that solution, and then retrospectively adjusts the strength of pheromone on the path it took proportionally to the fitness.

A procedure such as this, iterated over a population of ants, will provide two things: (a) a set of good solutions, and (b) a space of pheromones encoding global information.

Clearly (a) is very important. However, we will show that for real-world optimization of dynamic, uncertain, changing problems, (b) is of equal importance. It is for this reason that we have highlighted the ant algorithm as an important new approach which we believe is of great practical use. Briefly the importance of the space of pheromones is the following: if we change the problem in some small way (to reflect a change in demand, or a breakdown, delay or glitch in the manufacturing plant, or an addition to the plant), the information contained in the pheromones is still mostly relevant to the new problem. Hence a solution to the new/modified problem can be found considerably quicker. More traditional optimization procedures cannot do this and need a restart on the new problem.

The difficulty, as with applying all optimization algorithms, is in determining a suitable abstract representation: we need to define the space of pheromones and problem steps and choice procedure appropriately. It is known that for many algorithms (genetic algorithms, simulated annealing, etc.) the choice of representation is crucial in producing an effective optimization procedure. We do not expect ant algorithms to be any less representation-dependent in this regard.

An advantage of the ant algorithm, in this regard, is the relatively natural way in which expert rules of thumb may be encoded into the 'immediate reward' mentioned above. Many algorithms find it difficult to incorporate such hints or guidance into the search process. For ants algorithms this is quite straightforward, where the rules of thumb were encoded into the heuristic component. The bidding algorithm [13] is thus used for the calculation of the transition probabilities used by the ants in the solution construction procedure. This will be described in detail in the next section.

The algorithm presented here has also a number of enhancements with respect to the classic ACO. On one hand, to improve the search in the neighborhood of good solutions, a boost of pheromones is performed for high fitness solutions found; on the other, to enhance the exploration of very low probability steps in the solution construction, a random exploration is introduced.

Using these enhancements, the process of learning is constant and faster, and with proper choice of the parameter set, the optimal solutions are found within a few hundred iterations. We found that ants devote their resources very effectively to exploring a variety of high fitness solutions, while it is crucial to maintain the balance between learning and reinforcement.

3 Algorithm Scheme

The following section describes in detail the ant-bidding algorithm.

The optimization problem can be formulated as follows. Given:

1. M mixers, each having a production profile $\{V\}$ of the variants that can be done, producing each variant v of $\{V\}$ in a continuous manner at a rate $r_M(v)$, $v = 1, \ldots, V$ (we approximate batch mixing operations by assigning a rate equal to batch size and batch duration); we also define the variant changeover/cleanup times $t^M_{\text{variant}}(i, j)$ to change to variant j after variant i, $t^M_{\text{conc}}(i, j)$ and an additional changeover time due to some variant attributes (such concentration, or white/not white type); the maximum number of simultaneous connections to the tanks are also defined;

2. T storage tanks, each having a capacity of C_v, $v = 1, \ldots, V$, the changeover times $t^T_{\text{setup}}(i, j)$ to store variant j after variant i, the maximum number of simultaneous incoming connections from the mixers, and the maximum number of simultaneous connections to the packing lines;

3. PL packing lines, each having a packing profile $\{S\}$ of the SKU that can be packed, and for each a packing rate $r_{PL}(s)$, $s = 1, \ldots, S$, the variant changeover time $t^{PL}_{\text{variant}}(i, j)$ of packing a SKU having variant j after one having variant i, the pack-size changeover time $t^{PL}_{\text{size}}(i, j)$ of packing SKU j after i having a different pack-size, the maximum number of simultaneous connections from the tanks, the number n of scheduled maintenance operations with their intervals $[t_i, t_f]_k$, $k = 1, \ldots, n$ and the preferred attribute sequence of length m, $\{A^{PL}_1, \ldots, A^{PL}_m\}$, which is also PL dependent;

4. the demand profile of the S SKUs to complete for the period, where for each is specified a variant and pack size and the quantity to be done, $D(s), s = 1, \ldots, S$;

find the assignment and sequence of the V variants on the mixers and of SKUs on the packing lines that satisfy all the constraints (in particular limited number and capacity of the tanks, limited connectivity, preferred attribute sequence, clean/up changing times) and such to minimize the makespan, calculated as the time when the last SKU to be done is packed, since the beginning of the operations. In the following we will refer to *task* or *job* to indicate either the making of a variant or to the packing of an SKU.

In the ant algorithm scheme, each step is the completion of an assignment task. There are then two types of steps, 1. the making of a variant in a mixer (M-step), 2. the packing of a SKU on a packing line (P-step).

The problem is represented by two acyclic diagraphs $G(\mathbf{n}, \mathbf{e})$ where the nodes $\mathbf{n}$ represent the tasks to be completed and edges express a temporal sequence relation: $\mathbf{e}(i, j)$ = task j follows task i. The pheromones φ are represented with two distinct matrices relative to the two different types of tasks (making and packing). The element $\varphi(i, j)$ on each edge represents at any time the pheromone relative to the convenience of doing the job j immediately after job i on any resource. Moreover we have introduced at the beginning of the schedule on each resource (M mixers and PL packing lines) *virtual* tasks, to be able to store information also about the first (real) job, so that the first tasks scheduled are also associated with the pheromones (those connecting with the *virtual* tasks).

The dimension of the M-pheromones matrix is then $(V + M) \cdot V$ and that for the P-pheromones is $(S + PL) \cdot S$.

A. Initialization

The problem data is read in, the pheromones are set to a uniform value. The level of pheromones during the iterations is never allowed to drop below a minimum threshold and never be higher than a maximum level. This procedure allows the ants, on one hand, to never exclude zero-pheromone edges, and on the other, to never accumulate high amounts of pheromones on one edge, avoiding the formation of high concentration-pheromone trails. This may slow down the algorithm for very large problems, where it can be reasonable to remove those edges that are rarely used, being relative to task that are topologically distant (such as cities in the TSP).

B. Propagation of Ants

An iteration of the algorithm consists in the initialization and propagation of a number of artificial ants (usually 5 to 10). Each ant of the group completes a solution by choosing a sequence of steps, until no SKU is left to be done; since the problem size is $N = V + S$, this will involve N such steps for each ant, that is V M-steps and S P-steps during the construction procedure. Each ant maintains an updated taboo list, a Boolean list of $V + S$ elements to know what has been done and what needs to be completed. During propagation the pheromones are not changed, and only when the last ant of the group has finished, are the pheromones are updated.

Each artificial ant completes the following constructive procedure:

· **Choose a resource**

The ant progressively chooses which *resource* is available for the scheduling of the next job; with the term *resource* we indicate one of the following two:

1. a mixer and an empty tank,
2. a packing line and a not-empty tank.

The resource is chosen by examining among all the resources and jobs to complete, which one is available or will be available first, in consideration of the constraints; besides some tasks can be done on some *resource* but not on others (e.g. a mixer is available but it cannot handle any of the variants left to be done), in this case the resource is not considered as an available one. So, if a mixer and an empty tank are available, an M-step will be performed, or if a packing line and a not-empty tank are available, a P-step will be performed.

· **Choose a task**

M-step. Once a mixer and an empty tank have been identified, the bids for the variants (*MBids*) that still need to be done and that can be done on this resource, are calculated.

The bids of the variants are given by:

$$MBid(v) = \sum_{s=1,\ldots,S} PBid(s_v) \tag{1}$$

where v is the variant considered, s is the index for the SKUs s_v having this variant and *PBid* is the bid of SKU s over all the possible packing lines where the SKU can be done, finding the one with the highest value, as calculated by:

$$PBid(s_v) = \frac{D(s_v)}{T'} \tag{2}$$

where $D(s_v)$ is the demand of the SKU s, and the term T' includes the delay to wait for that PL to be available, the setup times, and the time to process the SKU: $D(s_v)/r_{PL}(s)$ where $r_{PL}(s)$ is the packing rate of SKU s on the packing line.

Note that in the term $PBid$ we have not taken into account any setup/changeover times for the mixer or tank, or other delays due to mixer and tank availability and the setup times for the packing lines: these quantities are not known at this time, being dependent on the schedule and on which packing line the specific SKU that will be scheduled.

The $MBids$ are finally scaled by a proper factor to be in the order of magnitude of the pheromones. The transition probabilities of making variant j after making the last variant i are given by:

$$P(j) \propto \frac{\varphi_V(i,j)^\alpha \cdot MBid(v_j)^\beta}{(1 + T_{\text{setup}}(i,j))^\gamma} \tag{3}$$

where φ_V are the M-pheromones, $T_{\text{setup}}(i,j)$ is the setup/changeover time due to the mixer and tank and it is computed as the maximum of the setup times $t^M_{\text{variant}}(i,j)$ + $t^M_{\text{conc}}(i,j)$ and $t^T_{\text{setup}}(i,j)$, considering the time when the mixer and tank become available respectively (they are usually not available at the same time, so overlapping in the setup times occurs); the factors α, β and γ are introduced to adjust the relative weights of these three components, the pheromones versus the heuristic factors (the bids and setup times).

The transition probabilities are used in the following ways:

1. Greedy step: choose the j that has the maximum value of probability
2. Probabilistic step: choose the j with a probability distribution given by (3)
3. Random step: choose j randomly.

Which type of step to make next is determined by two fixed cutoff parameters q_o and r_o with $q_o < r_o$. At each step, a random number r is generated in $[0,1]$; the ant makes a greedy step if $r < q_o$, else if $r < r_o$ the ant makes a probabilistic step, otherwise it makes a random step.

Once the step has been made (that is a task has been chosen) and the next task j has been selected, the necessary system variables are updated and the mixer starts to make the variant j, filling the tank(s) selected.

The parameters q_o and r_o are used to regulate the process exploration versus exploitation, that is to say, intensification versus diversification.

P-step. If the next available resource is a packing line and a not empty tank is found, the task will be to make a SKU on that packing line. In other words, the packing line has been chosen at this point, but the tank not yet, because there might be several not empty tanks. Before choosing the tank it is then necessary to calculate the probabilities for the SKUs, which will determine which variant (and then which tank) to use. Of course for the SKU to be considered in the bids, its corresponding variant must be present in some tank. Then the value of the bid is given by:

$$PBid(s) = \frac{D(s)}{T} \tag{4}$$

where $D(s)$ is the demand of the SKU s for the period, and T here is the resource allotment time, that is the total time required on the chosen resources to complete the SKU, and includes setup times of the packing lines, due to changes in the pack size, the presence of maintenance cycles (that will block the packing and shift its termination at a time subsequent to the end of the maintenance), the processing rates of the packing line for the SKU. In the calculation are included also the delays due to the following fact: another SKU with the same variant might have already been scheduled on one or more different packing line(s) and the variant quantity in the tank may not be enough to start packing an SKU using that variant. In the worst situation, it is necessary to introduce a delay to the start of the packing. The calculation of the exact delay is relatively expensive computationally (when there is multiple inflow and outflow to a single tank, for example), so in this calculation we only consider an approximate delay D, that guarantees the feasibility of the SKU.

The resource allotment time then given by:

$$T = \max(t_{\text{available}} - t, 0) + T_{\text{setup}} + t(s) + \Delta \tag{5}$$

where t is the current time, $t_{\text{available}}$ is the time when the packing line will finish the previous job and will be available, T_{setup} is the setup time (for changing pack size and/or variant), $t(s) = D(s)/r_{PL}(s)$ is the processing time of SKU s on the chosen PL; finally Δ is the approximate estimation of extra delays (approximate in the sense that the real delay can change due to availability of liquid in the tank and it will be computed only when the SKU has been chosen), and it includes also possible additional penalties due to the presence of maintenance during the packing task, since in this case, the packing has to be interrupted.

Then the transition probability for doing SKU j after SKU i, is then given by:

$$P(j) \propto \frac{\varphi_{SKU}(i,j)^{\alpha} \cdot PBid(s_j)^{\beta}}{(1 + T_{\text{setup}}(i,j))^{\gamma}} \tag{6}$$

where φ_{SKU} are the P-pheromones and $T_{\text{setup}}(i,j)$ is the adjusted setup time for doing j after i (that is, it is calculated considering variant and size changeovers from i to j). Again, the factors α, β and γ are introduced to adjust the relative weights of the heuristic components. And like before, which type of step to make next (that is if greedy, probabilistic or random) is set by the two parameters q_o and r_o.

C. Calculation of the Fitness

When all the SKUs in the demand profile have been completed (assigned to the packing lines), the schedule is complete. A complete solution is obtained and the fitness is calculated.

The fitness is the inverse of the maximum makespan MS over the packing lines, that is the time when the last packing line ends the last SKU:

$$f = \frac{q}{\max_{i=1,\dots,PL}(MS)} \tag{7}$$

where q is a scaling factor depending on the size of the problem, in order to maintain the fitness function (used to update the pheromones) in the appropriate order of magnitude.

D. Update of the Pheromones

The pheromone matrices are updated by evaporation and deposition processes: all pheromones evaporate at a rate $(1-\rho)$, while the pheromones on the solutions found (the paths of the ants) are augmented proportionally to the fitness (the goodness) of the solution found, according to the following equation:

$$\varphi(i,j) \leftarrow \rho \cdot \varphi(i,j) + \varepsilon \cdot \sum_{a=1\dots nAnts} f(a)|_{(i,j)\in S_a} \tag{8}$$

where $nAnts$ is the number of ants circulating at a time (between pheromones updates), S_a is the solution found by the ant a, $f(a)$ is the fitness of the solution S_a,

```
• Read the problem input data

• Initialize pheromones φSKU and φV

• for N_it times repeat:

  • Initialize and propagate the colony, repeating the following (nAnts) times:

    • initialize one ant a, and its taboo list.
      Repeat the following steps (tasks) completing the schedule:

      • Which step-type? Choose what resource will be available first.
          Do a M-Step if it is a (mixer+tank), do a P-Step if it is a
          packing (line+tank).
      • Calculate the bids, and then the transition probabilities.

      • Chose the step type, throwing a random number r:
          1. greedy step, if r<P_0
          2. probabilistic step, if r<r_0
          3. random step, otherwise.
      • Choose the task.
      • Take the step and update the system state.

    • Calculate the fitness f(a)
    • If fitness is greater than the best fitness, best fitness = f(a),
      store the solution.

  • evaporate the pheromones everywhere
  • increment the pheromones of the solution found.

• return the best solution found.
```

Fig. 2 Outline of the ants-bidding algorithm

which contributes to the increment on the edges $(i, j) \in S_a$. Here ϵ is constant, and ρ is also referred to as the pheromone persistency constant. If the fitness is close enough (usually 5%) or greater than the best fitness found so far, the pheromones on the edges of S_a are incremented by ϵ_{boost} instead of ϵ in Equation (8), a parameter that is usually set to one or two orders of magnitude larger than ϵ. In particular, if the fitness of a solution is larger than the fitness of the best solution found so far, the solution and its fitness are stored as the new best. The iteration then continues from b. with a new colony, until all the N_{it} iterations are completed. The scheme of the algorithm is presented in Fig. 2.

4 Further Enhancements

In this section we discuss some additional aspects/enhancements that have been included in the model.

A. Attribute Sequence

We may encounter additional constraints in production scheduling in the real world. This may arise due to the need to account for aspects of the problem that have not been explicitly modeled, such as to minimize clean up procedures in order to reduce waste and environmental impact. For example, we may impose an additional constraint that the SKU must follow a specified color sequence and/or pack size sequence, called attribute sequence, which depends on the packing line considered. In such cases, only the SKUs whose attribute is the correct one for the current attribute sequence of that packing line will have a non zero bid. When all the SKUs with that attribute are completed the attribute is allowed to change to the next value in the sequence. For the variant there is no direct attribute sequence on the mixers, but they will inevitably depend on the attribute sequence on the packing lines. In this way the only non-zero *MBids* calculated in Equation (1) are those for the variants associated with SKUs that are feasible in the current attribute sequence on at least a packing line. This is done simply setting to zero the *PBids* in Equation (2) that violate the sequence, so if there is no SKU in the actual sequence the resulting *MBid* will be zero.

B. Maintenance Times

On the packing lines periodic maintenance operations are usually scheduled on an ongoing basis. In the current implementation, we assume that the maintenance schedule is known at the beginning of the optimization. Maintenance is encoded in the following manner: 1. in the computation of the available resources, only those that are not under maintenance are considered; 2. if a line is selected with a maintenance upcoming, if the packing is not finished when the maintenance starts, the remainder of the packing continues after the maintenance has been completed. This will affect the factor *PBid* in Equation (2) because the time T' to complete the job will be increased by the duration of the maintenance.

If other maintenance cycle(s) are present, analogous checks are made and further interruption in the packing might happen.

C. Turning on/off the Mixers

During the production of a variant, it might happen that the mixer has a production rate that is not properly compensated on the other side by the packing lines, either because the overall packing rate is too low, or because the system is in a state where some or all the packing lines cannot pack the variant being made in the mixer. In this situation, since the tanks have a limited capacity, if production is not stopped, tank(s) can overflow. For this reason we have introduced a mechanism to turn off the mixers when this is about to happen. The mixer is turned back on when the level in the tank has dropped below a reasonable level, or there is a packing activity with an overall rate greater than the production rate.

Mixers are turned on and off several times. The estimation of the precise turn on/off times has proven to be an issue requiring some quite complex heuristic calculations as part of the optimization process.

In particular we have to compute the earliest possible time at which a mixer can be turned on while ensuring that all packing lines will be able to run and no tanks will overflow or become empty. Similarly we have to compute the earliest possible time at which a given job can begin on a packing line, again while ensuring that no tanks will overflow or become empty. In the general case, determining the *earliest time* is an optimization problem in its own right. Because of the time-criticality of this routine to the performance of the overall optimization, carefully constructed heuristics and a limited search are used for this calculation. These were designed and observed to yield good results (as will be seen), and not to create nonviable solutions. However, it is possible, under some circumstances, that the resulting solutions might be suboptimal. These algorithms must deal with mixers of differing speeds, potentially feeding a tank which feeds more than one packing line simultaneously (each of which may run and hence drain the tank at different speeds). Finding the earliest time such a job can start might require turning a mixer on earlier than before, but might also require turning off a mixer that was previously on. Such complicated permutations are very time-consuming to explore, given that these questions are asked in the innermost step of the optimization algorithm (i.e. are asked many thousands or millions of times). Therefore a number of heuristics were developed which considered separate cases such as that of a mixer being faster versus slower than the packing line(s) it feeds, and that of small versus large time ranges when the mixer is currently off, and that of packing jobs which extend across long versus short horizons. These heuristics were carefully evaluated and tested by statistically generating large numbers of artificial mixer/tank/packing-line scenarios. This gave us good confidence that the heuristics perform well. It is still theoretically possible that the heuristics don't return the optimal earliest time, however.

Split of the Making of the Variants

So far we have assumed that each variant is made at once and is completed in full, once its production starts, until the total necessary quantity (the sum of the demands of the SKUs having that variant) has been produced, even after an arbitrary number of on/off operations on the mixers. Nevertheless it could be more convenient to complete the production of a variant in two or more distinct phases.

We have implemented this enhancement by simply considering this fact. The attribute sequence might impose on the packing lines a sequence that will result in a suboptimal schedule if, for the SKU demand profile considered, there is no possibility of using the variant for contiguous/subsequent SKUs on the packing line. In other words, because of the attribute sequence, SKUs with another variant need to be packed before it is possible to complete the SKUs having the same variant. This situation will result in making and storage resources that are not fully exploited, since they are in a waiting state.

We then introduced the possibility of splitting a variant when this situation happens. The splitting is done at the beginning of the optimization by creating an additional variant in the optimization problem, whenever the presence is detected of one or more SKUs that will never be contiguous under the attribute sequence constraints on some packing line. A new variant is created relative to those SKUs, while the demand of the initial variant is adjusted.

With this enhancement, improvements with respect to the previous results are 2.91% on average, for the problems considered below in the next section (problems W19 to W22).

Finally one might want to consider also splitting the packing of the SKUs. We believe this procedure could also increase the quality of the best schedule found, but no implementation has been done at this time, since more complex criteria need to be introduced in this regard.

5 Results of the Optimization

We ran the model presented on problems of the following size: 30–40 SKUs, 10–12 variants, with usually 3 mixers, 5 tanks, 4–6 packing lines. These problems were given by Unilever for the scheduling of a real factory, based on data for fabric conditioner plants. Several tests were conducted. We present in this section only those for which we had an existing best solution as a benchmark. For each of these prob-

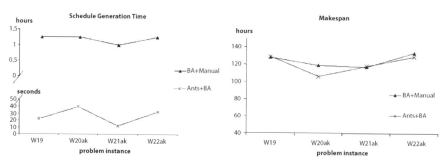

Fig. 3 Benchmark comparison of the new algorithm with the bidding algorithm combined with the best known. On the left, computation times are shown, while the right side shows the quality of the solutions found. The new algorithm finds solutions about two orders of magnitude faster (in minutes), and usually of better quality

lems, 5 optimization runs were repeated (that is starting from scratch, with uniform pheromones), each with 10,000 iterations, that are usually completed in about 20 seconds on a 2.66 GHz Pentium IV, using J2SE Runtime Environment 5.0. In most problems the best solution is found within the first 1000 iterations, and within 10,000 iterations the same absolute best solution is found in all 5 runs (except with problems W21ak and W22ak). The optimization was also run for 100,000 iterations in a search for further improvements.

We compared the results obtained in terms of computation time and quality with existing benchmarks of the bidding algorithm combined with human adjustments, that is manual methods based on visualization of the solution and adjustments by skilled trials moving/splitting jobs. This included the possibility of splitting SKUs as well, a possibility not implemented in the current version of the ants-bidding algorithm (in the figures, indicated Ants+BA). This method, however, takes times in the order of hours to generate the scheduling.

As can be seen, the Ants+BA algorithm can find optimized solutions much faster, with an improvement of about two orders of magnitude. Most importantly, the solutions are also improved in terms of optimization objective (shorter maximum makespan). The results of the runs conducted are compared with the known benchmark, and are shown in Table 1.

On average over all runs considered, an improvement of $< \Delta\% > = 3.29\%$ for the 10,000 iterations runs and $< \Delta\% > = 3.83\%$ for the 100,000 iterations runs is reached, which is significant over a week of operations on a single plant. Some negative signs are present for D, mainly due to the fact that the benchmark times given were rounded to the hour.

Besides this, it is also worth noticing that the new algorithm behaves in a quite different way with respect to the number of changeovers and setup times, as shown in Fig. 4. In particular variant changeovers on mixers and pack size changeovers are reduced, while variant changeovers on packing lines are increased. This result might be interpreted as due to the fact that usually pack size changeover durations are greater than variant changeover durations on packing lines.

Table 1 Results of optimization in terms of maximum makespan (in hours). The ant-bidding algorithm (Ants+BA) is compared with the best known solutions after 5 runs of 10,000 and 100,000 iterations, and the best solution found is adopted. The percentual improvement $\Delta\%$ is then calculated

Instance	Benchmark	Ants+BA (1.e5 iter.)	$\Delta\%$	Ants+BA (1.e6 iter.)	$\Delta\%$
W19	131	129.02	1.52	129.02	1.52
W20	101	99.36	1.64	99.36	1.64
W21	112	112.06	−0.05	112.06	−0.05
W22	103	96.79	6.22	96.79	6.22
W20ak	119	106.11	11.45	106.11	11.45
W21ak	117	117.15	−0.13	113.80	2.77
W22ak	133	129.88	2.37	128.76	3.24

Fig. 4 Comparison of the number of changeovers (**a**) making, (**b**) and (**c**) variant and SKU size on PL

The overall average improvement in the number of changeovers is (for the problems considered here) equal to 20.54%. We notice that there are several solutions having the same makespan, because minimizing the makespan leaves some degeneracy on the other packing lines, as they might finish at any time earlier than the maximum makespan and provide a viable solution of equal quality. This aspect could be useful when alternative ways of scheduling are needed.

A. Sensitivity to the Parameters

As in other optimization problems, ant algorithms demonstrate robustness with respect to changes in the instance of the problem with little or no tuning of the parameters required. The values for the parameters are set on some test data, to maximize the quality of the solutions found and the speed of their search. Values of the parameters that gave the best results are: $q_o = 0.4 - 0.5$ and $r_o = 0.99/0.98$ (one/two random step each 100 steps), $\alpha = 0.1$, $\beta = 1.0$, $\gamma = 2.0$, $r = 0.7$, $\epsilon = 1.0$ and $\epsilon_{boost} = 10.0$, with an offset of 0.01% from the best solution. In particular we observe that with respect to other optimization problems solved with ACO, the value of q_o used here seems quite low. This is due to the fact that the problem considered here is highly constrained, so it is better to favor exploration instead of exploitation.

The number of ants per iteration is always set to $nAnts = 1$, and does not seem to be a crucial parameter for these small–medium size problems.

B. Glitches on Packing Lines

In this section we analyze the robustness of the production schedule with respect to the effects of unforeseen glitches on the packing lines. This is particularly important, since there are many unexpected events in the real world that may require production rescheduling. The capacity to cope with glitches depends mainly on the *flexibility* of the plant. Notice that there are two situations that might affect the impact of a glitch:

a.   the glitch is on a PL that has little utilization.
b.   the glitch is on the PL that has high utilization.

To visualize the plant schedule and the effects of the glitch, we have developed a graphical user interface. In Fig. 5, two schedules are shown as a result of the optimization process. The horizontal axis is time, the top panel shows the mixer schedule, the middle panel the liquid quantities in the tanks and the bottom panel shows the packing line schedule. The maximum makespan is also indicated at the very top with the red line.

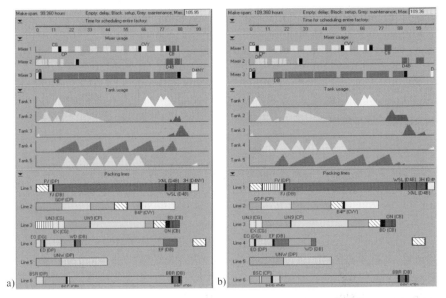

Fig. 5 (**a**) The glitch on packing line 3, is from 0 to 10 (in red) and happens to be on a PL which has some slack. In this case, there is no change in the makespan, since the duration of the glitch is less than the available slack. (**b**) Here the glitch (from 8 to 18) is on the *critical* packing line 1, the one with the maximum make-span and no flexibility to move other packing tasks to other lines. Hence the overall makespan increases

Variants (on mixers and tanks) and SKUs (on packing lines) are labeled with their names and colored with the respective color of the variant/SKU. The black rectangles represent setup times, while dark gray rectangles on the packing lines are the scheduled maintenance operations. In case **a**. the glitch (in red) occurs on a line with low utilization, so has no remarkable effects on overall production completion, while in case **b**. the glitch is on a line that is highly utilized and is the only one that can accomplish the long production of FJ(DB) (in blue) and the following packing tasks, so all production must shift, increasing the final makespan by the duration of the glitch.

The impact of the glitches are then essentially related to the flexibility of the plant. With dynamic scheduling the loss of time is never greater than the duration of the glitch.

C. Other Considerations

Because of the stochastic nature of this algorithm, to gather statistical consistency data the algorithm was run several times over repeated runs, each starting from scratch, with uniform pheromones distribution.

Over the runs, the average best fitness and the average system fitness as well as their standard deviations were calculated, to analyze the convergence process. By *system fitness* here we indicate the fitness of each ant (sampled every 10 iterations) averaged over the number of ants that are being considered in that interval, which is

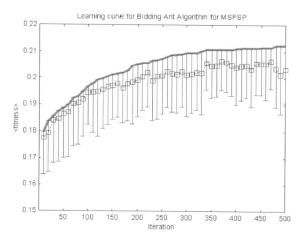

Fig. 6 Learning curve: the red solid line represents the best fitness found during iteration of the algorithm, the blue squares and bars represent the system fitness and its variation

related to the good solution discovery process. In Fig. 6, the red lines represent the average best fitness found so far (averaged over all runs), the blue squares the average system fitness (over the runs). The vertical bars represent the standard deviation of the average system fitness, to indicate how much variation has been present from one experiment to another. If, for example, the greediness is high (q_o) the squares will lie very close to the red line, with smaller error bars, indicating that the ants are performing less exploration.

It is also useful to evaluate the distribution of the average fitness and its cumulative distribution C, which are related to the probability P of finding a solution for

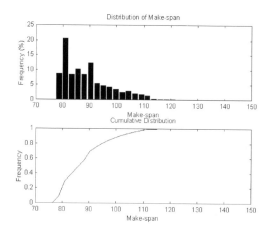

Fig. 7 Makespan distribution and its integral, of every agent that generates a solution during iteration of the algorithm

a given value of the fitness $(P = 1 - C)$ and to phase transitions in the system, as shown in Fig. 7.

Another relevant issue is how ants, once they have explored the space of solutions, are able to react to changes in the system. One type of change is to add, remove or split a job, perhaps as a consequence of demand changes. Others relate to plant operation, such as glitches or machine breakdowns. We can show that the pheromone memory allows the discovery of new solutions in a faster way. In Fig. 8, the red (lower) curve represents the fitness when the optimization is started as usual with a uniform distribution of pheromones. After 300 iterations the optimization is stopped and the problem is slightly changed by removing some tasks to be scheduled. The pheromones in this case were initialized with the distribution obtained at the end of the previous run. The blue (upper) curve shows not only a higher fitness (which actually depends on the fact that some jobs were removed, hence shortening the makespan), but most importantly a steeper rise.

6 Design

The optimization can also be run to test different plant configurations. Aided by data generated by an optimization system such as this, on the relative constrainedness of different parts of the manufacturing plant, a designer could understand and experiment with the implications of different types of design changes, and thereby understand which design features contribute most heavily to the creation of a factory which is *both efficient and robust* from the point of view of the operations/planning management.

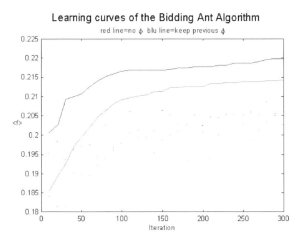

Fig. 8 Learning curves: the Y axis represent the fitness function (arbitrary unit), the red (lower) curve is a standard run of the algorithm. When the pheromones of the last iteration are used to initialize those for a modified problem, learning is much faster, as shown in the blue (upper) curve

It is possible to create several sets of experiments by changing each time the value of the demand for each SKU and by initializing the pheromones to the average calculated over the first set of experiments (i.e. those with unchanged demand). We conducted an analysis of the factory capability to absorb variations in the demand by introducing a uniform variation of 20% from period to period, in the demand of the SKUs and spread between multiple SKUs.

Figure 9 shows nine possible configurations for a plant, derived by variations of an original problem with 3 mixers ($M = 3$), 3 tanks ($T = 3$) and 5 packing lines ($PL = 5$). For the first row of plots, each plot is relative to an increase in the number of mixers, starting with $M = 2$, up to $M = 4$. The second row, each plot is relative to an increase in the number of tanks, starting from $T = 2$, up to $T = 4$. The third and last row, each plot is relative to the increase in the number of packing lines, starting from $PL = 4$ up to $PL = 6$.

In all the plots the z-axis is the makespan necessary to complete the schedule, while the x and y axes represent the incremental variation in the number of the other resources of the plant, e.g. in the first row the number of mixers is progressively increased, and each plot shows the resulting makespan as a variation of the number of PL versus the number of tanks, T. The green bars are proportional to the standard deviation, or variation for the time to complete the schedule.

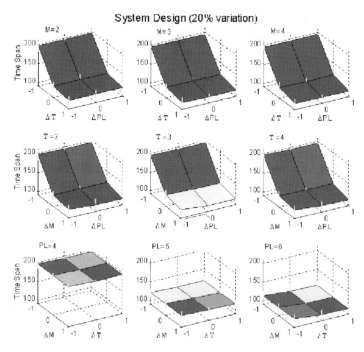

Fig. 9 Plots of the average time to complete a schedule with different plant designs: in each plot, we vary the number of resources available. Each row represents a different resource varying. First row mixers, second tanks, and third packing lines

We make two important observations here: on row 3, plot 1, we have that with $PL = 4$, even increasing the number of mixers to $M = 4$ ($\Delta M = 1$) and the number of tanks $T = 4(\Delta T = 1)$, no significantly better schedules are obtained, than just having $M = 2(\Delta M = -1)$, and $T = 2(\Delta T = -1)$. The situation of having just 4 packing lines is evidently a bottleneck for the system.

In other plots, as in row 2, plot 2, we notice that increasing PL and M is really beneficial if we move from $PL = 4(\Delta PL = -1)$ and $M = 2(\Delta M = -1)$, to $PL = 4(\Delta PL = -1)$ and $M = 3(\Delta M = 0)$, but not convenient at all to add an extra line, that is to $PL = 5(\Delta PL = 0)$ and $M = 4(\Delta M = 0)$ or from the initial point $PL = 4(\Delta PL = -1)$ and $M = 2(\Delta M = -1)$ to $PL = 5(\Delta PL = 0)$ and $M = 2(\Delta M = -1)$. We have added a packing line without obtaining any significant improvement.

The examples cited here support the following observations:

1. Tradeoffs exist between cost of design and robustness of design, in terms of the ability to robustly handle demand variations. Hence, robustness criteria need to be considered in deciding plant capacity.
2. Even if it is decided to increase capacity to improve design robustness to deal with demand uncertainty, the capacity increase needs to be decided on the basis of an analysis of plant bottlenecks.

A. Dynamic Capacity

We now examine phase transitions related behavior in this problem with a specific focus on capacity/flexibility related issues. We define in this section, a measure of the dynamic capacity of a plant based on the idea of considering the possible different ways (processes) to complete a finished product/job/task/object. We consider all the possible processes and distinguish among parallel and serial processes, respectively those that can happen at the same time, and those that follow one and other, like a sequence of operations, or sub-path.

The idea is to calculate for each finished product (here the SKUs) a measure of efficiency/flexibility or *utility* of the parallel and serial processes involved by first identifying each distinct way that the final product can be achieved with the corresponding sequence of operations (serial processes), and evaluating the speed at which they can complete the process. Then, those that represent alternative ways to accomplish the final product are added in a proper way. The calculation is repeated for all products. Finally all the utilities for all the products are combined taking into account also additional correction factors due to constraints on the maximum parallel processes present at any time.

In this way, for each possible final product, we consider all the feasible paths from start to completion. Each of these paths can be regarded as one set of operations or indefinitely many, and different finished products may share the same set, as in a typical multistage flowshop process, or have a quite different deconstruction into sets or simpler processes, as in a complex job-shop type process, or complex heterogeneous supply chain.

For each path we calculate, given the sequence of operations required, the minimum value of the utility U along the path, over either discrete steps or a continuous path, that is the minimum rate at which the task can be accomplished on this path. We repeat this calculation for all the pathways for this task, properly adding them together (some weighting might be needed). Adding up all the utilities for different paths for this task might not be sufficient, given that parallel processes might not be possible in reality because of constraints, such as limited simultaneous connections or limitations on resources. We need to examine for each of those paths how many of them can be effectively simultaneously possible and calculate the fraction of those that may be carried out at the same time. We can call this term the simultaneity constraint S, which corrects the utility U for the task t considered. The corrected utility is thus given by:

$$\tilde{U}(t) = S(t) \cdot \sum_{p \in \text{paths}} \min_{n \in \text{nodes}} U(p, n, t) \tag{9}$$

where the nodes represent the serial operations to complete a product on the path p, each characterized by a processing rate, and $\tilde{U}$ represents the corrected utility when taking into account constraints on simultaneous processing of tasks. The dynamic capacity is then given by:

$$C = \sum_{t \in \text{tasks}} \tilde{U}(t) \tag{10}$$

We have carried out the above procedure for the multistage flowshop problem (MS-FSP), which involes finding the dynamic capacity when the system is a plant with M mixers, producing V variants, each with different production profile and rate (which depends on the variant), T tanks, for temporary storage of different variants (one at a time), PL packing lines, characterized by packing profile (with SKU) and rate for each SKUs (a variant in a certain pack size). We assume that the first two stages are fully connected, while from tanks to packing lines there is a maximum number of simultaneous connections Nc at a time. A scheme of these settings is represented in Figure 10.

For each SKU s and each feasible path p (a combination of mixer and packing line) the utility can be defined as: $U(p, s) = \min r_M, r_{PL})$, where r_M and r_{PL} are respectively the mixing rate (for the variant) and the packing rate (for the SKU). The simultaneity constraint factor is taken into account in the following:

$$S(t) = \frac{\min(PL(s), N_c)}{\max(PL(s), N_c)} \tag{11}$$

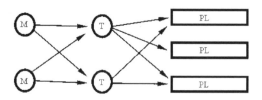

Fig. 10 Schematic representation of the liquid plant considered for the phase transition analysis

where *PL(s)* is the number of packing lines where the SKU s can be done and N_c the maximum number of connections at a time between the tank and the packing lines. Due to full connectivity from mixers to tanks, more correction factors are not necessary; we do not need this assumption in this context, but we take into account this aspect. Indeed the time to complete a task is governed by the minimum rate along the serial process, as explained in the text.

The total dynamic capacity is then given by:

$$C = \sum_{v \in \text{Variants}} S(s) \cdot \sum_{p \in \text{paths}} U(p, s(v)) \tag{12}$$

It should be noted that the dynamic capacity has been defined in this way to capture the flexibility of the plant, which is namely related to the various possible paths available to manufacture a product.

We generated a number of plant instances with varying degrees of connectivity and a number of demand profiles which are constantly increased by a finite amount. The results are shown in Fig. 11, where the z-axis is the time to complete the schedule (the makespan), the green bars represent the standard deviation over the demand variation, having added for each demand profile a 20% uniform random variation, and repeating this calculation over five runs. The red dots are missing data for plant profiles that have been interpolated.

We note that a phase transition is present in this case for C = 35,000, where a sudden drop occures in the time required to complete the schedule to a value that is fairly constant even with further increases of the capacity. The interpretation of this

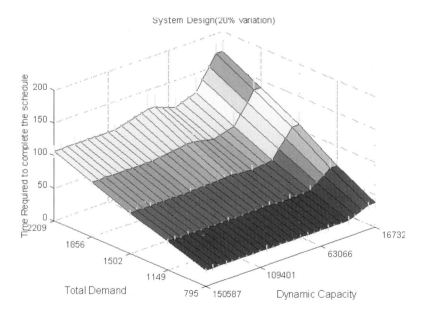

Fig. 11 Study of the factors affecting the time required to complete a production cycle, based on the total demand and dynamic capacity defined in Equation (12)

result is that, irrespective of the demand, the plant dynamic capacity needs to exceed a certain threshold to keep the plant makespans acceptable and robust to variable demands. This situation is represented also in Fig. 12, where the dynamic capacities of two existing plant configurations are computed and indicated by the two vertical lines. A great improvement could be accomplished if only these plants could increase their dynamic capacity and move after the phase transition, gaining about a 30% improvement. This improvement could be accomplished by simply increasing the number of connections between stages, with very little additional cost.

Interestingly enough, Unilever has anecdotal stories that relate to this analysis. Over the years a given factory (say a toothpaste factory) has changed from making 12 kinds of toothpaste, to 15 kinds, and then to 22, all without any substantial change in the efficiency of production. However, when the 23rd kind was introduced, suddenly the factory just couldn't cope with the production any more. Efficiency dropped drastically.

Further analysis and experiments need to be done varying the number of SKUs under different demand scenarios with a fixed plant capacity to examine any phase transition related effects. This is the subject of future work, although the order parameter we have introduced, and its phase transitions appear to be a likely explanation for these observations.

7 Conclusions

We have examined a number of aspects related to the scheduling of a multistage manufacturing plant. The use of an adaptive algorithm provides a better and more effective search in the solution space and copes with rescheduling in the case of malfunctioning or glitches, and variations in the demand. The improved speed in the

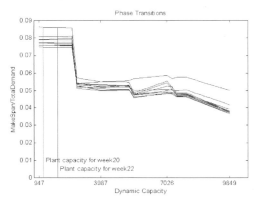

Fig. 12 Phase transition in the time necessary to complete a production cycle. An increase of the dynamic capacity can correspond a relevant decrease of the production times. The two vertical lines represent two plants from the data. Here it is shown that if bottlenecks are eliminated by improving connectivity among stages (e.g. little additional investment), the throughput of the plant can significantly increase

search for solutions allows one not only to find optimal solutions, but also to study several system configurations under many different conditions. The discovery of an order parameter and a deep theoretical analysis of the phase transitions improves the understanding of system design and its robustness and capability to respond to a changing environment.

Potential extensions of this work include applications to other flowshop and job-shop problems, but also to other complex/adaptive scheduling problems.

References

1. C. Y. Lee, G. Vairaktarakis (1994) Minimizing makespan in hybrid flowshops. Operations Research Letters 16:149–158
2. A. Babayan, D. He (2004) Solving the n-job 3-stage flexible flowshop scheduling problem using an agent-based approach. International Journal of Production Research 42(4):777–799
3. M. H. Bassett, P. Dave, F.J. Doyle, G.K. Kudva, J.F. Pekny, G.V. Reklaitis, S. Subrahmanyam, D. L. Miller, M.G. Zentner (1996) Perspectives on model based integration of process operations. Computers and Chemical Engineering 20(6-7):821
4. C. A. Floudas, X. Lin (2004) Continuous time versus discrete time approaches for scheduling of chemical processes – a review. Computers and Chemical Engineering 28(11):2109–2129
5. C. Charalambous, T. Tahmassebi, K. Hindi (2000) Modeling multi-stage manufacturing systems for efficient scheduling. European Journal of Operational Research 122(2):329
6. E. Nowicki, C. Smutnicki (1996) A fast tabu search algorithm for the permutation flow-shop problem. European Journal of Operational Research 91:160–175
7. T. Aldowaisan, A. Allahverdi (2003) New heuristics for no-wait flowshops to minimize makespan. Computers and Operations Research 30(8):1219–1231
8. M. Nawaz, E.E. Enscore Jr, I. Ham (1983) A heuristic algorithm for the m-machine, n-job flowshop sequencing problem. The International Journal of Management Sciences 11:91–95
9. N. G. Hall, C. Sriskandarajah (1996) A survey of machine scheduling problems with blocking and no-wait in process. Operations Research 44(3):510–525
10. V. Darley (1999) Towards a Theory of Autonomous, Optimizing Agents. PhD Thesis, Harvard
11. J. C. Beck, W. K. Jackson 1997, Constrainedness and the phase transition in job shop scheduling. Technical report TR 97-21, School of Computing Science, Simon Fraser University
12. H. Bauke, S. Mertens, A. Engel (2003) Phase transition in multiprocessor scheduling. Physical Review Letters 90(15):158701–158704
13. B. Ramachandran (1998) Automatic scheduling in plant design and operations. Internal Report, Unilever Research, Port Sunlight Laboratory
14. M. Dorigo, G. Di Caro, L. M. Gambardella (1999) Ant algorithms for discrete optimization. Artificial Life 5(2):137–172
15. M. Dorigo, V. Maniezzo, A. Colorni (1996) The ant system: optimization by a colony of cooperating agent. IEEE Transactions on Systems, Man and Cybernetics part B 26(1):29–41

16. L. M. Gambardella, M. Dorigo (1995) AntQ: A reinforcement learning approach to the traveling salesman problem. Proceedings of the Twelfth International Conference on Machine Learning ML95 252–260
17. L. M. Gambardella, E. D. Taillard, M. Dorigo (1999) Ant colonies for the quadratic assignment problem. Journal of the Operational Research Society 50:167–176
18. D. Costa, A. Hertz (1997) Ants can colour graphs. Journal of the Operational Research Society 48:295–305
19. L. M. Gambardella, M. Dorigo (2000) An ant colony system hybridized with a new local search for the sequential ordering problem. INFORMS Journal on Computing, 12(3):237–255
20. C. Rajendran, H. Ziegler (2004) Ant-colony algorithms for permutation flowshop scheduling to minimize makespan/totalflowtime of jobs. European Journal of Operation Research 155:426–438
21. A. Colorni, M. Dorigo, V. Maniezzo, M. Trubian (1994) Ant system for job-shop scheduling. Belgian Journal of Operations Research, Statistics and Computer Science 34:39–53
22. T. Stuetzle (1998) An ant approach for the flow shop problem. Proceedings of the 6th European Congress on Intelligent Techniques and Soft Computing (EUFIT 98), Vol. 3:1560–1564
23. E. Taillard (1993) Benchmarks for basic scheduling problems. European Journal of Operational Research 64:278–285
24. L. M. Gambardella, E. Taillard, G. Agazzi (1999) MACS-VRPTW: vehicle routing problem with time windows. In: M. Dorigo and F. Glover (eds) New Ideas in Optimization. Corne, McGraw-Hill, London
25. C. Blum, A. Roli (2003) Metaheuristics in combinatorial optimization: Overview and conceptual comparison. ACM Computing Surveys 35:268–308

Dynamic Load Balancing Using an Ant Colony Approach in Micro-cellular Mobile Communications Systems

Sung-Soo Kim[1], Alice E. Smith[2], and Soon-Jung Hong[3]

[1] Systems Optimization Lab. Dept. of Industrial Engineering, Kangwon National University, Chunchon, 200-701, Korea.
kimss@kangwon.ac.kr

[2] Industrial and Systems Engineering, Auburn University, AL 36849-5346, USA.
smithae@auburn.edu

[3] SCM Research & Development Team, Korea Integrated Freight Terminal Co., Ltd., Seoul, 100-101, Korea.
sjhong75@kift.kumho.co.kr

Abstract

This chapter uses an ant colony meta-heuristic to optimally load balance code division multiple access micro-cellular mobile communication systems. Load balancing is achieved by assigning each micro-cell to a sector. The cost function considers hand-off cost and blocked calls cost, while the sectorization must meet a minimum level of compactness. The problem is formulated as a routing problem where the route of a single ant creates a sector of micro-cells. There is an ant for each sector in the system, multiple ants comprise a colony and multiple colonies operate to find the sectorization with the lowest cost. It is shown that the method is effective and highly reliable, and is computationally practical even for large problems.

Key words: Load Balancing, Ant Colony Approach, Micro-cell Groupings

1 Introduction

In the last 15 years there has been substantial growth in micro-cellular mobile communication systems. It is imperative to provide a high level of service at minimum cost. With the substantial increase in cellular users, traffic hot spots and unbalanced call distributions are common in wireless networks. This decreases the quality of service and increases call blocking and dropping. One of the main design problems to be addressed in micro-cellular systems is location area management. This location area management problem can be generally stated as: For a given network of n cells, the objective is to partition the network into m location areas, without violating transmission constraints, and with minimum cost. This chapter addresses the problem of providing good quality of service at a reasonable level of cost for code division multiple access (CDMA) micro-cellular systems. To provide the best service for a given number of base stations and channels, the call load must be dynamically balanced considering the costs of call handoffs and call blockage. This is a location management optimization problem that can be accomplished through sectorization of the micro-cells. Figure 1 shows an example grouping that has one virtual base station (VBS) and three sectors. The maximum number of channel elements assigned to a VBS is termed hard capacity (HC). The maximum number of channel elements that a sector can accommodate is termed soft capacity (SC). HC is assumed to be 96 and SC is assumed to be 40 in this example. In Fig. 1(a) the total call demand is equal to HC (96) but, the total call demand in one sector is greater than 40 resulting in 30 blocked calls in that sector. Figure 1(b) has no blocked calls with the same HC and SC. Blocked calls are one consideration, while handoff calls are another. A disconnected grouping of micro-cells generates unnecessary handoffs between sectors as shown in Fig. 2(a). Therefore, the cells in a sector need to be connected compactly, as shown in Fig. 2(b).

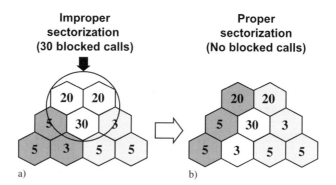

Traffic distribution and sectorization (HC: 96, SC: 40)

Fig. 1 Improper and proper groupings of micro-cells [14]

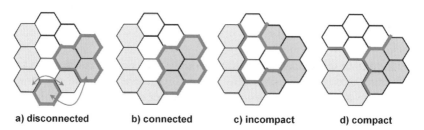

a) disconnected   b) connected   c) incompact   d) compact

Fig. 2 Examples of micro-cell groupings [14]

To minimize handoffs and interference among sectors, a measure of sector compactness, as Lee et al. [14] proposed, can be used. The following is a mathematical equation of the compactness index (CI):

$$CI = \frac{\sum_{i=1}^{n-1} \sum_{j=i+1}^{n} x_{ij} \times B_{ij}}{\sum_{i=1}^{n-1} \sum_{j=i+1}^{n} B_{ij}} \tag{1}$$

There are n cells. B_{ij} is 1 if cells i and j are adjacent, otherwise 0. If the sectors of cells i and j are the same, then $x_{ij} = 0$, otherwise 1. The CIs for Figs. 2(c) and (d) are $14/24 = 0.583$ and $9/24 = 0.375$, respectively. If 0.5 is chosen as the maximum CI, then Fig. 2(c) is infeasible.

The cells grouping problem is an NP-hard problem [11]. For load balancing of CDMA wireless systems previous research has explored the use of optimization heuristics. Kim and Kim [13] proposed a simulated annealing approach to minimize the cost of handoffs in the fixed part of a personal communication system network. Demirkol et al. [4] used SA to minimize handoff traffic costs and paging costs in cellular networks. Chan et al. [2] presented a genetic algorithm (GA) to reduce the cost of handoffs as much as possible while service performance is guaranteed. Lee et al. [14] used a GA to group cells to eliminate large handoff traffic and inefficient resource use. In their proposed sectorization, properly connected and compact sectors are considered to keep the handoffs as few as possible while satisfying the channel capacity in each sector. Brown and Vroblefski [1] altered the GA approach of [14] with less disruptive crossover and mutation operators, that is, operators that better maintain the structure of previous solutions in newly created solutions. They report improved results over the Lee et al. GA. The same authors used a grouping GA on a related problem to minimize location update cost subject to a paging boundary constraint [22]. Using the same fundamental problem formulation of [1] and [14], we propose a new heuristic based on an ant colony system for dynamic load balancing of CDMA wireless systems.

2 Ant Approach to Dynamic Load Balancing

The ant colony approach is one of the adaptive meta-heuristic optimization methods inspired by nature which include simulated annealing, GA and tabu search. The ant colony paradigm is distinctly different from other meta-heuristic methods in that it *constructs* an entire new solution set (colony) in each generation, while others focus on *improving* the set of solutions or a single solution from previous iterations. The ant optimization paradigm was inspired by the behavior of real ants. Ethnologists have studied how blind animals, such as ants, could establish shortest paths from their nest to food sources. The medium that is used to communicate information among individual ants regarding paths is pheromone. A moving ant lays some pheromone on the ground, thus marking the path. The pheromone, while gradually dissipating over time, is reinforced as other ants use the same trail. Therefore, efficient trails increase their pheromone level over time while poor ones reduce it to nil. Inspired by the behavior of real ants, Marco Dorigo introduced the ant colony optimization approach in his PhD Thesis in 1992 [5] and expanded it in further work, as summarized in [6–9]. The characteristics of ant colony optimization include:

1. a method to construct solutions that balances pheromone trails (characteristics of past solutions) with a problem-specific heuristic (normally, a simple greedy rule)
2. a method to both reinforce and evaporate pheromone.

Because of the ant paradigm's natural affinity for routing, there have been a number of ant algorithm approaches to telecommunications in previous research. Chu, et al. [3], Liu et al. [15], Sim and Sun [19], Gunes et al. [12] and Subing and Zemin [20] all used an ant algorithm for routing in telecommunications. Shyu et al. [17,18] proposed an algorithm based upon the ant colony optimization approach to solve the cell assignment problem. Subrata and Zomaya [21] used an ant colony algorithm for solving location management problems in wireless telecommunications. Montemanni et al. [16] used an ant colony approach to assign frequencies in a radio network. More recently, Fournier and Pierre [10] used an ant colony with local optimization to minimize handoff traffic costs and cabling costs in mobile networks.

Dynamic load balancing can be affected by grouping micro-cells properly and grouping can be developed through a routing mechanism. Therefore, we use ants and their routes to choose the optimum grouping of micro-cells into sectors for a given CDMA wireless system state.

2.1 Overview of the Algorithm

In our approach each ant colony (AC) consists of ants numbering the same as the number of sectors, and there are multiple colonies of ants (C colonies) operating simultaneously. That is, each ant colony produces one dynamic load balancing (sectoring) solution and the number of solutions per iteration is the number of colonies. Consider an example of accomplishing sectorization. There is one VBS and three sectors. In step 1, the ant system generates three ants, one for each of the three sectors. In step 2, a cell in each sector is chosen for the ant to begin in. In step 3, an ant chooses a cell to move to – moves are permitted to any adjacent cell that has not already been assigned to a sector. Step 4 continues the route formation of each ant, which results in sectorization of all micro-cells.

The flowchart in Figs. 3 and 4 gives the details of the algorithm. The variable *optimal* describes the best solution found so far (over all colonies and all iterations). The current available capacity of each VBS and each sector is calculated to determine which ant to move first for sectorization. The cell chosen for an ant to move to is based on the amount of handoff traffic (described in Sect. 2.4). When all cells are sectorized, CI is calculated using Equation (1). If CI is less than the specified level, the solution

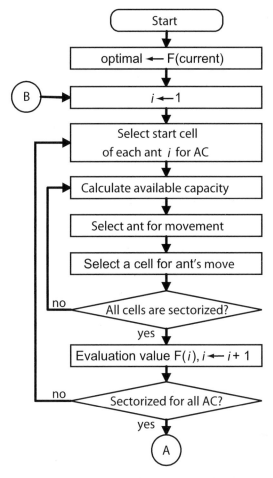

Fig. 3 Ant colony algorithm for dynamic load balancing

is feasible. Otherwise, it is infeasible (not compact enough) and discarded. After all feasible solutions are evaluated the minimum cost solution of an iteration is assigned to the variable *best*.

After all cells are sectorized by the ants in all colonies, the pheromone levels of each cell's possible assignment to each sector are updated using Equation (2). In this equation, $\tau_{ik}(t)$ is the intensity of pheromone of cell i for assignment to sector k at time t. $\Delta\tau_{ik}$ is an amount of pheromone added to cell i for assignment to sector k (we use a straightforward constant for this amount = 0.01). $\Delta\tau_{ik}^*$ is an elitist mechanism so that superior solutions deposit extra pheromone. If the best solution of the colonies 1 to C is also better than the current value of the variable *optimal*, we add a relatively large amount of pheromone = 10.0. If the best solution of the colonies 1 to C is worse than the current value of the variable *optimal* but the difference (GAP) between the

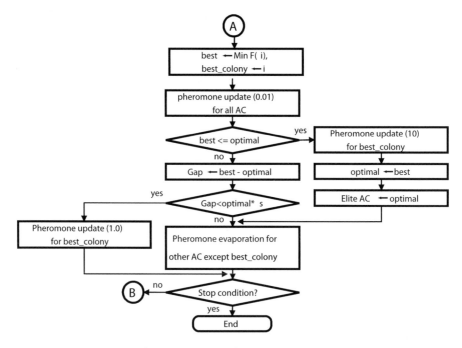

Fig. 4 Ant colony algorithm for dynamic load balancing Contd.

values of the variables *best* and *optimal* is less than the value of *optimal* · 0.05, that is, the objective function of the best solution in the colony is within 5% of the best solution yet found, we add an amount of pheromone = 1.0. ρ is a coefficient such that $(1 - \rho) \times \tau_{ik}(t)$ represents the evaporation amount of pheromone between times t and $t + 1$. We use $\rho = 0.5$.

$$\tau_{ik}(t + 1) = \rho \times \tau_{ik} + \sum_{j=1}^{C} \Delta \tau_{ikj} + \Delta \tau_{ik}^{*} \tag{2}$$

From Equation (2), it can be seen that the amount of pheromone change is elitist. That is, the pheromone deposited for the best ever solution is three orders of magnitude greater than an ordinary deposit of pheromone and the amount deposited for the best solution in the C colonies (if it meets the GAP criterion) is two orders of magnitude greater than usual. This elitism helps the ant system to converge relatively quickly.

2.2 Evaluation

The total cost is composed of the cost of blocked calls, the cost of soft and softer handoffs, and the cost of forced handoffs. Blocked calls are caused by exceeding HC or SC. When a mobile station with an ongoing call moves from one VBS to another, then a soft handoff occurs. When a mobile station with an ongoing call moves from one sector to another within a VBS, then a softer handoff occurs. When a cell changes its sector, all ongoing calls in the cell have to change sectors and a forced handoff occurs.

The cost of a micro-cellular system as proposed by Lee et al. [14] is used in this chapter and calculated based on the new grouping in time period $t + 1$ given the grouping of cells in time period t. There are M virtual base stations (BS_m, $m = 1, \ldots, M$); there is call demand of TD_i in each of the N cells, there is handoff traffic of h_{ij} from cell i to cell j, and there are K groupings (sectors) of micro-cells (SEC_k). The objective cost function [14] is

$$
\begin{aligned}
Min\ F\ =\ & c_1 \sum_m Max \left\{ \sum_{i \in BS_m} TD_i - HC_m, 0 \right\} \\
& + c_2 \sum_k Max \left\{ \sum_{i \in SEC_k} TD_i - SC_k, 0 \right\} \\
& + c_3 \sum_i \sum_j h_{ij} z_{ij} + c_4 \sum_i \sum_j h_{ij}(w_{ij} - z_{ij}) \\
& + c_5 \sum_i g_i TD_i
\end{aligned}
\tag{3}
$$

The first term is a summation over the M virtual base stations of the blocked calls due to hard capacity. The second term is a summation over the K sectors of the blocked calls due to soft capacity. The third term is the soft handoff traffic between adjacent cells with different VBSs. The fourth term is the softer handoff traffic between adjacent cells in different sectors within a VBS. The fifth term is the amount of forced handoff after sectorization (reconfiguration). z_{ij}, w_{ij}, and g_i are binary variables. z_{ij} is 1 if cells i and j are in different VBSs. w_{ij} is 1 if cells i and j are in different sectors. g_i is 1 if cell i changes sectors from the existing sectorization to the newly proposed one. c_1, c_2, c_3, c_4, and c_5 are weighting factors. The values of c_1, c_2, c_3, c_4, and c_5 are 10, 5, 2, 1, and 1 for examples in this chapter, as proposed by Lee et al. [14]. Larger weights are given to c_1 and c_2 because minimizing the blocked calls caused by hard and soft capacity is the first priority of sectorization.

2.3 Determination of Starting Cell for Each Ant

The following is the probability that cell i in sector k is selected for start.

$$
p(i, k)\ =\ \frac{TD_i}{\sum_{j \in SEC_k} TD_j}, \quad i \in SEC_k
\tag{4}
$$

Greater probability is given to cells that have large call demands to reduce forced handoff costs. We have one VBS and three sectors in the example shown in Fig. 5. Cell 4 in sector 1 has the highest probability (0.428) of starting. Cells 3 and 6 in sector 2 have the same highest probability (0.385) in sector 2. Cells 8 and 9 in sector 3 have the same highest probability (0.385) in sector 3.

2.4 Movement of Each Ant

The current available capacity of each VBS and each sector must be calculated. They are used to define ant movement. Capacities are calculated using following equations.

$$C\_BS_m = Max\left\{HC_m - \sum_{i \in BS_m} TD_i, L_{BS}\right\} \quad \text{for all} \quad m \tag{5}$$

$$C\_SEC_k = Max\left\{SC_k - \sum_{i \in SEC_k} TD_i, L_{SEC}\right\} \quad \text{for all} \quad k \tag{6}$$

Figures 6 and 7 are examples where $HC = 96$, $SC = 40$, lower bound of VBS (L_{BS}) = 3, and lower bound of sector (L_{SEC}) = 2. The available capacity for VBSs and sectors (C_BS_m and C_SEC_k) are calculated using Equations (5) and (6). We use the lower bounds of VBS and sectors (L_{BS} and L_{SEC}) to find the lowest total cost for sectorization. When searching for the optimal solution, we must consider that there are handoff costs and blocked calls. In other words, we might be able to save greater handoff costs even though we have some blocked calls in a VBS or sector. If

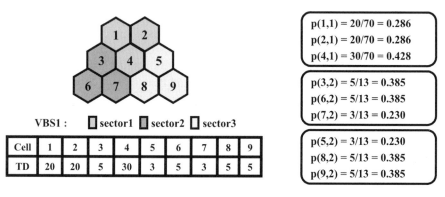

Fig. 5 Selection of starting cell for each ant

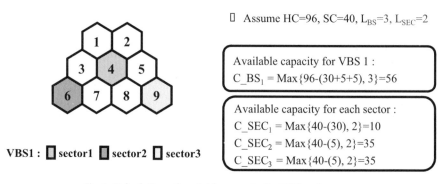

Fig. 6 Calculation of available capacity for VBS and sectors

cells 3, 4, and 8 are selected for sector 1 as shown in Fig. 7, sector 1 has no chance to be selected by an ant for sectorization because there is no current available capacity in sector 1 of VBS 1. To allow blocked calls in sector 1, a chance $(2/72 = 2.8\%)$ is given to sector 1 using the lower bound of sector 1. The value of the lower bound is given by the user based on expected blocked calls in the system. If we have a large lower bound, there is a high possibility of blocked calls.

If there is more than one VBS, a VBS for beginning movement must be chosen first. $P_{BS}(m)$ is the probability that VBS BS_m is selected to be moved from by an ant. After choosing VBS m', one of the sectors in VBS m' must be chosen. $P_{SEC}(k, m')$ is the probability that sector k in VBS m' is selected to be moved from by an ant. $P_{BS}(m)$ and $P_{SEC}(k, m')$ are calculated as follows:

$$P_{BS}(m) = \frac{C\_BS_m}{\sum_{u=1}^{M} C\_BS_u} \quad \text{for all} \quad m \tag{7}$$

$$P_{SEC}(k, m') = \frac{C\_SEC_k}{\sum_{l \in m'} C\_SEC_l} \quad \text{for all} \quad k \in BS_{m'} \tag{8}$$

The cell to be moved to by an ant is selected based on the amount of handoff traffic. $H_k(i)$ is the probability that cell i in N_k, is selected to move to first by an ant based on the amount of handoff traffic, h_{ij}. N_k is the set of cells which are not yet chosen for sector k and are adjacent to the cells of SEC_k.

$$H_k(i) = \frac{\sum_j (h_{ij} + h_{ji})}{\sum_i \sum_j (h_{ij} + h_{ji})}, \quad \text{for all} \quad i \in N_k, \quad \text{and} \quad j \in SEC_k \tag{9}$$

$phero(i, k)$ is the intensity of pheromone for cell i being assigned to sector k at time t which is $\tau_{ik}(t)$. This is indicative of the suitability of cell i for sector k. We set 0.001 for initial values of $phero(i, k)$ because the denominator of equation (10) cannot equal 0. $phero(i, k)$ is updated using Equation (2) from Sect. 2.1. $phero_k(i)$ is the probability of the suitability of cell i for sector k:

$$phero_k(i) = \frac{phero(i, k)}{\sum_{k=1}^{K} phero(i, k)} \quad \text{for all} \quad i \in N_k \tag{10}$$

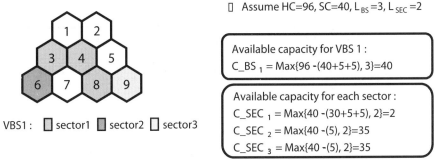

☐ Assume HC=96, SC=40, L_{BS} =3, L_{SEC} =2

Available capacity for VBS 1 :
C_BS $_1$ = Max{96 -(40+5+5), 3}=40

Available capacity for each sector :
C_SEC $_1$ = Max{40 -(30+5+5), 2}=2
C_SEC $_2$ = Max{40 -(5), 2}=35
C_SEC $_3$ = Max{40 -(5), 2}=35

VBS1 : ☐ sector1 ■ sector2 ☐ sector3

Fig. 7 Calculation of available capacity for VBS and sectors using lower bounds

Cell i is a cell adjacent to sector k. This cell has not been assigned to any sector yet. The probability that cell i will be assigned to sector k is

$$p_k(i) = \frac{\alpha H_k(i) + \beta\text{phero}_k(i)}{\sum_{l\in N_k}(\alpha H_k(l) + \beta\text{phero}_k(l))} \quad \text{for all} \quad i \in N_k \tag{11}$$

This probability considers both handoff traffic (termed the local heuristic in the ant colony literature) and pheromone. α and β are typical ant colony weighting factors where α weights the local heuristic and β weights the pheromone. For this chapter, $\alpha = 1$ and $\beta = 1$, giving equal weight to the local heuristic and the pheromone.

3 Experiments and Analysis

We consider three benchmarking problems from [14] (Table 1). We have recoded the GA proposed by Lee et al. [14] to compare the performance of our ant approach and the GA for these problems. 100 replications were performed of each algorithm for each problem. We use 10 ant colonies at each iteration, where each ant colony finds one solution. So, we have 10 different solutions at each iteration. We found the optimal solutions of the 12 and 19 cells problems using ILOG 5.1 to validate the performance of the heuristics. We terminate the ant system and the GA in these first two problems when an optimal solution is found and in the last problem (37 cells) by a CPU time of each replication of 3600 seconds. We define the convergence rate as how many times an optimal (or best found for the last problem) solution is obtained over 100 replications.

Table 1 Description of three benchmarking examples from Lee et al. [14]

	12 cells	19 cells	37 cells
Number of cells	12	19	37
number of VBSs	1	2	3
Number of sectors	3	6	9

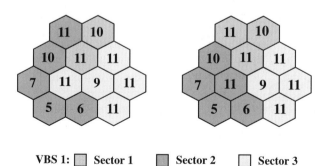

VBS 1: ▢ Sector 1   ■ Sector 2   ▢ Sector 3

Fig. 8 Comparison of groupings of 12 cells at time t and $t + 1$

For the 12 cells problem the objective function values of the old and the new groupings at times t and $t + 1$ are 255.604 and 217.842 as shown in Fig. 8. We have three ants in each colony because there are three sectors in one VBS. For the traffic distribution, we use an Erlang distribution with average traffic of 9. We set minimum CI to 0.5. We find an optimal solution with evaluation value of 217.842 using ILOG 5.1 with execution time = 4.42712 CPU seconds. The convergence rate of the ant approach to this optimal solution is 100% with 0.00711 CPU seconds per iteration while the convergence rate of GA is 98% with 0.02082 CPU seconds per iteration.

For the 19 cells problem the evaluation values of the old and the new groupings at times t and $t + 1$ are 601.388 and 284.597 as shown in Fig. 9 using an Erlang distribution with average traffic = 12. We have six ants in each colony because there are six sectors. We set minimum $CI = 0.65$. We find the optimal solution using ILOG 5.1 with an execution time 995.82 CPU seconds. The convergence rate of the ant ap-

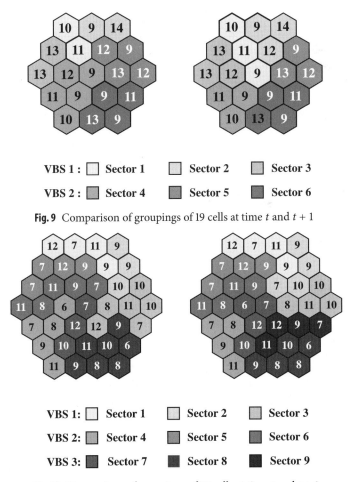

Fig. 9 Comparison of groupings of 19 cells at time t and $t + 1$

Fig. 10 Comparison of groupings of 37 cells at time t and $t + 1$

proach to this optimal solution is 100% with 0.06419 CPU seconds while the convergence rate of GA is 99% with 0.78378 CPU seconds.

For the large 37 cells problem, the evaluation values of the old and the new groupings at times t and $t + 1$ are 1091.18 and 726.288 as shown in Fig. 10 using an Erlang distribution with average traffic 9. We have nine ants in each ant colony because there are nine sectors. We set minimum $CI = 0.65$. Because this problem is too large to find

Table 2 Results of the ant colony approach and GA [14] for the 37 cells problem over 100 replications

Algorithm	Execution time	Objective minimum	maximum	average	Convergence rate
Ant System	5.0s	766.288	773.661	766.9409	73/100
	10.0s	766.288	773.661	766.9057	77/100
	20.0s	766.288	768.354	766.5772	86/100
	30.0s	766.288	768.354	766.5359	88/100
GA [14]	5.0s	766.288	888.258	798.7183	7/100
	10.0s	766.288	904.401	795.9874	12/100
	20.0s	766.288	874.574	785.0495	18/100
	30.0s	766.288	875.031	780.5263	18/100

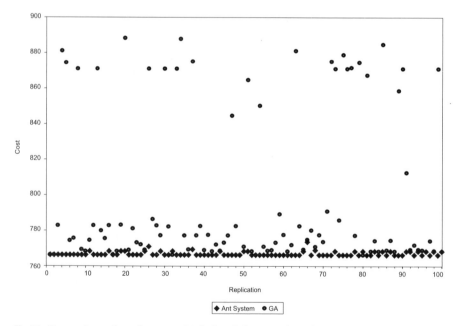

Fig. 11 Comparison of results using GA [13] and the ant colony for the 37 cells problem over 100 replications and execution time 5 seconds

the optimal solution exactly, we compare the performance of the ant approach and the GA using convergence rate within a limited CPU time. The convergence rates of 100 replications of the ant approach are 73, 77, 86, and 88% for computation times of 5, 10, 20, and 30 CPU seconds as shown in Table 2. Convergence rates of the GA are 7, 12, 18, and 18% for the same computation time. Not only does the ant approach far exceed the convergence rate to the best solution but the solutions found by the ant approach that are not the best are much closer to the best than those found by the GA (Figs. 11, 12, 13, and 14).

4 Conclusions

We have used the routing capability of the ant system paradigm to good effect in the problem of dynamic routing of micro-cellular systems. Our approach is computationally quick and reliable in terms of how close to optimal a given replication is likely to be. Using three test problems from the literature, we produced decidedly better results than the earlier published genetic algorithm approach and achieved optimality on the problems whose size allowed enumeration. There are some parameters to set for the ant system, but we chose straightforward ones and the method does not seem sensitive to their exact settings. The probabilities used for placement and movement of the ants were intuitively devised considering call traffic and available capacities.

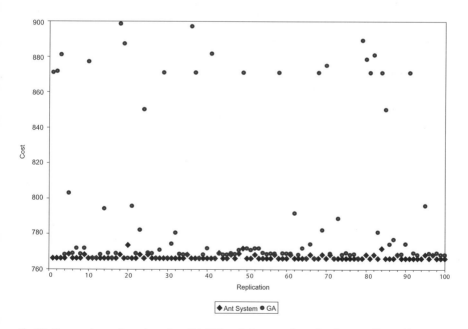

Fig. 12 Comparison of results using GA [13] and the ant colony for the 37 cells problem over 100 replications and execution time 10 seconds

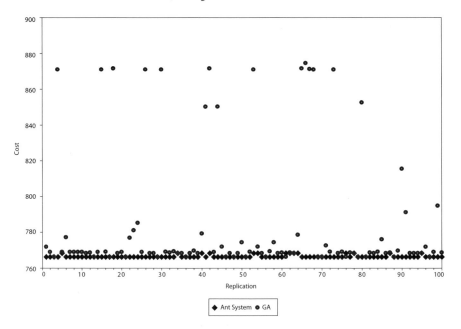

Fig. 13 Comparison of results using GA [13] and the ant colony for the 37 cells problem over 100 replications and execution time 20 seconds

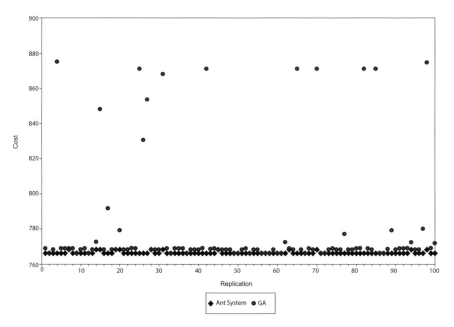

Fig. 14 Comparison of results using GA [13] and the ant colony for the 37 cells problem over 100 replications and execution time 30 seconds

References

1. Brown, E.C. and Vrobleski, M. (2004), A grouping genetic algorithm for the microcell sectorization problem, *Engineering Applications of Artificial Intelligence*, Vol. 17:589–598
2. Chan, T. M, Kwong, S, Man, K. F, and Tang, K. S (2002), Hard handoff minimization using genetic algorithms, *Signal Processing*, Vol. 82:1047–1058
3. Chu, C., JunHua Gu, J., Xiang Dan Hou, X., and Gu, Q. (2002), A heuristic ant algorithm for solving QoS multicast routing problem, *Proceedings of the 2002 Congress on Evolutionary Computation*, Vol. 2:1630–1635
4. Demirkol, I., Ersoy, C., Caglayan, M.U. and Delic, H. (2004), Location area planning and Dell-to-Switch assignment in cellular networks, *IEEE Transactions on Wireless Communications*, Vol. 3, No. 3:880–890
5. Dorigo, M. (1992), Optimization, Learning and Natural Algorithms, PhD Thesis, Politecnico di Milano, Italy
6. Dorigo, M. and Di Caro, G. (1999), The ant colony optimization meta-heuristic, in D. Corne, M. Dorigo and F. Glover (eds), New Ideas in Optimization, McGraw-Hill, 11–32
7. Dorigo, M. Di Caro, G., and Gambardella, L. M. (1999), Ant algorithms for discrete optimization, *Artificial Life*, Vol. 5, No. 2:137–172
8. Dorigo, M., Maniezzo, V. and Colorni, A., (1996), Ant system: optimization by a colony of cooperating gents, *IEEE Trans. on Systems, Man, and Cybernetics-Part B: Cybernetics*, Vol. 26, No 1:29–41
9. Dorigo, M., Gambardella, L. M. (1997), Ant colony system: a cooperative learning approach to the traveling salesman problem, *IEEE Trans. on Evolutionary Computation*, Vol. 1, No 1:53–66
10. Fournier, J.R.L. and Pierre, S. (2005), Assigning cells to switches in mobile networks using an ant colony optimization heuristic, *Computer Communication*, Vol. 28:65–73
11. Garey, M. R., Johnson, S. H., and Stockmeyer L. (1976), Some simplified NP-complete graph problems, *Theoretical Computer Science*, Vol. 1:237–267
12. Gunes, M., Sorges, U., and Bouazizi, I. (2002), ARA-the ant-colony based routing algorithm for MANETs, *Proceedings of International Conference on Parallel Processing Workshops*, 79–85
13. Kim, M. and Kim, J (1997), The facility location problems for minimizing CDMA hard handoffs, *Proceedings of Global Telecommunications Conference*, IEEE, Vol. 3:1611–1615
14. Lee, Chae Y., Kang, Hyon G., and Park, Taehoon (2002), A dynamic sectorization of micro cells for balanced traffic in CDMA: genetic algorithms approach, *IEEE Trans. on Vehicular Technology*, Vol.51, No.1:63–72
15. Liu, Y., Wu, J., Xu, K. and Xu, M. (2003), The degree-constrained multicasting algorithm using ant algorithm, *IEEE 10th International Conference on Telecommunications*, Vol. 1:370–374
16. Montemanni, R., Smith, D.H. and Allen, S. M. (2002), An ANTS algorithm for the minimum-span frequency assignment problem with multiple interference, *IEEE Trans. on Vehicular Technology*, Vol. 51, No. 5:949–953
17. Shyu, S.J., Lin, B.M.T., Hsiao, T.S. (2004), An ant algorithm for cell assignment in PCS networks, *IEEE International Conference on Networking, Sensing and Control*, Vol. 2:1081–1086
18. Shyu, S. J., Lin, B.M.T. and Hsiao, T.-S. (2006), Ant colony optimization for the cell assignment problem in PCS networks, *Computers & Operations Research*, Vol. 33:1713–1740
19. Sim, S.M. and Sun, W. H. (2003), Ant colony optimization for routing and load-balancing: survey and new directions, *IEEE Trans. on Systems, Man and Cybernetics, Part A*, Vol. 33, No. 5:560–572

20. Subing, Z and Zemin, L (2001), A Qos routing algorithm based on ant algorithm, *IEEE International Conference on Communications*, Vol. 5:1581–1585
21. Subrata, R. and Zomaya, A. Y. (2003), A comparison of three artificial life techniques for reporting cell planning in mobile computing, *IEEE Transactions on Parallel And Distributed Systems*, Vol. 14, No. 2:142–153
22. Vroblefski, M. and Brown, E. C. (2006), A grouping genetic algorithm for registration area planning, *Omega*, Vol. 34:220–230

New Ways to Calibrate Evolutionary Algorithms

Gusz Eiben and Martijn C. Schut

Department of Computer Science, Faculty of Science, VU University, Amsterdam, The Netherlands.
ae.eiben@few.vu.nl
mc.schut@few.vu.nl

Abstract

The issue of setting the values of various parameters of an evolutionary algorithm (EA) is crucial for good performance. One way to do it is by controlling EA parameters on-the-fly, which can be done in various ways and for various parameters. We briefly review these options in general and present the findings of a literature search and some statistics about the most popular options. Thereafter, we provide three case studies indicating a high potential for uncommon variants. In particular, we recommend focusing on parameters regulating selection and population size, rather than those concerning crossover and mutation. On the technical side, the case study on adjusting tournament size shows by example that global parameters can also be self-adapted, and that heuristic adaptation and pure self-adaptation can be successfully combined into a hybrid of the two.

Key words: Parameter Control, Self-adaptive, Selection, Population Size

1 Introduction

In the early years of evolutionary computing the common opinion was that evolutionary algorithm (EA) parameters are robust. The general claim was that the performance of an EA does not depend heavily on the right parameter values for the given problem instance at hand. Over the last two decades the EC community realised that setting the values of EA parameters is crucial for good performance. One way to calibrate EA parameters is by controlling them on-the-fly, which can be done in various ways and for various parameters [13,16,18]. The purpose of this chapter is to present a general description of this field, identify the main stream of research, and argue for alternative approaches that do not fall in the main stream. This argumentation is based on three case studies published earlier [7,15,17].

The rest of the chapter is organised as follows. Section 2 starts off with giving a short recap of the most common classification of parameter control techniques. Then we continue in Sect. 3 with an overview of related work, including some statistics on what types of parameter control are most common in the literature. Section 4 presents three case studies that substantiate our argument regarding the choice of the parameter(s) to be controlled. Section 5 concludes the chapter.

2 Classification of Parameter Control Techniques

In classifying parameter control techniques of an evolutionary algorithm, many aspects can be taken into account [1,13,16,18,53]. In this chapter we consider the three most important ones:

1. *What* is changed (e.g., representation, evaluation function, operators, selection process, mutation rate, population size, and so on)?
2. *How* the change is made (i.e., deterministic heuristic, feedback-based heuristic, or self-adaptive)?
3. *The evidence* upon which the change is carried out (e.g., monitoring performance of operators, diversity of the population, and so on)?

Each of these is discussed in the following.

2.1 *What* is Changed?

To classify parameter control techniques from the perspective of what component or parameter is changed, it is necessary to agree on a list of all major components of an evolutionary algorithm, which is a difficult task in itself. For that purpose, let us assume the following components of an EA:

- Representation of individuals
- Evaluation function
- Variation operators and their probabilities
- Selection operator (parent selection or mating selection)
- Replacement operator (survival selection or environmental selection)
- Population (size, topology, etc.)

Note that each component can be parameterised, and that the number of parameters is not clearly defined. For example, an offspring $\bar{v}$ produced by an arithmetical crossover of k parents $\bar{x}_1, \dots, \bar{x}_k$ can be defined by the following formula:

$$\bar{v} = a_1 \bar{x}_1 + \dots + a_k \bar{x}_k,$$

where $a_1, \dots, a_k$, and k can be considered as parameters of this crossover. Parameters for a population can include the number and sizes of subpopulations, migration rates, and so on for a general case, when more than one population is involved. Despite the somewhat arbitrary character of this list of components and of the list of parameters of each component, the "what-aspect" is one of the main classification features, since this allows us to locate where a specific mechanism has its effect.

2.2 *How* are Changes Made?

Methods for changing the value of a parameter (i.e., the "how-aspect") can be classified into: **parameter tuning** and **parameter control**. By parameter tuning we mean the commonly practised approach that amounts to finding good values for the parameters *before* the run of the algorithm and then running the algorithm using these values, which remain fixed during the run. Parameter control forms an alternative, as it amounts to starting a run with initial parameter values that are changed *during the run.*

We can further classify parameter control into one of the three following categories: deterministic, adaptive and self-adaptive. This terminology leads to the taxonomy illustrated in Fig. 1.

Deterministic parameter control This takes place when the value of a strategy parameter is altered by some deterministic rule. This rule fires at fixed moments, predetermined by the user (which explains the name "deterministic") and causes a predefined change without using any feedback from the search. Usually, a time-varying schedule is used, i.e., the rule is used when a set number of generations have elapsed since the last time the rule was activated.

Adaptive parameter control This takes place when there is some form of feedback from the search that serves as inputs to a mechanism used to determine the direction or magnitude of the change to the strategy parameter. The assignment of the value of the strategy parameter may involve credit assignment, based on the quality of solutions discovered by different operators/parameters, so that the updating mechanism can distinguish between the merits of competing strategies. Although the subsequent action of the EA may determine whether or not the new value persists or propagates throughout the population, the important point to note is that the updating mechanism used to control parameter values is externally supplied, rather than being part of the "standard" evolutionary cycle.

Self-adaptive parameter control The idea of the evolution of evolution can be used to implement the self-adaptation of parameters [6]. Here the parameters to be adapted are encoded into the chromosomes and undergo mutation and recombination. The better values of these encoded parameters lead to better individuals, which in turn are more likely to survive and produce offspring and hence propagate these better parameter values. This is an important distinction between adaptive and self-adaptive schemes: in the latter the mechanisms for the credit assignment and updating of different strategy parameters are entirely implicit, i.e., they are the selection and variation operators of the evolutionary cycle itself.

2.3 What *Evidence* Informs the Change?

The third criterion for classification concerns the evidence used for determining the change of parameter value [49, 51]. Most commonly, the progress of the search is

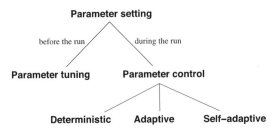

Fig. 1 Global taxonomy of parameter setting in EAs

monitored, e.g., by looking at the performance of operators, the diversity of the population, and so on. The information gathered by such a monitoring process is used as feedback for adjusting the parameters. From this perspective, we can make further distinction between the following two cases:

Absolute evidence We speak of absolute evidence when the value of a strategy parameter is altered by some rule that is applied when a predefined event occurs. The difference from deterministic parameter control lies in the fact that in deterministic parameter control a rule fires by a deterministic trigger (e.g., time elapsed), whereas here feedback from the search is used. For instance, the rule can be applied when the measure being monitored hits a previously set threshold – this is the event that forms the evidence. Examples of this type of parameter adjustment include increasing the mutation rate when the population diversity drops under a given value [38], changing the probability of applying mutation or crossover according to a fuzzy rule set using a variety of population statistics [37], and methods for resizing populations based on estimates of schemata fitness and variance [52]. Such mechanisms require that the user has a clear intuition about how to steer the given parameter into a certain direction in cases that can be specified in advance (e.g., they determine the threshold values for triggering rule activation). This intuition may be based on the encapsulation of practical experience, data-mining and empirical analysis of previous runs, or theoretical considerations (in the order of the three examples above), but all rely on the implicit assumption that changes that were appropriate to make on *another* search of *another* problem are applicable to *this* run of the EA on *this* problem.

Relative evidence In the case of using relative evidence, parameter values are compared according to the fitness of the offspring that they produce, and the better values get rewarded. The direction and/or magnitude of the change of the strategy parameter is not specified deterministically, but relative to the performance of other values, i.e., it is necessary to have more than one value present at any given time. Here, the assignment of the value of the strategy parameter involves credit assignment, and the action of the EA may determine whether or not the new value persists or propagates throughout the population. As an example, consider an EA using more crossovers with crossover rates adding up to 1.0 and being reset based on the crossovers performance measured by the quality of offspring they create. Such methods may be controlled adaptively, typically using "bookkeeping" to monitor performance and a user-supplied update procedure [11, 32, 45], or self-adaptively [5, 23, 35, 47, 50, 53] with the selection operator acting indirectly on operator or parameter frequencies via their association with "fit" solutions.

2.4 Summary

Our classification of parameter control methods is three-dimensional. The *component* dimension consists of six categories: representation, evaluation function, variation operators (mutation and recombination), selection, replacement, and population. The other dimensions have respectively three (deterministic, adaptive, self-adaptive) and two categories (absolute, relative). Their possible combinations are given in Table 1. As the table indicates, deterministic parameter control with relative evidence is impossible by definition, and so is self-adaptive parameter control with absolute evidence. Within the adaptive scheme both options are possible and are indeed used in practice.

3 Related Work

We conducted a literature review to get an overview of the work that has been done on the various parameters of evolutionary algorithms of the last decade. Our aim was not to deliver a fully annotated bibliography, but rather to illuminate some examples from the literature on this topic. The literature spans the conference proceedings of three major EC conferences: GECCO (1999–2006), CEC (1999–2006) and PPSN (1990–2006). In total we found 235 papers that were concerned, in any way (thus not necessarily (self-)adaptive), with one of the parameters of EAs mentioned above: representation, initialisation, evaluation function, variation operators, selection and population size. (In addition, we found 76 papers about adaptive EAs in general.) We categorised the 235 papers, the result of which is shown in Fig. 2. We consider this a preliminary overview giving some indication of the distribution of research effort spent on these issues. The histogram clearly shows that much research

Table 1 Refined taxonomy of parameter setting in EAs: types of parameter control along the type and evidence dimensions. The – entries represent meaningless (nonexistent) combinations

	Deterministic	Adaptive	Self-adaptive
Absolute	+	+	–
Relative	–	+	+

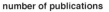

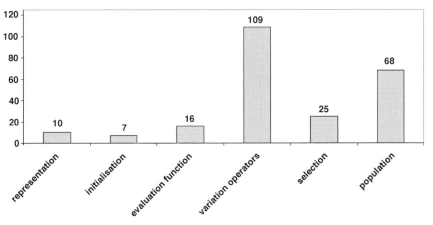

Fig. 2 Publication histogram

effort is spent on the variation operators (in general: 25, mutation: 54, crossover: 30). Also, the population parameter is well researched. However, we are aware of the fact that this number is biased, because it includes papers that are somewhat out of the scope of this chapter: for example, on population control in genetic programming, on the island-model of (sub)populations and on distributing (sub)populations in parallel evolutionary algorithms. We did not include papers on co-evolution.

We briefly discuss each EA parameter here, where we focus on the papers that explicitly look at (self-) adaptivity of the parameters. If possible, we make a distinction between deterministic, and self- adaptation within the discussion of a parameter.

3.1 Representation

Concerning representation, the *genome length* can be taken as a variable during an evolutionary run [28, 36, 43, 56]. Consider Ramsey et al. [43] who investigate a variable length genome under different mutation rates. To the suprise of the authors, the length of individuals self-adapts in direct response to the applied mutation rate. When tested with a broad range of mutation rates, the length tends to increase dramatically in the beginning and then decrease to a level corresponding to the mutation rate.

In earlier work, Harvey [28] presents an important property of variable-length genomes: "the absolute position of some symbols on the genotype can usually no longer be used to decide what feature those symbols relate to." Harvey sketches SAGA: a framework that was constructed to investigate the dynamics of a GA when genotype lengths are allowed to increase. The framework includes a particular crossover operator (SAGA cross) that has the requirement that the similarities are maximised between the two left segments that are swapped and between the two right segments that are swapped. This results in a computationally efficient algorithm where populations largely converge.

Stringer and Wu [56] show that a variable-length GA can evolve to shorter average size populations. This is observed when: (1) selection is absent from the GA, or (2) when selection focuses on some other property not influenced by the length of individuals. The model starts with an integer array of 100 elements, where each element represents an individual and the value denotes an individual's chromosome length. A simulated crossover produces children from two random parents, where the value of the first child equals the first (random) crossover point plus the value of the second parent less the second (random) crossover point; a similar procedure is used for the second child.

3.2 Initialisation

Although making the initialisation adaptive may seem contradictory (i.e., what should it adapt to initially?), there is a significant amount of literature dedicated to *dynamic restart strategies* for EAs [24,31,41,48]. This can be understood as (self-)adaptive initialisation.

Fukunga [24] shows how to find a good restart strategy in the context of resource-bounded scenarios: where the goal is to generate the best possible solution given a fixed amount of time. A new strategy is proposed that works based on a database of past performance of the algorithm on a class of problem instances. The resulting static restart strategy is competitive with those that are based on detection of lack of progress. This static strategy is an example of deterministic parameter control and it is surprising that it ourperforms the dynamic variant. According to the authors, one reason for this is that the dynamic variant can only consider local information of the current run.

Jansen [31] compares very simple deterministic strategies on a number of examples with different properties. The comparison is done in terms of a theoretical investigation on expected runtimes of various strategies. The strategies are more complex than fixing some point of time for a restart, but less complex than adaptive restart strategies. Two classes of dynamic restart strategies are presented and one additive and one multiplicative strategy is investigated in detail.

Finally, re-initialisation can also be considered in parallel genetic algorithms. Sekaj [48] researches this: the effect of re-initialisation is analysed with respect to convergence of the algorithm. In parallel genetic algorithms, (sub)populations may periodically *migrate* (as discussed later in the section about population). Sekaj lets re-initialisation happen when such migration took place. At re-initialisation, the current population was replaced by a completely new, randomly generated population. Additionally, two dynamic versions were presented: one where the algorithm after some number of generations compares the best individuals of each population and the worst population was re-initialised; and one which was based on the population diversity. Re-initialisation is able to remove differences between homogeneous and heterogeneous parallel GAs.

3.3 Evaluation Function

Regarding the evaluation function, some dynamic variants of this function are presented throughout the literature [19, 26, 34, 44].

The Stepwise Adaptation of Weights (SAW) technique has been introduced for problems that have a fitness function composed as a weighted sum of some atomic measure of quality. For instance, problems involving constraint satisfaction and data mining belong to this class, where the atomic measure can be the satisfaction of one given constraint or the correct classification of one given data record [77,80]. SAWing is an adaptive technique that adjusts the weights in the sum periodically at predefined intervals by increasing those that belong to "wrong" items, that is, to unsatisfied contraints or ill-classified records. Hereby SAWing effectively changes the fitness function and allows the algorithm to "concentrate" on the difficult cases. SAWing has been shown to work well on constraint satisfaction problems and data mining [78,79,81].

Reis et al. [44] propose and analyse a fractional-order dynamic fitness function: a fitness function based on fractional calculus. Besides the "default" fitness, the function has a component that represents the gain of the dynamical term. This dynamic function is compared with a static function that includes a discontinuity measurement. The comparison has been done in the domain of electronic circuit design. Both variants outperform the standard fitness algorithm.

Within the area of constraint optimisation problems, Kazarlis and Petridis [34] propose a technique where the problem constraints are included in the fitness function as penalty terms. During the GA evolution these terms are varied, such that the location of a global optimum is facilitated and local optima are avoided. In addition to a static penalty assigment method (which is more often used in these types of problems), an increasing function is introduced that depends on the evolution time. This function can be linear, exponential, square, cubic, quartic or 5-step. For the particular problems under research (Cutting and Packing and Unit Commitment Problem), it was found that the square function was the optimum increase rate of the penalty term (and not the linear function that was expected by the authors).

For the satisfiability problem, Gottlieb and Voss [26] compare three approaches based on adapting weights, where weights indicate the relative importance of a constraint in a particular satisfiability problem. Adaptation takes place after a fixed number of fitness evaluations. One of the three approaches yielded overall optimal performance that exploits SAT-specific knowledge.

It is noteworthy to mention that the dynamics of a fitness function can also be understood the other way around: where the fitness function is taken as being a dynamic one (because the problem is dynamic) and the EA has to deal with this. In such a case, the fitness function is thus not (self-)adaptive. For example, Eriksson and Olsson [19] propose a hybrid strategy to locate and track a moving global optimum.

3.4 Variation Operators

By far, most research effort in (self-)adaptive parameter control is spent on the variation operators: mutation and crossover. Although there are many papers about tuning the parameter values of the operator rates, a significant number look into (self-)adaptive parameter value control for mutation and crossover.

There are approximately a dozen papers that discuss the (self-)adaptive parameter control for *both* operators. Smith and Fogarty [50] use Kauffman's NK model to self-adapt the crossover and mutation rates. The method puts a mutation rate in the chromosome itself and lets the (global) mutation rate be based on some aggregated value of the mutation rates of the individuals. The authors compare their new method to a number of frequently used crossover operators with standard mutation rates. The results are competitive on simple problems, and significantly better on the most complex problems. Ho et al. [30] present a probabilistic rule-based adaptive model in which mutation and crossover rates are adapted during a run. The model works based on subpopulations that each use different (mutation or crossover or both) rates and good rates emerge based on the performance of the subpopulations. (Although called 'self-adaptive' by the authors, in our earlier mentioned terminology, this model is adaptive and not self-adaptive.) Finally, Zhang et al. [62] let the crossover and mutation rate adapt based on the application of K-means clustering. The population is clustered and rates are inferred based on the relative sizes of the clusters containing the best and worst chromosomes.

Regarding *mutation*, we see that the mutation operator is undoubtedly one of the most researched parameters according to our overview: 54 papers have been published about it. The topics of these papers range from introducing new mutation operators to the convergence of fitness gain effects of particular operators. Of these papers, 26 are specifically about (self-)adaptive mutation operators. We give some examples of these mutation operator studies. One of the earliest references in our overview is Bäck [4] who presents a self-adaptation mechanism of mutation rates. The rates are included in the individual itself. The methods enables a near-optimal schedule for the mutation rate. More recently, Katada et al. [33] looked at the domain of evolutionary robotics, where they tested a GA whose effective mutation rate changed according to the history of the genetic search. The individuals are neural networks that evolve over time. The control task was motion pattern discrimination. The variable mutation rate strategy shows better performance in this task, and this benefit was more pronounced with a larger genetic search space. Finally, Arnold [3] shows that *rescaled* mutations can be adaptively generated yielding robust and nearly optimal performance in problems with a range of noise strengths. Rescaling mutation has been suggested earlier in the literature, but this paper specifically discusses an adaptive approach for determining the scaling factor.

Concerning *crossover*, we briefly describe two studies. White and Oppacher [59] use automata to allow adaptation of the crossover operator probability during the run. The basic idea is to identify groups of bits within an individual that should be kept together during a crossover. An automaton encoded the probability that a given bit will be exchanged with the other parent under the crossover operators. First experiments show that the new crossover yields satisfactory results. The second study was undertaken by Vekaria and Clack [57] who investigate a number of biases to characterise adaptive recombination operators: directional – if alleles are either favoured or not for their credibility; credit – degree at which an allele becomes favoured; initialisation – if alleles are favoured without knowing their credibility; and hitchhiker – if alleles become favoured when they do not contribute to the fitness increase. Some experiments show, among other results, that initialisation bias (without mutation) does improve genetic search. Overall, the authors conclude that the biases are not always beneficial.

3.5 Selection

The majority of the 25 papers that we found with general reference to the selection mechanism of EAs are not about (self-)adaptive selection, but address rather a wide range of topics: e.g., without selection, effects of selection schemes, types of selection (clonal, anisotropic), and so forth. A stricter search shows that most studies that we categorised as being about (self-)adaptive selection actually refer to another EA parameter, for example, the mutation rate or the evaluation function. We only found one paper about (self-)adaptive survivor selection as defined in the terminology above.

Gorges-Schleuter [25] conducts a comparative study of global and local selection in evolution strategies. Traditionally, selection is a global parameter. In the so-called *diffusion model* for EAs, the individuals *only* exhibit local behaviour and the selection of partners for recombination and the selection individuals for survival are restricted to geographically nearby individuals. Local selection works then as follows: the first generated child is included in the next population whatsoever, each next child has to be better than its parent in order to be included in the next population.

3.6 Population

The population (size) parameter scores second-highest in our chapter overview. We already mentioned that this number is somewhat biased, because a number of the papers are about topics that are outside the scope of this chapter. General research streams that we identified regarding the population parameters with the 68 papers are: measuring population diversity, population size tuning, island model and migration parameter, ageing individuals, general effects of population size, population control in genetic programming, adaptive populations in particle swarm optimisation. For some topics, e.g., multi-populations or parallel populations, it is actually the *evaluation function* that is variable and not the population (size) itself – although at first sight the population size seems the varied parameter. This can also be said about co-evolutionary algorithms.

Approximately 10 of the 68 papers are specifically about (self-)adaptive population size. Later in this chapter, we discuss a number of these papers. Here, we briefly mention two other such papers. First, Lu and Yen [39] propose a dynamic multi-objective EA in which population growing and decline strategies are designed including a number of indicators that trigger the adaptive strategies. The EA is shown to be effective with respect to the population size and the diversity of the individuals. Secondly, Fernandes and Rosa [22] combine a dynamic reproduction rate based on population diversity, an ageing mechanism and a particular type of (macro-)mutation into one mechanism. The resulting mechanism is tested in a range of problems and shown to be superior in finding global optima.

4 Case Studies

We include three case studies that illustrate the benefits of (self-)adapting EA parameters. The first case study considers self-adaptive mutation and crossover and adaptive population size. The second study looks at on-the-fly population size adjustment. The third case study considers (self-)adaptive selection.

Throughout all case studies, we consider three important performance measures that reflect algorithm speed, algorithm certainty, and solution quality. The speed of optimization is measured by the Average number of Evaluations on Success (AES), showing how many evaluations it takes on average for the successful runs to find the optimum. The Success Rate (SR) shows how many of the runs were successful in finding the optimum. If the GA is somewhat unsuccessful ($SR < 100\%$), the measurement MBF (Mean Best Fitness) shows how close the GA can come to the optimum. If $SR = 100\%$, then the MBF will be 0, because every run found the optimum 0. (The MBF includes the data of all runs in it, the successful and the unsuccessful ones.)

4.1 An Empirical study of GAs "Without Parameters"

An empirical study on GAs "without parameters" by Bäck, Eiben and van der Vaart [7] can be considered as the starting point of the research this very chapter is based upon. The research it describes aims at eliminating GA parameters by making them (self-)adaptive while keeping, or even improving, GA performance. The quotes in the title indicate that this aim is only partly achieved, because the mechanisms for eliminating GA parameters can have parameters themselves. The paper describes methods to adjust

- the mutation rate (self-adaptive, by an existing method after [4]),
- the crossover rate (self-adaptive, by a newly invented method),
- the population size (adaptive, by an existing method after [2], [42, pp. 72–80])

on-the-fly, during a GA run.

The method to change mutation rate is self-adaptive. The mutation rate is encoded in extra bits at the tail of every individual. For each member in the starting population the rate is completely random within a given range. Mutation then takes place in two steps. First only the bits that encode the mutation rate are mutated and immediately decoded to establish the new mutation rate. This new mutation rate is applied to the main bits (those encoding a solution) of the individual. Crossover rates are also self-adaptive. A value between 0 and 1 is coded in extra bits at the tail of every individual (initialised randomly). When a member of the population is selected for reproduction by the tournament selection, a random number r below 1 is compared with the member's p_c. If r is lower than p_c, the member is ready to mate. If both selected parents are ready to mate two children are created by uniform crossover, mutated and inserted into the population. If it is not lower, the member will only be subject to mutation to create one child which undergoes mutation and survivor selection immediately. If both parents reject mating, the two children are created by mutation only. If one parent is willing to mate and the other one does not, then the parent that is not in for mating is mutated to create one offspring, which is inserted in the population immediately. The willing parent is on hold and the next parent selection round only picks one other parent. Population size undergoes changes through an adaptive scheme, based on the maximum-lifetime idea. Here every new individual is allocated a remaining lifetime (RLT) between the allowable minimum and maximum lifetime (MinLT and MaxLT) at birth. The RLT depends on the individual's fitness at the time of birth, related to other individuals in the population. In each cycle (roughly: generation), the remaining lifetime of all the members in the population is decremented by one. There is only one exception for the fittest member, whose

RLT is left unchanged. If the RLT of an individual reaches zero it is removed from the population.

The three methods to adjust parameters on-the-fly are then added to a traditional genetic algorithm and their effect on GA performance is investigated experimentally. The experimental comparison includes 5 GAs: a simple GA as benchmark, three GAs featuring only one of the parameter adjusting mechanisms and a GA that applies all three mechanisms and is therefore almost "parameterless". The experimental comparison is based on a test suite of five functions composed to comform to the guidelines in [8,60]: the sphere model, the generalised Rosenbrock function, the generalised Ackley function, the generalised Rastrigin function, and the fully deceptive six-bit function. All test functions are used with $n = 10$ dimensions and are scaled to have an optimal value of 0. We performed 30 runs for each condition.

In order to give a clear and compact overview of the performance of all GA variants we show the outcomes by ranking the GAs for each function in Table 2. To obtain a ranking, we award the best GA (fastest or closest to the minimum) one point, the second best GA two points, and so on, so the worst performing GA for a given function gets five points. If, for a particular function, two GAs finish very close to each other, we award them equally: add the points for both those rankings and divide that by two. After calculating these points for each function and each GA variant we add the points for all the functions to form a total for each GA. The GA with the least points has the best overall performance.

The overall competition ends in a close finish between the all-in GA as number one and AP-GA, the GA with adaptive population size, right on its heels. In this respect, the objective of the study has been achieved, using the all-in GA the user has fewer parameters[1] to set and gets higher performance than using the simple GA.

An unexpected outcome of this study is that adaptive population sizes proved to be the key feature to improve the benchmark traditional GA, TGA. Alone, or in combination with the self-adaptive variation operators, the mechanism to adjust the population size during a run causes a consequent performance improvement w.r.t. the benchmark GA. These outcomes give a strong indication that, contrary to past

Table 2 Ranking of the GAs (the labelling is obvious from the text)

	TGA	SAM-GA	SAX-GA	AP-GA	all-in GA
Sphere	2	5	3	1	4
Rosenbrock	3	5	4	4	2
Ackley	1	2.5	5	4	2.5
Rastrigin	2.5	5	4	2.5	1
Deceptive	4	5	2.5	2.5	1
Points	12.5	22.5	18.5	11	10.5
End rank	**3**	**5**	**4**	**2**	**1**

---

[1] This is not entirely true, since (1) *initial* values for those parameters must be given by the user and (2) the population adaptation method introduces two new parameters MinLT and MaxLT.

and present practice (where quite some effort is devoted to tuning or online controlling of the application rates of variance operators), studying control mechanisms for variable population sizes should be paid more attention.

After having published this paper, this conclusion has been generalised by distinguishing variation and selection as the two essential powers behind an evolutionary process [18, Chapter 2]. Here, variation includes recombination (crossover) and mutation; for selection we can further distinguish parent selection and survivor selection (replacement). Clearly, the population size is affected by the latter. From this perspective, the paper gives a hint that further to studying mechanisms for on-the-fly calibration of variation operators, the EC community should adopt a research agenda for on-the-fly selection control mechanisms, including those focusing on population size management.

4.2 Evolutionary Algorithms with On-the-Fly Population Size Adjustment

The investigation in [15] is a direct follow-up to [7] discussed in the previous section. As noted by Bäck et al. the population size is traditionally a rigid parameter in evolutionary computing. This is not only true in the sense that for the huge majority of EAs the population size remains constant over the run, but also for the EC research community that has not spent much effort on EAs with variable population sizes. However, there are biological and experimental arguments to expect that this would be rewarding. In natural environments, population sizes of species change and tend to stabilise around appropriate values according to factors such as natural resources and carrying capacity of the ecosystem. Looking at it technically, population size is the most flexible parameter in natural systems: it can be adjusted much more easily than, for instance, mutation rate.

The objective of this study is to perform an experimental evaluation of a number of methods for calibrating population size on-the-fly. Note that the paper does not consider (theory-based) strategies for *tuning* population size [65–68,70,74]. The inventory of methods considered for an experimental comparison includes the following algorithms from the literature. The Genetic Algorithm with Variable Population Size (GAVaPS) from Arabas [2], [42, pp. 72–80] eliminates population size as an explicit parameter by introducing the age and maximum lifetime properties for individuals. The maximum lifetimes are allocated at birth depending on fitness of the newborn, while the age (initialised to 0 at birth) is incremented at each generation by one. Individuals are removed from the population when their ages reach the value of their predefined maximum lifetime. This mechanism makes survivor selection unnecessary and population size an observable, rather than a parameter. The Adaptive Population size GA (APGA) is a variant of GAVaPS where a steady-state GA is used, and the lifetime of the best individual remains unchanged when individuals grow older [7]. In [27, 69] Harik and Lobo introduce a parameterless GA (PLGA) which evolves a number of populations of different sizes simultaneously. Smaller populations get more function evaluations, where population i is allowed to run four times more generations than the population $i + 1$. If, however, a smaller population converges, the algorithm drops it. The Random Variation of the Population

Size GA (RVPS) is presented by Costa et al. in [63]. In this algorithm, the size of the actual population is changed every N fitness evaluations, for a given N. Hinterding, Michalewicz and Peachey [29] presented an adaptive mechanism, in which three sub-populations with different population sizes are used. The population sizes are adapted at regular intervals (*epochs*) biasing the search to maximise the performance of the group with the mid-most size. The criterion used for varying the sizes is fitness diversity. Schlierkamp-Voosen and Mühlenbein [72] use a competition scheme between sub-populations to adapt the size of the sub-populations as well as the overall population size. There is a quality criterion for each group, as well as a gain criterion, which dictates the amount of change in the group's size. The mechanism is designed in such a way that only the size of the best group can increase. A technique for dynamically adjusting the population size with respect to the probability of selection error, based on Goldberg's research [67], is presented in [73]. Finally, the following methods have been selected for the experimental comparison.

1. GAVaPS from [2],
2. GA with adaptive population size (APGA) from [7],
3. the parameterless GA from [27],
4. the GA with Random Variation of Population Size (RVPS) from [63],
5. the Population Resizing on Fitness Improvement GA (PRoFIGA), newly invented for this paper.

The new method is based on monitoring improvements of the best fitness in the population. On fitness improvement the EA is made more explorative by increasing the population size. If the fitness is not improving (for a short while) the population is made smaller. However, if stagnation takes too long, then the population size is increased again. The intuition behind this algorithm is related to (a rather simplified view on) exploration and exploitation. The bigger the population size, the more it supports explorative search. Because in the early stages of an EA run fitness typically increases, population growth, hence exploration, will be more prominent in the beginning. Later on it will decrease gradually. The shrinking phase is expected to "concentrate" more on exploitation of fewer individuals after reaching the limits of exploration. The second kind of population growth is supposed to initiate renewed exploration in a population that ist stuck in local optima. Initial testing has shown that GAVaPS was very sensitive for the *reproduction ratio* parameter and the algorithm frequently increased the size of the population over several thousand individuals, which resulted in unreliable performance. For this reason it was removed from further experimentation.

When choosing the test suite for experimentation popular, but ad hoc collections of objective functions are deliberately avoided for reasons outlined in [14] and [18, Chapter 14]. Instead, the multimodal problem generator of Spears [54] is used for it has been designed to facilitate systematic studies of GA behavior. This generator creates random problem instances, i.e., fitness landscapes over bit-strings, with a controllable size (chromosome length) and degree of multi-modality (number of peaks). The test suite consists of 10 different landscapes for 100 bits, where the number of peaks ranges from 1 to 1000 through 1, 2, 5, 10, 25, 50, 100, 250, 500, and 1000.

We performed 100 runs for each condition. Here again, the performance of the algorithms is assessed by Success Rate (SR), Mean Best Fitness (MBF), and the Average number of Evaluations to a Solution (AES). SR is an effectivity measure that gives the percentage of runs in which the optimum (the highest peak) was found. MBF is also an effectivity measure showing the average of the best fitness in the last population over all runs. AES is a measure of efficiency (speed): it is the number of evaluations it takes on average for the successful runs to find the optimum. If a GA has no success ($SR = 0$) then the AES measure is undefined.

The main results are given in the graphs with a grid background in Fig. 3 and the left-hand side of Fig. 4.

The AES plots are shown in Fig. 3 (left). These graphs show clear differences between the algorithms. There are, however, no significant differences between the problem instances when only looking at the speed curves (except for the parameter-less GA). Apparently, finding a solution does not take more evaluations on a harder problem that has more peaks. (Although it should be noted that for harder problems the averages are taken over fewer runs, cf. the SR figures below, which reduces the reliability of the statistics.) This is an interesting artefact of the problem generator that needs further investigation. The increasing problem hardness, however, is clear from

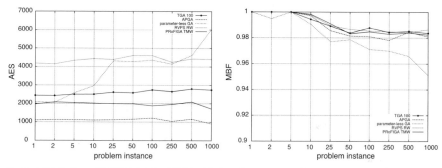

Fig. 3 AES (left) and MBF (right) of TGA, APGA, the parameterless GA, RVPS and PRoFIGA with max-eval = 10,000

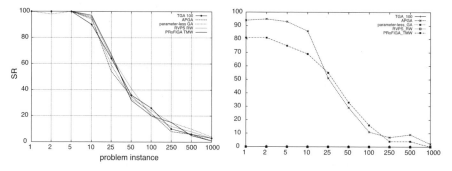

Fig. 4 SR of TGA, APGA, the parameterless GA, RVPS and PRoFIGA with max-eval = 10,000 (left) and with max-eval = 1500 (right)

the decreasing average quality of the best solution found (MBF), cf. Fig. 3 (right) and the decreasing probability of finding a solution (SR), cf. Fig. 4 (left).

We can rank the population (re)sizing methods based on the AES plots: APGA is significantly faster than the other methods, followed by PRoFIGA. The traditional GA comes third. The parameterless GA is only competitive for easy problems and the RVPS RW is clearly inferior to the other methods.

The SR and MBF results are quite homogeneous, with only one negative outlier, the parameterless GA. It seems that we cannot rank the algorithms by their effectivity. However, this homogeneity is a consequence of our choice of the maximum number of fitness evaluations in the termination criterion. Apparently it is "too" high allowing all contestants to reach the performance of the champions – be it slower. As a control experiment, we repeated all runs with the maximum number of fitness evaluations set to 1500. The resulting success rates are given in Fig. 4 (right), showing great differences. APGA and PRoFIGA obtain somewhat worse, but comparable SR results as before, but the other algorithms never find a solution yielding SR = 0 over all peaks.

Looking at the results we can conclude that adapting population sizes in an EA can certainly pay off. The gains in terms of efficiency, measured by the number of fitness evaluations needed to find a solution, can be significant: the winner of our comparison (APGA) achieves the same success rate and mean best fitness as the traditional GA with less than half of the work, and even the second best (PRoFIGA) needs 20% fewer evaluations. The second series of experiments shows that such an increase in speed can be converted into increased effectivity, depending on the termination condition. Here again, the winner is APGA, followed by PRoFIGA. It should be noted that two GAs from this comparison (the parameterless GA and RVPS RW) are much slower than the traditional GA. Hence, on-the-fly population (re)sizing is not necessarily better than traditional hand-tuning of a constant population size. The added value depends on the actual implementation, i.e., on *how* the population size is adjusted. Another observation made here is that the lifetime principle used in APGA eliminates explicit survivor selection and makes population size an observable instead of a user parameter. However, it should also be noted that using this idea does not mean that the number of EA parameters is reduced. In fact, it is increased in our case: instead of the population size N in the TGA, the APGA introduces two new parameters, *MinLT* and *MaxLT*.

These results can be naturally combined with those of Bäck et al. confirming the superiority of APGA on another test suite. Of course, highly general claims are still not possible about APGA. But these results together form a strong indication that incorporating on-the-fly population (re)sizing mechanisms based on the lifetime principle in EAs is a very promising design heuristic definitely worth trying and that APGA is a successful implementation of this general idea.

4.3   Boosting Genetic Algorithms with (Self-) Adaptive Selection

The paper [17] seeks an answer to the question whether it is feasible (i.e., possible and rewarding) to self-adapt selection parameters in an evolutionary algorithm? Note that the idea seems quite impossible considering that

- Self-adaptation manipulates parameters defined within an *individual*, hence the given parameter will have different values over different members of the population.
- Parameters regarding selection (e.g., tournament size in tournament selection or the bias in ranked biased selection) are inherently *global*, any given value holds for the whole population, not only for an individual.

This explains why existing approaches to controlling such parameters are either deterministic or adaptive.

The paper investigates self-adaptation of tournament size in a purely self-adaptive fashion and a variant that combines self-adaptation with a heuristic. The approach is based on keeping tournament size K as a globally valid parameter, but decomposing it. That is, to introduce local parameters k at the level of individuals that can be self-adapted and establish the global value through aggregating the local ones. Technically, this means extending the individual's chromosomes by an extra gene resulting in $\langle x, k \rangle$. Furthermore, two methods are required. One, to specify how to aggregate local k values to a global one. Two, a mechanism for variation (crossover and mutation) of the local k values.

The aggregation mechanism is rather straightforward. Roughly speaking, the global parameter will be the sum of the local votes of all individuals. Here we present a general formula applicable for any global parameter P and consider tournament size K as a special case.

$$P = \lceil \sum_{i=1}^{N} p_i \rceil \tag{1}$$

where $p_i \in [p_{\min}, p_{\max}]$, $\lceil \ \rceil$ denotes the ceiling function, and N is the population size.

Concerning variation of the extended chromosomes, crossover and mutation are distinguished. Crossover works on the whole $\langle x, k \rangle$, by whichever mechanism the user wishes. Mutation, however, is split. The x part of $\langle x, k \rangle$ is mutated by any suitable mutation operator, but for the k part a specific mechanism is used. A straightforward option would be the standard self-adaptation mechanism of σ values from Evolution Strategies. However, those σ values are not bounded, while tournament size is obviously bounded by zero and the population size. A possible solution is the self-adaptive mechanism for mutation rates in GAs as described by Bäck and Schütz [9]. This mechanism is introduced for $p \in (0, 1)$ and it works as follows:

$$p' = \left(1 + \frac{1-p}{p} \cdot e^{-\gamma \cdot N(0,1)} \right)^{-1} \tag{2}$$

where p is the parameter in question and γ is the learning rate, which allows for control of the adaptation speed. This mechanism has some desirable properties:

1. Changing $p \in (0, 1)$ yields a $p' \in (0, 1)$.
2. Small changes are more likely than large ones.

3. The expected change of p by repeatedly changing it equals zero (which is desirable, because natural selection should be the only force bringing a direction in the evolution process).
4. Modifying by a factor c occurs with the same probability as a modification by $1/c$.

The concrete mechanism for self-adaptive tournament sizes uses individual k values $k \in (0,1)$ and the formula of Equation (2) with $\gamma = 0.22$ (as recommended in [9]). Note that if a GA uses a recombination operator then this operator will be applied to the tournament size parameter k, just as it is applied to other genes. In practice this means that a child created by recombination inherits an initial k value from its parents and the definitive value k is obtained by mutation as described by Equation (2).

Besides the purely self-adaptive mechanism for adjusting tournament sizes the chapter also introduces a heuristic variant. In the self-adaptive algorithm as described above the direction (+ or −) as well as the extent (increment/decrement) of the change are fully determined by the random scheme. This is a general property of self-adaptation. However, in the particular case of regulating selection pressure we do have some intuition about the direction of change. Namely, if a new individual is better than its parents then it should try to increase selection pressure, assuming that stronger selection will be advantageous for him, giving a reproductive advantage over less fit individuals. In the opposite case, if it is less fit than its parents, then it should try to lower the selection pressure. Our second mechanism is based on this idea. Formally, we keep the aggregation mechanism from equation 1 and use the following rule. If $\langle x, k \rangle$ is an individual to be mutated (either obtained by crossover or just to be reproduced solely by mutation), then first we create x' from x by the regular bitflips, then apply

$$k' = \begin{cases} k + \Delta k & \text{if } f(x') \geq f(x) \\ k - \Delta k & \text{otherwise} \end{cases} \qquad (3)$$

where

$$\Delta k = \left| k - \left(1 + \frac{1-k}{k} e^{-\gamma N(0,1)}\right)^{-1} \right| \qquad (4)$$

with $\gamma = 0.22$.

This mechanism differs from "pure" self-adaptation because of the heuristic rule specifying the direction of the change. However, it could be argued that this mechanism is not a clean adaptive scheme (because the initial k values are inherited), nor a clean self-adaptive scheme (because the final k values are influenced by a user defined heuristic), but some hybrid form. For this reason we perceive and name this mechanism *hybrid self-adaptive* (HSA). All together this yields two new GAs: the GA with self-adaptive tournament size (GASAT) and the GA with hybrid self-adaptive tournament size (GAHSAT).

The experimental comparison of these GAs and a standard GA for benchmark is based on exactly the same test suite as the study in the previous section. The GAs are tested on the same landscapes in $\{0, 1\}^{100}$ with 1, 2, 5, 10, 25, 50, 100, 250, 500 and 1000 peaks obtained through the Multimodal Problem Generator of Spears [54]. We performed 100 runs for each condition. Also the performance measures are identical: the Mean Best Fitness (MBF) and its standard deviation (SDMBF), the Average number of Evaluations to a Solution (AES) and its standard deviation (SDAES), and the Success Rate (SR) are calculated, based on 100 independent runs. The results for the SGA, GASAT, and GAHSAT are shown in Table 3, Table 4, and Table 5, respectively.

The outcomes indicate that GASAT has better performance than SGA, but it is not as powerful as the hybrid self-adaptive mechanism. The initial research question about the feasibility of on-the-fly adjustment of K can be answered positively. It is interesting to remark here that the self-adaptive GAs are based on a simple mechanism (that was, nota bene, introduced for mutation parameters) and apply no sophisticated twists to it. Yet, they show very good performance that compares favorably with the best GA in [15]. The comparison between the former winner, APGA with adaptive population size, and GAHSAT is shown in Table 6. Note that the MBF results are omitted for they showed no significant difference. This comparison shows that the GAHSAT is very competitive, running out the APGA on the smoother landscapes.

Table 3 Results of SGA

Peaks	SR	AES	SDAES	MBF	SDMBF
1	100	1478	191	1.0	0.0
2	100	1454	143	1.0	0.0
5	100	1488	159	1.0	0.0
10	93	1529	168	0.9961	0.0142
25	62	1674	238	0.9885	0.0174
50	37	1668	221	0.9876	0.0140
100	22	1822	198	0.9853	0.0145
250	11	1923	206	0.9847	0.0137
500	6	2089	230	0.9865	0.0122
1000	5	2358	398	0.9891	0.0100

Table 4 End results of GASAT

Peaks	SR	AES	SDAES	MBF	SDMBF
1	100	1312	218	1.0	0.0
2	100	1350	214	1.0	0.0
5	100	1351	254	1.0	0.0
10	92	1433	248	0.9956	0.0151
25	62	1485	280	0.9893	0.0164
50	46	1557	246	0.9897	0.0128
100	21	1669	347	0.9853	0.0147
250	16	1635	336	0.9867	0.0130
500	3	1918	352	0.9834	0.0146
1000	1	1675	0	0.9838	0.0126

Table 5 End results of GAHSAT

Peaks	SR	AES	SDAES	MBF	SDMBF
1	100	989	244	1.0	0.0
2	100	969	206	1.0	0.0
5	100	1007	233	1.0	0.0
10	89	1075	280	0.9939	0.0175
25	63	1134	303	0.9879	0.0190
50	45	1194	215	0.9891	0.0127
100	14	1263	220	0.9847	0.0140
250	12	1217	166	0.9850	0.0131
500	7	1541	446	0.9876	0.0119
1000	4	1503	272	0.9862	0.0113

Table 6 Comparing GAHSAT and the winning APGA from [15]

	GAHSAT		APGA	
Peaks	SR	AES	SR	AES
1	100	989	100	1100
2	100	969	100	1129
5	100	1007	100	1119
10	89	1075	95	1104
25	63	1134	54	1122
50	45	1194	35	1153
100	14	1263	22	1216
250	12	1217	8	1040
500	7	1541	6	1161
1000	4	1503	1	910

5 Summary and Conclusions

The relevance of the above studies lies in the potential of on-the-fly adjustment of EA parameters that have not been widely considered in the past. The investigations reviewed here provide substantial evidence that on-line regulation of population size and selection can greatly improve EA performance. On the technical side, the case study on adjusting tournament size shows by example that global parameters can also be self-adapted, and that heuristic adaptation and pure self-adaptation can be successfully combined into a hybrid of the two.

On the general level, two things can be noted. First, we want to remark that parameter control in an EA can have two purposes. One motivation for controlling parameters on-the-fly is the assumption (observation) that in different phases of the search the given parameter should have different values for "optimal" algorithm performance. If this holds, then static parameter values are always inferior; for good EA performance one must vary this parameter. Another reason it can be done for is to avoid suboptimal algorithm performance resulting from suboptimal parameter values set by the user. The basic assumption here is that the algorithmic control mechanisms do this job better than the user could, or that they can do it approximately as good, but they liberate the user from doing it. Either way, they are beneficial.

The second thing we want to note is that making a parameter adaptive or self-adaptive does not necessarily mean that we have an EA with fewer parameters. For instance, in APGA the population size parameter is eliminated at the cost of introducing two new ones: the minimum and maximum lifetime of newborn individuals. If the EA performance is sensitive to these new parameters then such a parameter replacement can make things worse. But if the new parameters are less sensitive to accurate calibration, then the net effect is positive: the user can obtain a good algorithm with less effort spent on algorithm design. This latter is, however, hardly ever considered in evolutionary computing publications.

This phenomenon also occurs on another level. One could say that the mechanisms to adjust parameters are also (meta) parameters. For instance, the method that allocates lifetimes in APGA, or the function in Equation (2) specifying how the k values are mutated can be seen as high level parameters of the GA. It is in fact an assumption that these are well-chosen (smartly designed) and their effect is positive. Typically, there are more possibilities to obtain the required functionality, that is, there are possibly more well-working methods one can design. *Comparing different methods* implies experimental (or theoretical) studies very much like *comparing different parameter values* in a classical setting. Here again, it can be the case that algorithm performance is not so sensitive to details of this (meta) parameter, which can fully justify this approach.

Finally, let us place the issue of parameter control in a larger perspective. Over the last two decades the EC community has come to realise that EA performance, to a large extent, depends on well-chosen parameter values, which in turn may depend on the problem (instance) to be solved. In other words, it is now acknowledged that EA parameters need to be calibrated to specific problems and problem instances. Ideally, it should be the algorithm that performs the necessary problem-specific adjustments. Ultimately, it would be highly desirable to utilise the inherent adaptive power of an evolutionary process for calibrating *itself* to a certain problem instance, while solving that very problem instance. We believe that the extra computational overhead (i.e., solving the self-calibration problem additionally to the given technical problem) will pay off and hope to see more research on this issue.

References

1. P.J. Angeline. Adaptive and self-adaptive evolutionary computations. In *Computational Intelligence*, pages 152–161. IEEE Press, 1995
2. J. Arabas, Z. Michalewicz, and J. Mulawka. GAVaPS – a genetic algorithm with varying population size. In *Proceedings of the First IEEE Conference on Evolutionary Computation*, pages 73–78. IEEE Press, Piscataway, NJ, 1994
3. D.V. Arnold. Evolution strategies with adaptively rescaled mutation vectors. In *2005 Congress on Evolutionary Computation (CEC'2005)*, pages 2592–2599. IEEE Press, Piscataway, NJ, 2005
4. T. Bäck. The interaction of mutation rate, selection, and self-adaptation within a genetic algorithm. In Männer and Manderick [40], pages 85–94
5. T. Bäck. Self adaptation in genetic algorithms. In F.J. Varela and P. Bourgine, editors, *Toward a Practice of Autonomous Systems: Proceedings of the 1st European Conference on Artificial Life*, pages 263–271. MIT Press, Cambridge, MA, 1992
6. T. Bäck. Self-adaptation. In T. Bäck, D.B. Fogel, and Z. Michalewicz, editors, *Evolutionary Computation 2: Advanced Algorithms and Operators*, Chapter 21, pages 188–211. Institute of Physics Publishing, Bristol, 2000
7. T. Bäck, A.E. Eiben, and N.A.L. van der Vaart. An empirical study on GAs "without parameters". In Schoenauer et al. [46], pages 315–324
8. T. Bäck and Z. Michalewicz. Test landscapes. In T. Bäck, D.B. Fogel, and Z. Michalewicz, editors, *Handbook of Evolutionary Computation*, Chapter B2.7, pages 14–20. Institute of Physics Publishing, Bristol, and Oxford University Press, New York, 1997
9. Th. Bäck and M. Schütz. Intelligent mutation rate control in canonical genetic algorithms. In Zbigniew W. Ras and Maciej Michalewicz, editors, *Foundations of Intelligent Systems, 9th International Symposium, ISMIS '96, Zakopane, Poland, June 9-13, 1996, Proceedings*, volume 1079 of *Lecture Notes in Computer Science*, pages 158–167. Springer, Berlin, Heidelberg, New York, 1996
10. Y. Davidor, H.-P. Schwefel, and R. Männer, editors. *Proceedings of the 3rd Conference on Parallel Problem Solving from Nature*, number 866 in Lecture Notes in Computer Science. Springer, Berlin, Heidelberg, New York, 1994
11. L. Davis. Adapting operator probabilities in genetic algorithms. In J.D. Schaffer, editor, *Proceedings of the 3rd International Conference on Genetic Algorithms*, pages 61–69. Morgan Kaufmann, San Francisco, 1989
12. A.E. Eiben, T. Bäck, M. Schoenauer, and H.-P. Schwefel, editors. *Proceedings of the 5th Conference on Parallel Problem Solving from Nature*, number 1498 in Lecture Notes in Computer Science. Springer, Berlin, Heidelberg, New York, 1998
13. A.E. Eiben, R. Hinterding, and Z. Michalewicz. Parameter control in evolutionary algorithms. *IEEE Transactions on Evolutionary Computation*, 3(2):124–141, 1999.
14. A.E. Eiben and M. Jelasity. A critical note on experimental research methodology in EC. In *Proceedings of the 2002 Congress on Evolutionary Computation (CEC'2002)*, pages 582–587. IEEE Press, Piscataway, NJ, 2002

15. A.E. Eiben, E. Marchiori, and V.A. Valko. Evolutionary algorithms with on-the-fly population size adjustment. In X. Yao et al., editor, *Parallel Problem Solving from Nature, PPSN VIII*, number 3242 in Lecture Notes in Computer Science, pages 41–50. Springer, Berlin, Heidelberg, New York, 2004

16. A.E. Eiben, Z. Michalewicz, M. Schoenauer, and J.E. Smith. Parameter Control in Evolutionary Algorithms. In Lobo, Fernando G., Lima, Cláudio F. and Michalewicz, Zbigniew, editors, *Parameter Setting in Evolutionary Algorithms*, Studies in Computational Intelligence. Springer, 2007, pages 19–46

17. A.E. Eiben, M.C. Schut, and A.R. deWilde. Boosting genetic algorithms with (self-) adaptive selection. In *Proceedings of the IEEE Conference on Evolutionary Computation*, 2006, pages 1584–1589

18. A.E. Eiben and J.E. Smith. *Introduction to Evolutionary Computing*. Springer, Berlin, Heidelberg, New York, 2003

19. R. Eriksson and B. Olsson. On the performance of evolutionary algorithms with life-time adaptation in dynamic fitness landscapes. In *2004 Congress on Evolutionary Computation (CEC'2004)*, pages 1293–1300. IEEE Press, Piscataway, NJ, 2004

20. L.J. Eshelman, editor. *Proceedings of the 6th International Conference on Genetic Algorithms*. Morgan Kaufmann, San Francisco, 1995

21. H.-G. Beyer et al., editor. *Proceedings of the Genetic and Evolutionary Computation Conference (GECCO-2005)*. ACM, 2005

22. C. Fernandes and A. Rosa. Self-regulated population size in evolutionary algorithms. In Th.-P. Runarsson, H.-G. Beyer, E. Burke, J.-J. Merelo-Guervos, L. Darell Whitley, and X. Yao, editors, *Parallel Problem Solving from Nature – PPSN IX*, number 4193 in Lecture Notes in Computer Science, pages 920–929. Springer, Berlin, Heidelberg, New York, 2006

23. D.B. Fogel. *Evolutionary Computation*. IEEE Press, 1995

24. A.S. Fukunga. Restart scheduling for genetic algorithms. In Eiben et al. [12], pages 357–366

25. M. Gorges-Schleuter. A comparative study of global and local selection in evolution strategies. In Eiben et al. [12], pages 367–377

26. J. Gottlieb and N. Voss. Adaptive fitness functions for the satisfiability problem. In Schoenauer et al. [46], pages 621–630

27. Georges R. Harik and Fernando G. Lobo. A parameter-less genetic algorithm. In Wolfgang Banzhaf et al., editor, *Proceedings of the Genetic and Evolutionary Computation Conference*, volume 1, pages 258–265. Morgan Kaufmann, 1999

28. I. Harvey. The saga-cross: the mechanics of recombination for species with variable-length genotypes. In Männer and Manderick [40], pages 269–278

29. R. Hinterding, Z. Michalewicz, and T.C. Peachey. Self-adaptive genetic algorithm for numeric functions. In Voigt et al. [58], pages 420–429

30. C.W. Ho, K.H. Lee, and K.S. Leung. A genetic algorithm based on mutation and crossover with adaptive probabilities. In *1999 Congress on Evolutionary Computation (CEC'1999)*, pages 768–775. IEEE Press, Piscataway, NJ, 1999

31. T. Jansen. On the analysis of dynamic restart strategies for evolutionary algorithms. In J.J. Merelo Guervos, P. Adamidis, H.-G. Beyer, J.-L. Fernandez-Villacanas, and H.-P. Schwefel, editors, *Proceedings of the 7th Conference on Parallel Problem Solving from Nature*, number 2439 in Lecture Notes in Computer Science, pages 33–43. Springer, Berlin, Heidelberg, New York, 2002

32. B.A. Julstrom. What have you done for me lately?: Adapting operator probabilities in a steady-state genetic algorithm. In Eshelman [20], pages 81–87

33. Y. Katada, K. Okhura, and K. Ueda. An approach to evolutionary robotics using a genetic algorithm with a variable mutation rate strategy. In Yao et al. [61], pages 952–961

34. S. Kazarlis and V. Petridis. Varying fitness functions in genetic algorithms: studying the rate of increase of the dynamics penalty terms. In Eiben et al. [12], pages 211–220

35. N. Krasnogor and J.E. Smith. Emergence of profitable search strategies based on a simple inheritance mechanism. In Spector et al. [55], pages 432–439

36. C.-Y. Lee and E.K. Antonsson. Adaptive evolvability via non-coding segment induced linkage. In Spector et al. [55], pages 448–453

37. M. Lee and H. Takagi. Dynamic control of genetic algorithms using fuzzy logic techniques. In S. Forrest, editor, *Proceedings of the 5th International Conference on Genetic Algorithms*, pages 76–83. Morgan Kaufmann, San Francisco, 1993

38. J. Lis. Parallel genetic algorithm with dynamic control parameter. In *Proceedings of the 1996 IEEE Conference on Evolutionary Computation*, pages 324–329. IEEE Press, Piscataway, NJ, 1996

39. H. Lu and G.G. Yen. Dynamic population size in multiobjective evolutionary algorithm. In *2002 Congress on Evolutionary Computation (CEC'2002)*, pages 1648–1653. IEEE Press, Piscataway, NJ, 2002

40. R. Männer and B. Manderick, editors. *Proceedings of the 2nd Conference on Parallel Problem Solving from Nature*. North-Holland, Amsterdam, 1992

41. K.E. Mathias, J.D. Schaffer, L.J. Eshelman, and M. Mani. The effects of control parameters and restarts on search stagnation in evolutionary programming. In Eiben et al. [12], pages 398–407

42. Z. Michalewicz. *Genetic Algorithms + Data Structures = Evolution Programs*. Springer, Berlin, Heidelberg, New York, 3rd edition, 1996

43. C.L. Ramsey, K.A. de Jong, J.J. Grefenstette, A.S. Wu, and D.S. Burke. Genome length as an evolutionary self-adaptation. In Eiben et al. [12], pages 345–353

44. C. Reis, J.A. Tenreiro Machado and J. Boaventura Cunha. Fractional dynamic fitness functions for ga-based circuit design. In Beyer et al. [21], pages 1571–1572

45. D. Schlierkamp-Voosen and H. Mühlenbein. Strategy adaptation by competing subpopulations. In Davidor et al. [10], pages 199–209

46. M. Schoenauer, K. Deb, G. Rudolph, X. Yao, E. Lutton, J.J. Merelo, and H.-P. Schwefel, editors. *Proceedings of the 6th Conference on Parallel Problem Solving from Nature*, number 1917 in Lecture Notes in Computer Science. Springer, Berlin, Heidelberg, New York, 2000

47. H.-P. Schwefel. *Numerische Optimierung von Computer-Modellen mittels der Evolutionsstrategie*, volume 26 of *ISR*. Birkhaeuser, Basel/Stuttgart, 1977

48. I. Sekaj. Robust parallel genetic algorithms with re-initialisation. In Yao et al. [61], pages 411–419

49. J.E. Smith. *Self Adaptation in Evolutionary Algorithms*. PhD Thesis, University of the West of England, Bristol, UK, 1998

50. J.E. Smith and T.C. Fogarty. Adaptively parameterised evolutionary systems: Self adaptive recombination and mutation in a genetic algorithm. In Voigt et al. [58], pages 441–450

51. J.E. Smith and T.C. Fogarty. Operator and parameter adaptation in genetic algorithms. *Soft Computing*, 1(2):81–87, 1997

52. R.E. Smith and E. Smuda. Adaptively resizing populations: Algorithm, analysis and first results. *Complex Systems*, 9(1):47–72, 1995

53. W.M. Spears. Adapting crossover in evolutionary algorithms. In J.R. McDonnell, R.G. Reynolds, and D.B. Fogel, editors, *Proceedings of the 4th Annual Conference on Evolutionary Programming*, pages 367–384. MIT Press, Cambridge, MA, 1995

54. W.M. Spears. *Evolutionary Algorithms: the Role of Mutation and Recombination*. Springer, Berlin, Heidelberg, New York, 2000

55. L. Spector, E. Goodman, A. Wu, W.B. Langdon, H.-M. Voigt, M. Gen, S. Sen, M. Dorigo, S. Pezeshk, M. Garzon, and E. Burke, editors. *Proceedings of the Genetic and Evolutionary Computation Conference (GECCO-2001)*. Morgan Kaufmann, San Francisco, 2001

56. H. Stringer and A.S. Wu. Behavior of finite population variable length genetic algorithms under random selection. In Beyer et al. [21], pages 1249–1255

57. K. Vekaria and C. Clack. Biases introduced by adaptive recombination operators. In W. Banzhaf, J. Daida, A.E. Eiben, M.H. Garzon, V. Honavar, M. Jakiela, and R.E. Smith, editors, *Proceedings of the Genetic and Evolutionary Computation Conference (GECCO-1999)*, pages 670–677. Morgan Kaufmann, San Francisco, 1999

58. H.-M. Voigt, W. Ebeling, I. Rechenberg, and H.-P. Schwefel, editors. *Proceedings of the 4th Conference on Parallel Problem Solving from Nature*, number 1141 in Lecture Notes in Computer Science. Springer, Berlin, Heidelberg, New York, 1996

59. T. White and F. Oppacher. Adaptive crossover using automata. In Davidor et al. [10], pages 229–238

60. D. Whitley, K. Mathias, S. Rana, and J. Dzubera. Building better test functions. In Eshelman [20], pages 239–246

61. X. Yao, E. Burke, J.A. Lozano, J. Smith, J.-J. Merelo-Guervos, J.A. Bullinaria, J. Rowe, P. Tino, A. Kaban, and H.-P. Schwefel, editors. *Parallel Problem Solving from Nature – PPSN-VIII*, number 3242 in Lecture Notes in Computer Science. Springer, Berlin, Heidelberg, New York, 2004

62. J. Zhang, S.H. Chung, and J. Zhong. Adaptive crossover and mutation in genetic algorithms based on clustering technique. In Beyer et al. [21], pages 1577–1578

63. J. Costa, R. Tavares, and A. Rosa. An experimental study on dynamic random variation of population size. In *Proc. IEEE Systems, Man and Cybernetics Conf.*, volume 6, pages 607–612, Tokyo, 1999. IEEE Press

64. S. Forrest, editor. *Proceedings of the 5th International Conference on Genetic Algorithms*. Morgan Kaufmann, San Francisco, 1993

65. D.E. Goldberg. Optimal population size for binary-coded genetic algorithms. TCGA Report No. 85001, 1985

66. D.E. Goldberg. Sizing populations for serial and parallel genetic algorithms. In J.D. Schaffer, editor, *Proceedings of the 3rd International Conference on Genetic Algorithms*, pages 70–79. Morgan Kaufmann, San Francisco, 1989

67. D.E. Goldberg, K. Deb, and J.H. Clark. Genetic Algorithms, Noise, and the Sizing of Populations. IlliGAL Report No. 91010, 1991

68. N. Hansen, A. Gawelczyk, and A. Ostermeier. Sizing the population with respect to the local progress in $(1,\lambda)$-evolution strategies – a theoretical analysis. In *Proceedings of the 1995 IEEE Conference on Evolutionary Computation*, pages 80–85. IEEE Press, Piscataway, NJ, 1995

69. F.G. Lobo. *The parameter-less Genetic Algorithm: rational and automated parameter selection for simplified Genetic Algorithm operation*. PhD Thesis, Universidade de Lisboa, 2000

70. C.R. Reeves. Using genetic algorithms with small populations. In Forrest [64], pages 92–99

71. J. Roughgarden. *Theory of Population Genetics and Evolutionary Ecology*. Prentice-Hall, 1979

72. D. Schlierkamp-Voosen and H. Mühlenbein. Adaptation of population sizes by competing subpopulations. In *Proceedings of the 1996 IEEE Conference on Evolutionary Computation*. IEEE Press, Piscataway, NJ, 1996

73. R.E. Smith. Adaptively resizing populations: An algorithm and analysis. In Forrest [64]

74. R.E. Smith. *Population Sizing*, pages 134–141. Institute of Physics Publishing, 2000
75. J. Song and J. Yu. *Population System Control*. Springer, 1988
76. V.A. Valkó. Self-calibrating evolutionary algorithms: Adaptive population size. Master's Thesis, Free University Amsterdam, 2003
77. B. Craenen and A.E. Eiben. Stepwise adaptation of weights with refinement and decay on constraint satisfaction problems. In L. Spector, E. Goodman, A. Wu, W.B. Langdon, H.-M. Voigt, M. Gen, S. Sen, M. Dorigo, S. Pezeshk, M. Garzon, and E. Burke, editors, *Proceedings of the Genetic and Evolutionary Computation Conference*, pages 291–298. Morgan Kaufmann, 2001
78. B. Craenen, A.E. Eiben, and J.I. van Hemert. Comparing evolutionary algorithms on binary constraint satisfaction problems. *IEEE Transactions on Evolutionary Computation*, 7(5):424–444, 2003
79. J. Eggermont, A.E. Eiben, and J.I. van Hemert. Adapting the fitness function in GP for data mining. In R. Poli, P. Nordin, W.B. Langdon, and T.C. Fogarty, editors, *Genetic Programming, Proceedings of EuroGP'99*, Volume 1598 of *LNCS*, pages 195–204. Springer-Verlag, 1999
80. A.E. Eiben, B. Jansen, Z. Michalewicz, and B. Paechter. Solving CSPs using self-adaptive constraint weights: how to prevent EAs from cheating. In D. Whitley, D. Goldberg, E. Cantu-Paz, L. Spector, I. Parmee, and H.-G. Beyer, editors, *Proceedings of the Genetic and Evolutionary Computation Conference*, pages 128–134. Morgan Kaufmann, 2000
81. A.E. Eiben and J.I. van Hemert. SAW-ing EAs: adapting the fitness function for solving constrained problems. In D. Corne, M. Dorigo, and F. Glover, editors, *New Ideas in Optimization*, Chapter 26, pages 389–402. McGraw-Hill, London, 1999

Divide-and-Evolve: a Sequential Hybridization Strategy Using Evolutionary Algorithms

Marc Schoenauer[1], Pierre Savéant[2], and Vincent Vidal[3]

[1] Projet TAO, INRIA Futurs, LRI, Bât. 490, Université Paris Sud, 91405 Orsay, France.
marc.schoenauer@inria.fr

[2] Thales Research & Technology France, RD 128, 91767 Palaiseau, France.
pierre.saveant@thalesgroup.com

[3] CRIL & Université d'Artois, rue de l'université - SP16, 62307 Lens, France.
vidal@cril.univ-artois.fr

Abstract

Memetic algorithms are hybridizations of evolutionary algorithms (EAs) with problem-specific heuristics or other meta-heuristics, that are generally used within the EA to locally improve the evolutionary solutions. However, this approach fails when the local method stops working on the complete problem. *Divide-and-Evolve* is an original approach that evolutionarily builds a sequential slicing of the problem at hand into several, hopefully easier, sub-problems: the embedded (meta-)heuristic is only asked to solve the 'small' problems, and *Divide-and-Evolve* is thus able to globally solve problems that are intractable when directly fed into the heuristic. The *Divide-and-Evolve* approach is described here in the context of temporal planning problems (TPPs), and the results on the standard Zeno transportation benchmarks demonstrate its ability to indeed break the complexity barrier. But an even more prominent advantage of the *Divide-and-Evolve* approach is that it immediately opens up an avenue for multi-objective optimization, even when using single-objective embedded algorithm.

Key words: Hybrid Algorithms, Temporal Planning, Multi-objective Optimization

1 Introduction

Evolutionary Algorithms (EAs) are bio-inspired meta-heuristics crudely borrowing from the Darwinian theory of natural evolution of biological populations (see [7] for the most recent comprehensive introduction, or [6] for a brief introduction of the basic concepts). In order to solve the optimization problem at hand, EAs evolve a population of individuals (tuples of candidate solutions) relying on two main driving forces to reach the optimal solution: *natural selection* and *blind variations*. Natural selection biases the choices of the algorithm toward good performing individuals (w.r.t. the optimization problem at hand) at reproduction time and at survival time (*survival of the fittest*). Blind variation operators are stochastic operators defined on the search space, which create new individuals from *parents* in the current population, independently of their performance (hence the term "blind"). They are usually categorized into *crossovers*, producing *offspring* from two parents, and *mutations* that create one offspring from a single parent. Whereas the natural selection part of an EA is (almost) problem independent, the choice of the search space (the *representation*) and the corresponding variation operators has to be done anew for each application domain, and requires problem-specific expertise.

This has been clearly demonstrated in the domain of "combinatorial optimization", where it is now well-known (see Grefenstette's seminal paper [12]) that generic EAs alone are rarely efficient. However, the flexibility of EAs allows the user to easily add domain knowledge at very different levels of the algorithm, from representation [23] to *ad hoc* variation operators [31] to explicit use of other optimization techniques within the EA. The most successful of such hybridizations use other heuristics or meta-heuristics to locally improve all individuals that are created by the EA, from the initial population to all offspring that are generated by the variation operators. Such algorithms have been termed "Memetic Algorithms" or "Genetic Local Search" [22]. Those methods are now the heart of a very active research field, as witnessed by the yearly WOMA series (Workshops on Memetic Algorithms), journal special issues [13] and edited books [14].

However, most memetic approaches are based on finding local improvements of candidate solutions proposed by the evolutionary search mechanism using dedicated local search methods that have to tackle the complete problem. Unfortunately, in many combinatorial domains, this simply proves to be impossible when reaching some level of complexity. This chapter proposes an original hybridization of EAs with a domain-specific solver that addresses this limitation in domains where the task at hand can be sequentially decomposed into a series of (hopefully) simpler tasks. Temporal planning is such a domain, that will be used here to instantiate the *Divide-and-Evolve* paradigm.

Artificial Intelligence Planning is a form of general problem solving task which focuses on problems that map into *state models* that can be defined by a state space S, an initial state $s_0 \subseteq S$, a set of goal states $S_G \subseteq S$, a set of actions $A(s)$ applicable in each state S, and a transition function $f(a, s) = s'$ with $a \in A(s)$, and $s, s' \in S$. A solution to this class of models is a sequence of applicable actions mapping the initial state s_0 to a goal state that belongs to S_G.

An important class of problems is covered by temporal planning, which extends classical planning by adding a duration to actions and by allowing concurrent actions in time [11]. In addition, other metrics are usually needed for real-life problems to ensure a good plan, for instance a cost or a risk criterion. A common approach is to aggregate the multiple criteria, but this relies on highly problem-dependent features and is not always meaningful. A better solution is to compute the set of optimal non-dominated solutions – the so-called Pareto front.

Because of the high combinatorial complexity and the multi-objective features of Temporal Planning Problems (TPPs), EAs seem to be good general-purpose candidate methods.

However, there have been very few attempts to apply EAs to planning problems and, as far as we know, not any to temporal planning. Some approaches use a specific representation (e.g. dedicated to the battlefield courses of action [24]). Most of the domain-independent approaches see a plan as a program and rely on genetic programming and on the traditional blocks-world domain for experimentation (starting with the Genetic Planner [27]). A more comprehensive state of the art on genetic planning can be found in [2] where the authors experimented with a variable length chromosome representation. It is important to notice that all these works search the space of (partial) plans.

In this context, the *Divide-and-Evolve* approach, borrowing from the Divide-and-Conquer paradigm, tries to slice the problem at hand into a sequence of problems that are hopefully easier to solve by the available OR or local methods. The solution to the original problem is then obtained by a concatenation of the solutions to the different sub-problems.

Note that the idea to divide the plan trajectory into small chunks has been studied with success in [15]. The authors have shown the existence of landmarks, i.e. sets of facts that must be true at some point during execution of any solution plan, and the impact of ordering them on search efficiency. They also prove that deciding if a fact is a landmark and finding ordering relations is PSPACE-complete. In this way, *Divide-and-Evolve* can be seen as an attempt to generate ordered sets of landmarks using a stochastic approach. However, the proposed approach is not limited to finding sets of facts that must absolutely be true within every solution to the initial problem. In particular, it also applies to problems that have no landmark *per se*, for simple symmetry reasons: there can be several equivalent candidate landmarks, and only one of them can and must be true at some point.

The chapter is organized as follows: the next section presents an abstract formulation of the *Divide-and-Evolve* scheme, and starting from its historical (and pedagogical) root, the TGV paradigm. Generic representation and variation operators are also introduced. Section 3 introduces an actual instantiation of the *Divide-and-Evolve* scheme to TPPs. The formal framework of TPPs is first introduced, then the TPP-specific issues for the *Divide-and-Evolve* implementation are presented and discussed. Section 4 is devoted to experiments on the TPP transportation Zeno benchmark for both single and multi-objective cases. The local problems are solved using the exact temporal planner CPT [28], a freely-available optimal temporal planner, for its temporal dimension. The last section highlights the limitations of the present work and sketches further directions of research.

Note that an initial presentation of *Divide-and-Evolve* was published at the Evo-COP'06 conference [25], in which only very preliminary results were presented in the single-objective case. Those results are here validated more thoroughly on the three instances Zeno10, Zeno12 and Zeno14: the new experiments demonstrate that the *Divide-and-Evolve* approach can repeatedly find the optimal solution on all three instances. Moreover, the number of backtracks that are needed by the optimal planner

CPT to solve each sub-problem is precisely analyzed, and it is demonstrated that it is possible to find the optimal solution even when limiting the number of backtracks that CPT is allowed to perform for each sub-problem, thus hopefully speeding up the complete optimization. On the other hand, however, the multi-objective results that are presented here are the same as those of [25], and are recalled here for the sake of completeness, as they represent, as far as we know, the very first results of Pareto multi-objective optimization in temporal planning, and open up many avenues for further research.

2 The *Divide-and-Evolve* Paradigm

This section presents the *Divide-and-Evolve* scheme, an abstraction of the TGV paradigm that can be used to solve a planning problem when direct methods face a combinatorial explosion due to the size of the problem. The TGV approach might be a way to break the problem into several sub-problems, hopefully easier to solve than the initial global problem.

2.1 The TGV Metaphor

The *Divide-and-Evolve* strategy springs from a metaphor on the route planning problem for the French high-speed train (TGV). The original problem consists in computing the shortest route between two points of a geographical landscape with strong bounds on the curvature and slope of the trajectory. An evolutionary algorithm was designed [5] based on the fact that the only local search algorithm at hand was a greedy deterministic algorithm that could solve only very simple (i.e. short distance) problems. The evolutionary algorithm looks for a split of the global route into small consecutive segments such that a local search algorithm can easily find a route joining their extremities. Individuals represent sets of intermediate train stations between the station of departure and the terminus. The convergence toward a good solution was obtained with the definition of appropriate variation and selection operators [5]. Here, the state space is the surface on which the trajectory of the train is defined.

Generalization

Abstracted to planning problems, the route is replaced by a sequence of actions and the "stations" become intermediate states of the system. The problem is thus divided into sub-problems and "to be close" becomes "to be easy to solve" by some local algorithm $\mathcal{L}$. The evolutionary algorithm plays the role of an oracle pointing at some imperative states worth going through.

2.2 Representation

The problem at hand is an abstract AI planning problem as described in the introduction. The representation used by the EA is a variable length list of states: an individual is thus defined as $(s_i)_{i\in[1,n]}$, where the length n and all the states s_i are unknown and subject to evolution. States s_0 and $s_{n+1} \equiv s_G$ will represent the initial state and the goal of the problem at hand, but will not be encoded in the genotypes. By reference to the original TGV paradigm, each of the states s_i of an individual will be called a *station*.

Requirements

The original TGV problem is purely topological with no temporal dimension and reduces to a planning problem with a unique action: moving between two points. The generalization to a given planning domain requires being able to:

1. define a distance between two different states of the system, so that $d(S, T)$ is somehow related to the difficulty for the local algorithm $\mathcal{L}$ to find a plan mapping the initial state S to the final state T;
2. generate a chronological sequence of virtual "stations", i.e. intermediate states of the system, that are close to one another, s_i being close to s_{i+1};
3. solve the resulting "easy" problems using the local algorithm $\mathcal{L}$;
4. "glue" the sub-plans into an overall plan of the problem at hand.

2.3 Variation Operators

This section describes several variation operators that can be defined for the general *Divide-and-Evolve* approach, independently of the actual domain of application (e.g. TPPs, or the original TGV problem).

Crossover

The crossover operation exchanges stations between two individuals. Because of the sequential nature of the fitness, it seems a good idea to try to preserve sequences of stations, resulting in straightforward adaptations to variable-length representation of the classical 1- or 2-point crossover operators.

Suppose you are recombining two individuals $(s_i)_{1 \le n}$ and $(T_i)_{1 \le m}$. The 1-point crossover amounts to choosing one station in each individual, say s_a and T_b, and exchanging the second part of the lists of stations, obtaining the two offspring $(s_1, \dots, s_a, T_{m+1}, \dots T_b)$ and $(T_1, \dots, T_b, s_{n+1}, \dots, s_n)$ (2-point crossover is easily implemented in a similar way). Note that in both cases, the length of each offspring is likely to differ from those of the parents.

The choice of the crossover points s_a and T_b can be either uniform (as done in all the work presented here), or distance-based, if some distance is available: pick the first station s_a randomly, and choose T_b by, for example, a tournament based on the distance from s_a (this is ongoing work).

Mutation

Several mutation operators can be defined. Suppose individual $(s_i)_{1 \le n}$ is being mutated:

* **At the individual level**, the *Add* mutation simply inserts a new station s_{new} after a given station (s_a), resulting in an $n + 1$-long list, $(s_1, \dots, s_a, s_{new}, s_{a+1}, \dots, s_n)$. Its counterpart, the *Del* mutation, removes a station s_a from the list.

Several improvements on the pure uniform choice of s_a can be added and are part of ongoing work, too: if the local algorithm fails to successfully join all pairs of successive stations, the last station that was successfully reached by the local algorithm can be preferred for station s_a (in both the *Add* and *Del* mutations). If all partial problems are solved, the most difficult one (e.g. in terms of number of backtracks) can be chosen.

- **At the station level**, the definition of each station can be modified – but this is problem-dependent. However, assuming there exists a station-mutation operator μ_S, it is easy to define the individual-mutation M_{μ_S} that will simply call μ_S on each station s_i with a user-defined probability p_{μ_S}. Examples of operators μ_S will be given in Sect. 3, while simple Gaussian mutation of the (x, y) coordinates of a station were used for the original TGV problem [5].

3 Application to Temporal Planning

3.1 Temporal Planning Problems

Domain-independent planners rely on the Planning Domain Definition Language (PDDL) [20], inherited from the STRIPS model [8], to represent a planning problem. In particular, this language is used for a competition[1] which has been held every two years since 1998 [1,16,19,21]. The language has been extended to represent TPPs in PDDL2.1 [10]. For the sake of simplicity, and because the underlying temporal planner that we use, CPT [28, 29], does not strictly conform to PDDL2.1, the temporal model is often simplified as explained below [28].

A *Temporal PDDL Operator* is a tuple $o = \langle pre(o), add(o), del(o), dur(o) \rangle$ where $pre(o)$, $add(o)$ and $del(o)$ are sets of ground atoms that respectively denote the preconditions, add effects and del effects of o, and $dur(o)$ is a rational number that denotes the *duration* of o. The operators in a PDDL input can be described with variables, used in predicates such as (at ?plane ?city). The variables ?plane and ?city are then replaced by CPT with the objects of a particular problem in an initial *grounding* process.

A *Temporal Planning Problem* is a tuple $P = \langle A, I, O, G \rangle$, where A is a set of atoms representing all the possible facts in a world situation, I and G are two sets of atoms that respectively denote the initial state and the problem goals, and O is a set of ground PDDL operators.

As is common in partial order causal link (POCL) planning [30], two dummy actions are also considered, *Start* and *End* with zero durations, the first with an empty precondition and effect I; the latter with precondition G and empty effects. Two actions a and a' interfere when one deletes a precondition or positive effect of the other. The simple model of time in [26] defines a valid plan as a plan where interfering actions do not overlap in time. In other words, it is assumed that the preconditions need to hold until the end of the action, and that the effects also hold at the end and cannot be deleted during the execution by a concurrent action.

---

[1] see http://ipc.icaps-conference.org/

A *schedule* P is a finite set of action occurrences $\langle a_i, t_i \rangle$, $i = 1, \ldots, n$, where a_i is an action and t_i is a non-negative integer indicating the starting time of a_i (its ending time is $t_i + dur(a_i)$). P must include the *Start* and *End* actions, the former with time tag 0. The same action (except for these two) can be executed more than once in P if $a_i = a_j$ for $i \neq j$. Two action occurrences a_i and a_j *overlap* in P if one starts before the other ends; namely if $[t_i, t_i + dur(a_i)] \cap [t_j, t_j + dur(a_j)]$ contains more than one time point.

A schedule P is a *valid plan* iff interfering actions do not overlap in P and for every action occurrence $\langle a_i, t_i \rangle$ in P its preconditions $p \in pre(a)$ are true at time t_i. This condition is inductively defined as follows: p is true at time $t = 0$ if $p \in I$, and p is true at time $t > 0$ if either p is true at time $t - 1$ and no action a in P ending at t deletes p, or some action a' in P ending at t adds p. The *makespan* of a plan P is the time tag of the *End* action.

3.2 CPT: an Optimal Temporal Planner

An optimal temporal planner computes valid plans with minimum makespan. Even though an optimal planner was not mandatory (as discussed in Sect. 5), we have chosen *CPT* [28], a freely-available optimal temporal planner, for its temporal dimension and for its constraint-based approach, which provides a very useful data structure when it comes to gluing the partial solutions (see Sect. 2.2). Indeed, since in temporal planning actions can overlap in time, the simple concatenation of sub-plans, though providing a feasible solution, obviously might produce a plan that is not optimal with respect to the total makespan, even if the sequence of actions is the optimal sequence. However, thanks to the causal links and order constraints maintained by CPT, an improved global plan can be obtained by shifting sub-plans as early as possible in a final state of the algorithm.

Another argument for choosing CPT was the fact that it is a sound and complete planner in the following sense: a valid plan with makespan equal to a given bound B on the number of allowed backtracks is found if and only if one such plan exists. There are then many strategies for adjusting the bound B so that an optimal makespan is produced; e.g., the bound may be increased until a plan is found, or can be decreased until no plan is found, etc.

Indeed, because one motivation for the *Divide-and-Evolve* approach is to tackle large instances that are too complex to be directly solved by the local algorithm (CPT in our case), it is important to be able to launch only limited searches by CPT: a bound on the number of allowed backtracks could be added to all CPT calls, and the fitness penalized when this bound is reached (CPT stops without giving a result in that case). More details will be given in Sect. 4.1.

One final reason for originally choosing CPT, before the collaboration among the authors of this work started, was that a binary version was freely available on the third author's Web page. However, even though a tighter collaboration rapidly became effective, it proved nevertheless intractable to call CPT as a subroutine, for technical reasons (CPT-2 was written in CLAIRE). Hence data had to be passed through files, and CPT launched anew each time, resulting in a huge waste of CPU resources. This drawback will be solved by switching to CPT-3, the most recent version of CPT (on-going work).

3.3 Rationale for using *Divide-and-Evolve* for Temporal Planning

The reasons for the failure of standard OR methods addressing TPPs come from the exponential complexity of the number of possible actions when the number of objects involved in the problem increases. It has been known for a long time that taking into account the interactions between sub-goals can decrease the complexity of finding a plan, in particular when these sub-goals are independent [18]. Moreover, computing an ideal ordering on sub-goals is as difficult as finding a plan (PSPACE-hard), as demonstrated in [17]. which proposes an algorithm for computing an approximation of such an ordering. The basic idea when using the *Divide-and-Evolve* approach is that each local sub-plan ("joining" stations s_i and s_{i+1}) should be easier to find than the global plan (joining the station of departure s_0 and the terminus s_{n+1}). This will now be demonstrated on the Zeno transportation benchmark (see http://ipc.icaps-conference.org/).

Table 1 illustrates the decomposition of a relatively difficult problem in the Zeno domain (zeno14 from IPC-3 benchmarks), a transportation problem with 5 planes (plane1 to plane5) and 10 persons (person1 to person10) to travel among 10 cities (city0 to city9). A plane can fly at two different speeds. Flying fast requires more fuel. A plane has a fuel level and might be refueled when empty. A person is either at a city or in a plane and requires to be boarded and disembarked.

Analyzing the optimal solution found by CPT-3, it was possible (though not triv-ial) to manually divide the optimal "route" of this solution in the state space into four intermediate stations between the initial state and the goal. It can be seen that very few moves (plane or person) occur between two consecutive stations (the ones in bold in each column of Table 1). Each sub-plan is easily found by CPT, with a maxi-mum of 1 backtrack and 1.87 seconds of search time. It should be noted that most of the time spent by CPT is on pre-processing: this operation is actually repeated each time CPT is called, but could be factorized at almost no cost ... except coding time.

Note that the final step of the process is the compression of the five sub-plans (see Sect. 2.2: it is performed in 0.10 second (plus 30.98 seconds for pre-processing) without any backtracking, and the overall makespan of the plan is 476, much less than the sum of the individual makespans of each sub-plan (982).

To summarize, the recomposed plan, with a makespan of 476, required a total running time of 193.30 seconds (including only 7.42 s of pure search) and only one backtrack, whereas a plan with the same optimal makespan of 476 was found by CPT in 2692.41 seconds and 566,681 backtracks. Sect. 5 will discuss this issue.

3.4 Description of the State Space

Non-temporal States
A natural state space for TPPs, as described at the beginning of this section, would be the actual space of all possible time-stamped states of the system. Obviously, the

Table 1 State decomposition of the Zeno14 instance. (The new location of moved objects appears in bold)

Objects	Init (station 0)	Station 1	Station 2	Station 3	Station 4	Goal (station 5)
plane 1	city 5	city 5	city 5	**city 6**	city 6	city 6
plane 2	city 2	**city 0**	city 0	**city 2**	**city 3**	city 3
plane 3	city 4	**city 7**	**city 9**	city 7	city 7	**city 9**
plane 4	city 8	city 8	city 8	**city 7**	**city 5**	city 5
plane 5	city 9	**city 6**	**city 1**	city 1	**city 8**	city 8
person 1	city 9	city 9	city 9	city 9	city 9	city 9
person 2	city 1	city 1	city 1	city 1	**city 8**	city 8
person 3	city 0	city 0	city 0	**city 2**	city 2	city 2
person 4	city 9	city 9	city 9	**city 7**	city 7	city 7
person 5	city 6	city 6	**city 1**	city 1	city 1	city 1
person 6	city 0	city 0	city 0	**city 6**	city 6	city 6
person 7	city 7	city 7	city 7	city 7	**city 5**	city 5
person 8	city 6	city 6	**city 1**	city 1	city 1	city 1
person 9	city 4	**city 7**	city 7	city 7	**city 5**	city 5
person 10	city 7	city 7	**city 9**	city 9	city 9	city 9
Makespan	150		203	150	276	203
Backtracks	0		0	0	1	0
Search time	1.34		1.27	1.32	1.87	1.52
Total time	32.32		32.25	32.30	32.85	32.50

	Compression		Global Search
Makespan	476		476
Backtracks	0		566, 681
Search time	0.10		2660.08
Total time	31.08 (total : 193.30)		2692.41

size of such a space is far too big and we simplified it by restricting the stations to non-temporal states. However, even with this simplification, not all "non-temporal" states can be considered in the description of the "stations".

Limiting the Possible States

First, the space of all possible states grows exponentially with the size of the problem. Second, not all states are consistent w.r.t. the planning domain. For instance, an object cannot be located at two places at the same time in a transportation problem – and inferring such state invariants is feasible but not trivial [9]. Note also that determining plan existence from a propositional STRIPS description has been proved to be PSPACE-complete [3].

A possible way to overcome this difficulty would be to rely on the local algorithm to (rapidly) check the consistency of a given situation, and to penalize unreachable stations. However, this would clearly be a waste of computational resources, possibly leading to a far too difficult problem to solve for the EA (it would have to "discover" again and again that one object cannot be at the same time at two different locations, without a way to generalize and save these across different situations).

On the other hand, introducing domain knowledge into EAs has been hailed for some time as the royal road to success in evolutionary computation [12]. Hence, it seems a more promising approach to add state invariants to the description of the state space in order to remove as many of the inconsistent states as possible. The good thing is that it is not necessary to remove *all* inconsistent states since, in any case, the local algorithm is there to help the EA to spot them – inconsistent stations will be given poor fitness, and will not survive the following selection steps. In particular, only state invariants involving a single predicate have been implemented in the present work.

3.5 Representation of Stations

It was decided to describe the stations using **only the predicates that are present in the goal** of the overall problem, and to maintain the state invariants based on the semantics of the problem.

A good example is given in Table 1: the goal of this benchmark instance is to move the persons and planes in cities listed in the last column. No other predicate than the corresponding (at objectN cityM) predicates is present in the goal. Through a user-supplied file, the algorithm is told that only the at predicates will be used to represent the stations, with the syntactic restrictions that within a given station, the first argument of an at predicate can appear only once (at is said to be *exclusive* with respect to its first argument). The state space that will be explored by the algorithm thus amounts to a vector of 15 fluents (instantiated predicates) denoting that an item is located in a city (a column of Table 1). In addition, the actual implementation of a station includes the possibility to "remove" (in fact, comment out) a predicate of the list: the corresponding object will not move during this sub-plan.

Distance

The distance between two stations should reflect the difficulty the local algorithm has in finding a plan joining them. At the moment, a purely syntactic domain-independent distance is used: the number of different predicates not yet reached. The difficulty can then be estimated by the number of backtracks needed by the local algorithm. It is reasonable to assume that most local problems where only a few predicates need to be changed from the initial state to the goal will be easy for the local algorithm – though this is certainly not true in all cases.

Random Generation of Stations

Thanks to the state invariants described above, generating random stations now amounts to choose among consistent stations, and is thus rather simple for a single station. Nevertheless, the generation of initial individuals (sequences of stations $(s_i)_{i\in[1,n]}$ such that all local problems (s_i, s_{i+1}) are simple for the local algorithm) remains an issue. A very representation-specific method has been used for TPPs, and will be described in the next section.

3.6 Representation-specific Operators

The initialization of an individual (see Sect. 3.5) and the station-mutation operator (see Sect. 2.3) will be described for the chosen problem-specific representation for TPPs.

Initialization

First, the number of stations is chosen uniformly in a user-supplied interval. The user also enters a maximal distance d_{max} between stations: two consecutive stations will not differ by more that d_{max} predicates (the minimal number of stations is eventually adjusted in order to meet the requirement, according to the distance between the initial state and the goal state). A matrix is then built, similar to the top lines of Table 1: each line corresponds to one of the goal predicates, each column is a station. Only the first and last columns (corresponding to initial state and goal) are filled with values. A number of "moves" are then randomly added in the matrix, at most d_{max} per column, and at least one per line. Additional moves are then added according to another user-supplied parameter, and without exceeding the d_{max} limit per column. The matrix is then filled with values, starting from both ends (init and goal), constrained column-wise by the state invariants, as described in Sect. 3.4 and line-wise by the values in the init and goal states. If some station proves to be inconsistent at some point, it is rejected and a new one is generated. A final sweep on all predicates comments out some of the predicates with a given probability.

Station Mutation

Thanks to the simplified representation of the states (a vector of fluents with a set of state invariants), it is straightforward to modify one station randomly: with a given probability, a new value for the non-exclusive arguments is chosen among the possible values respecting all constraints (including the distance constraints with previous and next stations). In addition, each predicate might be commented out from the station with a given probability, as in the initialization phase.

4 First Experiments

4.1 Single Objective Optimization

Our main playground to validate the *Divide-and-Evolve* approach is that of transportation problems, and started with the zeno domain as described in Sect. 3.3. As can be seen in Table 1, the description of the stations in zeno domain involves a single predicate, at, with two arguments. It is *exclusive* w.r.t. its first argument. Three instances have been tried, called zeno10, zeno12 and zeno14, from the simplest to the hardest.

The simple zeno10 (resp. zeno12) instance can be solved very easily by CPT-2 alone, in less than 2 seconds (resp. 125 seconds), finding the optimal plans with makespan 453 (resp. 549) using 154 (resp. 27560) backtracks. On the other hand, the zeno14 instance could not be solved at all by CPT-2. However, as described in Table 1, the new version CPT-3 could solve it, with a makespan of 476 and using 566,681 backtracks.

Algorithmic Settings

The EA that was used for the first implementation of the *Divide-and-Evolve* paradigm uses standard algorithmic settings at the population level:

- population size was set to 10 to limit the CPU cost;
- both $(10+10) - ES$ and $(10, 70) - ES$ evolution engines were used: the 10 parents give birth to either 10 or 70 children, and the best 10 among the 10 children plus the 10 parents $((10 + 10) - ES)$ or among the 70 children $((10, 70) - ES)$ become the parents of next generation;
- 1-point crossover is applied to 25% of the individuals;
- the other 75% undergo mutation: 25% of the mutations are the *Add* (resp. *Del*) generic mutations (Sect. 2.3). The remaining 50% of the mutations are called problem-specific station mutations. Within a station mutation, a predicate is randomly changed in 75% of the cases and a predicate is removed (resp. restored) in each of the remaining 12.5% cases. (see Sect. 3.6);
- Initialization is performed using initial size in $[2, 10]$, maximum distance of 3 and probability to comment out a predicate is set to 0.1.

Note that at the moment, no lengthy parameter tuning was performed for those proof-of-concept experiments, and the above values were decided based upon a very limited set of initial experiments.

The Fitness

The target objective is here the total makespan of a plan – assuming that a global plan can be found, i.e. that all problems (s_i, s_{i+1}) can be solved by the local algorithm. If one of the local problems cannot be solved, the individual is declared *infeasible* and is penalized in such a way that all infeasible individuals are worse than any feasible one. Moreover, this penalty is proportional to the number of remaining stations (relative to the total number of stations) after the failure, in order to provide a nice slope for the fitness landscape towards feasibility. For feasible individuals, an average of the total makespan and the sum of the makespans of all partial problems is applied. The latter is needed in order to promote smaller individuals. When only the total makespan is used, the algorithm tends to generate some lengthy individuals with useless intermediate stations, slowing down the whole run because of all the respective calls to CPT.

Results on Zeno 10

For zeno10, all runs found the optimal solution in the very early generations, for both evolution engines $(10 + 10) - ES$ and $(10, 70) - ES$ (rather often, in fact, the initialization procedure produced a feasible individual that CPT could compress to the optimal makespan).

As already mentioned (see Sect. 3.2), the number of backtracks used by CPT was in all cases limited to a large number to avoid endless searches. However, in order to precisely investigate the simplification due to *Divide-and-Evolve*, we took a closer look at the number of backtracks used by CPT alone, and noticed two things: first, when forbidden to use any backtrack, CPT nevertheless found a suboptimal solution with makespan 915); second, when searching for the optimal solution without limit on the number of backtracks, CPT never used more than 33 backtracks on a single iteration, with a total of 154 altogether. It was hence decided to try different limits on the number of backtrack per iteration during *Divide-and-Evolve* procedure, from 35 (above the actual number that is necessary for CPT alone) to 1 (0 was not actually possible there). The results are presented in Table 2 and demonstrate that, in all cases, *Divide-and-Evolve* was able to drive CPT toward the optimal solution, even though no backtracks could actually be used by CPT. Note, however, that for the most difficult case (limit set to 1 backtrack), the $(10 + 10) - ES$ engine did not perform very well, while the $(10, 70) - ES$ case was much more robust.

Here again, when forbidden to use any backtrack, CPT nevertheless found a suboptimal solution with makespan 815. When searching for the optimal solution without limit on the number of backtracks, CPT never used more than 8066 backtracks on a single iteration, with a total of 27,560 altogether. As with zeno10, different limits on the number of backtrack were run, from 8070 (slightly above the actual number that is necessary for CPT alone) to 1. The results are presented in Table 2 and again demonstrate that *Divide-and-Evolve* was indeed able to drive CPT toward the optimal solution, though not when allowed no backtrack at all. Also, for the most difficult case (limit set to 10 backtracks), the $(10 + 10) - ES$ engine could not find the optimal solution, while the $(10, 70) - ES$ case could, with a much smaller number of stations. Note that the $(10, 70) - ES$ engine was able to find an optimal solution with no backtrack only once in 11 runs, with more than 50 stations . . .

Results on Zeno 12

For zeno12 (see Table 3), most runs (around 80% on average) found the optimal solution. The running time for one generation of the $(10, 70) - ES$ engine (70 evalu-

Table 2 Performance of *Divide-and-Evolve* on zeno10 using the $(10+10)-ES$ evolution engine (except last line) when the number of backtracks allowed for CPT is limited

Limit	Makespan	Maximal # BKT	# Stations	# Success
35	453	33	5	5/11
20	453	2	20	5/11
10	453	1	5	5/11
1	453	1	6	1/11 (10,10)-ES
			6	9/11 (10,70)-ES

Table 3 Performance of *Divide-and-Evolve* on zeno12 using the $(10+10)-ES$ evolution engine (except last line) when the number of backtracks allowed for CPT is limited

Limit	Makespan	Maximal # BKT	# Stations	# Success
8,070	549	5,049	9	8/11
8,000	549	4,831	13	4/11
100	549	5	21	2/11
50	549	28	3	2/11
20	549	3	20	3/11
10	549	2 − 1	7 − 11	6/11 (10,70)-ES

ations) was about 30 minutes on a 3.4 GHz Pentium IV processor (because different individuals can have very different numbers of stations, all running times are rough averages over all runs performed for those experiments). All solutions were found in less than 20 generations.

Results on Zeno 14

A more interesting case is that of zeno14: remember that the present *Divide-and-Evolve* EA uses CPT-2, that is unable to find any solution to zeno14: the results given in Table 1 have been obtained using CPT-3. But whereas it proved unable to solve the full problem, CPT-2 could nevertheless be used to solve the hopefully small instances of zeno14 domain that were generated by the *Divide-and-Evolve* approach – though taking a huge amount of CPU time for that (on average, 90 minutes for one generation of 70 evaluations). Note that here, setting a limit on the number of backtracks allowed for CPT was in any case mandatory, to prevent CPT from exploring the "too complex" cases that would have resulted in a never-returning call (as does a call to the full problem).

The optimal solution (makespan 476) was found in 3 out of 11 runs, with a limit on the number of backtracks set to 120,000. Note that *Divide-and-Evolve* was unable to find the optimal solution when using a smaller number of backtracks, though it repeatedly found feasible solutions (see Table 4) even with the lowest limit of one backtrack (while CPT is not able to find any feasible solution alone, whatever the number of backtracks allowed).

Table 4 Performance of *Divide-and-Evolve* on zeno14 with limited number of backtracks using the $(10, 70) - ES$ evolution engine

Limited no BKT	Best Makespan
120,000	656
10,000	892
1,000	603
1	868

Discussion of Single-objective Results

The main conclusion of these experiments is the proof-of-concept of the *Divide-and-Evolve* approach. Not only has *Divide-and-Evolve* been able to find an optimal solution to zeno14 using a version of CPT that was unable to do so, but it also has demonstrated that it could find optimal solutions to a given problem using a very limited setting for CPT. Although, due to the huge overload of CPT calls through Unix forks, it was not possible to see a statistically significant decrease in the CPU time needed for different settings of the number of backtracks allowed for CPT, there is no doubt that limiting CPT will allow *Divide-and-Evolve* to progress more quickly. Further experiments (using CPT3) are, however, needed to more precisely quantify the gain, and determine the best tradeoff.

4.2 A Multi-objective Problem

Problem Description
In order to test the feasibility of the multi-objective approach based on the *Divide-and-Evolve* paradigm, we extended the zeno benchmark with an additional criterion, that can be interpreted either as a cost, or as a risk: in the former case, this additional objective is an additive measure, whereas in the latter case (risk) the aggregation function is the max operator.

The problem instance is shown in Fig. 1: the only available routes between cities are displayed as edges, only one transportation method is available (plane), and the duration of the transport is shown on the corresponding edge. Risks (or costs) are attached to the cities (i.e. concern any transportation that either lands or takes off from that city). In the initial state, the three persons and the two planes are in City 0, and the goal is to transport them to City 4.

As can be easily computed (though there is a little trick here), there are three remarkable Pareto-optimal solutions, corresponding to traversing only one of the

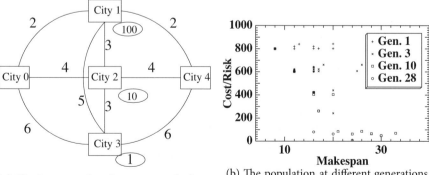

(a) The instance: durations are attached to edges, costs/risks are attached to cities (in gray circles).

(b) The population at different generations for a successful run on the cost (additive) instance of the zeno mini-problem of Figure 1(a)

Fig. 1 The multi-objective Zeno benchmark

three middle cities. Going through City 1 is fast, but risky (costly), whereas going through City 3 is slow and safe and cheap.

When all persons go through respectively City 1, City 2 and City 3, the corresponding values of the makespans and costs in the additive case are (8, 800), (16, 80) and (24, 8), whereas they are, in the max case, (8, 100), (16, 10) and (24, 1).

Problem Complexity

It is easy to compute the number of possible virtual stations: each one of the three persons can be in one of the five cities, or not mentioned (absent predicate). Hence there are $3^6 = 729$ possible combinations, and 729^n possible lists of length n. So even when n is limited to 6, the size of the search space is approx. 10^{17} …

The Algorithm

The EA is based on the standard NSGA-II multi-objective EA [4]: standard tournament selection of size 2 and deterministic replacement among parents + offspring, both based on the Pareto ranking and crowding distance selection; a population size of 100 evolves over 30 generations. All other parameters were as those used for the single objective case.

Fitnesses

The problem has two objectives: one is the the total makespan (as in the single-objective case), the other is either the **risk** (aggregated using the **max** operator) or the **cost** (an **additive** objective). Because the global risk only takes three values, there is no way to have any useful gradient information when used as fitness in the max case. However, even in the additive case, the same arguments than for the makespan apply (Sect. 4.1), and hence, in all cases, the second objective is the sum of the overall risk/cost and the average (not the sum) of the values for all partial problems – excluding from this average those partial problems that have a null makespan (when the goal is already included in the initial state).

Results

For the additive (**cost**) case, the most difficult Pareto optimum (going through city 3 only) was found four times out of 11 runs. However, the two other remarkable Pareto optima, as well as several other points in the Pareto front were also repeatedly found by all runs. Figure 1(b) shows different snapshots of the population at different stages of the evolution for a typical successful run: at first (+), all individuals have a high cost (above 800); at generation 3 ('×'), there exist individuals in the population that have cost less than 600; at generation 10 (squares), many points have a cost less than 100. But the optimal (24,8) solution is only found at generation 28 (circles).

The problem in the **risk** context (the max case) proved to be, as expected, slightly more difficult. All three Pareto optima (there exist no other point of the true Pareto front in the max case) were found only in two runs out of 11. However, all runs found both the two other Pareto optima, as well as the slightly suboptimal solution that goes only through city 3 but did not find the little trick mentioned earlier, resulting in a (36,1) solution.

In both cases, those results clearly validate the *Divide-and-Evolve* approach for multi-objective TPPs – remember that CPT has no knowledge of the risk/cost in its optimization procedure – it only aggregates the values *a posteriori*, after having computed its optimal plan based on the makespan only – hence the difficulty in finding the third Pareto optimum going only through `city3`.

5 Discussion and Further Work

First, note that any planner can be used to solve the local problems. In particular, both exact and suboptimal planners are suitable. Some experiments will be made using planners other than CPT. However, because the final goal is to find an optimal plan which joins the station of departure and the terminus, using an optimal planner might be mandatory, and, at least, most probably makes things easier for the EA. Because CPT is developed and maintained by one of the authors, we will more likely contnue to use it in the future.

A primary theoretical concern is the existence of a decomposition for any plan with optimal makespan. At the moment, because of the restriction of the representation to the predicates that are in the goal, some states become impossible to describe. If one of these states is mandatory for all optimal plans, the evolutionary algorithm may be unable to find the optimal solution. In the Zeno domain for instance, it can be necessary to link a specific person to a specific plane. This may happen when two persons are boarded on two planes, which thus play a symmetrical role between two given stations, but do not play a symmetrical role w.r.t. the overall goal of the problem. The `in` predicate should then be taken into account when splitting the optimal solution. The main difficulty, however, is to add the corresponding state invariant between `at` and `in` (a person is either `at` a location or `in` a plane).

The results presented in Sect. 4.1, although demonstrating that *Divide-and-Evolve* can solve problems that cannot be directly solved by CPT, they also show that the search capabilities of the proposed algorithm should be improved for more robustness.

There is plenty of room for improvements, e.g. on the variation operators: at the moment, both are completely blind, without any use of any domain knowledge. Of course, this is compliant with a "pure" evolutionary approach ... that is also known to be completely inefficient in the case of combinatorial optimization problems. Crossover can be improved by choosing the crossing station in the second parent such that it is close to that of the first parent, at least at that moment, according to the syntactic distance. The *Add* mutation that randomly adds a station in the list will be improved by building the new station in such a way that it is half-way from both surrounding stations. The choice of station to be deleted in the *Del* mutation will be biased toward the stations that are very easy to reach (in terms of actual number of backtracks used).

Also, all parameters of the algorithm will be carefully fine-tuned.

Of course the *Divide-and-Evolve* scheme has to be experimented on more examples. The International Planning Competition provides many instances in several domains that are good candidates. Preliminary results on the `driver` problem showed very similar results to those reported here on the `zeno` domain. However, other domains, such as the `depot` domain, or many real-world domains, involve (at least) two predicates in their goal descriptions (e.g. `in` and `on` for `depot`). It is hence necessary to increase the range of allowed expressions in the description of individuals.

Other improvements will result from the move to CPT-3, the new version of CPT, entirely rewritten in C. It will be possible to call CPT from within the EA, and hence to perform all grounding, pre-processing and CSP representation only once: at the moment, CPT is launched anew for each partial computation, and a quick look at Table 1 shows that on the `zeno14` problem, for instance, the run-time per individual will decrease from 193 to 8 seconds. Though this will not *per se* improve the quality of the results, it will allow us to tackle more complex problems than even `zeno14`. Along the same lines, other planners, in particular suboptimal planners, will also be tried in lieu of CPT, as maybe the *Divide-and-Evolve* approach could find optimal results using suboptimal planners (as done in some sense in the multi-objective case; see Sect. 4.2).

Greater improvements will be possible after that move, with respect to problem representation. Because *Divide-and-Evolve* will have access to all exclusions among predicates that are derived and maintained by CPT, exclusions among predicates might be automatically derived, including exclusions across predicates, such as those involving predicates `in` and `on` in the `depot` domain. Second, and maybe more important, the expressive power of the representation of the stations will be increased: at the moment, only predicates that are listed in the overall goal are considered in the intermediate stations. And the example of `zeno14` clearly shows that, though the *DAE* approach can indeed break the complexity barrier, solving instances that CPT could not directly solve, it will not be able to reach the global optimum with such restriction (one can construct examples where `in` predicates are necessary to actually optimally break the problem). It is hence planned to allow other predicates to be used to represent intermediate stations. Of course, this will also increase the size of the search space, and some detailed analysis will be needed to somehow determine the minimal set of predicates that are needed for a given problem in order that the *DAE* approach can find the global optimum.

A last but important remark about the results is that, at least in the single objective case, the best solution found by the algorithm was always found in the early generations of the runs (35 at most for `Zeno14`): it could be the case that the simple splits of the problem into smaller sub-problems that are done during the initialization are the main reasons for the good results. Detailed investigations will show whether or not an EA is actually useful in that context!

Nevertheless, we do believe that using evolutionary computation is mandatory in order to solve multi-objective optimization problems, as witnessed by the results of Sect. 4.2, which are, to the best of our knowledge, the first ever results of Pareto optimization for TPPs, and are enough to justify the *Divide-and-Evolve* approach.

Acknowledgement. Vincent Vidal is supported by the ANR "Planevo" project n°JC05_41940.

References

1. F. Bacchus. The 2000 AI Planning Systems Competition. *Artificial Intelligence Magazine*, 22(3):47–56, 2001
2. A.H. Brie and P. Morignot. Genetic Planning Using Variable Length Chromosomes. In *15th Int. Conf. on Automated Planning and Scheduling*, 2005
3. T. Bylander. The Computational Complexity of Propositional STRIPS planning. *Artificial Intelligence*, 69(1-2):165–204, 1994
4. K. Deb, S. Agrawal, A. Pratab, and T. Meyarivan. A Fast Elitist Non-Dominated Sorting Genetic Algorithm for Multi-Objective Optimization. In M. Schoenauer et al., editor, *PPSN'2000*, pages 849–858. Springer-Verlag, LNCS 1917, 2000
5. C. Desquilbet. Détermination de trajets optimaux par algorithmes génétiques. Rapport de stage d'option B2 de l'Ecole Polytechnique. Palaiseau, France, Juin 1992 Advisor: Marc Schoenauer. In French
6. A.E. Eiben and M. Schoenauer. Evolutionary Computing. *Information Processing Letter*, 82(1):1–6, 2002
7. A.E. Eiben and J.E. Smith. *Introduction to Evolutionary Computing*. Springer Verlag, 2003
8. R. Fikes and N. Nilsson. STRIPS: A New Approach to the Application of Theorem Proving to Problem Solving. *Artificial Intelligence*, 1:27–120, 1971
9. M. Fox and D. Long. The Automatic Inference of State Invariants in TIM. *Journal of Artificial Intelligence Research*, 9:367–421, 1998
10. M. Fox and D. Long. PDDL2.1: An Extension to PDDL for Expressing Temporal Planning Domains. *Journal of Artificial Intelligence Research*, 20:61–124, 2003
11. H. Geffner. Perspectives on Artificial Intelligence Planning. In *Proc. AAAI-2002*, pages 1013–1023, 2002
12. J. J. Grefenstette. Incorporating Problem Specific Knowledge in Genetic Algorithms. In Davis L., editor, *Genetic Algorithms and Simulated Annealing*, pages 42–60. Morgan Kaufmann, 1987
13. W.E. Hart, N. Krasnogor, and J.E.Smith, editors. *Evolutionary Computation – Special Issue on Memetic Algorithms*, Volume 12:3. MIT Press, 2004
14. W.E. Hart, N. Krasnogor, and J.E. Smith, editors. *Recent Advances in Memetic Algorithms*. Studies in Fuzziness and Soft Computing, Vol. 166. Springer Verlag, 2005
15. J. Hoffmann, J. Porteous, and L. Sebastia. Ordered Landmarks in Planning. *Journal of Artificial Intelligence Research*, 22:215–278, 2004
16. Jörg Hoffmann and Stefan Edelkamp. The Deterministic Part of IPC-4: An Overview. *Journal of Artificial Intelligence Research*, 24:519–579, 2005
17. J. Koehler and J. Hoffmann. On Reasonable and Forced Goal Orderings and their Use in an Agenda-Driven Planning Algorithm. *Journal of Artificial Intelligence Research*, 12:338–386, 2000
18. R. Korf. Planning as Search: A Quantitative Approach. *Artificial Intelligence*, 33:65–88, 1987

19. D. Long and M. Fox. The 3rd International Planning Competition: Results and Analysis. *Journal of Artificial Intelligence Research*, 20:1–59, 2003
20. D. McDermott. PDDL – The Planning Domain Definition language. At http://ftp.cs. yale.edu/pub/mcdermott, 1998
21. D. McDermott. The 1998 AI Planning Systems Competition. *Artificial Intelligence Magazine*, 21(2):35–56, 2000
22. P. Merz and B. Freisleben. Fitness Landscapes and Memetic Algorithm Design. In David Corne, Marco Dorigo, and Fred Glover, editors, *New Ideas in Optimization*, pages 245–260. McGraw-Hill, London, 1999
23. N. J. Radcliffe and P. D. Surry. Fitness variance of formae and performance prediction. In L. D. Whitley and M. D. Vose, editors, *Foundations of Genetic Algorithms 3*, pages 51–72. Morgan Kaufmann, 1995
24. J.L. Schlabach, C.C. Hayes, and D.E. Goldberg. FOX-GA: A Genetic Algorithm for Generating and Analyzing Battlefield Courses of Action. *Evolutionary Computation*, 7(1):45–68, 1999
25. Marc Schoenauer, Pierre Savéant, and Vincent Vidal. Divide-and-Evolve: a new memetic scheme for domain-independent temporal planning. In J. Gottlieb and G. Raidl, editors, *Proc. EvoCOP'06*. Springer Verlag, 2006
26. D. Smith and D. S. Weld. Temporal Planning with Mutual Exclusion Reasoning. In *Proc. IJCAI-99*, pages 326–337, 1999
27. L. Spector. Genetic Programming and AI Planning Systems. In *Proc. AAAI 94*, pages 1329–1334. AAAI/MIT Press, 1994
28. V. Vidal and H. Geffner. Branching and Pruning: An Optimal Temporal POCL Planner based on Constraint Programming. In *Proc. AAAI-2004*, pages 570–577, 2004
29. V. Vidal and H. Geffner. Branching and Pruning: An Optimal Temporal POCL Planner based on Constraint Programming (to appear). *Artificial Intelligence*, 2005
30. D. S. Weld. An Introduction to Least Commitment Planning. *AI Magazine*, 15(4):27–61, 1994
31. D. Whitley, T. Starkweather, and D. Fuquay. Scheduling problems and traveling salesman: The genetic edge recombination operator. In J. D. Schaffer, editor, *Proc. 3rd Intl Conf. on Genetic Algorithms*. Morgan Kaufmann, 1989

Local Search Based on Genetic Algorithms

Carlos García-Martínez[1] and Manuel Lozano[2]

[1] Dept. of Computing and Numerical Analysis, University of Córdoba, 14071 Córdoba, Spain.
cgarcia@uco.es

[2] Dept. of Computer Science and Artificial Intelligence, University of Granada, 18071 Granada, Spain.
lozano@decsai.ugr.es

Abstract

Genetic Algorithms have been seen as search procedures that can quickly locate high performance regions of vast and complex search spaces, but they are not well suited for fine-tuning solutions, which are very close to optimal ones. However, genetic algorithms may be specifically designed to provide an effective *local search* as well. In fact, several genetic algorithm models have recently been presented with this aim. In this chapter, we call these algorithms *Local Genetic Algorithms*.

In this chapter, first, we review different instances of local genetic algorithms presented in the literature. Then, we focus on a recent proposal, the Binary-coded Local Genetic Algorithm. It is a *Steady-state Genetic Algorithm* that applies a *crowding replacement method* in order to keep, in the population, groups of chromosomes with high quality in different regions of the search space. In addition, it maintains an external solution (*leader chromosome*) that is crossed over with individuals of the population. These individuals are selected by using *Positive Assortative Mating*, which ensures that these individuals are very similar to the leader chromosome. The main objective is to orientate the search in the nearest regions to the leader chromosome.

We show an empirical study comparing a *Multi-start Local Search* based on the binary-coded local genetic algorithm with other instances of this metaheuristic based on local search procedures presented in the literature. The results show that, for a wide range of problems, the multi-start local search based on the binary-coded local genetic algorithm consistently outperforms multi-start local search instances based on the other local search approaches.

Key words: Local Genetic Algorithms, Local Search Procedures, Multi-start Local Search

1 Introduction

Local Search Procedures (LSPs) are optimisation methods that maintain a solution, known as *current solution*, and explore the search space by steps within its *neighbourhood*. They usually go from the current solution to a better *close* solution, which is used, in the next iteration, as current solution. This process is repeated till a stop condition is fulfilled, e.g. there is no better solution within the neighbourhood of the current solution.

Three important LSPs are:

- *First Improvement Local Search* [3]: Replaces the current solution with a randomly chosen neighbouring solution with a better fitness value.
- *Best Improvement Local Search* [3]: Replaces the current solution with the best among all the neighbouring solutions.
- *Randomised K-opt LSP* (RandK-LS) [26, 27, 34]: Looks for a better solution by altering a variable number of k components of the current solution per iteration, i.e. the dimension of the explored neighbourhood is variable.

The interest on LSPs comes from the fact that they may effectively and quickly explore the *basin of attraction* of optimal solutions, finding an optimum with a high degree of accuracy and within a small number of iterations. In fact, these methods are a key component of metaheuristics that are *state-of-the-art* for many optimisation problems, such as *Multi-Start Local Search* (MSLS) [4], *Greedy Randomised Adaptive Search Procedures* (GRASP) [8, 41], *Iterated Local Search* (ILS) [30], *Variable Neighbourhood Search* (VNS) [35], and *Memetic Algorithms* (MAs) [36].

Genetic Algorithms (GAs) [17, 24] are optimisation techniques that use a *population of candidate solutions*. They explore the search space by *evolving* the population through four steps: *parent selection, crossover, mutation,* and *replacement.* GAs have been seen as search procedures that can locate high performance regions of vast and complex search spaces, but they are not well suited for fine-tuning solutions [14, 29]. However, the components of the GAs may be *specifically designed* and their parameters *tuned,* in order to provide an *effective local search behaviour.* In fact, several GA models have recently been presented with this aim [29, 31]. In this chapter, these algorithms are called *Local Genetic Algorithms* (LGAs).

LGAs have some advantages over classic LSPs. Most LSPs lack the ability to follow the proper path to the optimum on complex search landscapes. This difficulty becomes much more evident when the search space contains very narrow paths of arbitrary direction, also known as *ridges.* That is because LSPs attempt successive steps along orthogonal directions that do not necessarily coincide with the direction of the ridge. However, it was observed that LGAs are capable of following ridges of arbitrary direction in the search space regardless of their direction, width, or even, discontinuities [29]. Thus, the study of LGAs is a promising way to design more effective metaheuristics based on LSPs [13, 22, 29, 31, 40, 47].

The aim of this chapter is to analyse LGAs in depth. In order to do this:

- First, we introduce the LGA concept and identify its main properties.
- Second, we review different LGA instances presented in the literature.
- Finally, we focus on a recent LGA example, the *Binary-coded LGA* (BLGA) [12]. We describe an empirical study comparing a MSLS based on the BLGA with other instances of this metaheuristic based on LSPs proposed in the literature. The results show that, for a wide range of problems, the MSLS instance based on the BLGA consistently outperforms the MSLS instances based on the other local search approaches.

The chapter is organised as follows. In Sect. 2, we outline three LSPs that have been often considered in the literature to build metaheuristics based on LSPs. In Sect. 3, we introduce a brief overview about GAs. In Sect. 4, we inspect the LGA concept and review several LGA instances found in the literature. In Sect. 5, we describe an example of LGA, the BLGA. In Sect. 6, we show an empirical study comparing the performance of the MSLS based on the BLGA with other instances of the MSLS based on LSPs presented in Sect. 2. Finally, in Sect. 7, we provide some conclusions and future research directions.

2 Local Search Procedures in the Literature

LSPs are improvement heuristics that maintain a solution, known as *current solution* (X^C), and search its *neighbourhood* ($N(X^C)$) for a better one. If a better solution $S \in N(X^C)$ is found, S becomes the new X^C and the neighbourhood search starts again. If no further improvement can be made, i.e. $\nexists\, S \in N(X^C)$ such as S improves X^C, then, a local or global optimum has been found.

The interest in LSPs comes from the fact that they may effectively and quickly explore the *basin of attraction* of optimal solutions, finding an optimum with a high degree of accuracy and within a small number of iterations. The reasons for this high exploitative behaviour are:

- LSPs usually keep as X^C the best found solution so far, and
- $N(X^C)$ is composed of solutions with minimal differences from X^C, i.e. LSPs perform a *local* refinement on X^C.

Three important LSPs are:

- *First Improvement Local Search* (First-LS) [3]: Works by comparing X^C with neighbouring solutions. When a neighbouring solution appears better, X^C is replaced and the process starts again. If all the neighbouring solutions are worse than X^C, then, the algorithm stops. In First-LS, $N(X^C)$ is usually defined as the set of solutions with minimal differences from X^C, i.e. in binary-coded problems, S differs from X^C only in one bit, $\forall S \in N(X^C)$.
- *Best Improvement Local Search* (Best-LS) [3]: Generates and evaluates all the neighbouring solutions of X^C. Then, the best one replaces X^C if it is better. Otherwise, the algorithm stops. In Best-LS, $N(X^C)$ is usually defined as in First-LS.
- *Randomised K-opt LSP* (RandK-LS) [26, 27, 34]: Is a variation of the *K-opt* LSP presented in [33]. That was specifically designed to tackle binary-coded problems. Its basic idea is to find a solution by flipping a variable number of k bits in the solution vector per iteration. In each step, n (n is the dimension of the problem) solutions ($X^1, X^2, \ldots, X^n$) are generated by flipping one bit of the previous solution, i.e. solution X^{i+1} is obtained by flipping one bit of the solution X^i (X^1 is generated from X^C). A candidate set is used to assure that each bit is flipped no more than once. Then, the best solution in the sequence is accepted as X^C for the next iteration, if it is better, otherwise the algorithm stops and returns X^C.

LSPs are a key component of many metaheuristics. In order to perform a *global search*, these metaheuristics look for *synergy* between the exploitative behaviour of the LSP and *explorative components*. In this way, while the explorative components ensure that different promising search zones are focused upon, the LSP obtains the best possible accurate solutions within those regions. Examples of metaheuristics based on LSPs are:

- *MSLS* [4]: Iteratively applies the LSP to random solutions.
- *GRASP* [8,41]: Generates randomised heuristic solutions to the specific problem, and applies the LSP to them.
- *ILS* [30]: The LSP is initially applied to a random solution. In the following iterations, the LSP is applied to solutions generated by altering previous ones.
- *VNS* [35]: The idea is similar to that of ILS. The main difference is that the solutions are lightly or strongly altered depending on whether or not the new solutions improve the best one so far.
- *MAs* [36]: They are evolutionary algorithms that apply LSPs in order to refine the individuals of the population.

3 Genetic Algorithms

GAs are general purpose search algorithms that use principles inspired by *natural genetic populations* to evolve solutions to problems [17,24]. The basic idea is to maintain a *population of chromosomes* that represent candidate solutions to the concrete problem. The GA evolves the population through a process of competition and controlled variation. Each chromosome in the population has an associated *fitness* to determine which ones are used to form new chromosomes in the competition process, which is called *parent selection*. The new ones are created using genetic operators such as *crossover* and *mutation*.

GAs have had a great measure of success in search and optimisation problems. The reason for a great part of their success is their ability to exploit the information accumulated about an initially unknown search space in order to bias subsequent searches into useful subspaces. This is their key feature, particularly in large, complex, and poorly understood search spaces, where classical search tools (enumerative, heuristic, ...) are inappropriate, offering a valid approach to problems requiring efficient and effective search techniques.

Two of the most important GA models are the *Generational* GA and the *Steady-state* GA:

- The Generational GA [17] creates new offspring from the members of an old population, using the genetic operators, and places these individuals in a new population that becomes the old population when the whole new population is created.
- The Steady-state GA (SSGA) [44,49] is different from the generational model in that there is typically one single new member inserted into the new population at any one time. A *replacement/deletion* strategy defines which member in the current population is forced to perish (or vacate a slot) in order to make room for the new offspring to compete (or, occupy a slot) in the next iteration. The basic algorithm step of the SSGA is shown in Fig. 1.

```
SSGA()

    initialise P;
    evaluate P;

    while (stop-condition is not fulfilled)
        parents ← select two chromosomes from P;
        offspring ← combine and mutate the chromosomes in parents;
        evaluate offspring;
        R ← select an individual from P; //replacement strategy
        decide if offspring should replace R;
```

Fig. 1 Structure of a SSGA.

4 Local Genetic Algorithms

There are two primary factors in the search carried out by a GA [49]:
- *Selection pressure.* In order to have an effective search there must be a search criterion (the fitness function) and a selection pressure that gives individuals with higher fitness a higher chance of being selected for reproduction, mutation, and survival. Without selection pressure, the search process becomes random and promising regions of the search space would not be favoured over non-promising regions.
- *Population diversity.* This is crucial to a GA's ability to continue fruitful exploration of the search space.

Selection pressure and population diversity are inversely related:
- increasing selection pressure results in a faster loss of population diversity, while
- maintaining population diversity offsets the effect of increasing selection pressure.

Traditionally, GA practitioners have carefully designed GAs in order to obtain a balanced performance between selection pressure and population diversity. The main objective was to obtain their beneficial advantages simultaneously: to allow the most promising search space regions to be reached (*reliability*) and refined (*accuracy*).

Due to the flexibility of the GA architecture, it is possible to design GA models specifically aimed to provide *effective local search*. In this way, their unique objective is to obtain *accurate* solutions. In this chapter, these algorithms are named *Local Genetic Algorithms*.

LGAs present some advantages over classic LSPs. Most LSPs lack the ability to follow the proper path to the optimum on complex search landscapes. This difficulty becomes much more evident when the search space contains very narrow paths of arbitrary direction, also known as *ridges*. This is because LSPs attempt successive steps along orthogonal directions that do not necessarily coincide with the direction of the ridge. However, it was observed that LGAs are capable of following ridges of arbitrary direction in the search space regardless of their direction, width, or even, discontinuities [29]. Thus, the study of LGAs becomes a promising way to allow the design of more effective metaheuristics based on LSPs. In fact, some LGAs were considered for this task [13, 22, 29, 31, 40, 47].

In the following sections, we explain the features of different LGA instances that may be found in the literature. In addition, we cite the corresponding metaheuristic models in which LGAs were integrated:

- μGAs [29] (Section 4.1).
- The Crossover Hill-climbing [31] (Section 4.2).
- LGAs based on female and male differentiation [13] (Section 4.3).
- LGAs as components of distributed GAs [22, 40, 47] (Section 4.4).

4.1 μGAs Working as LGAs

In [29], a *Micro-Genetic Algorithm* (μGA) (GA with a small population and short evolution) is used as LGA within a memetic algorithm. Its mission is to refine the solutions given by the memetic algorithm. It evolves a population of *perturbations* (P^i), whose aptitude values depend on the solution given by the memetic algorithm. Its main features are the following:

- It is an *elitist* Generational GA that uses roulette wheel parent selection, a ten-point crossover, and bit mutation with adaptive probabilities. In addition, by using a small population (five individuals), the μGA may achieve high selection pressure levels, which allows accurate solutions to be reached.
- The perturbation space is defined in such a way that the μGA explores a small region centred on the given solution. Thus, it offers local improvements to the given solution.

The memetic algorithm based on the μGA was tested against 12 different evolutionary algorithm models, which include a simple GA and GAs with different hill-climbing operators, on five hard constrained optimisation problems. The simulation results revealed that this algorithm exhibits good performance, outperforming the competing algorithms in all test cases in terms of solution accuracy, feasibility rate, and robustness.

We should point out that μGAs have been considered as LGAs by other authors:

- Weicai et al. [48] propose a *Multi-agent GA* that makes use of a μGA.
- Meloni et al. [32] insert a μGA in a *multi-objective evolutionary algorithm* for a class of sequencing problems in manufacturing environments.
- Papadakis et al. [38] use a μGA within a GA-based fuzzy modelling approach to generate TSK models.

4.2 A Real-coded LGA: Crossover Hill-climbing

Lozano et al. [31] propose a *Real-coded Memetic Algorithm* that uses *Crossover Hill-climbing* (XHC) as LGA. Its mission is to obtain the best possible accuracy levels to lead the population towards the most promising search areas, producing an effective refinement on them. In addition, an adaptive mechanism is employed to determine the probability with which solutions are refined with XHC.

The XHC is a Real-coded Steady-state LGA that maintains a pair of parents and performs crossover repeatedly on this pair until some number of offspring, n_{off}, is reached. Then, the best offspring is selected and replaces the worst parent, only if it is better. The process iterates n_{it} times and returns the two final current parents.

The XHC proposed may be conceived as a *Micro Selecto-Recombinative Real-coded LGA* model that employs the minimal population size necessary to allow the crossover to be applicable, i.e. two chromosomes. Although XHC can be instantiated with any crossover operator, the authors used a *self-adaptive real-parameter operator* that generates offspring according to the current distribution of the parents. If the parents are located close to each other, the offspring generated by the crossover might be distributed densely around them. On the other hand, if the parents are located far away from each other, then the offspring will be sparsely distributed.

Experimental results showed that, for a wide range of problems, the real-coded memetic algorithm with XHC operator consistently outperformed other real-coded memetic algorithms appearing in the literature.

Other studies have considered some variants of the XHC algorithm [6,7,37].

4.3 LGAs Based on Female and Male Differentiation

Parent-Centric Crossover Operators (PCCOs) is a family of real-parameter crossover operators that use a probability distribution to create offspring in a restricted search region marked by one of the parent, *the female parent*. The range of this probability distribution depends on the distance among the female parent and the other parents involved in the crossover, *the male parents*.

Traditionally, PCCO practitioners have assumed that every chromosome in the population may become either a female parent or a male parent. However, it is very important to emphasise that female and male parents have two differentiated roles [13]:

- female parents *point* to the search areas that will receive sampling points, whereas
- male parents are used to determine the *extent* of these areas.

With this idea in mind, García-Martínez et al. [13] propose applying a *Female and Male Differentiation* (FMD) process before the application of a PCCO. The FMD process creates two different groups according to two tuneable parameters (N_F and N_M):

- G_F with the N_F best chromosomes in the population, which can be female parents; and
- G_M with the N_M best individuals, which can be selected as male parents.

An important feature of this FMD process is that it has a strong influence on the degree of *selection pressure* kept by the GA:

- On the one hand, when N_F is low, high selection pressure degrees are achieved, because the search process is very focused in the best regions.
- On the other hand, if N_F is high, the selection pressure is softened, providing extensive sampling on the search areas represented in the current population.

The authors argue that the two parameters associated with the FMD process, N_F and N_M, may be adequately adjusted in order to design *Local Real-coded GAs* that reach *accurate* solutions:

- On the one hand, with low N_F values ($N_F = 5$), the GA keeps the best solutions found so far in a similar way to which LSPs keep the best solution found so far in X^C.
- On the other hand, PCCOs sample the neighbourhood of the N_F best solutions as LSPs sample the neighbourhood of X^C, i.e. PCCOs perform a *local* refinement on the N_F best solutions.

In addition, the authors argue that *Global Real-Coded GAs* can also be obtained by adequately adjusting N_F and N_M. Global Real-coded GAs offer *reliable* solutions when they tackle multimodal and complex problems.

Finally, with the aim of achieving *robust operation*, García-Martínez et al. followed a simple *hybridisation technique* to put together a Global Real-coded GA and a Local Real-coded GA.

Empirical studies confirmed that this hybridisation was very competive with *state-of-the-art on metaheuristics for continuous optimisation*.

4.4 LGAs as Components of Distributed GAs

Distributed GAs keep in parallel, several independent *subpopulations* that are processed by a GA [45]. Periodically, a *migration mechanism* produces a chromosome exchange between the subpopulations. Making distinctions between the subpopulations by applying GAs with different configurations, we obtain *Heterogeneous Distributed Genetic Algorithms* (HDGAs). These algorithms represent a promising way for introducing correct exploration/exploitation balance in order to avoid premature convergence and reach accurate final solutions.

Next, we describe three HDGA models that assign to every subpopulation a different exploration or exploitation role. In this case, the exploitative subpopulations are LGAs whose mission is to refine the solutions that have been migrated from explorative subpopulations:

- *Gradual Distributed Real-coded GAs* [22] (Sect. 4.4.1).
- *GA Based on Migration and Artificial Selection* [40] (Sect. 4.4.2).
- *Real Coded GA with an Explorer and an Exploiter Population* [47] (Sect. 4.4.3).

4.4.1 Gradual Distributed Real-coded GAs

The availability of crossover operators for real-coded GAs [21, 23] that generate different exploration or exploitation degrees makes the design of Heterogeneous Distributed Real-coded GAs based on these operators feasible. Herrera et al. [22] propose *Gradual Distributed Real-coded GAs* (GD-RCGAs) that apply a different crossover operator to each subpopulation. These operators are differentiated according to their associated exploration and exploitation properties and the degree thereof. The effect achieved is a *parallel multiresolution* with regard to the action of the crossover operators. Furthermore, subpopulations are suitably connected to exploit this multiresolution in a *gradual* way.

GD-RCGAs are based on a hypercube topology with three dimensions (Fig. 2). There are two important sides to be differentiated:

- The *front side* is devoted to exploration. It is made up of four subpopulations $E_1, \ldots, E_4$, to which exploratory crossover are applied. The exploration degree increases clockwise, starting at the lowest E_1, and ending at the highest E_4.
- The *rear side* is for exploitation. It is composed of subpopulations $e_1, \ldots, e_4$ that undergo exploitative crossover operators. The exploitation degree increases clockwise, starting at the lowest e_1, and finishing at the highest e_4. Notice that the $e_1, \ldots, e_4$ populations are LGAs that achieve different exploitation levels.

The connectivity of the GD-RCGA allows a *gradual refinement* when migrations are produced from an exploratory subpopulation toward an exploitative one, i.e., from E_i to e_i, or between two exploitative subpopulations from a lower degree to a higher one, i.e. from e_i to e_{i+1}.

Experimental results showed that the GD-RCGA consistently outperformed sequential real-coded GAs and homogeneous distributed real-coded GAs, which are equivalent to them, and other real-coded evolutionary algorithms reported in the literature.

4.4.2 GA Based on Migration and Artificial Selection

In [40], a distributed GA, called GAMAS, was proposed. GAMAS uses four subpopulations, denoted as *species I–IV*, which supply different exploration or exploitation levels by using different mutation probabilities:

- Species II is a subpopulation used for exploration. For this purpose, it uses a high mutation probability ($p_m = 0.05$).
- Species IV is a subpopulation used for exploitation. This way, its mutation probability is low ($p_m = 0.003$). Species IV is an LGA that attempts to achieve high exploitation by using a low mutation probability.
- Species III is an exploration and exploitation subpopulation with $p_m = 0.005$.

GAMAS selects the best individuals from species II–IV, and introduces them into species I whenever those are better than the elements in this subpopulation. The mis-

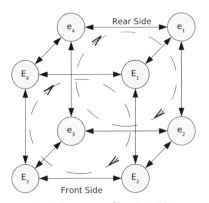

Fig. 2 Structure of a GD-RCGA

sion of species I is to preserve the best chromosomes appearing in the other species. At predetermined generations, its chromosomes are reintroduced into species IV by replacing all of the current elements in this species.

Experimental results showed that GAMAS consistently outperforms simple GAs and alleviates the problem of *premature convergence*.

4.4.3 Real-coded Genetic Algorithm with an Explorer and an Exploiter Population

Tsutsui et al. [47] propose a GA with two populations whose missions are well differentiated: one is aimed to explore the search space, whereas the other is an LGA that searches the neighbourhood of the best solution obtained so far. Both of them are generational GAs. However, the LGA uses a *fine-grained mutation* and a population of half the size of the explorer population. This way, the LGA performs a high exploitation over the best solution so far.

The proposed technique exhibited performance significantly superior to standard GAs on two complex highly multimodal problems.

5 Binary-coded Local Genetic Algorithm

In this section, we describe a recent LGA example, the Binary-coded LGA (BLGA) [12] that may be used to design metaheuristics based on LSPs. The aim of BLGA is two-fold:

- On the one hand, BLGA has been specifically designed to perform an effective local search in a similar way to LSPs. BLGA optimises locally the solutions given by the metaheuristic, by steps within their neighbourhoods.
- On the other hand, while BLGA performs the local search, its population (P) acquires information about the location of the best search regions. Then, BLGA can make use of the knowledge in P in order to guide the search. This kind of information cannot be used by LSPs.

BLGA is a SSGA (Sect. 3) that uses a *crowding replacement method* (*restricted tournament selection* [19]) that favours the formation of *niches* (groups of chromosomes of high quality located in different and scattered regions of the search space) in P. In addition, BLGA maintains an external chromosome, the *leader chromosome* (C^L), which plays the same role as X^C in classical LSPs:

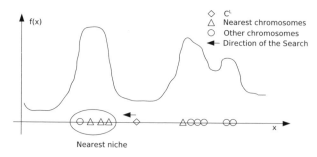

Fig. 3 Niches considered to guide the local search

- BLGA samples new solutions within the neighbourhood of C^L in a similar way to LSPs with X^C, by means of a *multi-parent* version of the *uniform crossover operator* [44]. In addition, BLGA directs the sampling operation towards the closest niches to C^L (Fig. 3) by selecting parents with *positive assortative mating* [9].
- BLGA keeps the best sampled solution in C^L just as LSPs keep the best solution obtained so far in X^C.

5.1 General Scheme of the BLGA

Let's suppose that a particular metaheuristic applies the BLGA as LSP. When the metaheuristic calls the BLGA to refine a particular solution, the BLGA considers this solution as C^L. Then, the following steps are carried out during each iteration (Fig. 4):

1. *Mate selection.* m chromosomes $(Y^1, Y^2, ..., Y^m)$ are selected from the population by applying the positive assortative mating m times (Sect. 5.2).
2. *Crossover.* C^L is crossed over with $Y^1, Y^2, ..., Y^m$ by applying the multi-parent uniform crossover operator, generating an offspring Z (Sect. 5.3).
3. *To update the leader chromosome and replacement.* If Z is better than C^L, then C^L is inserted into the population using the restricted tournament selection (Sect. 5.4) and Z becomes the new C^L. Otherwise, Z is inserted in the population using this replacement scheme.

All these steps are repeated until a termination condition is achieved (Sect. 5.5).

5.2 Positive Assortative Mating

Assortative mating is the natural occurrence of mating between individuals of similar phenotype more or less often than expected by chance. Mating between individuals with similar phenotype more often is called positive assortative mating and less often is called negative assortative mating. Fernandes et al. [9] implement these ideas to design two mating selection mechanisms. A first parent is selected by the roulette wheel method and n_{ass} chromosomes are selected with the same method (in BLGA all the candidates are selected at random). Then, the similarity between each of these

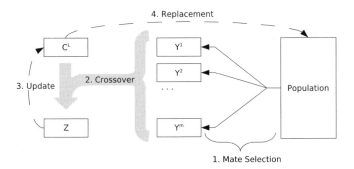

Fig. 4 Model of the BLGA

chromosomes and the first parent (C^L in the BLGA) is computed (similarity between two binary-coded chromosomes is defined as the Hamming distance between them). If assortative mating is negative, then the one with less similarity is chosen. If it is positive, the genome more similar to the first parent is chosen to be the second parent. In the case of BLGA, the first parent is C^L and the method is repeated m times, in order to obtain m parents.

Since positive assortative mating selects similar individuals to C^L, it helps BLGA to achieve the two main objectives:

- Positive assortative mating helps BLGA to perform a local refinement on C^L because similar parents make the crossover operator to sample near to C^L.
- Positive assortative mating probabilistically guides the search according to the information kept in P, because it probabilistically selects chromosomes from the nearest niches to C^L (see Fig. 3).

5.3 Multi-parent Uniform Crossover Operator

Since the main aim of the BLGA is to *fine-tune* C^L, it should sample new points near it. *Uniform crossover* (UX) [44] creates an offspring from two parents by choosing the genes of the first parent with the probability p_f. If it uses a high p_f value, it will generate the offspring near to the first parent. The BLGA uses a *multi-parent UX* that will be defined below.

During application of the crossover operator, the BLGA uses a *short term memory* mechanism to avoid the generation of any offspring previously created. It remembers the genes of C^L that have been flipped when generating an offspring Z^k. Then, it avoids flipping those genes of C^L, in order to prevent the creation of Z^k once again. In order to do that, this mechanism maintains a mask, $M = (M_1, \ldots, M_n)$, where $M_i = 1$ indicates that the ith gene of C^L (C_i^L) cannot be flipped in order to create a new offspring. Initially, and when C^L is updated with a better solution, any gene can be flipped, so M_i is set to 0 for all $i \in \{1, \ldots, n\}$.

The pseudocode of the crossover operator with short term memory is shown in Fig. 5, where $U(0, 1)$ is a random number in $[0, 1]$, $RI(1, m)$ is a random integer in $\{1, 2, \ldots, m\}$, and p_f is the probability of choosing genes from C^L. It creates the offspring Z as follows:

- Z_i is set to C_i^L for all $i = 1, \ldots, n$ with $M_i = 1$.
- If $M_i = 0$, then Z_i is set to C_i^L with probability p_f. Otherwise, Z_i is set to the ith gene of a randomly chosen parent Y^j. The mask is updated if Z_i is different from C_i^L.
- Finally, if the Z obtained is equal to C^L, then a gene i with $M_i = 0$ chosen at random, is flipped and the mask is updated.

Tabu Search [15] also uses a short term memory. Tabu search stores in that memory the last movements that were used to generate the current solution. It forbides those movements in order to avoid sampling previous solutions. In tabu search, each forbidden movement in the short term memory has a *tabu tenure* that indicates when it should be removed from the memory. When the tabu tenure expires, the movement is permitted again.

```
multiparent_UX(C^L, Y^1, ... , Y^m, M, p_f)

   For (i = 1, ..., n)

      If (M_i = 1  OR  U(0,1) < p_f) //short term memory mechanism
         Z_i ← C_i^L ;

      Else
         k ← RI(1, m);
         Z_i ← Y_i^k ;

         If (Z_i ≠ C_i^L)
            M_i ← 1  ; //update the mask

   If (Z = C^L)
      j ← RI(1, n)  such as  M_i = 0;
      M_i ← 1  ; //update the mask
      Z_i ← 1 - Z_i ;

   Return Z;
```

Fig. 5 Pseudocode of the Multi-parent Uniform Crossover Operator with short term memory

5.4 Restricted Tournament Selection

BLGA considers *Restricted Tournament Selection* (RTS) [19] as its *crowding replacement method*. The application of RTS together with the use of high population size may favour the creation of groups of chromosomes with high quality in P, which become located in different and scattered regions of the search space (*niches*). In this way, the population of the BLGA acquires knowledge about the location of the best regions of the search space. The aim of the BLGA is to use this information to guide future searches.

The pseudocode of the RTS is shown in Fig. 6. Its main idea is to replace the closest chromosome R to the one being inserted in the population, from a set of n_T randomly selected ones.

```
RTS(Population, solution)

   G_T ← Select randomly n_T individuals from Population;
   R ← Choose from G_T the most similar
       chromosome to solution;

   If (solution is better than R)
      replace R with solution;
```

Fig. 6 Pseudocode of restricted tournament selection

5.5 Stop Condition

It is important to notice that when every component of the mask of the short term memory (Sect. 5.3) is equal to 1, then, C^L will not be further improved, because the crossover operator will create new solutions exactly equal to C^L. Thus, this condition will be used as the stop condition for the BLGA, and the BLGA will return C^L to the metaheuristic.

6 Experiments: Comparison with Other LSPs

This section reports on an empirical comparative study between the BLGA method and other LSPs for binary-coded problems presented in the literature: First-LS [3]; Best-LS [3]; and RandK-LS [26, 27, 34].

The study compares four instances of the simplest LSP based metaheuristic, the *Multi-start Local Search* [4], each one with a different LSP. The pseudocode of the MSLS metaheuristic is shown in Fig. 7.

The four MSLS instances are defined as follows:

- MS-First-LS: MSLS with the First-LS.
- MS-Best-LS: MSLS with the Best-LS.
- MS-RandK-LS: MSLS with the RandK-LS.
- MS-BLGA: MSLS with the BLGA.

We have chosen the MSLS metaheuristic in order to avoid possible synergies between the metaheuristic and the LSP. In this way, comparisons among the LSPs are fairer. All the algorithms were executed 50 times, each with a maximum of 100,000 evaluations.

The BLGA uses 500 individuals as the population size, $p_f = 0.95$ and $m = 10$ mates for the crossover operator, $n_{ass} = 5$ for the positive assortative mating, and

```
multistart_LS (LSP)

    S^best ← generate random solution;

    While (stop-condition is not fulfilled)
        S ← generate a random solution;
        S' ← perform LSP on S;

        If (S' is better than S^best)
            S^best ← S';

    Return S^best;
```

Fig. 7 Pseudocode of the MSLS metaheuristic

$n_T = 15$ for restricted tournament selection (parameter values from [12]). The population of the BLGA does not undergo initialisation after the iterations of the MSLS, i.e. the initial population of the BLGA at the jth iteration of the MS-BLGA is the last population of the $(j-1)$th iteration. On the other hand, the leader chromosome is given by the MSLS, i.e. it is generated at random, at the beginning of the iterations of the metaheuristic.

We used the 19 test functions described in Appendix A. Table 1 indicates their name, dimension, optimisation criteria, and optimal fitness value.

The results for all the algorithms are included in Table 2. The performance measure is the average of the best fitness function found over 50 executions. In addition, a two-sided *t-test* at 0.05 level of significance was applied in order to ascertain if the differences in performance of the MS-BLGA are significant when compared with those for the other algorithms. We denote the direction of any significant differences as follows:

Table 1 Test problems

Name	Dimension	Criterion	f^*
Onemax(400)	400	minimisation	0
Deceptive(13)	39	minimisation	0
Deceptive(134)	402	minimisation	0
Trap(1)	36	maximisation	220
Trap(4)	144	maximisation	880
Maxcut(G11)	800	maximisation	572.7[1]
Maxcut(G12)	800	maximisation	621[2]
Maxcut(G17)	800	maximisation	Not known
Maxcut(G18)	800	maximisation	1063.4[1]
Maxcut(G43)	1000	maximisation	7027[2]
M-Sat(100,1200,3)	100	maximisation	1[3]
M-Sat(100,2400,3)	100	maximisation	1[3]
NkLand(48,4)	48	maximisation	1[3]
NkLand(48,12)	48	maximisation	1[3]
BQP('gka')	50	maximisation	3414[4]
BQP(50)	50	maximisation	2098[4]
BQP(100)	100	maximisation	7970[4]
BQP(250)	250	maximisation	45,607[4]
BQP(500)	500	maximisation	116,586[4]

[1] Upper bounds presented in [11].

[2] Upper bounds presented in [10].

[3] 1 is the maximum possible fitness value, however an optimal solution with that fitness value may not exist, depending on the current problem instance.

[4] Best known values presented in [1].

Table 2 Comparison of the MS-BLGA with other MSLS instances

	MS-First-LS	MS-Best-LS	MS-RandK-LS	MS-BLGA
Onemax(400)	0 ~	0 ~	0 ~	0
Deceptive(13)	8.68 ~	3.36−	14.32+	8.68
Deceptive(134)	177.6−	128.4−	201.6+	185.84
Trap(1)	213.12+	219.1 ~	201.86+	218.38
Trap(4)	790.08+	828.92+	781.78+	869.3
Maxcut(G11)	437.36+	349.6+	441+	506.64
Maxcut(G12)	425.6+	335.16+	431.32+	497.36
Maxcut(G17)	2920.82+	2824.66+	2946.58+	2975.7
Maxcut(G18)	849.86+	628.32+	873.82+	898.08
Maxcut(G43)	6427.44+	5735.84+	6463.1 ~	6463.18
M-Sat(100,1200,3)	0.9551+	0.9526+	0.9563 ~	0.9566
M-Sat(100,2400,3)	0.9332 ~	0.9314+	0.9335 ~	0.9338
NkLand(48,4)	0.7660+	0.7647+	0.7694+	0.7750
NkLand(48,12)	0.7456 ~	0.7442 ~	0.7493 ~	0.7468
BQP("gka")	3414 ~	3414 ~	3414 ~	3414
BQP(50)	2098 ~	2094.08 ~	2096.72 ~	2098
BQP(100)	7890.56+	7831.7+	7881.52+	7927.56
BQP(250)	45,557.16 ~	45,171.38+	45,504.22+	45,510.96
BQP(500)	115,176.88 ~	108,588.26+	115,335.34 ~	115,256.3

- A plus sign (+): the performance of MS-BLGA is better than that of the corresponding algorithm.
- A minus sign (−): the algorithm improves the performance of MS-BLGA.
- An approximate sign (~): no significant differences.

We have introduced Fig. 8 in order to facilitate the analysis of these results. It shows the percentage improvements, reductions, and non-differences, according to the t-test, obtained when comparing MS-BLGA with the other algorithms on all the test problems.

From Fig. 8, we can say that MS-BLGA performs better than all the other algorithms for more than the 50% of the test problems, and better than or equivalent to

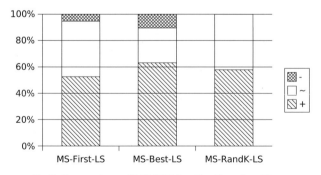

Fig. 8 Comparison of MS-BLGA with other algorithms

almost 90%. Thus, we may conclude that the BLGA is a very promising algorithm for dealing with binary-coded optimisation problems.

On the other hand, Fig. 9 shows the percentage improvements, reductions and non-differences obtained when using MS-BLGA for each test problem (with regard to the other algorithms). Two remarks are worth mentioning regarding Fig. 9:

- MS-BLGA is one of the best algorithms for almost 90% of the test functions. Specifically, MS-BLGA achieves better or equivalent results to those of the other algorithms for all functions, except the two Deceptive ones.
- MS-BLGA returns the best results on four of the five Max-Cut problems.

To sum up, we may conclude that the BLGA, working within the MSLS metaheuristic, is very competitive with classic LSPs, because it obtains better or equivalent results for almost all the test problems considered in this study.

7 Conclusions

In this chapter, we have shown that GAs may be specifically designed with the aim of performing an effective local search: we called these GAs Local GAs. First, we surveyed different LGA instances appearing in the literature. Then, we focused on the BLGA, a recent LGA proposal. BLGA incorporates a specific mate selection mechanism, the crossover operator, and a replacement strategy to direct the local search towards promising search regions represented in the proper BLGA population.

An experimental study, including 19 binary-coded test problems, has shown that when we incorporate the BLGA into a MSLS metaheuristic, this metaheuristic improves results compared with the use of other LSPs that are frequently used to implement it. The good performance of the LGAs reviewed and the satisfactory results

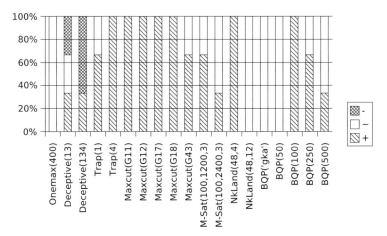

Fig. 9 Performance of MS-BLGA on each test problem

given by the BLGA indicate that further study of these GAs is a topic of major interest. We currently intend to:

- analyse the behaviour of LGAs when they are used by different metaheuristics based on LSPs [4, 8, 30, 35, 36, 41]. Specifically, we are interested in the BLGA.
- extend our investigation to different test-suites (other coding schemes) and real-world problems.

Acknowledgement. This research has been supported by the Spanish MEC project TIN2005-08386-C05-01.

A Appendix. Test Suite

The test suite used for the experiments consists of 19 binary-coded test problems (n is the dimension of the problem). They are described in the following sections.

A.1 Onemax Problem

This is a minimisation problem that applies the following formula:

$$f(X) = n - \sum_{i=1}^{n} X_i \tag{1}$$

We have denoted as Onemax(n) an instance of the Onemax problem with n decision variables: we used Onemax(200).

A.2 Deceptive Problem

In deceptive problems [16] there are certain schemata that guide the search towards a solution that is not globally competitive. It is due to, the schemata that have the global optimum do not bear significance and so, they may not proliferate during the genetic process. The deceptive problem used consists of the concatenation of k sub-problems of length 3. The fitness for each 3-bit section of the string is given in Table 3. The overall fitness is the sum of the fitnesses of these deceptive subproblems. To obtain an individual's fitness, the value of this function is subtracted from the maximum value ($30k$). Therefore, the optimum has a fitness of zero.

We denoted as Deceptive(k) an instance of the Deceptive problem with k sub-problems of length 3. We used two instances: Deceptive(13) and Deceptive(134).

Table 3 Deceptive order-3 problem

Chromosomes	000	001	010	100	110	011	101	111
Fitness	28	26	22	14	0	0	0	30

A.3 Trap Problem

Trap problem [46] consists of misleading subfunctions of different lengths. Specifically, the fitness function $f(X)$ is constructed by adding subfunctions of length 1 (F_1), 2 (F_2), and 3 (F_3). Each subfunction has two optima: the optimal fitness value is obtained for an all-ones string, while the all-zeroes string represents a local optimum. The fitness of all other strings in the subfunction is determined by the number of zeroes: the more zeroes the higher the fitness value. This causes a large basin of attraction towards the local optimum. The fitness values for the subfunctions are specified in Table 4, where the columns indicate the number of ones in the subfunctions F_1, F_2, and F_3. The fitness function $f(X)$ is composed of four subfunctions F_3, six subfunctions F_2, and 12 subfunctions F_1. The overall length of the problem is thus 36. $f(X)$ has 2^{10} optima of which only one is the global optimum: the string with all ones having a fitness value of 220.

$$f(X) = \sum_{i=0}^{3} F_3(X_{[3i:3i+2]}) + \sum_{i=0}^{5} F_2(X_{[2i+12:2i+13]}) + \sum_{i=0}^{11} F_1(X_{24+i}) \qquad (2)$$

We used two instances of the Trap problem:

- Trap(1), which coincides exactly with the previous description. And,
- Trap(4), which applyies Trap(1) to a chromosome with four groups of 36 genes. Each group is evaluated with Trap(1), and the overall fitness of the chromosomes is the sum of the fitnesses of each group.

A.4 Max-Sat Problem

The satisfiability problem in propositional logic (SAT) [42] is the task of deciding whether a given propositional formula has a model. More formally, given a set of m clauses $\{C_1, \ldots, C_m\}$ involving n Boolean variables $X_1, \ldots, X_n$ the SAT problem is to decide whether an assignment of values to variables exists such that all clauses are simultaneously satisfied.

Max-Sat is the optimisation variant of SAT and can be seen as a generalisation of the SAT problem: given a propositional formula in conjunctive normal form (CNF), the Max-Sat problem then is to find a variable assignment that maximises the number of satisfied clauses. It returns the percentage of satisfied clauses.

We used two sets of instances of the Max-Sat problem with 100 variables, three variables by clause, and 1200 and 2400 clauses, respectively. They were obtained using

Table 4 Fitness values of the subfunctions F_i of length i; the columns represent the number of bits in the subfunction that are equal to one

	0	1	2	3
F_3	4	2	0	10
F_2	5	0	10	
F_1	0	10		

the random generator in [43] ([5]). They are denoted as M-Sat(n, m, l, $seed$), where l indicates the number of variables involved in each clause, and $seed$ is a parameter needed to randomly generate the Max-Sat instance. Each execution of each algorithm used a different $seed$, i.e. the ith execution of every algorithm used the same $seed_i$, whereas the jth execution used $seed_j$.

A.5 NK-Landscapes

In the NK model [28], N represents the number of genes in a haploid chromosome and K represents the number of linkages each gene has to other genes in the same chromosome. To compute the fitness of the entire chromosome, the fitness contribution from each locus is averaged as follows:

$$f(X) = \frac{1}{N} \sum_{i=1}^{N} f(locus_i) \tag{3}$$

where the fitness contribution of each locus, $f(locus_i)$, is determined by using the (binary) value of gene i together with values of the K interacting genes as an index into a table T_i of size 2^{K+1} of randomly generated numbers uniformly distributed over the interval $[0, 1]$. For a given gene i, the set of K linked genes may be randomly selected or consists of the immediately adjacent genes.

We used two sets of instances of the NK-Landscape problem: one with $N = 48$ and $K = 4$, and another with $N = 48$ and $K = 12$. They are denoted as NKLand (N, K, $seed$), where $seed$ is a parameter needed to randomly generate the NK-Landscape instance. They were obtained using the code offered in [39] ([5]). Each execution of each algorithm used a different $seed$, i.e. the ith executions of all the algorithms used the same $seed_i$, whereas the jth executions used $seed_j$.

A.6 Max-Cut Problem

The Max-Cut problem [25] is defined as follows: let an undirected and connected graph $G = (V, E)$, where $V = \{1, 2, \ldots, n\}$ and $E \subset \{(i, j) : 1 \le i < j \le n\}$, be given. Let the edge weights $w_{ij} = w_{ji}$ be given such that $w_{ij} = 0 \ \forall (i, j) \notin E$, and in particular, let $w_{ii} = 0$. The Max-Cut problem is to find a bipartition (V_1, V_2) of V so that the sum of the weights of the edges between V_1 and V_2 is maximised.

We used five instances of the Max-Cut problem (G11, G12, G17, G18, G43), obtained by means of the code in [50] ([20]).

A.7 Unconstrained Binary Quadratic Programming Problem

The objective of the Unconstrained Binary Quadratic Programming (BQP) [1, 18] is to find, given a symmetric rational $n \times n$ matrix $Q = (Q_{ij})$, a binary vector of length n that maximises the following quantity:

$$f(X) = X^t Q X = \sum_{i=1}^{n} \sum_{j=1}^{n} q_{ij} X_i X_j, \quad X_i \in \{0, 1\} \tag{4}$$

We used five instances with different values for n. They were taken from the OR-Library [2]. They are the first instances of the BQP problems in the files 'bqpgka', 'bqp50', 'bqp100', 'bqp250', 'bqp500'. They are called BQP('gka'), BQP(50), BQP(100), BQP(250), and BQP(500), respectively.

References

1. Beasley JE (1998) Heuristic algorithms for the unconstrained binary quadratic programming problem. Tech. Rep., Management School, Imperial College, London, UK.
2. Beasley JE (1990) The Journal of the Operational Research Society 41(11):1069–1072. (http://people.brunel.ac.uk/ mastjjb/jeb/info.html)
3. Blum C, Roli A (2003) ACM Computing Surveys 35(3):268–308
4. Boese KD, Muddu S (1994) Operations Research Letters 16:101–113
5. De Jong K, Potter MA, Spears WM (1997) Using problem generators to explore the effects of epistasis. In: Bäck T (ed) Proc. of the Seventh International Conference on Genetic Algorithms. Morgan Kaufmann
6. Dietzfelbinger M, Naudts B, Van Hoyweghen C, Wegener I (2003)IEEE Transactions on Evolutionary Computation 7(5):417–423
7. Elliott L, Ingham DB, Kyne AG, Mera NS, Pourkashanian M, Wilson CW (2004) An informed operator based genetic algorithm for tuning the reaction rate parameters of chemical kinetics mechanisms. In: Deb K, Poli R, Banzhaf W, Beyer H-G, Burke EK, Darwen PJ, Dasgupta D, Floreano D, Foster JA, Harman M, Holland O, Lanzi PL, Spector L, Tettamanzi A, Thierens D, Tyrrell AM (eds) Proc. of the Genetic and Evolutionary Computation Conference, LNCS 3103. Springer, Berlin Heidelberg
8. Feo T, Resende M (1995) Journal of Global Optimization 6:109–133.
9. Fernandes C, Rosa A (2001) A study on non-random mating and varying population size in genetic algorithms using a royal road function. Proc. of the 2001 Congress on Evolutionary Computation, IEEE Press, Piscataway, New Jersey
10. Festa P, Pardalos PM, Resende MGC, Ribeiro CC (2002) Optimization Methods and Software 17(6):1033–1058
11. Fischer I, Gruber G, Rendl F, Sotirov R (2006) Mathematical Programming 105(2–3): 451–469
12. García-Martínez C, Lozano M, Molina D (2006) A Local Genetic Algorithm for Binary-coded Problems. In: Runarsson TP, Beyer H-G, Burke E, Merelo-Guervós JJ, Whitley LD, Yao X (eds) 9th International Conference on Parallel Problem Solving from Nature, LNCS 4193. Springer, Berlin Heidelberg
13. García-Martínez C, Lozano M, Herrera F, Molina D, Sánchez AM (2007) Global and local real-coded genetic algorithms based on parent-centric crossover operators. European Journal of Operational Research. In Press, Corrected Proof, Available online 18 October 2006
14. Gendreau M, Potvin J-Y (2005) Annals of Operations Research 140(1):189–213
15. Glover F, Laguna M (1999) Operational Research Society Journal 50(1):106–107
16. Goldberg DE, Korb B, Deb K (1989) Complex Systems 3:493–530
17. Goldberg DE (1989) Genetic Algorithms in Search, Optimization, and Machine Learning. Addison-Wesley, Reading, MA.
18. Gulati VP, Gupta SK, Mittal AK (1984) European Journal of Operational Research 15:121–125

19. Harik G (1995) Finding multimodal solutions using restricted tournament selection. In: Eshelman LJ (ed) Proc. of the 6th International Conference on Genetic Algorithms. Morgan Kaufmann, San Mateo, California
20. Helmberg C, Rendl F (2000) Siam Journal of Optimization 10(3):673–696
21. Herrera F, Lozano M, Verdegay JL (1998) Artificial Intelligence Revue 12(4):265–319
22. Herrera F, Lozano M (2000) IEEE Trans. on Evolutionary Computation 4(1):43–63
23. Herrera F, Lozano M, Sánchez AM (2003) International Journal of Intelligent Systems 18(3):309–338
24. Holland JH (1992) Adaptation in Natural and Artificial Systems. The MIT Press Cambridge, MA, USA
25. Karp RM (1972) Reducibility among combinatorial problems. In: Miller R, Thatcher J (eds), Complexity of Computer Computations. Plenum Press, New York
26. Katayama K, Tani M, Narihisa H (2000) Solving large binary quadratic programming problems by effective genetic local search algorithm. In: Whitley D, Goldberg D, Cantu-Paz E, Spector L, Parmee I, Beyer H-G (eds) Proc. of the 2000 Genetic and Evolutionary Computation Conference. Morgan Kaufmann
27. Katayama K, Narihisa H (2001) Trans. IEICE (A) J84-A(3):430–435
28. Kauffman SA (1989) Lectures in the Sciences of Complexity 1:527–618
29. Kazarlis SA, Papadakis SE, Theocharis JB, Petridis V (2001) IEEE Transactions on Evolutionary Computation 5(3):204–217
30. Lourenço HR, Martin O, Stützle T (2002) Iterated local search. In: Glover F, Kochenberger G (eds) Handbook of Metaheuristics. Kluwer Academic, Boston, MA, USA
31. Lozano M, Herrera F, Krasnogor N, Molina D (2004) Evolutionary Computation Journal 12(3):273–302
32. Meloni C, Naso D, Turchiano B (2003) Multi-objective evolutionary algorithms for a class of sequencing problems in manufacturing environments. Proc. of the IEEE International Conference on Systems, Man and Cybernetics 1
33. Merz P (2002) Nk-fitness landscapes and memetic algorithms with greedy operators and k-opt local search. In: Krasnogor N (ed) Proc. of the Third International Workshop on Memetic Algorithms (WOMA III)
34. Merz P, Katayama K (2004) Bio Systems 79(1–3):99–118
35. Mladenovic N, Hansen P (1997) Computers in Operations Research 24:1097–1100
36. Moscato P (1999) Memetic algorithms: a short introduction. In: Corne D, Dorigo M, Glover F (eds), New Ideas in Optimization. McGraw-Hill, London
37. Nasimul N, Hitoshi I (2005) Enhancing differential evolution performance with local search for high dimensional function optimization. In: Beyer HG, O'Reilly UM, Arnold DV, Banzhaf W, Blum C, Bonabeau EW, Cantu-Paz E, Dasgupta D, Deb K, Foster JA, De Jong ED, Lipson H, Llora X, Mancoridis S, Pelikan M, Raidl GR, Soule T, Tyrrell AM, Watson J-P, Zitzler E (eds) Proc. of the Genetic and Evolutionary Computation Conference. ACM Press, New York
38. Papadakis SE, Theocharis JB (2002) Fuzzy Sets and Systems 131(2):121–152
39. Potter MA. http://www.cs.uwyo.edu/~wspears/nk.c
40. Potts JC, Giddens TD, Yadav SB (1994) IEEE Transactions on Systems, Man, and Cybernetics 24:73–86
41. Resende MGC, Ribeiro CC (2003) International Series in Operations Research and Management Science 57:219–250
42. Smith K, Hoos HH, Stützle T (2003) Iterated robust tabu search for MAX-SAT. In: Carbonell JG, Siekmann J (eds) Proc. of the 16th conference on the Canadian Society for Computational Studies of Intelligence, LNCS 2671. Springer, Berlin Heidelberg

43. Spears WM. http://www.cs.uwyo.edu/~wspears/epist.html
44. Sywerda G (1989) Uniform crossover in genetic algorithms. In: Schaffer JD (ed) Proc. of the Third International Conference on Genetic Algorithms, Morgan Kaufmann, San Francisco, CA, USA
45. Tanese R (1987) Parallel genetic algorithms for a hypercube. In: Grefenstette JJ (ed) Proc. of the Second International Conference on Genetic Algorithms Applications. Hillsdale, NJ, Lawrence Erlbraum
46. Thierens D (2004) Population-based iterated local search: restricting neighborhood search by crossover. In: Deb K, Poli R, Banzhaf W, Beyer H-G, Burke EK, Darwen PJ, Dasgupta D, Floreano D, Foster JA, Harman M, Holland O, Lanzi PL, Spector L, Tettamanzi A, Thierens D, Tyrrell AM (eds) Proc. of the Genetic and Evolutionary Computation Conference, LNCS 3103. Springer, Berlin Heidelberg
47. Tsutsui S, Ghosh A, Corne D, Fujimoto Y (1997) A real coded genetic algorithm with an explorer and an exploiter population. In: Bäck T (ed) Proc. of the Seventh International Conference on Genetic Algorithms. Morgan Kaufmann Publishers, San Francisco
48. Weicai Z, Jing L, Mingzhi X, Licheng J (2004) IEEE Transactions on Systems, Man, and Cybernetics - Part B: Cybernetics 34(2):1128–1141
49. Whitley D (1989) The GENITOR algorithm and selection pressure: why rank-based allocation of reproductive trials is best. In: Schaffer JD (ed) Proc. of the Third International Conference on Genetic Algorithms, Morgan Kaufmann, San Francisco, CA, USA
50. Ye Y. http://www.stanford.edu/ yyye/yyye/Gset/

Designing Efficient Evolutionary Algorithms for Cluster Optimization: A Study on Locality

Francisco B. Pereira[1,3], Jorge M.C. Marques[2], Tiago Leitão[3], and Jorge Tavares[3]

[1] Instituto Superior de Engenharia de Coimbra, Quinta da Nora, 3030-199 Coimbra, Portugal.
xico@dei.uc.pt

[2] Departamento de Química, Universidade de Coimbra, 3004-535 Coimbra, Portugal.
qtmarque@ci.uc.pt

[3] Centro de Informática e Sistemas da Universidade de Coimbra, 3030 Coimbra, Portugal.
tleitao@dei.uc.pt, jast@dei.uc.pt

Abstract

Cluster geometry optimization is an important problem from the Chemistry area. Hybrid approaches combining evolutionary algorithms and gradient-driven local search methods are one of the most efficient techniques to perform a meaningful exploration of the solution space to ensure the discovery of low energy geometries. Here we perform a comprehensive study on the locality properties of this approach to gain insight to the algorithm's strengths and weaknesses. The analysis is accomplished through the application of several static measures to randomly generated solutions in order to establish the main properties of an extended set of mutation and crossover operators. Locality analysis is complemented with additional results obtained from optimization runs. The combination of the outcomes allows us to propose a robust hybrid algorithm that is able to quickly discover the arrangement of the cluster's particles that correspond to optimal or near-optimal solutions.

Key words: Cluster Geometry Optimization, Hybrid Evolutionary Algorithms, Locality, Potential Energy

1 Introduction

A cluster is an aggregate of between a few and millions of atoms or molecules, which may present distinct physical properties from those of a single molecule or bulk matter. The interactions among those atoms (or molecules) may be described by a multi-dimensional function, designated as Potential Energy Surface (PES), whose knowledge is mandatory in the theoretical study of the properties of a given chemical system. The arrangement of particles corresponding to the lowest potential energy (i.e., the global minimum on the PES) is an important piece of information, needed to understand the properties of real clusters. Usually, for systems with many particles (such as clusters), the PES is approximately written in an analytical form as a sum of all pair-potentials (i.e., functions that depend on the distance connecting each pair of atoms or molecules). Due to their simplicity, both Lennard-Jones [1,2] and Morse [3] potentials are among the most widely applied as pair-wise models in the study of clusters. In particular, Morse functions may be used to describe either long-range interactions, such as in the alkali metal clusters, or the short-range potentials arising between, for example, C_{60} molecules. From the point of view of global optimization, Morse clusters (especially the short-range ones) are considered to be more challenging than those described by the Lennard-Jones potential [4,5]. Indeed, short ranged Morse clusters tend to present a rough energy landscape due to the great number of local minima and their PESs are more likely to have a multiple-funnel topography [4].

Since the early 1990s Evolutionary Algorithms (EAs) have been increasingly applied to global optimization problems from the Chemistry/Biochemistry area. Cluster geometry optimization is a particular example of one of these problems [5–12]. Nearly all the existing approaches rely on hybrid algorithms combining EAs with local search methods that use first-order derivative information to guide the search into the nearest local optimum. State of the art EAs adopt real-valued representations codifying the Cartesian coordinates of the atoms that compose the cluster [12]. The performance of evolutionary methods can be dramatically increased if local optimization is used to improve each individual that is generated. Hybrid approaches for this problem were first proposed by Deaven and Ho [6] and, since then, have been used in nearly all cluster optimization situations. Typically, local methods perform a gradient-driven local minimization of the cluster potential, allowing the hybrid algorithm to efficiently discover the nearest local optimum.

Locality is an essential requirement to ensure the efficiency of search and has been widely studied by the evolutionary computation community [13–18]. Locality indicates that small variations in the genotype space imply small variations in the phenotype one. A locally strong search algorithm is able to efficiently explore the neighborhood of the current solutions. When this condition is not satisfied, the exploration performed by the EA is inefficient and tends to resemble random search.

The goal of our analysis is to perform an empirical study on the locality properties of the hybrid algorithm that is usually adopted for cluster optimization. The analysis adopts the framework proposed by Raidl and Gottlieb [15]. In this model a set of inexpensive static measures is used to characterize the interactions between representation and genetic operators and assess how they influence the interplay among the genotype/phenotype space. We extend this framework to deal with an optimization situation where the joint efforts of an EA and a gradient driven local method are combined during exploration of the search space. The study considers a broad set of genetic operators, suitable for a real valued representation. Furthermore, two distance measures are defined and used: fitness based and structural distance.

Mutation is the most frequent operator considered in locality studies. In a previous paper we presented a detailed analysis concerning its properties when applied in this evolutionary framework [19]. Here, we briefly review the main conclusions and extend the work to consider crossover. We believe that to obtain a complete characterization of the hybrid EA search competence, crossover must also be taken into account. Regarding this operator, locality should measure its ability to generate descendants by preserving and combining useful features of both parents.

Results allow us to gain insight about the degree of locality induced by genetic operators. With regard to mutation, results establish a clear hierarchy in the locality strength of different types of mutation. As for crossover, the analysis shows that one of the operators is able to maintain/promote diversity, even if similar individuals compose the population. The other two crossover operators considered in this study require mutation to maintain diversity. To the best of our knowledge, this is the first time that a comprehensive locality analysis has been used to study hybrid algorithms for cluster geometry optimization. Results help to provide a better understanding of the role played by each one of the components of the algorithm, which may be important for future applications of EAs to similar problems from the Chemistry area. For the sake of completeness, the empirical locality study is complemented with additional results obtained from real optimization experiments. The outcomes confirm the main conclusions of the static analysis.

The structure of the chapter is the following: in Section 2 we briefly describe Morse clusters. In Section 3 we present the main components of the hybrid algorithm used in the experiments. A brief report of some optimization results is presented in Section 4. Section 5 comprises a detailed analysis on the locality properties of the algorithm. In Section 6 we present the outcomes of the optimization of a large cluster to confirm the results of the locality analysis and, finally, Section 7 gathers the main conclusions.

2 Morse Clusters

Morse clusters are considered a benchmark for testing the performance of new methods for cluster structure optimization. The energy of such a cluster is represented by the N-particle pair-wise additive potential [3] defined as

$$V_{\text{Morse}} = \epsilon \sum_{i}^{N-1} \sum_{j>i}^{N} \left\{ \exp\left[-2\beta(r_{ij} - r_0)\right] - 2\exp\left[-\beta(r_{ij} - r_0)\right] \right\} \qquad (1)$$

where the variable r_{ij} is the distance between atoms i and j in the cluster structure. The bond dissociation energy ϵ, the equilibrium bond length r_0 and the range exponent of the potential β are parameters defined for each individual pair-wise Morse interaction. Usually, these are assumed to be constant for all interactions in a cluster formed by only one type of atom. The potential of (1) is a scaled version [20] of the Morse function with non-atom-specific interactions, where ϵ and r_0 have both been set to 1 and β has been fixed at 14, which corresponds to a short-range interaction. Global optimization is particularly challenging for short-range Morse clusters, since they have many local minima and a "noisy" PES [4]. This simplified potential has already been studied by other authors [5, 9, 11, 20], and the minima are well established for many values of N [21].

The application of local minimization methods, as described below, requires the specification of the analytical gradient of the function to be optimized. In Cartesian coordinates, the generic element n of the gradient of the Morse cluster potential may be given by

$$g_n = -2\beta\epsilon \sum_{i\neq n}^{N} \left(\frac{x_{ni}}{r_{ni}}\right) \{\exp[-2\beta(r_{ni} - r_0)] - \exp[-\beta(r_{ni} - r_0)]\} \qquad (2)$$

where $x_{ni} = x_n - x_i$. Similar expressions apply for the y and z directions.

3 EAs for Morse Clusters Optimization

EAs have been used since 1993 for cluster geometry optimization. A comprehensive review of these efforts, including an outline of state-of-the art applications, can be found in [8]. Regarding the application of EAs to Morse clusters, the most important works are from Johnston and collaborators [5, 11]. In our analysis we adopt an experimental model similar to the one used by these researchers. Its main components have been proposed and evaluated by different teams [5–7, 12].

3.1 Chromosome Representation and Evaluation

An individual must specify the location of the atoms that compose the cluster. For aggregates with N atoms, a solution is composed of $3 \times N$ real values specifying the Cartesian coordinates of each of the particles. The scheme presented in Fig. 1 illustrates the chromosome format. Zeiri proposed this representation in 1995 [12] and, since then, it has become the most widely used in this context [5, 7]. The coordinate values range between 0 and λ. We set λ to $N^{1/3}$, as this interval ensures that the aggregate volume scales correctly with N [11].

There is another parameter δ that specifies the minimum distance that must exist between atoms belonging to the same cluster. It is useful to prevent the generation of aggregates with particles that are excessively close to each other. This avoids possible numerical problems (if two particles are too close, then the pair-wise potential will tend to infinity) and reduces the size of the search space. This parameter is used in the generation of the initial population and during the application of genetic operators.

To assign fitness to an individual we just have to calculate its potential energy using (1).

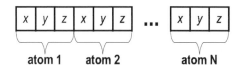

Fig. 1 Structure of a chromosome

3.2 Population Model and Genetic Operators

A generational model is adopted and the standard set of variation operators is used. For crossover, three different operators are analyzed in this research study: uniform, cut and splice and generalized cut and splice crossover. The purpose of all of them is to exchange sub-clusters between parents when generating descendants. In this context, a sub-cluster is defined as a subset of the atoms that compose the cluster. In uniform crossover, the atoms that will compose the offspring are randomly selected from those of the parents. More specifically, the parent chromosomes are scanned from left to right (atoms 1 through N) and, in each position, the child inherits the atom from one of the parents with equal probability. When combining particles from two parents to create a descendant, uniform crossover does not consider the spatial distribution of the atoms. It just cares about the ordering of atoms in the chromosome, which is not related to their positioning in 3D space.

Cut and splice crossover (C&S crossover), proposed by Deaven and Ho in 1995 [6], was specifically designed for cluster geometry optimization. Unlike generic operators, such as uniform crossover, C&S is sensitive to the semantic properties of the structures being manipulated and therefore it is able to arrange a more suitable combination of the parents' features. Since its proposal, it has become widely used and several authors confirm that it enhances the performance of the algorithm [8, 11]. When generating two descendants $D1$ and $D2$ from parents $P1$ and $P2$, C&S determines the sub-clusters to be exchanged in the following way:

1. Apply random rotations to $P1$ and $P2$.
2. Define a random horizontal cutting plane (parallel to the xy plane) for $P1$. This plane splits $P1$ in two complementary parts (X atoms below the plane and $N - X$ atoms above it);
3. Define a horizontal cutting plane (parallel to the xy plane) for $P2$, in such a way that X atoms stay below the plane and $N - X$ are above it.
4. Generate $D1$ and $D2$ by combining complementary parts of each one of the parents.

Special precautions are taken when merging sections from different parents to ensure that the distance between two atoms is never smaller than δ.

Unlike uniform crossover, C&S ensures that the contribution of each parent is formed by a set of atoms that are close together (they are above or below a randomly determined plane). The sub-clusters will tend to have low-energy[1], therefore increasing the likelihood of combining useful building blocks that compose good quality solutions. In addition to these two existing operators, here we propose and analyze a generalization of C&S. This will help us to perform a more detailed study concerning the locality properties of crossover operators used in evolutionary cluster optimization. The new operator, which we will identify as generalized cut and splice (GenC&S), acts in a way that resembles standard C&S crossover. The most relevant difference is related to the way it determines the sub-clusters to be exchanged. With

---

[1] Potential energy is directly related to the distance between pairs of atoms.

GenC&S, subsets of atoms that are close together in the parent clusters will form the building blocks used to create descendants. The constraint that exists in the original cut and splice operator (atoms above/below the plane) is removed and Euclidian distance is the only criterion used to select atoms.

More specifically, GenC&S creates a descendant $D1$ from parents $P1$ and $P2$ in the following way (the other descendant $D2$ is created swapping the role played by the parents):

1. Select a random atom CP from $P1$.
2. Select a random number $X \in [1, N - 2]$, where N is the number of atoms that compose the cluster.
3. From $P1$, copy CP and the X atoms closer to it, to $D1$.
4. Select $N - (X + 1)$ atoms from $P2$ to complete $D1$. Give preference to atoms that, in the 3D space, are closer to the original location of CP. Skip atoms that are too close (i.e., at a distance smaller than δ) to particles already belonging to $D1$.

Depending on the atom distribution, in a small number of situations it might be impossible to select enough particles from $P2$ to complete $D1$. This can happen because too many atoms are skipped due to the distance constraint. If this situation occurs, $D1$ is completed with atoms randomly placed.

There is another difference between C&S and GenC&S: in the second operator no random rotations are applied to the parents before the genetic material is mixed. This action is not necessary in the generalized version because we removed the constraint that forces the cutting plane defining the sub-clusters to be parallel to the xy plane and therefore there is no bias associated with this operator.

Two mutation operators were tested in this work: Sigma mutation and Flip mutation. We consider that mutation is performed on atoms, i.e., when applied it modifies the value of the three coordinates that specify the position of a particle in 3D space. The first operator is an evolutionary strategy (ES) like mutation and acts in the following way: when undergoing mutation, the new value v_{new} for each one of the three coordinates of an atom (x, y, z) is obtained from the old value v_{old} through the expression:

$$v_{\text{new}} = v_{\text{old}} + \sigma N(0, 1) \qquad (3)$$

where $N(0, 1)$ represents a random value sampled from a standard Normal distribution and σ is a parameter from the algorithm. The new value must be between 0 and λ.

Flip mutation works in the following way: when applied to an atom, it assigns new random values to each one of its coordinates, i.e., it moves this atom to a random location.

3.3 Local Optimization

Local optimization is performed with the Broyden–Fletcher–Goldfarb–Shanno limited memory quasi-Newton method (L-BFGS) of Liu and Nocedal [22, 23]. L-BFGS is a powerful optimization technique that aims to combine the modest storage and computational requirements of conjugate gradient methods with the superlinear convergence exhibited by full memory quasi-Newton methods (when sufficiently close to a solution, Newton methods are quadratically convergent). In this limited memory algorithm, the function to be minimized and its gradient must be supplied, but knowledge about the corresponding Hessian matrix is not required *a priori*.

L-BFGS is applied to every generated individual. During local search, the maximum number of iterations that can be performed is specified by a parameter of the algorithm, the Local Search Length (LSL). However, L-BFGS stops as soon as it finds a local optimum, so the effective number of iterations can be smaller than the value specified by LSL.

4 Optimization Results

The main goal of the research reported here is to study the locality of different genetic operators. This will be carried out in the next section. Nevertheless, to establish an appropriate background, we first present some experimental results.

Aggregates ranging from 19 to 50 atoms compose the standard instances used when Morse clusters are adopted as a benchmark for assessing the efficiency of evolutionary algorithms. The original research conducted by Johnston et al. [11] revealed that the hybrid algorithm is efficient and reliable, as it was able to find nearly all known best solutions. The only exception was the cluster with 30 atoms, where the current putative optimum was only reported in a subsequent paper by the same authors [5]. The algorithm used in the experiments relied on C&S crossover and flip mutation as genetic operators. Other details concerning the optimization can be found in [5,11].

In 2006, we developed a hybrid algorithm to be used in locality analysis. In order to confirm its search competence, we repeated the experiments of searching for the optimal geometry of Morse clusters ranging from 19 to 50 atoms. C&S crossover and the two mutation operators previously described were used in the tests. Results obtained confirmed the efficiency of the hybrid approach, as it was able to find all known best solutions. Regardless of this situation, a more detailed analysis of the outcomes revealed that there are some differences in results achieved by different mutation operators. While experiments performed with Sigma mutation achieve good results for all instances, tests done with Flip mutation reveal a less consistent behavior. For small clusters (up to 30 atoms), the results achieved are analogous to those obtained by Sigma mutation. As the clusters grow in size, the algorithm shows signs of poor scalability and its performance starts to deteriorate. For clusters with more than 36 atoms, most times it fails to find the optimum. A detailed description and analysis of the results and a complete specification of the parameter settings used in the experiments may be found in [19].

5 Locality Analysis

The cluster with 50 atoms (the largest instance that was considered in the previous optimization experiments) was selected to perform all the tests concerning the locality properties of the different genetic operators. When appropriate, the empirical analysis is complemented with experimental results.

5.1 Related Work

Many approaches have been proposed to estimate the behavior of EAs when applied to a given problem. Some of these techniques adopt measures that are, to some extent, similar to the locality property adopted in this work. In this section we highlight the most relevant ones.

The concept of fitness landscapes, originally proposed by Wright [24], establishes a connection between solution candidates and their fitness values and it has been widely used to predict the performance of EAs. Several measures for fitness landscapes were defined for this task. Jones and Forrest proposed fitness distance correlation as a way to determine the relation between fitness value and distance to the optimum [25]. If fitness values increase as distance to the optimum decreases, then search is expected to be easy for an EA [26].

An alternative way to analyze the fitness landscape is to determine its ruggedness. Some autocorrelation measures help to determine how rugged a landscape is. Weinberger [27] proposed the adoption of autocorrelation functions to measure the correlation of all points in the search space at a given distance. Another possibility to investigate the correlation structure of a landscape is to perform some random walks. The value obtained with the random walk correlation function can then be used to determine the correlation length, a value that directly reflects the ruggedness of the landscape [24]. In general, smoother landscapes are highly correlated, making the search for an EA easier. More rugged landscapes are harder to explore in a meaningful way.

Sendhoff et al. studied the conditions for strong causality on EAs [18]. A search process is said to be locally strongly causal if small variations in the genotype space imply small variations in the phenotype space. In the above-mentioned work, variations in genotypes are caused by mutation (crossover is not applied). Fitness variation is used to access distances in the phenotype space. A probabilistic causality condition is proposed and studied in two situations: optimization of a continuous mathematical function and optimization of the structure of a neural network. They conclude that strong causality is essential, as it allows for controlled small steps in the phenotype space that are provoked by small steps in the genotype space.

The empirical framework to study locality that we adopt in our research was proposed by Raidl and Gottlieb [15]. The model is useful to study how the adopted representation and the genetic operators are related and how this interplay influences the performance of the search algorithm. The analysis is based on static measures applied to randomly generated individuals that help to quantify the distance between solutions in the search space and how it is linked to the similarity among corresponding phenotypes. This model allows the calculation of three features, which are essential for good performance: locality, heritability and heuristic bias. Locality was already defined in the introduction. Heritability refers to the ability of crossover operators to create children that combine meaningful features of both parents. Heuristic bias concerns the genotype–phenotype mapping function. Some functions that favor the mapping towards phenotypes with higher fitness might help to increase performance. This effect is called heuristic bias. The authors suggest that these properties can be studied either in a static fashion or can be dynamically analyzed during actual optimization runs. They also claim that the results achieved provide a reliable basis for assessing the efficiency of representations and genetic operators. In the above mentioned work, this framework is used to compare different representations for the multidimensional knapsack problem.

5.2 Definitions

When performing studies with an evolutionary framework it is usual to consider two spaces: the genotype space Φ_g and the phenotype space Φ_p [28]. Genetic operators work on Φ_g, whereas the fitness function f is applied to solutions from Φ_p: f: $\Phi_p \to \mathbb{R}$. A direct representation is adopted in this chapter. Since there is no maturation or decoder function, genetic operators are directly applied to phenotypes. Then, it is not necessary to perform an explicit distinction between the two spaces and, from now on, we will refer to individuals or phenotypes to designate points from the search space.

To calculate the similarity between two individuals from Φ_p, a phenotypic distance has to be defined. This measure captures the semantic difference between two solutions and is directly related to the problem being solved. We determine phenotypic distance in the two following ways.

Fitness Based Distance
Determining the fitness distance between two phenotypes A, B is straightforward:

$$d_{fit}(A, B) = |f(A) - f(B)| \tag{4}$$

In cluster optimization, it calculates the difference between the potential energy of the two solutions.

Structural Distance
According to (1), the basic features that influence the quality of an N-atom cluster are the $N \times (N-1)/2$ interactions occurring between particles forming the aggregate. The interaction between atoms i and j depends only on the distance r_{ij} between them. We implement a simple method to approximate the structural shape of a cluster. First, all the $N \times (N - 1)/2$ distances between atoms are calculated. Then, they are separated into several sets according to their values. We consider 10 sets S_i. The limits for each S_i, $i = 1, \ldots, 10$, are defined as follows:

$$\left[\frac{i - 1}{10} \times \mu, \frac{i}{10} \times \mu \right[, i = 1, \ldots, 10 \tag{5}$$

where μ is the maximum distance between two atoms. Considering the parameter λ, μ is equal to $\sqrt{3\lambda^2}$. Structural distance captures the dissimilarity between two clusters

A and B in terms of the distances between all pairs of atoms. It is measured in the following way:

$$d_{\text{struct}}(A, B) = \frac{1}{10} \sum_{i=1}^{10} |\#S_i(A) - \#S_i(B)| \tag{6}$$

where $\#S_i(A)$ (likewise, $\#S_i(B)$) is the cardinality of subset S_i for cluster A (likewise, for cluster B).

5.3 Mutation Innovation

To analyze the effect of mutation on locality we adopt the innovation measure proposed by Raidl and Gottlieb [13]. The distance between the individuals involved in a mutation is used to predict the effect of the application of this operator. Let X be a solution and X^m the result of applying mutation to X. The mutation innovation MI is measured as follows:

$$MI = \text{dist}(X, X^m) \tag{7}$$

Distance can be calculated using either fitness based or structural distance. MI illustrates how much innovation the mutation operator introduces, i.e., it aims to determine how much this operator modifies the semantic properties of an individual. Locality is directly related to this measure. The application of a locally strong operator implies a small modification in the phenotype of an individual (i.e., there will be a small phenotypic distance between the two involved solutions). Conversely, operators with weak locality allow large jumps in the search space, complicating the task of the search algorithm. To determine the MI, 1000 random individuals were generated and then, a sequence of mutations was applied to each one of them. In each one of the 1000 mutation series, distance is measured between the original individual and the solution created after $k \in \{1, 2, 3, 4, 5, 10, 25, 50, 100\}$ successive mutation steps. To conform with the adopted optimization framework, local search is considered as part of this genetic operator, i.e., L-BFGS is applied after each mutation and distance is measured using the solution that results from this operation.

We will study the locality properties of two mutation operators: Sigma and Flip. For the first operator, three values for σ are tested: $\{0.01 \times \lambda, 0.1 \times \lambda, 0.25 \times \lambda\}^2$. We expect that the variation in the value of σ will provide insight to the effect of this parameter on the performance of the algorithm.

Fitness Based Distance

To simplify the analysis, distances between the original solution and the successive mutants are grouped in different sets. Given a d_{fit} fitness distance between two solutions, set $\mathbf{G_i}$ to which d_{fit} is assigned, is determined in the following way:

---

[2] The σ value used in Sigma mutation is proportional to λ to ensure that its effect is comparable for clusters of different size. Nevertheless, in the text we will adopt the simplified notation Sigma0.01 instead of Sigma0.01×λ (alike for other values of σ).

{**G0** : $0 \le d_{\text{fit}} < 1$; **G1** : $1 \le d_{\text{fit}} < 5$; **G2** : $5 \le d_{\text{fit}} < 10$; **G3** : $10 \le d_{\text{fit}} < 20$; **G4** : $20 \le d_{\text{fit}} < 30$; **G5** : $30 \le d_{\text{fit}} < 50$; **G6** : $50 \le d_{\text{fit}} < 100$; **G7** : $100 \le d_{\text{fit}} < 250$; **G8** : $250 \le d_{\text{fit}} < 500$; **G9** : $500 \le d_{\text{fit}}$}.

The specific values selected to determine intervals are arbitrary. The relevant information to obtain here is the distribution of the fitness distances through the sets. Situations where values tend to be assigned to higher order sets (i.e., large variations), suggest that the locality is low. In Figs. 2 and 3 we present, for the number of mutations considerd, the distances between the original solutions and the mutants iteratively generated. The three parts of Fig. 2 show results from experiments performed with Sigma, and Fig. 3 shows results obtained with Flip mutation. In each column (corresponding to a given mutation step), we show the distribution of the 1000 distances for the 10 $\mathbf{G_i}$ sets. It is clear that there are important differences in the locality of mutation. Experiments combining Sigma0.01 with L-BFGS exhibit the highest locality and, even after 100 steps, nearly all fitness distances belong to sets **G0**–**G2** (more than 50% are in cluster **G0**). This shows that there is still a clear relation, maybe even excessive, with the departure point. On the contrary, experiments with Sigma0.25 and Flip mutation evidence lower locality. The modification they induce is substantial (large displacement of one atom strongly modifies the fitness of the solution) and local search might not be able to make an appropriate repair. Sigma0.1 is between these two scenarios. In the beginning it shows signs of reasonable locality, but after some steps it approaches the distribution exhibited by Sigma0.25 and Flip. This is a behavior that is more in accordance with what one should expect from a mutation operator. Individuals that are just a few steps away should have similar phenotypic properties. Moreover, distance should gradually increase, as more mutations are applied to one of them.

Structural Distance

Analysis of MI with structural distance confirms the conclusions from the previous section. In Fig. 4 we show the structural distance for the same operators and settings. Once again, it is clear that Sigma0.01 has high locality, Sigma0.25 and Flip have both low locality and Sigma0.1 is between the two extremes. Results with both phenotypic distances reveal that, as σ increases, the effect on the locality of Sigma mutation tends to approach that of Flip mutation.

5.4 Crossover Innovation

Crossover innovation measures the ability of this operator to create descendants that are different from their parents. Let C be a child resulting from the application of crossover to parents $P1$ and $P2$. Following the definition proposed by Raidl and Gottlieb [15], crossover innovation CI can be measured as follows:

$$CI = \min \{\text{dist}(C, P1), \text{dist}(C, P2)\} \qquad (8)$$

According to (8), CI measures the phenotypic distance between a child and its phenotypically closer parent. In general, we expect CI to be directly related to the distance

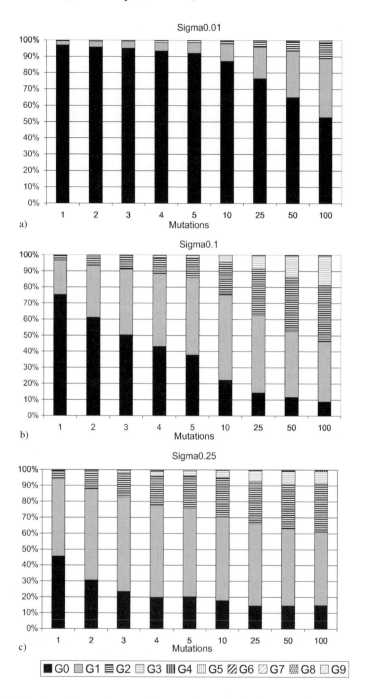

Fig. 2 Distribution of fitness distances between the original solutions and the mutants iteratively generated for Sigma mutation: **a)** Sigma0.01; **b)** Sigma0.1; **c)** Sigma0.25

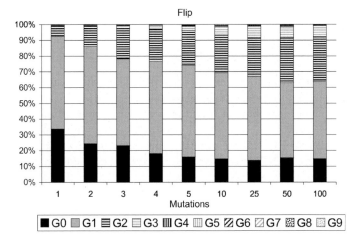

Fig. 3 Distribution of fitness distances between the original solutions and the mutants iteratively generated for Flip mutation

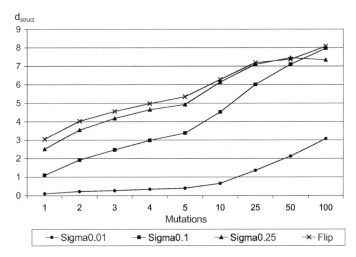

Fig. 4 Average structural distances between the original solution and the mutants iteratively generated

that exists between parents involved in crossover. Similar parents tend to create descendants that are also close to both of them. On the contrary, dissimilar parents tend to originate larger crossover innovations.

Nevertheless, under the same circumstances (i.e., when applied to the same pair of individuals), different crossover operators might induce distinct levels of innovation. This disparity reflects diverse attitudes on how the genetic material is combined. When exploring the search space, it is important to rely on crossover operators that maintain a moderately high value of CI. This will help to preserve population diversity and ensure that an appropriate exploration of the space is performed. Anyway, it is also important that CI is not too high because this might prevent the preservation and combination of useful features that are inherited from the parents. It is a well-known fact that mixing properties from the parents in a meaningful way is one of the most important roles of crossover [29].

The CI of the different crossover operators was empirically analyzed. To study how the parental distance affects this measure, we adopted the following procedure: parent $P1$ was randomly generated and then kept unchanged throughout the experiments, while parent $P2$ was obtained from $P1$ through the application of a sequence of mutations. In the experiments performed, we measured CI after $k \in \{1, 2, 3, 4, 5, 10, 25, 50, 100\}$ successive mutation steps. Local optimization is applied to the original solution and also after each mutation. Sigma0.1 was the operator chosen to generate the sequence of mutated individuals that act as parent $P2$. As can be confirmed from Figs. 2 and 4, at the beginning of the sequence (i.e., when the number of mutations is small), the parents will be similar. When the number of successive mutations increases, difference between parents tends to increase steadily ($P1$ remains unchanged while $P2$ accumulates mutations). To conform with the adopted optimization framework, the child that is generated is locally improved before CI is determined.

CI was used to study the locality properties of three crossover operators: $\{Uniform, C\&S, GenC\&S\}$. The procedure described was repeated 1000 times for each of the operators. Results obtained with the two distance measures will be analyzed separately.

Fitness Based Distance

The 1000 fitness distances were grouped in the same 10 sets **G0–G9** previously described. In Fig. 5 we present, for the three crossover operators, the distances between a child and its phenotypically closer parent. Each column corresponds to a given distance between the two parents: in the first column from the left, parents are just one mutation away, while in the last one they are 100 mutations away. In every one of these columns we show the distribution of the 1000 distances for the 10 G_i sets.

It is clear from the figures that the combined application of crossover and subsequent local search establishes a process with fairly high locality. Even with parents that are 100 mutations away, the child maintains a clear relation, in terms of the potential energy of the cluster, with at least one of its parents. Nearly all fitness distances lie in sets **G0**, **G1** and **G2**, meaning that the variation in potential energy does not exceed 10.

Another relevant outcome is the noteworthy difference between the results achieved with C&S and the results obtained with the other two crossover operators. C&S seems to be insensitive to the difference that exists between parents. The distribution of the fitness distances is similar, whether the parents are almost identical or have large dissimilarities. This result suggests that the diversity level of the population is irrelevant when C&S is applied. The justification for the results might be related to the way C&S acts. Before cutting parents, it applies a random rotation to each one of them. As the rotations are independent, even if the two parents are identical the descendants might be distinct. This is an unusual behavior because most

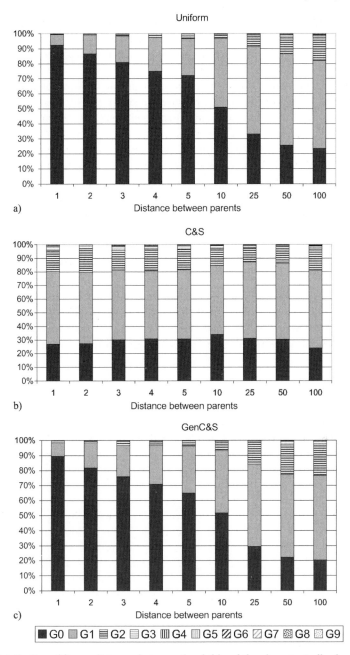

Fig. 5 Distribution of fitness distances between the child and the phenotypically closer parent:
a) Uniform; **b)** C&S; **c)** GenC&S

crossover operators are unable to introduce any novelty into the population when
they are applied to an identical pair of solutions. On the contrary, C&S is capable of

adding diversity to the population. As a consequence, existing diversity might not be as relevant as it is in other situations. Finally, this outcome also suggests that mutation might be less important in experiments with C&S than in experiments that rely on other crossover operators.

Figures displaying the *CI* distribution of uniform and GenC&S operators present a pattern that is more in compliance with the expected behavior of crossover. When parents are similar, the innovation is small. As the distance between parents increases, average *CI* also increases.

Structural Distance

Results achieved with structural distance are presented in the chart from Fig. 6. They are in agreement with the information provided by fitness based distance. While C&S crossover is insensitive to the distance that exists between parents, both uniform and GenC&S tend to generate more innovative children as this distance increases.

5.5 Additional Tests

To complement our study, and to determine if the empirical study is confirmed by experimental results, we performed an extended set of optimization experiments using the same Morse instance with 50 atoms that was selected for the locality analysis. The study focused on the behavior of the search algorithm when using different genetic operators. The settings are the following: Evaluations: 3,000,000; Population Size: 100; Elitist Strategy; Tournament Selection with tourney size 10; Crossover operators: {Uniform, C&S, GenC&S}; Crossover rate: 0.7; Mutation operators: {Sigma, Flip}; $\sigma = \{0.01 \times \lambda, 0.1 \times \lambda, 0.25 \times \lambda\}$; Mutation rate: {0.0, 0.01, 0.05, 0.1, 0.2, 0.3}; LSL: 200; δ: 0.5.

Each iteration performed by L-BFGS counts as one evaluation. The initial population is randomly generated and for every set of parameters we performed 30 runs.

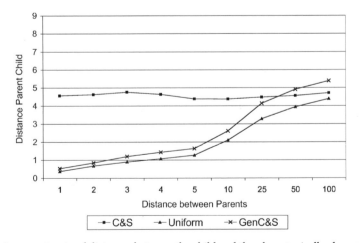

Fig. 6 Average structural distances between the child and the phenotypically closer parent

When appropriate, statistical significance of the results is accessed with a t-test (level of significance 0.01).

In Tables 1 to 3 (respectively for uniform, C&S and GenC&S), we present an overview of the results achieved. For each one of the settings we present the average of the best fitness over the 30 runs (MBF). In brackets we also present the Gap, defined as the distance between the MBF and the putative optimum value for the potential energy of a Morse cluster with 50 atoms (gap value expressed in percentage).

If we combine the information obtained with the locality analysis and the optimization results it is possible to infer some conclusions concerning the behavior

Table 1 Optimization results of the 50-atom Morse cluster using uniform crossover

		Mutation rate			
Mutation	0.01	0.05	0.1	0.2	0.3
Sigma 0.01	−188.247	−189.247	−190.449	−191.197	−191.894
	(5.1)	(4.6)	(4.0)	(3.7)	(3.3)
Sigma 0.1	−192.831	−193.831	−194.086	−193.561	−194.348
	(2.8)	(2.3)	(2.2)	(2.5)	(2.1)
Sigma 0.25	−194.433	−194.745	−193.548	−187.483	−183.937
	(2.0)	(1.9)	(2.5)	(5.5)	(7.3)
Without mutation	−182.427 (8.1)				

Table 2 Optimization results of the 50-atom Morse cluster using C&S crossover

		Mutation rate			
Mutation	0.01	0.05	0.1	0.2	0.3
Sigma 0.01	**-195.656**	**-195.066**	**-195.246**	**-195.223**	**-195.547**
	(1.4)	(1.7)	(1.6)	(1.6)	(1.5)
Sigma 0.1	−194.505	**-195.329**	**-194.971**	−194.132	−194.455
	(2.0)	(1.6)	(1.8)	(2.2)	(2.0)
Sigma 0.25	−193.995	−193.464	−191.784	−187.481	−183.347
	(2.2)	(2.5)	(3.4)	(5.5)	(7.6)
Without mutation	**-194.816** (1.8)				

Table 3 Optimization results of the 50-atom Morse cluster using GenC&S crossover

		Mutation rate			
Mutation	0.01	0.05	0.1	0.2	0.3
Sigma 0.01	−193.708	−194.313	−193.967	−194.054	−194.760
	(2.4)	(2.1)	(2.3)	(2.2)	(1.9)
Sigma 0.1	**-194.626**	−194.789	−194.634	**-195.225**	**-196.177**
	(1.9)	(1.8)	(1.9)	(1.6)	(1.2)
Sigma 0.25	**-195.254**	**-195.272**	**-195.275**	−190.498	**-186.093**
	(1.6)	(1.6)	(1.6)	(4.0)	(6.2)
Without mutation	−192.913 (2.8)				

of the search algorithm. Experiments performed with uniform crossover obtain results of inferior quality to those achieved by the other two operators. This is true for all settings adopted during the tests. The analysis presented in the previous section showed that the locality properties of uniform crossover are comparable to those of GenC&S. In contrast, optimization results suggest that knowing the locality of an operator is not sufficient to predict its efficiency. The difference in performance between uniform crossover and the other two operators shows that, when solving difficult optimization problems, it is essential to rely on specific operators to explore the search space. Specific operators are sensitive to the structure of individuals being manipulated and, therefore, increase the probability of exchanging genetic material in a meaningful way.

Results in bold in Tables 1 to 3 identify the best crossover operator for each specific setting. There is a clear separation between the settings where C&S had the best performance and settings where GenC&S was better. When Sigma0.01 mutation is used, experiments performed with C&S always achieve the best results. On the contrary, when Sigma0.25 is adopted, experiments performed with the new crossover operator always exhibit the best performance. When Sigma0.1 is used, differences in performance between these two operators are small (with the exception of the experiments performed with a mutation rate of 0.3). Table 4 reviews the statistical analysis performed. The symbol '$\star$' identifies settings where there are significant differences between the results achieved by experiments performed with C&S and GenC&S. Results show that they occur in two situations:

(i) They are visible when the effect of mutation is almost irrelevant. This happens in the test performed without mutation and also in the experiment using Sigma0.01 with rate 0.01. In these scenarios C&S crossover is clearly more efficient than GenC&S. Locality analysis results help to explain why this happens. C&S does not require distinct parents to generate original descendants. It is therefore able to maintain and promote the diversity level of the population. The addition of a mutation operator with the ability to perform considerable changes in the individuals might lead to too large disruptions in the solutions preventing an efficient exploration of the search space.

(ii) Significant differences also occur when mutation plays a major role in the optimization process. More specifically, differences appear in nearly all experiments performed with Sigma0.25 (the only exception being the situation with a mutation rate of 0.01) and also in the test performed with Sigma0.1 with rate 0.3. In all these situations, GenC&S was more efficient than the original C&S operator. This result is

Table 4 C&S *vs* GenC&S: significant differences (50-atom Morse cluster)

Mutation	Mutation rate				
	0.01	0.05	0.1	0.2	0.3
Sigma 0.01	$\star$				
Sigma 0.1					$\star$
Sigma 0.25		$\star$	$\star$	$\star$	$\star$
Without mutation	$\star$				

also in accordance with the locality analysis. GenC&S is sensitive to the diversity level of the population and therefore it requires different parents to create children with a substantial level of innovation. Just as experimental results show, its performance is enhanced if the mutation operator helps to maintain an appropriate level of diversity. In the experiments that are between these two extremes, optimization results achieved by the two crossover operators can be considered similar.

Results from the tables also confirm the robustness of the hybrid EA. In most cases (particularly in experiments performed with crossover operators that are sensitive to the structures being manipulated), the gap to the putative global optimum is small, ranging between 1.5 and 2.0%.

6 Optimization of a Larger Cluster

We performed a final set of tests on the optimization of a Morse cluster with 80 atoms. Our aim was twofold: first, we wanted to determine if the results achieved in a difficult optimization situation confirm our locality analysis. Also, we intended to collect some statistics during the runs to measure the diversity of the population throughout the optimization. In the previous section, the locality of genetic operators was studied separately. Now, by collecting these statistics on the fly, we expected to gain insight on how the combination of different genetic operators with other algorithmic components influence search dynamics. We will also verify whether these results confirm the static empirical analysis that was performed earlier.

6.1 Optimization Results

The settings of the experiments performed were as follows: Number of runs: 30; Evaluations: 3,000,000; Population size: 100; Elitist strategy; Tournament selection with tourney size 10; Crossover operators: {C&S, GenC&S} with rate 0.7; Sigma0.1 mutation with rate {0.0, 0.01, 0.05, 0.1, 0.2}; LSL: 200; δ: 0.5.

Here, we did not aim to conduct an all-inclusive study. Our goal was just to obtain additional verification whether the locality analysis could find support in optimization results. That is why we maintained the settings selected for the optimization of the 50-atom cluster, even though we were aware that the number of evaluations should be increased to enable an appropriate exploration of the search space. Nevertheless, 3 million evaluations should be enough to provide some hints concerning the search performance of different genetic operators. Also, we selected just a subset of the genetic operators previously considered. We did not perform experiments with uniform crossover. Results achieved on the optimization of the 50-atom cluster show that it is clearly less efficient than the other two crossover operators and so it was not considered in this last step of the research. As for mutation, we selected Sigma0.1, as it proved to be the most balanced operator.

In Table 5 we show an overview of the results achieved. For each one of the selected settings we present the mean best fitness calculated over the 30 runs and the gap to the putative global optimum expressed in percentage (value in brackets). The first row presents results from experiments performed with C&S and the second shows results achieved in tests done with GenC&S. Values in bold highlight settings where the MBF is significantly better than that achieved by the other test performed with the same mutation rate.

A brief overview of the optimization results shows that experiments performed with GenC&S achieved better results than those that used original C&S crossover. The only exception is when mutation is absent. Here, the MFB is better in the experiment performed with the original C&S operator. Actually, results show that this is the best setting for C&S crossover. As soon as mutation is added, the efficiency of the operator decreases. This is true even for small rates, even though the increase of the mutation rate amplifies the effect. This situation was visible in tests performed with the 50-atom cluster, but here it is more evident.

Experiments performed with GenC&S crossover achieve better results when mutation is used. It is also clear that fairly high mutation rates are required to enhance the performance of the algorithm when this type of crossover is used. The only experiment that was able to discover the putative global optimum for this instance was the one that combined GenC&S crossover and Sigma0.1 mutation with a rate of 0.2. Even though the relevance of this finding should be handled with care (just as mentioned at the beginning of this section, the number of evaluations might be too small to allow a proper exploration of the search space), it nevertheless is a pointer to the efficiency of the different operators. Anyway, the gaps that exist between the MFB and the global optimum are more relevant to assess the efficiency of the search algorithm. When using GenC&S crossover and Sigma mutation, gaps range between 1.1 and 1.9%. These low values demonstrate the competence of the optimization method, showing that promising areas of the search space can be discovered even with a limited number of evaluations.

We performed a brief statistical analysis to confirm the validity of our conclusions. Values in bold in Table 5 show that there is a significant difference between the MBF attained by C&S and GenC&S when no mutation is used. When a moderate mutation rate is adopted (0.01 and 0.05), GenC&S outperforms C&S even though differences are not statistically significant. As the mutation rate increases, differences in MBF become more evident. When 0.1 or 0.2 are used, there is a significant difference between the results achieved by the two crossover operators. We also studied whether there are significant differences between experiments that used the same crossover operator and different mutation rates. Tables 6 and 7 summarize these results. Cells

Table 5 Optimization results of the 80-atom Morse cluster

	Mutation rate				
	0.0	0.01	0.05	0.1	0.2
C&S	**-334.318**	−332.984	−332.783	−331.433	−328.197
	(1.9)	(2.3)	(2.4)	(2.7)	(3.7)
GenC&S	−329.520	−334.318	−336.083	**-335.755**	**-337.043**
	(3.3)	(1.9)	(1.4)	(1.5)	(1.1)

marked with '$\star$' identify a situation where a significant difference exists. The results are once again in compliance with our previous analysis. When using C&S crossover (Table 6), all experiments performed with a mutation rate of 0.2 achieve significantly poorer results than experiments performed with other mutation rates. The only exception is when experiments with 0.1 and 0.2 rates are compared (in this situation, the difference is not significant). As for GenC&S (Table 7), significant differences are observed when comparing experiments with and without mutation. All MBFs obtained in tests performed without mutation are significantly worse that those achieved by experiments that rely on mutation to promote diversity.

Graphs showing the optimization of the 80-atom cluster are presented in Fig. 7. It shows the evolution of the best solution (averaged over 30 runs) during the opti-

Table 6 C&S significant differences (80-atom Morse cluster)

	0.0	0.01	0.05	0.1	0.2
	0.6				
0.0	0.6				$\star$
	0.6				
	0.6	0.6			
0.01	0.6	0.6			$\star$
	0.6	0.6			
	0.6	0.6	0.6		
0.05	0.6	0.6	0.6		$\star$
	0.6	0.6	0.6		
	0.6	0.6	0.6	0.6	
0.1	0.6	0.6	0.6	0.6	
	0.6	0.6	0.6	0.6	
	0.6	0.6	0.6	0.6	0.6
0.2	0.6	0.6	0.6	0.6	0.6
	0.6	0.6	0.6	0.6	0.6

Table 7 GenC&S significant differences (80-atom Morse cluster)

	0.0	0.01	0.05	0.1	0.2
	0.6				
0.0	0.6	$\star$	$\star$	$\star$	$\star$
	0.6				
	0.6	0.6			
0.01	0.6	0.6			
	0.6	0.6			
	0.6	0.6	0.6		
0.05	0.6	0.6	0.6		
	0.6	0.6	0.6		
	0.6	0.6	0.6	0.6	
0.1	0.6	0.6	0.6	0.6	
	0.6	0.6	0.6	0.6	
	0.6	0.6	0.6	0.6	0.6
0.2	0.6	0.6	0.6	0.6	0.6
	0.6	0.6	0.6	0.6	0.6

mization for four selected settings: two experiments without mutation and two experiments using both operators (mutation rate 0.2). For the sake of clarity, lines obtained from experiments performed with other mutation rates (0.01, 0.05 and 0.1) are not shown. However, results presented are representative of the behavior of the optimization algorithm. When C&S crossover is used, there is a clear advantage if the mutation operator is not present. After the initial stage, when both lines exhibit a similar pattern, the experiment combining C&S and Sigma mutation starts to stagnate, suggesting that it is unable to identify promising areas of the search space. When GenC&S is used, two completely different patterns emerge. If this operator is used in combination with mutation, there is continuous improvement of the best solution found. If mutation is not considered, then the algorithm quickly converges to a suboptimal solution and is completely unable to escape from it. This is an expected result, given the locality analysis of GenC&S that was presented in Sect. 5.

6.2 Search Dynamics

The optimization results for the 80-atom cluster are in accordance with the locality analysis that was performed earlier. This agreement is another sign suggesting that locality analysis is a useful tool to predict the performance of evolutionary algorithms when solving difficult optimization problems.

Before concluding the chapter, we present a last set of results that hopefully will contribute to a full clarification of the behavior of the algorithm when exploring the coordinate space. Locality analysis is performed in static environments, where a single operator is applied at a time. In a real optimization situation, interactions that exist between the genetic operators and selection might influence the behavior of the algorithm. We now present several measures that illustrate how different configurations of the search algorithm, in terms of the genetic operators used and parameter settings adopted, are able to promote and maintain diversity in a population.

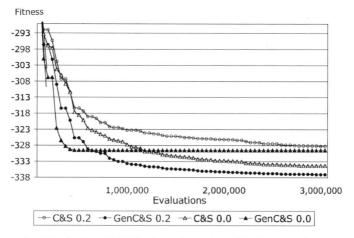

Fig. 7 Evolution of MBF in the optimization of the 80-atom cluster

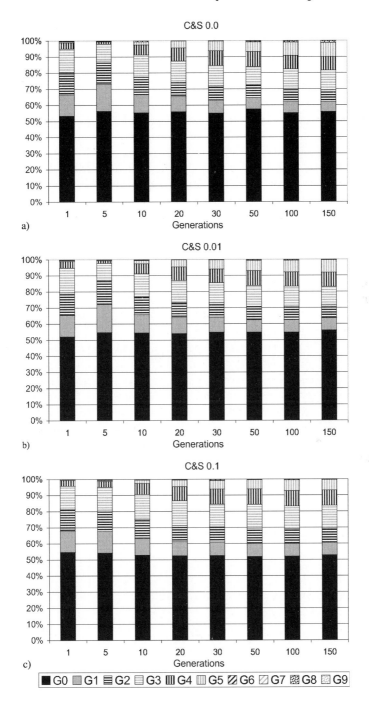

Fig. 8 Distribution of fitness distances in the population throughout the optimization of the 80-atom cluster: **a)** C&S 0.0; **b)** C&S 0.01; **c)** C&S 0.1

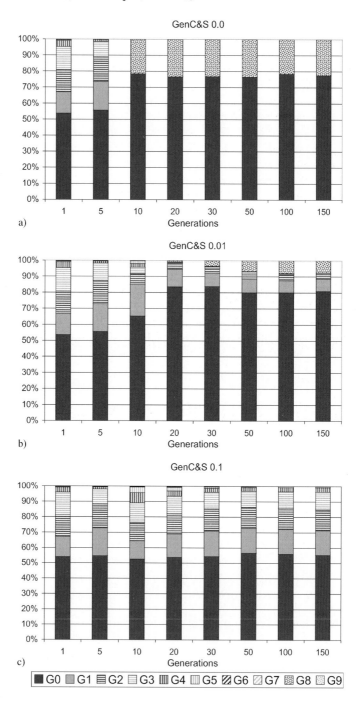

Fig. 9 Distribution of fitness distances in the population throughout the optimization of the 80-atom cluster: **a)** GenC&S 0.0; **b)** GenC&S 0.01; **c)** GenC&S 0.1

When the algorithm was searching for good solutions for the 80-atom cluster, we collected in several predetermined generations, $\{1, 5, 10, 20, 30, 50, 100, 150\}$, the average distance between all pairs of individuals from the population. The two distance measures previously defined were used. In Figs. 8 and 9 we show the distribution of the fitness distances obtained with different settings. The three parts of Fig. 8 display results achieved in experiments performed with C&S crossover, and Fig. 9 presents the outcomes from tests performed with GenC&S. We show results only from optimization experiments performed with the following mutation rates: $\{0, 0.01, 0.1\}$. Experiments performed with other mutation rates follow the same pattern.

Two clear distinct configurations emerge. When C&S crossover is used, the diversity of the population is similar throughout the optimization. Also, the mutation rate (and even the absence of mutation at all) does not influence the diversity level. Real optimization results thus confirm the static analysis. C&S crossover is able to maintain a high level of diversity, independently of the difference that exists between parents.

In experiments performed with GenC&S there is a clear distinction between mutation used or not used. If crossover acts alone then the algorithm quickly converges to a situation where approximately 80% of the individuals are identical and the other 20% are distinct[3]. This result confirms the locality study that showed how GenC&S is unable to inject diversity in the population. As soon as mutation is added, the problem is reduced. When a mutation rate of 0.01 is adopted, a small level of diversity

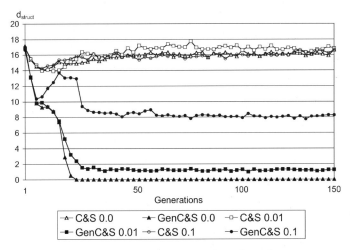

Fig. 10 Average structural distances in the population throughout the optimization of the 80-atom cluster

---

[3] The 20% of descendants that have a considerable different potential energy from that of its parents is a consequence of the way GenC&S acts: when two mates are nearly identical, it might be impossible to select enough atoms from the parents in such a way that the minimum distance constraint is satisfied. If this happens the descendant is completed with atoms placed at random locations.

is already visible, even though it is still clearly below the one that is verified in experiments performed with C&S. If the mutation rate is raised to 0.1, the diversity is comparable to the one that is visible in tests performed with the other crossover operator. This dependence on mutation to promote the diversity of the population was already predicted during the locality analysis of GenC&S. Fig. 10 presents the results obtained for structural distance in the same optimization experiments. They confirm all the analysis. Actually, they are even more evident as they perfectly rank the diversity level achieved by different combinations of crossover and mutation.

7 Conclusions

In this chapter we studied the locality properties of the hybrid evolutionary algorithm usually applied in cluster geometry optimization. Several Morse clusters instances, a well-known NP-hard benchmark model system, were selected for the analysis.

Two distance measures, required to determine the semantic difference between two solutions, were used to conduct a comprehensive analysis concerning the locality strength of an extended set of mutation and crossover operators.

In what concerns mutation, the empirical study showed that there are important differences in the locality level induced by different operators. Sigma mutation is an appropriate variation operator, but a moderate standard deviation should be selected to ensure the preservation of a reasonable correlation between individuals before and after the application of this operator. Conversely, flip mutation has low locality. This operator tends to perform large jumps in the search space, complicating the task of the exploration algorithm.

As for crossover, interesting patterns emerge from the analysis. C&S, which is the most widely used operator for cluster optimization, showed a remarkable innovation capacity. This operator is able to generate original descendants, even when the diversity of the population is low. This can be considered as an interesting behavior, but it also suggests that C&S might have difficulties in performing a meaningful identification and combination of important features that exist in parents. The other two operators considered in the analysis, GenC&S and uniform crossover, exhibit a behavior that is more in compliance with what is expected from crossover. If the mating parents are similar, the descendants tend to be analogous to them. If the distance between parents is high, the probability of generating a child with distinct features increases.

To validate our analysis we performed several optimization experiments using different settings and distinct combinations of genetic operators. Experimental outcomes support the most relevant results of the locality analysis and confirm the role played by different genetic operators.

This study is part of an ongoing project concerning the application of EAs to optimization problems from the Chemistry area. In the near future, we plan to extend our model to consider heritability and heuristic bias, the other two features that compose the original framework proposed by Raidl and Gottlieb. Results obtained with this analysis will be valuable for the future development of enhanced methods to employ in optimization problems with properties similar to the ones addressed in this research.

Acknowledgement. This work was supported by Fundação para a Ciência e Tecnologia, Portugal, under grant POSC/EIA/55951/2004.

We are grateful to the John von Neumann Institut für Computing, Jülich, for the provision of supercomputer time on the IBM Regatta p690+ (Project EPG01).

References

1. J. E. Jones. On the Determination of Molecular Fields. II. From the Equation of State of a Gas. Proc. Roy. Soc. A, 106, 463–477, 1924
2. J. E. Lennard-Jones. Cohesion. Proc. Phys. Soc., 43, 461–482, 1931
3. P. Morse. Diatomic Molecules According to the Wave Mechanics. II. Vibrational Levels. Phys. Rev., 34, 57–64, 1929
4. J. P. K. Doye, R. Leary, M. Locatelli and F. Schoen. Global Optimization of Morse Clusters by Potential Energy Transformations, Informs Journal on Computing, 16, 371–379, 2004
5. R. L. Johnston. Evolving Better Nanoparticles: Genetic Algorithms for Optimising Cluster Geometries, Dalton Transactions, 22, 4193–4207, 2003
6. D. M. Deaven and K. Ho. Molecular Geometry Optimization with a Genetic Algorithm, Phys. Rev. Lett. 75, 288–291, 1995
7. B. Hartke. Global Geometry Optimization of Atomic and molecular Clusters by Genetic Algorithms, In L. Spector et al. (Eds.), Proceedings of the Genetic and Evolutionary Computation Conference (GECCO-2001), 1284–1291
8. B. Hartke. Application of Evolutionary Algorithms to Global Cluster Geometry Optimization, In R. L. Johnston (Ed.), Applications of Evolutionary Computation in Chemistry, Structure and Bonding, 110, 33–53, 2004
9. F. Manby, R. L. Johnston and C. Roberts. Predatory Genetic Algorithms. Commun. Math. Comput. Chem. 38, 111–122, 1998
10. W. Pullan. An Unbiased Population-Based Search for the Geometry Optimization of Lennard-Jones Clusters: $2 \leq N \leq 372$. Journal of Computational Chemistry, 6(9), 899–906, 2005
11. C. Roberts, R. L. Johnston and N. Wilson (2000). A Genetic Algorithm for the Structural Optimization of Morse Clusters. Theor. Chem. Acc., 104, 123–130, 2000
12. Y. Zeiri. Prediction of the Lowest Energy Structure of Clusters Using a Genetic Algorithm, Phys. Rev., 51, 2769–2772, 1995
13. J. Gottlieb and C. Eckert. A Comparison of Two Representations for the Fixed Charge Transportation Problem, In M. Schoenauer et al. (Eds.), Parallel Problem Solving from Nature (PPSN VI), 345–354, Spinger-Verlag LNCS, 2000
14. J. Gottlieb and G. Raidl. Characterizing Locality in Decoder-Based EAs for the Multi-dimensional Knapsack Problem, In C. Fonlupt et al. (Eds.), Artificial Evolution: Fourth European Conference, 38–52, Springer-Verlag LNCS, 1999
15. G. Raidl and J. Gottlieb. Empirical Analysis of Locality, Heritability and Heuristic Bias in Evolutionary Algorithms: A Case Study for the Multidimensional Knapsack Problem. Evolutionary Computation, 13(4), 441–475, 2005

16. F. Rothlauf and D. Goldberg, Prüfernumbers and Genetic Algorithms: A Lesson on Hoe the Low Locality on an Encoding Can Harm the Performance of Gas, In M. Schoeneauer et al. (Eds.), Parallel Problem Solving from Nature PPSN VI, 395–404, 2000

17. F. Rothlauf. On the Locality of Representations, In E. Cantú-Paz et al. (Eds.), Proceedings of the Genetic and Evolutionary Computation Conference (GECCO-2003), Part II, 1608–1609, 2003

18. B. Sendhoff, M. Kreutz and W. Seelen. A Condition for the Genotype-Phenotype Mapping: Causality. In T. Bäck (Ed.), Proceedings of the 7th International Conference on Genetic Algorithms (ICGA-97), 73–80, 1997

19. F. B. Pereira, J. M. C. Marques, T. Leitão, J. Tavares. Analysis of Locality in Hybrid Evolutionary Cluster Optimization. In G. Yen et al. (Eds.), Proceedings of the IEEE Congress on Evolutionary Computation (CEC-2006), pp. 8049–8056, 2006

20. J. P. K. Doye and D. J. Wales. Structural Consequences of the Range of the Interatomic Potential. A Menagerie of Clusters. J. Chem. Soc. Faraday Trans. 93, 4233–4243, 1997

21. D. J. Wales et al. The Cambridge Cluster Database, URL: http://www-wales.ch.cam. ac.uk/CCD.html, accessed on January 2007

22. D. C. Liu and J. Nocedal. On the Limited Memory Method for Large Scale Optimization, Mathematical Programming B, 45, 503–528, 1989

23. J. Nocedal. Large Scale Unconstrained Optimization, In A. Watson and I. Duff (Eds.), The State of the Art in Numerical Analysis, 311–338, 1997

24. S. Wright. The Roles of Mutation, Inbreeding, Crossbreeding and Selection in Evolution. In Proceedings of the 6th International Conference on Genetics, Vol. 1, 356–366, 1932

25. T. Jones and S. Forrest. Fitness Distance Correlation as a Measure of Problem Difficulty for Genetic Algorithms. In L. Eshelman (Ed.), Proceedings of the 6th International Conference on Genetic Algorithms (ICGA-95), 184–192, 1995

26. P. Merz. Memetic Algorithms for Combinatorial Optimization Problems: Fitness Landscapes and Effective Search Strategies. Ph.D. Thesis, Department of Electrical Engineering and Computer Science, University of Siegen, Germany, 2000

27. E. D. Weinberger. Correlated and Uncorrelated Fitness Landscapes and How to Tell the Difference, Biological Cybernetics, 63, 325–336, 1990

28. W. Hart, T. Kammeyer, R. Belew. The Role of Development in Genetic Algorithms. In D. Whitley and M. Vose, (Eds.), Foundations of Genetic Algorithms 3, Morgan Kaufmann, pp. 315–332, 1995

29. D. Thierens, D. Goldberg. Mixing in Genetic Algorithms. In S. Forrest (Ed.), Proceedings of the Fifth International Conference on Genetic Algorithms (ICGA-93), Morgan Kaufmann, pp. 38–45, 1993

Aligning Time Series with Genetically Tuned Dynamic Time Warping Algorithm

Pankaj Kumar[1], Ankur Gupta[1], Rajshekhar[1], V.K. Jayaraman[2], and B.D. Kulkarni[3]

[1] Summer Trainee, Chemical Engineering Department, IIT Kharagpur, India.

[2] Chemical Engineering Division, National Chemical Laboratory, Pune, India.
vk.jayaraman@ncl.res.in

[3] Chemical Engineering Division, National Chemical Laboratory, Pune, India.
bd.kulkarni@ncl.res.in

Abstract

It is well known that Dynamic Time Warping (DTW) is superior to Euclidean distance as a similarity measure in time series analyses. Use of DTW with the recently introduced warping window constraints and lower bounding measures has significantly increased the accuracy of time series classification while reducing the computational expense required. The warping window technique learns arbitrary constraints on the warping path while performing time series alignment. This work utilizes genetic algorithms to find the optimal warping window constraints which provide a better classification accuracy. Performance of the proposed methodology has been investigated on two problems from diverse domains with favorable results.

Key words: Time Series Alignment and Classification, Dynamic Time Warping, Euclidean Distance, Lower Bounding, Genetic Algorithms

1 Introduction

The superiority of Dynamic Time Warping (DTW) over Euclidean distance as a similarity measure is a well established fact [1]. This leads to an obvious next step to use DTW instead of Euclidean distance as the proximity measure while dealing with different time series for classification or indexing purposes. The application of DTW in time series analysis has already been investigated in several domains. It was introduced by the speech recognition community [2] but has spread to other fields of application, such as bioinformatics [3], chemical engineering [4], biometric data [5], signature analysis [6], indexing of historical documents [7] and robotics [8] to name only a few. However, direct application of DTW has been limited due to its quadratic time complexity [9].

A recently introduced constrained window based warping method [1] for time series alignment has given rise to significant enhancement in the accuracy of the DTW method and application of a lower bounding measure has led to reduced computation time, making the methodology more desirable. The warping window concept is a framework that learns arbitrary constraints on the warping path while performing sequence alignment. The use of a lower bounding measure [1, 10] has significantly reduced the computational time required by pruning off the unnecessary computationally expensive calculations.

Ratanamahatana and Keogh [1] obtained the constraints on the warping path by formulating the problem as a classical search problem and implemented generic heuristic search techniques. As the use of heuristic search techniques like backward and forward hill climbing search methodologies perform poorly in obtaining a global optimum solution of the problem, application of genetic algorithms (GA) can be of significant help in the current scenario. In this work, we employ the widely used GA to obtain the warping path constraints.

The rest of the chapter is organized as follows: Sect. 2 provides a short resume of earlier research work on DTW and related work. Section 3 highlights details of lower bounding measures and their utility. Section 4 explains the heuristic method employed by Ratanamahatana and Keogh [1] for learning the proposed bands for different classes and briefly outlines the genetic algorithm and explains how it has been employed in the current framework. Details of the experimental data are provided in Sect. 5. In Sect. 6, we discuss our experimental results. Finally, Sect. 7 provides conclusions drawn from our work.

2 Dynamic Time Warping (DTW)

Consider two multivariate time series Q and C, of length n and m respectively,

$$Q = q_1, q_2, q_3, \ldots, q_i, \ldots, q_n \tag{1}$$

$$C = c_1, c_2, c_3, \ldots, c_j, \ldots, c_m \tag{2}$$

where, q_i and c_j are both vectors such that, $q_i, c_j \in \mathfrak{R}^p$ with $p \geq 1$. To align these two sequences using DTW, we construct a n-by-m matrix where the $(i\text{th}, j\text{th})$ element of the matrix corresponds to the squared distance,

$$d(i, j) = \sum_{r=1}^{r=p} (q_{i,r} - c_{j,r})^2 \tag{3}$$

To find the best match between these two sequences, one finds a path through the matrix that minimizes the total cumulative distance between them. Such a path will be called a warping path. A warping path, W, is a contiguous set of matrix elements that characterizes a mapping between Q and C. The kth element of W is defined as $W(k) = (i, j)_k$. The time-normalized distance [11] between the two time series is defined over the path as,

$$DTW(Q, C) = \min_{W} \left[\sqrt{\dfrac{\sum\limits_{k=1}^{k=K} d(W(k)) \cdot \phi(k)}{\sum\limits_{k=1}^{k=K} \phi(k)}} \right] \tag{4}$$

where, $\phi(k)$ is the non-negative weighting coefficient K and is the length of the warping path W, which satisfies the condition,

$$\max(m, n) \leq K \leq m + n - 1 \tag{5}$$

The normalization is done to compensate for K, the number of steps in the warping path, W, which can be different for different cases; i.e., number of steps will be very low if too many diagonal steps are followed and hence the objective function in Equation (4) will prefer such paths. Using path normalization, the path selection procedure is made independent of number of steps taken. The symmetric normalization [11] has been used for the purpose, given as,

$$\phi(k) = (i(k) - i(k - 1)) + (j(k) - j(k - 1)) \tag{6}$$

Note that for a horizontal or a vertical step, the value of $\phi(k)$ is 1, but for the diagonal step, it is 2. The path, however, is constrained to the following conditions [1]:

Boundary conditions: The path must start at $W(1) = (1, 1)$ and end at $W(K) = (n, m)$ that is, the warping path has to start at the top right and end at the bottom left of the distance matrix.

Continuity and monotonic condition: Every point in the two time-series must be used in the warping path, and both i and j indexes can only increase by 0 or 1 on each step along the path. In other words, if we take a point (i, j) from the matrix, the next point must be either of $(i, j + 1), (i + 1, j + 1)$ and $(i + 1, j)$.

Warping Window condition: The path as we know intuitively, should not wander too far from the diagonal. The concept of band has been introduced to keep the movement of path close to the diagonal. This band has been illustrated in Fig. 1 by the shaded region.

Mathematically, the band can be defined as,

$$j - R_i \leq i \leq j + R_i \tag{7}$$

where, $R_i = d$ such that $0 \leq d \leq m$ and $1 \leq i \leq m$. R_i is the permissible range of warping above and to the right of the diagonal. The band will be symmetric with respect to the diagonal. Even Euclidean distance can be defined in the case where the bandwidth $R_i = 0$. In general, any arbitrary shaped band can be formed using the above definition of the band. Now, the optimization problem given by Equation (4)

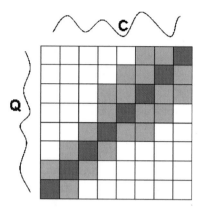

Fig. 1 A sample warping window band

can be effectively solved by dynamic programming technique as done in [1, 11]. For this the cumulative distance matrix is calculated [10–12],

$$D(i, j) = \min \begin{cases} D(i-1, j) + d(i, j) \\ D(i-1, j-1) + 2 \cdot d(i, j) \\ D(i, j-1) + d(i, j) \end{cases} \tag{8}$$

This takes care of the continuity and monotonic conditions along with the path normalization. Using this recursive formula for dynamic programming we can find the warping path W. Now, along this warping path W the values of $d(i, j)$ can be summed up, and the final value of DTW distance can be found as given by Eq. (4). However, in cases where the warping window constraint is used, only those points are considered in Eq. (8) that satisfy Eq. (7). This reduces the number of paths that need to be considered.

3 Lower-bounding Measures

Calculation of the DTW distance takes up a lot of computation time, and a classification algorithm that uses DTW as a distance measure is bound to be computationally inefficient. To solve this problem, a fast lower bounding measure is used to prune-off unnecessary calculations [10, 13].

3.1 Lower-bounding Measure (LB_{measure})

Let us consider the two time-series Q and C defined in Eqs. (1) and (2). Using the global constraints on the warping path given by Eq. (7), we construct an envelope across Q bounded by two time-series U and L, given as,

$$u_{j,r} = \max(q_{j-R,r} : q_{j+R,r}) \tag{9}$$

$$l_{j,r} = \min(q_{j-R,r} : q_{j+R,r}) \tag{10}$$

Using the above definitions of U and L, we define $LB_{measure}$ as,

$$LB\_measure(Q,C) = \sqrt{\frac{1}{(m+n-1)} \sum_{j=1}^{j=m} \sum_{r=1}^{r=p} \begin{cases} (c_{j,r}-u_{j,r})^2 & \text{if } c_{j,r} > u_{j,r} \\ (l_{j,r}-c_{j,r})^2 & \text{if } c_{j,r} < l_{j,r} \\ 0 & ,otherwise \end{cases}} \quad (11)$$

It is important to note that the above defined $LB_{measure}$ for two time series will always be lower than the corresponding DTW measure. This point can be understood very easily by considering the fact that while finding $LB_{measure}$, we need not consider continuity and monotonic path constraints, which have to be followed strictly in the case of DTW measure evaluation. A mathematical proof of this has been provided in [10]. The above conclusion can be employed in classification problem by finding the $LB_{measure}$ before evaluating the DTW measure and in the process, saving a lot of computation effort while identifying the nearest neighbor.

4 Learning Multiple Bands

4.1 Learning Multiple Bands for Classification using Heuristics

The next step involves finding bands for different classes. In the approach employed by Ratanamahatana and Keogh [1], it has been posed as a search problem, and generic heuristic techniques were used to obtain bands. The search can be performed as either a forward or backward hill-climbing search algorithm. Forward search begins with a uniform bandwidth of zero width, while backward search begins with maximum bandwidth m above and to the right of the diagonal.

In backward search, one begins with the maximum possible bandwidth, evaluates the accuracy with this bandwidth, then reduces the bandwidth of the complete segment, and re-evaluates the accuracy. If any improvement is obtained, one continues reducing the bandwidth of the entire segment, otherwise the segment is split into two equal halves, and the bandwidth of the two segments recursively reduced separately. This is done until one reaches zero bandwidth for the segment or a threshold value for the length of the segment is reached, or no further improvement can be made. Forward search is very similar, except that one starts off with zero bandwidth and increases the bandwidth in every step. For the heuristic search one needs a heuristic function to measure the quality of operation and to determine whether an improvement has been made or not. Ratanamahatana and Keogh [1] used an accuracy metric as heuristic function. The accuracy metric used was the number of correct classifications using the DTW distance and the band.

The threshold condition is used as terminating conditions to prevent any over fitting of data. Ratanamahatana and Keogh [1] employed the threshold condition

$$end - start + 1 \leq threshold \quad (12)$$

The value of the threshold has been taken as $\frac{1}{2}\sqrt{m}$. Theoretically, the value of the threshold can be as low as a single cell.

4.2 Learning Multiple Bands for Classification using GA

Since their introduction by Holland [14] in 1975, genetic algorithms have evolved to a great extent. Several variants of GA have been developed [15] and have been employed as well equipped tools for tackling problems from several domains.

Genetic algorithms are a particular class of evolutionary algorithm that uses techniques inspired by evolutionary biology and the Darwinian concept of "survival of the fittest". These are implemented as computer simulations in which first, a population of abstract representations of candidate solutions (individuals) is randomly generated. Each individual represents a potential solution to the problem. This population of potential solutions goes through a simulated process of evolution by creating new individuals through the breeding of the present individuals. Genetic algorithms do not require continuity of parameters or existence of derivatives in the problem domain. They can effectively handle multi-modal and multi-parameter type optimization problems, which are extremely difficult to handle using classical optimization methods.

Based on the complexity of DTW calculations involved, we utilized the basic form of genetic algorithm, the structure of which is shown in Table 1. The details of the components involved are:

Individual representation: Each individual is represented in the form of a binary string. This string evolves from the binary encoding of the parameters involved in the optimization problem. Although, there are a variety of ways in which a candidate solution can be represented, bit-encoding is the most common, owing to its precision and ease of use [14].

Fitness function: Each individual represents a potential solution to the problem with varying quality measured in terms of fitness function value. The most obvious function to be employed to evaluate the fitness for the current purpose is the classification accuracy of different individuals. However, in the present scenario it is common for many individuals to have the same accuracy. To sort out this problem, we include a parameter, bandwidth along with accuracy, to differentiate between the fitness of individuals having the same accuracy. Hence, if two individuals have the same accuracy, the one with the narrower bandwidth will be regarded as fitter.

Table 1 Algorithmic structure of GA

1. Randomly initialize the population of candidate solutions
2. Evaluate the fitness of each candidate solution
3. Perform elitism
4. Select candidate solutions for reproduction stage
5. Perform crossover on selected candidate solutions
6. Perform mutation on individuals obtained from Step 4
7. Obtain the new generation from individuals resulting from Step 3, 5 and 6
8. Check for termination criterion, if not satisfied Go to Step 2

Initial generation: A pool of candidate solutions is initialized in the form of randomly generated binary strings. The size of the pool, i.e. the population size considered in this work, is 100.

Termination criteria: The generational progress continues until some termination criteria are satisfied. In the current work, when the number of generations reaches a maximum defined *a priori*, evolution of new generations is stopped. The maximum number of generations has been kept at 50.

Selection procedure: This selection is done stochastically so that fitter solutions have more probability of being selected for the next generation. A tournament selection procedure is used for the purpose. A set of three individuals is randomly selected from the current generation, and the one with the maximum value of fitness function is passed to the reproduction stage, where genetic operators are employed.

Crossover operator: Crossover simulates the natural process of reproduction, where a pair of parent solutions is selected for breeding, and two new child solutions are created which share many characteristics of their parents. A higher probability of crossover is required for an effective local search and convergence to a solution. The crossover probability is taken as 0.8. As is common practice, the operator is implemented on parent solutions by swapping the parts of corresponding binary strings at random location

Mutation operator: Mutation introduces new individuals by changing an arbitrary bit of the binary strings corresponding to randomly selected individuals. It is like a random walk through the search space and signifies a global search. Mutation produces random solutions in the search space, and thus a high mutation probability will result in a highly unstable system, and hence will slow down the convergence. A low value of mutation probability will maintain diversity, yet keep the search space stable. The mutation probability is taken as 0.15.

Elitism operator: Elitism is a technique for ensuring that the best solutions are not lost during crossover and mutation. It copies a very few individuals with the highest fitness values to the next generation without any change, thus safeguarding against loss of elite individuals. This is fulfilled by retaining the best 5% of the population from the current to the next generation.

Following the algorithmic structure provided in Table 1, GA can be very easily implemented to obtain the optimum warping window once the bands are represented in binary string format.

5 Data Sets

Performance of the proposed methodology was investigated on problems from two very different domains. In the first application the methodology was employed in the bioinformatics field; the second application was to fault diagnosis in a chemical engineering system.

5.1 Gene Expression Profiles

Modeling and analysis of temporal aspects of gene expression data are gaining increasing importance [16,17]. Classification of these gene expression profiles has great potential in identifying the functions of genes in various cellular processes and gene regulations. It can be of great help in disease diagnosis and drug treatment evaluation [18]. The approach is also beneficial in grouping together similar kinds of genes, and in the process can help one to understand the functions and characteristics of various unknown genes [19].

We took the four-class classification problem of identifying the four gene clusters obtained by Iyer et al. [20]. Referring to the original work, the genes belonging to different clusters are: cluster 1: genes 263–296; cluster 2: genes 301–343; cluster 3: genes 394–407; cluster 4: genes 493–517. There are a total of 116 expression profiles with each profile measured at 19 different time points. Figure 2 provides sample time series from each class.

5.2 CSTR Batch Profiles

The continuously stirred tank reactor (CSTR) is an integral equipment in chemical engineering operations. In the second case study, we performed fault diagnosis of a jacketed CSTR in which an irreversible, exothermic, first-order reaction $(A \rightarrow B)$ is taking place. The system is equipped with three control loops, controlling the outlet temperature, the reactor holdup and the outlet concentration. A detailed mathematical model has been provided by Luyben [21]. This model has been further explored by Venkatasubramanian et al. [22] in their work on the application of neural networks to fault diagnosis. They introduced different ranges of operating parameters resulting in one normal operation and six abnormal operations. Table 2 lists these malfunctions.

Generation of the historical database The inlet stream state values were varied in each run to simulate abnormal operations. Each abnormal operating condition was character-

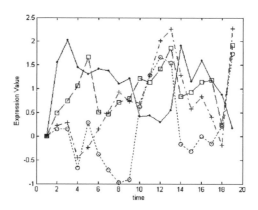

Fig. 2 Microarray gene expression, time series from each class

Table 2 Selected malfunctions for CSTR fault diagnosis

Malfunction (fault)	Variation from normal operation (%)
High inlet flow rate	$+(5-15)$
Low inlet flow rate	$-(5-15)$
High inlet concentration of reactant	$+(5-15)$
Low inlet concentration of reactant	$-(5-15)$
High inlet temperature of reactant	$+(5-15)$
Low inlet temperature of reactant	$-(5-15)$

ized by an abnormal value of an input variable. The normal operations were simulated by varying any of the input variables in the neighborhood of the perfectly normal operation. This variation was kept in the range 2.0%. The magnitudes of the input variable were varied randomly in each simulation. The duration of each simulation was 4 hours and the sampling interval was randomly varied from 2 minutes. Each of the seven operating conditions was simulated 50 times to provide a historical database of 350 batches. Gaussian measurement noise was added to the measured variables so that the signal-to-noise ratio for the CSTR profile was approximately equal to ten.

6 Results and Discussion

In this section, we discuss the performance of the proposed methodology for supervised classification of the data sets obtained from the case studies discussed in the previous section. The accuracy of classification has been measured using the "1-nearest-neighbour" with "leaving-one-out" approach. The Ratanamahatana and Keogh [1] methodology is different from the classical DTW in two senses: they employed the lower bounding technique to enhance computational efficiency and used the warping window concept to increase the accuracy. They employed heuristic hill-climbing search algorithms to train the warping window. In this work, we investigated the classification accuracy of constrained window based DTW, based on a warping window trained with classical genetic algorithms and compared its performance with that obtained by hill-climbing algorithms. The classification results obtained by the different methodologies are listed in Table 3.

As was expected, application of a genetic algorithm to train the warping window in DTW is highly effective in classifying the gene expression profiles. The classification accuracy is more than 98% with use of the genetic algorithm.

Table 3 Comparison of results

Case Studies	Classification Accuracy (%)		
	Hill climbing (Forward)	Hill climbing (Backward)	Genetic Algorithms
Gene expression Classification	95.690	96.552	98.276
CSTR fault diagnosis	93.714	94.286	94.286

However, the classification performance for the fault diagnosis of CSTR does not show much improvement. The classification accuracy of the backward hill climbing search algorithm and the GA remains the same at 94.286%. This observation may be attributed to the fact that attaining a higher accuracy becomes really difficult if application of DTW is unable to achieve any higher accuracy with change in the warping window. The classification accuracy observed using the forward hill climbing algorithm is slightly lower at 93.714%.

7 Conclusions and Future Work

In this work a genetically tuned DTW method with a constrained window based warping method [1] for time series alignment is introduced. In the original formulation of Ratanamahatana and Keogh [1] the constraints on the warping path were obtained by formulating the problem as a classical search problem. Further, generic heuristic search techniques were employed by them to get the optimal bands. With the help of two case studies we have shown that the application of a GA can significantly enhance performance and produce improved solutions.

Acknowledgement. We gratefully acknowledge financial assistance provided by the Department of Biotechnology, New Delhi, India.

References

1. Ratanamahatana CA and Keogh E (2004) Making time-series classification more accurate using learned constraints. In Proceedings of SIAM International Conference on Data Mining (SDM '04), Lake Buena Vista, Florida, pp. 11–22
2. Itakura F (1975) Minimum prediction residual principle applied to speech recognition. IEEE Trans Acoustics Speech Signal Process ASSP 23:52–72
3. Aach, J. and Church, G. (2001) Aligning gene expression time series with time warping algorithms, Bioinformatics 17:495–508
4. Chu S, Keogh E, Hart D and Pazzani M (2002) Iterative deepening dynamic time warping for time series. In: Proc 2nd SIAM International Conference on Data Mining
5. Gavrila DM and Davis LS (1995) Towards 3-d model-based tracking and recognition of human movement: a multi-view approach. In International Workshop on Automatic Face- and Gesture-Recognition. pp. 272–277
6. Munich M and Perona P (1999) Continuous dynamic time warping for translation-invariant curve alignment with applications to signature verification. In Proceedings of 7th International Conference on Computer Vision, Korfu, Greece, pp. 108–115
7. Rath T and Manmatha R (2002) Word image matching using dynamic time warping. In Proceedings of the Conference on Computer Vision and Pattern Recognition (CVPR 03), Vol. II:521–527
8. Schmill M, Oates T and Cohen P (1999) Learned models for continuous planning. In 7th International Workshop on Artificial Intelligence and Statistics
9. Berndt D and Clifford J (1994) Using dynamic time warping to find patterns in time series. AAAI-94 Workshop on Knowledge Discovery in Databases, pp. 229–248

10. Rath TM and Manmatha R (2002) Lower-Bounding of Dynamic Time Warping Distances for Multivariate Time Series. Technical Report MM-40, Center for Intelligent Information Retrieval, University of Massachusetts Amherst

11. Sakoe H and Chiba S (1978) Dynamic-programming algorithm optimization for spoken word recognition. IEEE Transactions on Acoustics, Speech, and Signal Processing 26:43–49

12. Henk-Jan Ramaker et al. (2003) Dynamic time warping of spectroscopic BATCH data, Analytica Chimica Acta, Volume 498, Issues 1–2, 28:133–153

13. Keogh E (2002) Exact indexing of dynamic time warping. In: Proceedings of the 28th Very Large Databases Conf. (VLDB), Hong Kong, China, August 20–23, 406–417

14. Holland J (1975) Adaptation in Natural and Artificial Systems, University of Michigan Press, Ann Arbor, MI.

15. Larrañaga P, Kuijpers CMH, Murga RH, Inza I and Dizdarevic S (1999) Genetic algorithms for the travelling salesman problem: A review of representations and operators, Artificial Intelligence Review 13:129–170

16. Brown P and Botstein D (1999) Exploring the new world of the genome with DNA microarrays, Nature Genetics supplement 21:33–37

17. Schena M, et al. (1995) Quantitative monitoring of gene expression patterns with a cDNA microarray. Science 270:467–470

18. Golub TR, Slonim DK, Tamayo P, Huard C, Gassenbeek M, Mesirov JP, Coller H, Loh ML, Downing JR, Caligiuri MA, Bloomfield DD and Lander ES (1999) Molecular classification of cancer: class discovery and class prediction by gene expression monitoring, Science, 286(15):531–537

19. Eisen M, et al. (1998) Cluster analysis and display of genome-wide expression patterns. Proceedings of the National Academy of Sciences, USA, 95:14863–14868

20. Iyer VR, et al. (1999) The transcriptional program in the response of human fibroblasts to serum. Science, 283(5398):83–87

21. Luyben William L. (1973), Process Modeling, Simulation and Control for Chemical Engineers, McGraw Hill, New York

22. Venkatasubramanian V, Vaidyanathan R and Yamamoto Y (1990) Process fault detection and diagnosis using neural networks-i. Steady-state processes, Computers Chem. Engg., 14(7):699–712

Evolutionary Generation of Artificial Creature's Personality for Ubiquitous Services

Jong-Hwan Kim[1], Chi-Ho Lee[1], Kang-Hee Lee[2], and Naveen S. Kuppuswamy[1]

[1] Korea Advanced Institute of Science and Technology (KAIST), 373-1, Guseong-Dong, Yuseong-Gu, Daejeon, Republic of Korea
jhkim, chiho, naveen@rit.kaist.ac.kr

[2] Telecommunictaion R&D Center, Telecommunication Network Business, Samsung Electronics Co., Ltd., Yeongtong-gu, Suwon-si, Gyeoggi-do, Republic of Korea
khanghee76.lee@samsung.com

Abstract

Ubiquitous robot systems represent the state-of-the-art in robotic technology. This paradigm seamlessly blends mobile robot technology (Mobot) with distributed sensor systems (Embot) and overseeing software intelligence (Sobot), for various integrated services. The wide scope for research in each component area notwithstanding, the design of the Sobot is critical since it performs the dual purpose of overseeing intelligence as well as user interface. The Sobot is hence modeled as an Artificial Creature with autonomously driven behavior. This chapter discusses the evolutionary generation of an artificial creature's personality. The artificial creature has its own genome in which each chromosome consists of many genes that contribute to defining its personality. The large number of genes also allows for a highly complex system. However, it becomes increasingly difficult and time-consuming to ensure reliability, variability and consistency for the artificial creature's personality while manually assigning gene values for the individual genome. One effective approach to counter these problems is the Evolutionary Generative Process for an Artificial Creature's Personality (EGPP) which forms the focus of this chapter. EGPP evolves a genome population such that it customizes the genome, meeting a simplified set of personality traits desired by the user. An evaluation procedure for each genome of the population is carried out in a virtual environment using tailored perception scenarios. Effectiveness of this scheme is demonstrated by using an artificial creature Sobot, 'Rity' in the virtual 3D world created in a PC, designed to be a believable interactive software agent for human–robot interaction.

Key words: Ubiquitous Robot, Genetic Robot, Evolvable Artificial Creature, Artificial Genome, Artificial Personality

1 Introduction

Ubiquitous robotic systems are emerging. These systems negate the necessity for personal robotic systems to limit themselves to the conventional notion of a stand-alone robot platform. Brady defined robotics as the 'Intelligent connection of perception to action' [1]. Ubiquitous robotics lends itself to that description, by allowing us to redefine the interconnection between the three components, intelligence, perception and action, by manifesting them individually as the intelligent Software Robot–Sobot, the perceptive Embedded Robot–Embot and the physically active Mobile Robot–Mobot, respectively, as described in [2–7]. The interconnection is therefore created through the network and the integration is carried out using the middleware in the ubiquitous space (u-space). This can be conceptualized as a networked cooperative robot system. The core intelligence of this system is comprised of software robots. Distributed Embots ensure that the Sobots possess context aware perceptive capabilities. Lastly, Mobots act upon the service requests in the physical domain. Networking technology such as the IPv6 format and broadband wireless systems is the key leveraging these advancements. Ubiquitous robots will thus be able to understand what the user needs, even without the issuance of a direct command, and be able to supply continuous and seamless service. Following the general concepts of ubiquitous computing, ubiquitous robotic systems are seamless, calm, and context aware. The Sobot can connect to, and be transmitted to any device, at any time and at any place within the u-space, by maintaining its own unique IP address. It is context aware and can automatically and calmly provide services to the user. Embots collect and synthesize sensory information through the detection, recognition and authentication of users and other robots. Mobots proceed to act by providing the general users with integrated services. Middleware enables the Ubibot to interact and manage data communication reliably without disrupting the protocols in the u-space. From this description it is evident that the design of the intelligent Sobot is critical to the functioning of the ubiquitous robot system.

Sobots by their very nature of being software agents capable of seamless transition, are eminentlt suitable an interface for users to communicate with the entire ubiquitous robot system. However, in order to do this effectively, they must present an interactive interface. It is for this reason that Sobots are modeled as artificial creatures.

Artificial creatures, also referred to variously as interactive creatures, autonomous agents, synthetic characters, software robots, or 3D avatars, have been developed to entertain humans in real-time interaction. In general, an artificial creature has various virtual and physical sensors, which influence internal states such as motivation, homeostasis, emotion, etc., and then lead to a different behavior externally according to the internal state. These software agents have significant potential for application in the entertainment industry. In particular, such agents have applications in the development of interfaces with human–robot interaction systems, where the human user may have meaningful interactions with the creature. Most previous work, however, dealt with behavior selection and learning mechanisms to improve the output performance [8–16].

The concept of genetic encoding and evolutionary mechanisms emerged to improve the output performance of game characters in video gaming. The brain structures were encoded as a genome to activate proper behavior patterns when it interacted with a user in real time [17, 18]. Genes in a creature's genomes code for such structures as chemo-receptors, reactions and brain lobes in the neural network-based brain model, rather than outward phenomena such as disease-resistance, fearlessness, curiosity, or strength. Also, in strategy games and simulators, neuroevolution was employed in training artificial neural networks as agent controllers, where fitness was determined by game play [19–23]. In [24, 25], key parameters of the game character's behavior were genetically encoded for evolution by which they were made to be persistently competitive and to generate novel behaviors.

Along with the genetic encoding and evolutionary mechanism for the artificial creature, its personality should also be emphasized to create a believable one. The significance of having a diverse personality was noted in [26]: 'Personality is a determiner of, not merely a summary statement of, behavior.' It was claimed that the personality was crucial in building a reliable emotional agent. To build such a truly believable artificial creature, it is required to provide an evolutionary generative process that has the power to represent its personality by composing genetic codes of parameters related by internal states and outward behaviors. The evolutionary generative process includes implementations of the artificial creature, virtual environment, perception scenario, personality model and evolutionary algorithm for a specific personality desired by a user. The evolutionary algorithm requires investigations of genetic representation, fitness function, genetic operators and how they affect the evolutionary process in the simulated environment.

This chapter focuses on evolving an artificial creature's personality as desired by using its computer-coded genome in a virtual environment. The primary application is that of providing a believable and interactive agent for personal use. The agent would reside in hardware devices such as personal computer and mobile phone in order to provide ubiquitous services. Video clips of this scenario maybe seen at http://rit.kaist.ac.kr/research.php (under the links 'Intelligent Robots', 'Software Robot' under the title "Geney"). The genome is composed of multiple artificial chromosomes each of which consists of many genes that contribute to defining the creature's personality. It provides primary advantages for artificial reproduction, the ability to evolve, and the reusability among artificial creatures [27, 28]. The large number of genes also allows for a highly complex system. However, it is difficult and time-consuming to manually assign values to them to ensure reliability, variability and consistency of the artificial creature's personality. The evolutionary generative process for an artificial creature's personality (EGPP) which forms the crux of this chapter, effectively deals with these issues providing a powerful tool for artificial creature personality generation and evolution.

EGPP is a software system to generate a genome as its output, which characterizes an artificial creature's personality in terms of various internal states and their concomitant behaviors. Initialization of the genome population is carried out by setting some parameters in the graphical user interface (GUI). EGPP acts on the initial population for evolution to have a plausible creature given the personality preference set through the GUI by the user. In a control architecture, to implement the plausible creature, a stochastic voting mechanism is employed for behavior selection. Also, gene masking operators are introduced to forbid unreasonable behaviors. The evaluation procedure for each individual (genome) is carried out in a virtual environment by applying a series of perceptions (perception scenario) to the artificial creature with the corresponding genome and then by measuring its fitness. The proposed EGPP has been validated by implanting the evolved genome into the artificial creature. In experiments the artificial creature, Rity, developed in a 3D virtual world, is used as a test agent in a personal computer environment.

This chapter is organized as follows: Sect. 2 describes the general concept of ubiquitous robot system incorporating Sobot, Embot and Mobot. Section 3 introduces an artificial creature, Rity, its internal control architecture, and the structure of the genome, which is composed of a set of chromosomes including the fundamental genes, the internal state related genes, and the behavior related genes. Section 4 describes the proposed EGPP along with the personality model, evolutionary algorithm, perception scenario and fitness function. Experiments are carried out to demonstrate its performance and effectiveness in Sect. 5. Concluding remarks follow in Sect. 6.

2 The Ubiquitous Robot System

The ubiquitous robot system is comprised of networked and integrated cooperative robot systems existing in the ubiquitous world. It includes Software Robot Sobots, Embedded Robot Embots and Mobile Robot Mobots in their various forms. The ubiquitous robot is created and exists within a u-space which provides both its physical and virtual environment. It is anticipated that in the years to come the world will consist of many such u-spaces, each being based on the IPv6 protocol or a similar system and interconnected through wired or wireless broadband network in real time.

The primary advantage of ubiquitous robot systems is that they permit abstraction of intelligence from the real world by decoupling it from perception and action capabilities. Sensory information is standardized along with motor or action information and this permits the abstract intelligence to proceed with the task of providing services in a seamless, calm and context aware manner. Ubiquitous robots provide us with services through the network at anytime and anywhere in a u-space through distributed capabilities provided by the component Sobot, Embot and Mobot. Each robot, however, has specific individual intelligence and roles, and can communicate information through networks. The development of effective Middleware is critical to this system since it serves to overcome difficulties due to heterogeneous protocols and network difficulties.

Some of the integrated services and solutions offered by the ubiquitous robot technology include ubiquitous home services for security and safety, location-based services like GIS, health services in telemedicine, ubiquitous learning systems and ubiquitous commerce services.

As mentioned earlier, the ubiquitous robot system incorporates three kinds of robot systems: Sobots, Embots and Mobots under the ambit of the following definitions.

2.1 Software Robot: Sobot

Sobots are the intelligent component of the Ubibot system whose domain lies wholly within the software realm of the network. It can easily traverse through the network to connect with other systems irrespective of temporal and geographical limitations. Sobots are capable of operating as intelligent entities without help from other ubiquitous robots and are typically characterized by self-learning, context-aware intelligence and, calm and seamless interaction abilities. Within the u-space, Sobots try and recognize the prevailing situation and often make decisions on the course of action and implement them without directly consulting the user each time. They are proactive and demonstrate rational behavior and show capabilities to learn new skills. It is also totally pervasive in its scope and thus is able to provide seamless services throughout the network.

The Sobot intelligence also has an important subsidiary function to perform, which is for the interface with the user. Coupled with its ability to seamlessly transmit itself, it must additionally present an interactive, believable interface for user comfort and convenience. It is for this reason that the Sobot is developed as an artificial creature, with autonomous behaviors, as described in Sect. 3.

2.2 Embedded Robot: Embot

The embedded robots, as the name implies, are implanted within the environment or upon Mobots. They together comprise the perceptive component of the ubiquitous robot system. Utilizing a wide variety of sensors in a sensor network, Embots detect and monitor the location of a user or a Mobot, authenticating them and also integrate assorted sensory information thus comprehending the current environmental situation. Embots are also networked and equipped with processing capabilities and thus may deliver information directly or under the Sobot's instructions to the user. Embots are characterized by their calm sensing, information processing and information communication capabilities. Embots offer great functionality by being able to sense features such as human behavior, status, relationships and also environmental conditions impacting human behavior. They also possess abilities to perform data mining, which can enhance information search processes.

2.3 Mobile Robot : Mobot

Mobots offer a broad range of services for general users specifically within the physical domain of the u-space. Mobility is a key property of Mobots, as well as the general capacity to provide services in conjunction with Embots and Sobots. The Mobot is usually in continous communication with the Sobot in order to provide practical services based on information given by the Embot. Alternatively, Mobots serve Embots as a platform for data gathering. Mobots are typically multi-purpose service robots with functionalities extending to home, office and public facilities.

2.4 Middleware

Middleware allows communication within and among ubiquitous robots using a variety of network interfaces and protocols. Middleware usually varies from one vendor to the next depending upon a variety of factors. The selected middleware allows conversion of the constituent entities of the ubiquitous robot system into specific components with respect to the developer, thereby making it convenient to update functions, maintain resources and perform power management. The Middleware structure for a ubiquitous robot system must contain at least one interface and one broker. The interfaces refer to the hardware level interfaces of the communication protocols such as Bluetooth and Ethernet and the software level interfaces like HTTP and FTP. The broker enables the system to make an offer of service irrespective of the operating structure, position and type of interface. This thus enables Sobots to receive information from a wide variety of Embots and to communicate with the Mobots.

3 Sobot as an Artificial Creature

This section introduces a Sobot as an artificial creature, Rity, in a 3D virtual world, its internal control architecture and genome composed of a set of chromosomes [13].

3.1 Artificial Creature, Rity

An artificial creature is defined as an agent which behaves autonomously driven by its internal states such as motivation, homeostasis and emotion. It should also be able to interact with humans and its environment in real time. Rity is designed to fulfill the requirements for an artificial creature. It represents itself visually on the screen as a dog and may interact with humans based on information through a mouse, a camera or a microphone.

The internal control architecture in this chapter is composed of four primary modules, that is, perception module, internal state module, behavior selection module and motor module. All the modules are embodied in Rity, which is a 3D virtual pet with 12 degrees of freedom. It is developed in Visual C++ 6.0 and OpenGL and works well on Pentium III machines or above. Figure 1 illustrates both the internal control architecture and a screen shot of the computer screen showing Rity, in a virtual 3D environment. A brief explanation of each module in the internal architecture follows.

Perception module The perception module can recognize and assess the environment and subsequently send the information to the internal state module. Rity has several virtual sensors for light, sound, temperature, touch, vision, orientation and time.

Internal state module
The internal state module defines the creature's internal state with the motivation unit, the homeostasis unit and the emotion unit. In Rity, motivation is composed of six states: curiosity, intimacy, monotony, avoidance, greed and the desire to control.

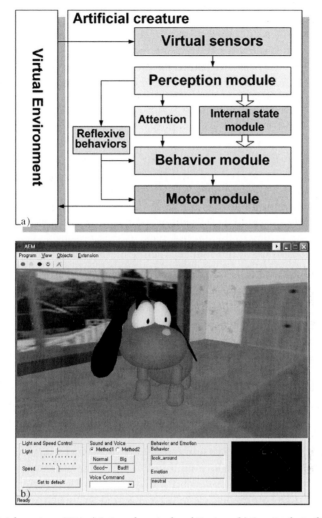

Fig. 1 Artificial creature, Rity. **a**) Internal control architecture. **b**) Screen shot of Rity in a 3D virtual world

Homeostasis includes three states: fatigue, hunger and drowsiness. Emotion includes five states: happiness, sadness, anger, fear and neutral. In general, the number of internal states depends on an artificial creature's architecture. Each internal state is updated by its own weights, which connect the stimulus vector to itself and are also represented as a vector. For instance, motivation vector $\mathbf{M}$ is defined as

$$\mathbf{M}(t) = [m_1(t), m_2(t), \dots, m_6(t)]^T \tag{1}$$

where $m_k(t)$ is the kth state in the internal state module. Each motivation state is updated by

$$m_k(t+1) = m_k(t) + \{\lambda_k(\overline{m_k} - m_k(t)) + \mathbf{S}^T \cdot \mathbf{W}_k^M(t)\}, \quad k = 1, 2, \dots, 6 \tag{2}$$

where $\mathbf{S}$ is the stimulus vector, $\mathbf{W}_k^M$ is a weight matrix connecting $\mathbf{S}$ to the kth state in the internal state module, $\overline{m_k}$ is the constant to which the internal state converges without any stimuli, and λ_k is the discount factor between 0 and 1. Similarly, the following update equations are defined for the homeostasis unit using state vector $\mathbf{H}(t)$ and weight matrix $\mathbf{W}_k^H$, and also the emotion unit using state vector $\mathbf{E}(t)$ and weight matrix $\mathbf{W}_k^E$, respectively:

$$h_k(t+1) = h_k(t) + \{\lambda_k(\overline{h_k} - h_k(t)) + \mathbf{S}^T \cdot \mathbf{W}_k^H(t)\}, \quad k = 7,8,9 \tag{3}$$

$$e_k(t+1) = e_k(t) + \{\lambda_k(\overline{e_k} - e_k(t)) + \mathbf{S}^T \cdot \mathbf{W}_k^E(t)\}, \quad k = 10,11,\ldots,14, \tag{4}$$

where $\mathbf{H}(t) = [h_7(t), h_8(t), h_9(t)]^T$ and $\mathbf{E}(t) = [e_{10}(t), e_{11}(t), \ldots, e_{14}(t)]^T$.

Behavior selection module

The behavior selection module is used to choose a proper behavior based on Rity's internal state. According to the internal state, various reasonable behaviors can be selected probabilistically by introducing a voting mechanism [14,15], where each behavior has its own voting value. The procedure of behavior selection is as follows:

1. Determine the temporary voting vector, $\mathbf{V}_{temp}$ using $\mathbf{M}$ and $\mathbf{H}$.
2. Calculate voting vector $\mathbf{V}$ by applying attention and emotion masks to $\mathbf{V}_{temp}$.
3. Calculate a behavior selection probability, $p(b)$, using $\mathbf{V}$.
4. Select a proper behavior b with $p(b)$ among various behaviors.

Each step is described in the following.

1. The temporary voting vector is defined as follows:

$$\mathbf{V}_{temp}^T = \left(\mathbf{M}^T \mathbf{D}^M + \mathbf{H}^T \mathbf{D}^H\right)$$
$$= [v_{t1}, v_{t2}, \ldots, v_{tz}] \tag{5}$$

where a superscript T represents a transpose of a vector and z represents the number of behaviors provided for Rity. v_{tr}, $r = 1,\ldots,z$, is the temporary voting value. As there are 6 motivation states and 3 homeostasis states for Rity, $6 \times z$ matrix $\mathbf{D}^M$ and $3 \times z$ matrix $\mathbf{D}^H$ are the behavioral weight matrices connecting motivation and homeostasis to behaviors, respectively.

2. Two masking matrices for attention and emotion assist Rity in selecting a natural behavior by masking unreasonable behaviors. An attention masking matrix $\mathbf{Q}^A(a)$ is obtained by the attentional percept, a, which has its own masking value. The matrix is defined as a diagonal matrix with diagonal entries $q_1^a(a), \ldots, q_z^a(a)$ where $q_r^a(\cdot)$, $r = 1,\ldots,z$, is the masking value, and $0 \le q_r^a(\cdot) \le 1$. Similarly, an emotion masking matrix, $\mathbf{Q}^E(e)$, where e is the dominant emotion, is defined. From these two masking matrices and the temporary voting vector, the behavior selector obtains a final voting vector as follows:

$$\mathbf{V}^T = \mathbf{V}_{temp}^T \mathbf{Q}^A(a) \mathbf{Q}^E(e)$$
$$= [v_1, v_2, \ldots, v_z] \tag{6}$$

where v_r, $r = 1, 2, \ldots, z$, is the rth behavior's voting value.

3. The selection probability $p(b_k)$ of a behavior, b_k, $k = 1, 2, \ldots, z$, is calculated from the voting values as follows:

$$p(b_k) = v_k / \sum_{r=1}^{z} (v_r).$$

4. By employing the probability proportional selection mechanism, a reasonable and natural behavior is selected.

In addition to behaviors, Rity has five facial expressions for happiness, sadness, anger, fear and neutral state. It is chosen for the dominant emotional state.

Motor module

The motor module incorporates virtual actuators to execute the selected behavior in the virtual 3D environment.

3.2 Genetic Representation

This section presents a genetic representation for an artificial creature that would be capable of animal-like evolution. Due to the existence of the pleiotypic and polygenic nature of the genotype, a single gene influences multiple phenotypic characters (pleiotypic nature) and a single phenotypic character is directly inspired by multiple genes (polygenic nature). To reflect this complexity to Rity's chromosomal coding, a sophisticated internal architecture is encoded to avoid a purely mechanistic response.

An artificial creature is made up of genome, a set of chromosomes, $\mathbf{C}_k$, $k = 1, 2, \ldots, c$, which has the capability of passing its traits to its offspring. Each chromosome $\mathbf{C}_k$ is composed of three gene vectors: the Fundamental gene vector (F-gene vector), $\mathbf{x}_k^F$, the Internal state related gene vector (I-gene vector), $\mathbf{x}_k^I$, and the Behavior related gene vector (B-gene vector), $\mathbf{x}_k^B$, and is defined as

$$\mathbf{C}_k = \begin{bmatrix} \mathbf{x}_k^F, \\ \mathbf{x}_k^I, \\ \mathbf{x}_k^B \end{bmatrix}, \quad k = 1, 2, \ldots, c$$

with

$$\mathbf{x}_k^F = \begin{pmatrix} x_{1k}^F \\ x_{2k}^F \\ \vdots \\ x_{wk}^F \end{pmatrix}, \mathbf{x}_k^I = \begin{pmatrix} x_{1k}^I \\ x_{2k}^I \\ \vdots \\ x_{yk}^I \end{pmatrix}, \mathbf{x}_k^B = \begin{pmatrix} x_{1k}^B \\ x_{2k}^B \\ \vdots \\ x_{zk}^B \end{pmatrix},$$

where w, y, and z are the sizes of the F-gene vector, I-gene vector, and B-gene vector, respectively. F-genes represent fundamental characteristics of Rity, including genetic information such as volatility, initial values, constant values ($\overline{m_k}$ in (2), $\overline{h_k}$ in (3) and $\overline{e_k}$ in (4)) and the discounting factor (λ_k of (2)–(4)). The volatility gene in $\mathbf{C}_k$ determines whether the kth internal state is volatile or non-volatile when reset. I-genes include genetic codes representing their internal state by setting the weights of $\mathbf{W}_k^M(t)$ in (2), $\mathbf{W}_k^H(t)$ in (3) and $\mathbf{W}_k^E(t)$ in (4). These genes shape the relationship between perception and internal state. B-genes include genetic codes related to output behavior by setting the weights of $\mathbf{D}^M$ and $\mathbf{D}^H$ in (5), and $\mathbf{Q}^E$ in (6).

An artificial genome, $\mathbf{G}$, composed of a chromosomal set, is defined as

$$\mathbf{G} = \left[\, \mathbf{C}_1 \mid \mathbf{C}_2 \mid \ldots \mid \mathbf{C}_c \, \right],$$

where c is the number of chromosomes in the genome.

Rity is implemented by $w = 4$, $y = 47$, $z = 77$, and $c = 6 + 3 + 5 = 14$. These values are equivalent to the ability of perceiving 47 different types of percepts and of outputting 77 different behaviors as response. Figure 2 shows the 14 chromosomes, where the first six $\mathbf{C}_1$–$\mathbf{C}_6$ are related to motivation: curiosity ($\mathbf{C}_1$), intimacy ($\mathbf{C}_2$), monotony ($\mathbf{C}_3$), avoidance ($\mathbf{C}_4$), greed ($\mathbf{C}_5$), and desire to control ($\mathbf{C}_6$), the next three $\mathbf{C}_7$–$\mathbf{C}_9$ are related to homeostasis: fatigue ($\mathbf{C}_7$), drowsiness ($\mathbf{C}_8$), and hunger ($\mathbf{C}_9$), and the last five $\mathbf{C}_{10}$–$\mathbf{C}_{14}$ are related to emotion: happiness ($\mathbf{C}_{10}$), sadness ($\mathbf{C}_{11}$), anger ($\mathbf{C}_{12}$), fear ($\mathbf{C}_{13}$), and neutral ($\mathbf{C}_{14}$). As each chromosome is represented by 4 F-genes, 47 I-genes and 77 B-genes, Rity has 1792 genes in total. The genes in Fig. 2 are originally represented by real numbers: initial and constant values of F-genes range from 0 to 500, I-genes from −500 to 500, and B-genes from 0 to 1000. F- and B-genes are normalized to brightness values from 0 to 255, which are expressed as black-and-white rectangles. The darker the color, the higher its value. In addition to positive normalization, I-genes may have negative values and are normalized as red-and-black rectangles in the same manner.

The 2D genetic representation has advantages of representing essential characteristics of three types of genes intuitively, reproducing the evolutionary characteristics of living creatures, and enabling users to easily insert or delete other types of chromosomes and genes related to an artificial creature's personality and other information.

Procedure EA
begin

$\qquad t \leftarrow 0$

i)   initialize $P(t)$

ii)   gene-mask $P(t)$

iii)   evaluate $P(t)$

iv)   store the best genome $b(t)$ among $P(t)$

v)   **while (not** termination-condition) **do**

$\qquad$ **begin**

$\qquad\qquad t \leftarrow t + 1$

vi)   select $P(t)$ from $P(t − 1)$

vii)   alter $P(t)$

viii)   gene-mask $P(t)$

ix)   evaluate $P(t)$

x)   store the best genome $b(t)$ among $b(t − 1)$ and $P(t)$

$\qquad$ **end**

end

Fig. 2 Artificial genome of Rity

4 Evolutionary Generative Process for a Personality (EGPP)

To build a truly believable artificial creature, it is required to design an evolutionary generative process to generate a genome representing its personality. The process includes an implementation of the artificial creature and its virtual environment, its personality model, and an evolutionary algorithm for a desired personality.

4.1 Personality Model

Big five personality dimensions are employed and Rity's internal traits are classified for the corresponding personalty dimension. They are classified as follows: extroverted (as opposed to introverted), agreeable (as opposed to antagonistic), conscientious (as opposed to negligent), openness (as opposed to closedness), and neuroticism (as opposed to emotional stability) [29]. From these, in this chapter agreeable and antagonistic personalities are engineered for Rity to demonstrate the feasibility of the EGPP. By comparing the performance for the two contrasting personalities, the evaluation can be easily made concerning its ability to provide consistent (the ability to exhibit reliably expectant behaviors) and uniquely distinct personality. The agreeable personality assumes strength in curiosity, intimacy and happiness, and weakness in greed, desire to control, avoidance, anger and fear. In contrast, the antagonistic personality assumes weakness in curiosity, intimacy and happiness, and strength in greed, desire to control, avoidance, anger and fear.

Considering the personality traits, the preference values of the agreeable and the antagonistic personality models are assigned between 0 and 1 as in Table 1, where ψ_k^I and ψ_k^B are preference values for the kth internal state and behavior group, respectively. These values mean user's desired preference and will be used for defining the fitness function. Preference values in the table are denoted by

$$\Psi = \begin{bmatrix} \psi_1^I & \psi_2^I & \cdots & \psi_c^I \\ \psi_1^B & \psi_2^B & \cdots & \psi_c^B \end{bmatrix}. \tag{7}$$

Table 1 Preference values for the agreeable and antagonistic personalities

Internal State			Assigned preference values			
			Agreeable personality		Antagonistic personality	
Mode	State	k	ψ_k^I	ψ_k^B	ψ_k^I	ψ_k^B
	Curiosity	1	0.5	0.5	0.2	0.2
	Intimacy	2	**0.8**	**0.8**	0.2	0.2
Motivation	Monotony	3	0.5	0.5	0.5	0.5
	Avoidance	4	0.2	0.2	**0.8**	**0.8**
	Greed	5	0.2	0.2	**0.8**	**0.8**
	Control	6	0.1	0.1	0.7	0.7
	Fatigue	7	0.2	0.2	0.2	0.2
Homeostasis	Drowsiness	8	0.2	0.2	0.2	0.2
	Hunger	9	0.2	0.2	0.2	0.2
	Happiness	10	**0.8**	**0.8**	0.2	0.2
	Sadness	11	0.5	0.5	0.5	0.5
Emotion	Anger	12	0.2	0.2	**0.8**	**0.8**
	Fear	13	0.2	0.2	**0.8**	**0.8**
	Neutral	14	0.5	0.5	0.2	0.2

The preference values are set by using slider bars in a graphical user interface (GUI) (see Fig. 6).

4.2 Procedure of Evolutionary Algorithm

EGPP includes an evolutionary algorithm, which maintains a population of genomes, G_i^t, $i = 1, 2, \ldots, n$, with the form of a two-dimensional matrix, $P(t) = \{G_1^t, G_2^t, \ldots, G_n^t\}$ at generation t, where n is the size of the population. G_i^t is defined as

$$G_i^t = \left[\, C_{i1}^t \,\middle|\, C_{i2}^t \,\middle|\, \ldots \,\middle|\, C_{ic}^t \,\right] = \begin{bmatrix} x_i^{Ft} \\ x_i^{It} \\ x_i^{Bt} \end{bmatrix}$$

$$= \begin{bmatrix} x_{i1}^{Ft} & x_{i2}^{Ft} & \ldots & x_{ic}^{Ft} \\ x_{i1}^{It} & x_{i2}^{It} & \ldots & x_{ic}^{It} \\ x_{i1}^{Bt} & x_{i2}^{Bt} & \ldots & x_{ic}^{Bt} \end{bmatrix}$$

Figure 3 illustrates the procedure of the evolutionary algorithm in the following manner:

(i) Preference values are assigned via a GUI and a population of genomes is initialized (Sect. 4.3)

(ii) and (viii) Each G_i^t is masked by I-gene and B-gene maskings, in order to generate reasonable behaviors. Masked genomes replace the original ones (Sect. 4.4)

(iii) and (ix) The masked genome is implanted to the artificial creature, a series of perceptions is applied to it, and then its fitness is evaluated (Sects. 4.5 and 4.6).

(iv) and (x) The best genome is selected among the genomes in $P(t)$, and stored in $b(t)$ which is the best solution for generation t.

Procedure EA
begin

$t \leftarrow 0$
i)   initialize $P(t)$
ii)   gene-mask $P(t)$
iii)   evaluate $P(t)$
iv)   store the best genome $b(t)$ among $P(t)$
v)   **while (not** termination-condition) **do**
        begin
            $t \leftarrow t + 1$
vi)   select $P(t)$ from $P(t-1)$
vii)   alter $P(t)$
viii)   gene-mask $P(t)$
ix)   evaluate $P(t)$
x)   store the best genome $b(t)$ among $b(t-1)$ and $P(t)$
        end
end

Fig. 3 Procedure of evolutionary algorithm for an artificial creature's personality

(v) Steps (vi)–(x) are repeated until the termination condition is reached. The maximum number of generations is used as a termination condition.

(vi) In the **while** loop, a new population (iteration $t + 1$) is formed by fitness proportional selection in that generation.

(vii) Some members of the new population undergo transformations by means of the crossover operators, Θ_χ^F, Θ_χ^I and Θ_χ^B, and the mutation operators Θ_μ^F, Θ_μ^I and Θ_μ^B, to form new genomes (Sect. 4.7).

4.3 Initialization Process

In the step 'initialize $P(t)$', x_{pk}^{F0}, x_{qk}^{I0}, x_{rk}^{B0}, $p = 1, 2, \ldots, w$, $q = 1, 2, \ldots, y$, $r = 1, 2, \ldots, z$, $k = 1, 2, \ldots, c$, of all $\mathbf{G}_i^0 = \mathbf{G}_i^t|_{t=0}$, $i = 1, 2, \ldots, n$, in $P(0) = \{\mathbf{G}_1^0, \mathbf{G}_2^0, \ldots, \mathbf{G}_n^0\}$ at generation $t = 0$, are initialized. As there are many genes to be evolved, their characteristics should be considered in the initialization process for computational efficiency and performance improvement. F-genes of all $\mathbf{G}_i^0 = \mathbf{G}_i^t|_{t=0}$ are initialized individually, since the format and the scale of genes are distinctly different from each other. $x_{1k}^{F0}, x_{2k}^{F0}, x_{3k}^{F0}$ and x_{4k}^{F0} represents volatility for genes inheritance, initial value, constant value and discount factor in order. x_{1k}^{F0} is given as either 0 or 1 when the corresponding state is volatile or not. x_{2k}^{F0} and x_{3k}^{F0} are generated in $U[0, F_{\max}]$ (uniformly distributed random variables on $[0, F_{\max}]$), where $F_{\max}$ is the upper bound for the two gene values. The discounter factor, $x_{4k}^{F0} \in U(0, 1]$.

I- and B-genes are randomly initialized in between lower and upper bounds considering user's preference values in (7). These values are used for center of the range gene values at initialization. I-genes, x_{qk}^{I0}, of all $\mathbf{G}_i^0 = \mathbf{G}_i^t|_{t=0}$ are randomly initialized as follows:

$$x_{qk}^{I0} = m_{qk}^I \dot{U}[\ \max[0, I_{\max} \cdot (\psi_k^I - \delta_I)],$$
$$I_{\max} \cdot (\psi_k^I + \delta_I)\ ] \tag{8}$$

where m_{qk}^I is the qth I-gene mask of the kth chromosome, $I_{\max}$ is the upper bound of I-gene values and δ_I is a constant for deciding the range of initialization of I-genes. B-genes, x_{rk}^{B0}, of $\mathbf{G}_i^0 = \mathbf{G}_i^t|_{t=0}$ are similarly initialized as follows:

$$x_{rk}^{B0} = U[\ \max[0, B_{\max} \cdot (\psi_k^B - \delta_B)],$$
$$B_{\max} \cdot (\psi_k^B + \delta_B)\ ] \tag{9}$$

where $B_{\max}$ is the upper bound of B-gene values and δ_B is a constant for the range of B-gene values.

4.4 Gene Masking

To build a truly believable artificial creature with a specific personality, the artificial creature is required to have a proper genome that leads to the generation of plausible internal states and behaviors. In this regard, a gene masking process is needed to isolate unnecessary genes. The masking process is divided into I-gene masking

(I-masking), and B-gene masking (B-masking). I-masking enables EGPP to provide the artificial creature with reasonable internal states. The I-masking $\Theta_m^I(\mathbf{x}^I)$ for I-gene vector $\mathbf{x}_k^I$, $k = 1, 2, \ldots, c$, is defined as

$$\Theta_m^I(\mathbf{x}^I) = \left(\mathbf{m}_1^I abs(\mathbf{x}_1^I) \ \mathbf{m}_2^I abs(\mathbf{x}_2^I) \ \ldots \ \mathbf{m}_c^I abs(\mathbf{x}_c^I) \right)$$

where $\mathbf{m}_k^I$, $k = 1, 2, \ldots, c$, is an I-masking matrix and $abs(\mathbf{x}_k^I)$ is a resultant vector after the absolute value of each element is taken. Each element in the I-masking matrix has one of three masking values, -1, 0, or $+1$, which represent negative masking, zero masking, and positive masking, respectively. With I-masking, I-genes $\mathbf{x}^I$ are replaced by their masked equivalents. Similarly, B-masking is required such that the artificial creature may select more appropriate behaviors given a specified internal state and perception. The B-masking $\Theta_m^B(\mathbf{x}^B)$ for B-gene vector $\mathbf{x}_k^B$, $k = 1, 2, \ldots, c$, is defined as

$$\Theta_m^B(\mathbf{x}^B) = \left(\mathbf{m}_1^B \mathbf{x}_1^B \ \mathbf{m}_2^B \mathbf{x}_2^B \ \ldots \ \mathbf{m}_c^B \mathbf{x}_c^B \right)$$

where $\mathbf{m}_k^B$, $k = 1, 2, \ldots, c$, is a B-masking matrix. In the same manner as I-genes, all B-genes are replaced by their masked equivalents. Elements of the B-masking matrix take values either 0 or 1. Zero masking prevents the behavior from being selected, while positive masking retains the voting values for the behaviors relevant to the perception.

4.5 Perception Scenario

A series of randomly generated perceptions is applied to the artificial creature and its internal states and behaviors are observed. These internal and external responses are used to evaluate its genome for a specific personality at every generation. The perception scenario is designed using stimuli from the environment. The manner in which the stimuli may be applied is customizable and is formalized as follows: each step in the perception scenario is characterized by an event. For the user, the event represents a stimulus applied to the artificial creature. For the creature, it is a perceived event. Based on this formalization, the perception scenario is defined as the permutation of perceivable information of an artificial creature, for the given perception scenario time. Ideally, it is desired that the user inputs scenarios in a feedback fashion, analyzing the response to each input, but this is an intricate and exhaustive process, requiring extensive analysis on the part of the user, which is inappropriate for a home-user scenario and hence the randomly generated sequence is utilized. However, care is taken to ensure that illogically sequenced perception scenarios do not result, such as the situation where an obstacle must exist prior to the perception 'SUDDEN_DISAPPEAR' activating as a possibility. Perception scenario is characterized as follows:

$$(\mathbb{A}, \tilde{\mathbb{A}}, \check{\mathbb{P}}, t_s, T_s) \tag{10}$$

where $\mathbb{A} = \{A_1, A_2, \ldots, A_q\}$ is the set of all perception groups representing similar perceived events and q is the number of perception groups. $\tilde{\mathbb{A}} = \{\tilde{A}_1, \tilde{A}_2, \ldots, \tilde{A}_q\}$, for $\tilde{A}_i \subseteq A_i (i = 1,\ldots,q)$, which is used in a scenario for which a user does not wish to or cannot feasibly expose the artificial creature to a complete set of stimuli. $\tilde{\mathbb{P}} = \{\tilde{p}_1, \tilde{p}_2, \ldots, \tilde{p}_q\}$ is the set of the generation probabilities for the different percepts associated with each $\tilde{A}_i \in \tilde{\mathbb{A}}$. Events occur at discrete time intervals with random variable time step, $t_s \in [t_{min}, t_{max}]$, where the minimum of t_{min} is the sampling time, ΔT. T_s is the duration length of the scenario, and is called the perception scenario time.

Rity can perceive 47 perceptions and a set of all the perception groups in Table 2 is $\mathbb{A} = \{A_{po}, A_{ob}, A_{bt}, A_{ph}, A_{so}, A_{of}, A_{ba}\}$, where A_{po} is the perception group related to posture, A_{ob} obstacle, A_{bt} brightness/temperature, A_{ph} pat/hit, A_{so} sound, and A_{of} object/face, and A_{ba} battery. Perception scenarios are generated through the GUI as shown in Fig. 5, where the user can set both the scenario time (T_s) and voting values for the generation probabilities of perception groups. Once 'Start' button is pressed, a scenario is automatically generated as shown on the right side of the figure, where the number represents the event occurrence time. The parameters, $t_{min} = 0.1$ s, $t_{max} = 10$ s, $T_s = 500$ s, $\tilde{\mathbb{A}} = \{A_{po}, A_{ob}, A_{bt}, A_{ph}, A_{so}, A_{of}, A_{ba}\}$ and voting values $\{\tilde{v}_1, \tilde{v}_2, \ldots, \tilde{v}_q\} = \{0.5, 0.5, 0.5, 0.7, 0.5, 0.7, 0.5\}$ were used. The set of generation probabilities $\tilde{\mathbb{P}}$ can be calculated by $\tilde{p}_i = \tilde{v}_i / \sum_{r=1}^{q} (\tilde{v}_r)$. Figure 4 shows its outcomes. Perception scenario 1 (Fig. 4) is used for evaluating genomes and perception scenario 2 (Fig. 4) is for verifying the selected genome by EGPP. The upper parts of Fig. 4 show timing diagrams of the 47 types of perceptions for 500 s. The lower parts show histograms representing the frequency of each perception.

4.6 Fitness Function

Considering the diverse range of personalities, a well-designed fitness function is needed to evaluate genomes for a specific personality. The evaluation procedure has the following three steps:

- Step 1: A genome is imported to the artificial creature.
- Step 2: A series of random stimuli in a perception scenario is applied to the artificial creature in a virtual environment.
- Step 3: A fitness is calculated by evaluating its internal states and behaviors.

Table 2 List of perceptions and their groups

Group	Perception	Group	Perception	Group	Perception	Group	Perception
Posture (A_{po})	POWER_ON	Bright-ness/ (A_{bt})	SUDDEN_LIGHTNESS	Pat/Hit (A_{ph})	HEAD_PATTED	Object/ Face (A_{of})	OBJECT1_DETECTED
	SHAKEN		SUDDEN_DARKNESS		HEAD_HITTED		OBJECT1_CLOSE
	DANDLED	Tempe-rature (A_{bt})	GLARING		BODY_PATTED		OBJECT2_DETECTED
	SHOCKED		NORMAL		BODY_HITTED		OBJECT2_CLOSE
	LIFTED		DARK	Sound (A_{so})	SOUND_NOISY		OBJECT3_DETECTED
	FALLEN		VERY_DARK		SOUND_NORMAL		OBJECT3_CLOSE
	CORRECT_POSTURE		MILD		SOUND_CALM		FACE_DETECTED
Obstacle (A_{ob})	OBSTACLE_EXIST		HOT		SUDDEN_LOUD		FACE_CLOSE
	DISTANCE_NEAR		COLD		SUDDEN_CALM	Battery (A_{ba})	BATTERY_LOW
	DISTANCE_MID				VOICE		BATTERY_NORMAL
	DISTANCE_FAR				VOICE_GOOD		BATTERY_FULL
	SUDDEN_APPEAR				VOICE_BAD		
	SUDDEN_DISAPPEAR				VOICE_HELLO _OR_BYE		
	CLIFF						

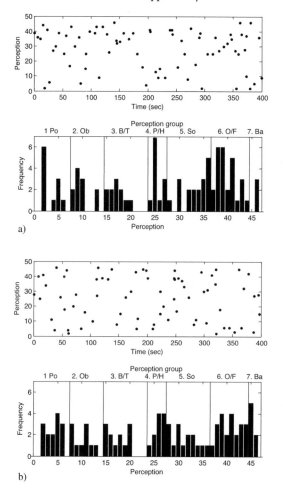

Fig. 4 Generated perception scenarios for 47 perceptions and 7 perception groups. **a**) Perception scenario 1 for evaluation of genomes. **b**) Perception scenario 2 for verification of the selected genome

In Step 2, dependent on the imported genome, it generates internal states and relevant behaviors in response to stimuli. The fitness function can be designed by using the difference between the user's preference and the following two evaluation functions: one to evaluate internal states and the other to evaluate behaviors (see (13)).

Evaluation function for internal states

Internal state vector, $\mathbf{I}(t, \mathbf{G}) = [\alpha_1(t, \mathbf{G}), \dots, \alpha_c(t, \mathbf{G})]^T$ is defined at time t for a genome $\mathbf{G}$, where c is a number of internal states.

Rity has $\mathbf{M}(t, \mathbf{G}) = [m_1(t, \mathbf{G}), \dots, m_6(t, \mathbf{G})]^T = [\alpha_1(t, \mathbf{G}), \dots, \alpha_6(t, \mathbf{G})]^T$ in ((1), $\mathbf{H}(t, \mathbf{G}) = [h_7(t, \mathbf{G}), \dots, h_9(t, \mathbf{G})]^T = [\alpha_7(t, \mathbf{G}), \dots, \alpha_9(t, \mathbf{G})]^T$, and

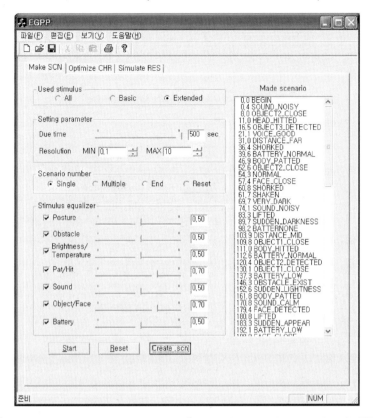

Fig. 5 Perception scenario time setting and perception scenario generation via GUI

$\mathbf{E}(t, \mathbf{G}) = [e_{10}(t, \mathbf{G}), \ldots, e_{14}(t, \mathbf{G})]^T = [\alpha_{10}(t, \mathbf{G}), \ldots, \alpha_{14}(t, \mathbf{G})]^T$ from (2), (3) and (4). For fitness, one evaluates the possession ratio of each internal state in response to stimuli in a perception scenario for perception scenario time T_s. The possession ratio of the kth ($k = 1, 2, \ldots, c$) internal state for T_s, $\Phi_{pk}^I(T_s, \mathbf{G})$, is defined as

$$\Phi_{pk}^I(T_s, \mathbf{G}) = \left(\sum_{j=1}^{T_s/\Delta T} \alpha_k(T_s, \mathbf{G}) \right) / \Phi_p^I(T_s, \mathbf{G}), \tag{11}$$

where $\Phi_p^I(T_s, \mathbf{G})$ is the sum of the possession value of all internal states defined by

$$\Phi_p^I(T_s, \mathbf{G}) = \sum_{j=1}^{T_s/\Delta T} \sum_{k=1}^{c} \alpha_k(T_s, \mathbf{G}). \tag{(11).a}$$

Evaluation function for behaviors

Given a set of behavior groups $\mathbf{B_c}^T = [\beta_1, \beta_2, \ldots, \beta_c]$, one examines the frequency of each behavior group (Table 3) for T_s. The frequency of the kth behavior group for T_s is defined as

$$\Phi_{fk}^{BG}(T_s, \mathbf{G}) = f_k^{BG}(T_s, \mathbf{G})/n_{BG}, \tag{12}$$

where the data set consists of $n_{BG} = \sum_{k=1}^{c} f_k^{BG}(T_s, \mathbf{G})$ observations, with the behavior group β_k appearing $f_k^{BG}(T_s, \mathbf{G})$ times for k ($k = 1, 2, \ldots, c$). As shown in Table 3, the behavior groups are classified on the basis of how each behavior group is

Table 3 List of state related behavior groups

Related internal mode	Index	Internal state	Internal state related behavior group
Motivation	1	Curiosity	approach, observe, touch, chase, search_looking_around
	2	Intimacy	touch, eyecontact, glad, approach, show_me_face, hug_me, lean, chase, once_again, handshake
	3	Monotony	shake_arm_and_leg, snooze, look_around, stretch
	4	Avoidance	turn_and_ignore, look_away, move_backward, cover_eyes, kick, roar, head_down, hide_head, resist
	5	Greed	look_around, search_looking_around, search_wandering, collect, piss, roar, chase
	6	Control	show_me_face, hug_me, give_me_ball, once_again, complain_for_food
Homeostasis	7	Fatigue	refuse, rest, crouch, recharge, faint, head_down
	8	Drowsiness	yawn, sleep, snooze, lie, rest
	9	Hunger	complain_for_food, rumble, search_wandering
Emotion	10	Happiness	hurrah, dance_with_arms, mouth_open, shake_arms
	11	Sadness	hit_ground, head_down, be_ill, hug_me, weep
	12	Anger	shake_head, hit_head, roar, turn_and_ignore, look_away
	13	Fear	resist, hide_head, tremble, faint, move_backward
	14	Neutral	stop, lie, look_around

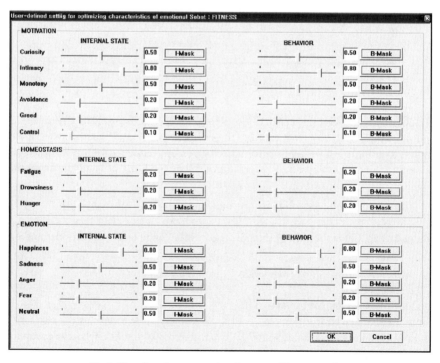

Fig. 6 Users' preference setting for agreeable personality

closely related to each internal state, where each behavior has the advantage of good consistency with each internal state.

The user sets the relevant preference values ψ_k^I and ψ_k^B in (7) for Rity's personality according to his/her preference (Table 1) through the GUI (Fig. 6), where each preference value is assigned between 0 and 1. Since the user's preference corresponds to the desired personality, EGPP finds the I-genes and B-genes to meet the preference by utilizing the preference in the fitness function. Agreeable personality model in Table 1 is set by the GUI in Fig. 6. If one clicks either the 'I-Mask' or 'B-Mask' button in Fig. 6, the gene masking process can be implemented.

Using (11) and (12), the fitness function is defined as

$$\Phi(T_s, \mathbf{G}) = N - \rho \left[\sum_{k=1}^{c} (1/\tilde{\psi}_k^I) |\tilde{\psi}_k^I - \Phi_{pk}^I(T_s, \mathbf{G})| + \sum_{k=1}^{c} (1/\tilde{\psi}_k^B) |\tilde{\psi}_k^B - \Phi_{fk}^{BG}(T_s, \mathbf{G})| \right]$$

(13)

with the normalized preference value, ψ_k^I and ψ_k^B, defined as

$$\tilde{\psi}_k^I = \psi_k^I / \sum_{l=1}^{c} \psi_l^I, \qquad \tilde{\psi}_k^B = \psi_k^B / \sum_{l=1}^{c} \psi_l^B$$

(13.a)

where $\Phi_{pk}^I(T_s, \mathbf{G})$, is the possession ratios of the kth internal state, $\Phi_{fk}^{BG}(T_s, \mathbf{G})$ is the frequency of the kth behavior group in $\mathbf{B_c}$. N is a constant number and ρ a scale factor for the difference terms.

4.7 Genetic Operators

EGPP uses two genetic operators for crossover and mutation. Crossover operator Θ_χ is divided into the F-crossover operator, Θ_χ^F, I-crossover operator, Θ_χ^I, and B-crossover operator, Θ_χ^B. These operations are performed only between parental genes of the same kind, length, and chromosomal order. For example, there are two parental artificial genomes $\mathbf{G}_1^t$ and $\mathbf{G}_2^t$. In the case of Θ_χ^F, $\mathbf{x}_{1k}^{Ft}$ can crossover only with $\mathbf{x}_{2k}^{Ft}$ of the same F-genes. Based on this policy, two kinds of crossover are possible. In the first method, Θ_χ^F, Θ_χ^I and Θ_χ^B operate together between two arbitrary parents at crossover rate p_χ, where two offspring are generated from two parents. In the second method, Θ_χ^F, Θ_χ^I and Θ_χ^B operate independently between two arbitrary parents at three different kinds of gene-dependent crossover rates p_χ^F, p_χ^I and p_χ^B. Consequently two offspring are generated from six parents. In the same manner, the mutation operator Θ_μ is divided into operators Θ_μ^F, Θ_μ^I, and Θ_μ^B and the two methods outlined above can be applied.

Although this section shows only two kinds of crossover and mutation operators, as examples, it should be noted that various crossover and mutation operators can be developed for the genomes of 2D representation.

5 Experiments

The agreeable and antagonistic personality models were chosen to validate the proposed EGPP and the fittest genome obtained was implanted into Rity to verify the feasibility of EGPP. The parameter settings of EGPP were applied equally in both cases of agreeable and antagonistic personalities. The population size was 20 and the number of generations was 1000. Crossover and mutation operate independently between two arbitrary parents at three different kinds of gene-dependent crossover and mutation rates. The crossover and mutation rates for the I- and B-genes were set to $(0.1, 0.05)$ and $(0.2, 0.05)$, respectively. F-crossover and F-mutation rates were set to 0.0 to keep the assigned fundamental characteristics.

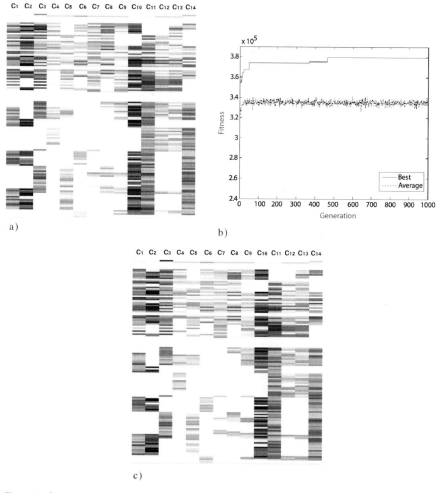

a)

b)

c)

Fig. 7 Evolutionary generation of the agreeable genome by EGPP. **a)** An initial genome. **b)** Evolution process by EGPP. **c)** Final genome

5.1 Generating Genomes by EGPP

Figure 7 shows the evolution process for generating a genome for an agreeable personality. The initial genomes were generated using the process as outlined in Sect. 4.4.3 One of them is shown in Fig. 7, which has chromosomes for intimacy (C_2) and happiness (C_{10}) with strong I- and B-genes while chromosomes for avoidance (C_4), greed (C_5), desire to control (C_6), anger (C_{12}) and fear (C_{13}) have weak I-genes and B-genes. Despite providing intuitive bounds for random initialization, these settings did not always guarantee a viable set of genomes for an agreeable personality in a complex test environment. They do, however, provide a heuristic starting point for optimizing the genome for a more consistent personality with EGPP. The

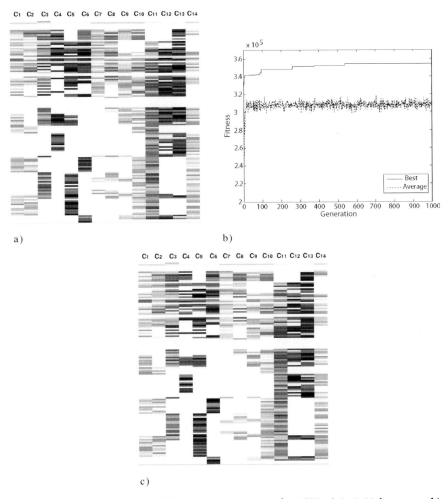

a)

b)

c)

Fig. 8 Evolutionary generation of the antagonistic genome by EGPP. **a)** An initial genomes. **b)** Evolution process by EGPP. **c)** Final genome

progress of the EGPP is shown in Fig. 7, where a steady improvement in performance is quantitatively demonstrated. Figure 7 illustrates the genome after EGPP, indicating significant changes to the genome's structure. Similarly, Fig. 8 illustrates the same process of EGPP for genomes that were evolved for an antagonistic personality. The optimized genomes for the agreeable and antagonistic personalities can be seen in Table 4.

The normalized preference column in the table indicates the normalized user assigned preference. A feature which can be seen, however, is the discrepancy between the normalized value and that from the optimized genome. This discrepancy exists for a number of reasons, such as complexity introduced by the large parameter set, the constraints enforced by the masking process, and the dependency on the scenario chosen for optimization. Thus, despite an exact match being near impossible due to the random nature of these factors, the algorithm generates values which are nevertheless close to the normalized values.

Table 4 Optimized genomes for the agreeable and antagonistic personalities

| | | | Agreeable personality | | | | | |
| | | | Internal State | | | Behavior | | |
Group	State	k	Assigned preference	Normalized preference	Optimized genome	Assigned preference	Normalized preference	Optimized genome
Motivation	Curiosity	1	0.5	0.098	0.165	0.5	0.098	0.170
	Intimacy	2	0.8	0.157	0.166	0.8	0.157	0.207
	Monotony	3	0.5	0.098	0.100	0.5	0.098	0.148
	Avoidance	4	0.2	0.039	0.031	0.2	0.039	0.044
	Greed	5	0.2	0.039	0.057	0.2	0.039	0.059
	Control	6	0.1	0.020	0.000	0.1	0.020	0.022
Homeostasis	Fatigue	7	0.2	0.039	0.056	0.2	0.039	0.030
	Drowsiness	8	0.2	0.039	0.043	0.2	0.039	0.037
	Hunger	9	0.2	0.039	0.042	0.2	0.039	0.037
Emotion	Happiness	10	0.8	0.157	0.066	0.8	0.157	0.237
	Sadness	11	0.5	0.098	0.122	0.5	0.098	0.000
	Anger	12	0.2	0.039	0.043	0.2	0.039	0.007
	Fear	13	0.2	0.039	0.100	0.2	0.039	0.000
	Neutral	14	0.5	0.098	0.007	0.5	0.098	0.001
			Antagonistic personality					
Group	State	k	Assigned preference	Normalized preference	Optimized genome	Assigned preference	Normalized preference	Optimized genome
Motivation	Curiosity	1	0.2	0.039	0.032	0.2	0.039	0.041
	Intimacy	2	0.2	0.039	0.001	0.2	0.039	0.052
	Monotony	3	0.5	0.098	0.030	0.5	0.098	0.047
	Avoidance	4	0.8	0.157	0.179	0.8	0.157	0.337
	Greed	5	0.8	0.157	0.199	0.8	0.157	0.047
	Control	6	0.7	0.137	0.086	0.7	0.137	0.012
Homeostasis	Fatigue	7	0.2	0.039	0.071	0.2	0.039	0.029
	Drowsiness	8	0.2	0.039	0.000	0.2	0.039	0.035
	Hunger	9	0.2	0.039	0.061	0.2	0.039	0.006
Emotion	Happiness	10	0.2	0.039	0.019	0.2	0.039	0.029
	Sadness	11	0.5	0.098	0.079	0.5	0.098	0.076
	Anger	12	0.8	0.157	0.138	0.8	0.157	0.169
	Fear	13	0.8	0.157	0.106	0.8	0.157	0.122
	Neutral	14	0.2	0.039	0.000	0.2	0.039	0.000

5.2 Verification of Evolved Genomes

This section verifies the effectiveness of EGPP by implanting the agreeable genome A (Fig. 7(c)) and the antagonistic genome B (Fig. 8(c)) into two artificial creatures, Rity A and Rity B, respectively, and by observing their internal states and behaviors when perception scenario 2 (Fig. 4(b)) is applied to them. The verified results are shown in Table 5. It can be seen that, similar to the optimized results, the discrepancy between the normalized value and that from the evaluated genome, is caused by the large parameter set and masking process constraints, as mentioned earlier, in addition to the use of a different perception scenario from that for the optimization process.

Verification on internal state responses

Figures 9 and 10 show the experimental results for internal state responses when the perception scenario 2 was applied to agreeable Rity A and antagonistic Rity B,

Table 5 Verification of genomes by perception scenario 2, which is shown in Fig. 4 for verification of the optimized genome)

Group	State	k	Agreeable personality					
			Assigned preference	Normalized preference	Evaluated genome	Assigned preference	Normalized preference	Evaluated genome
Motivation	Curiosity	1	0.5	0.098	0.154	0.5	0.098	0.172
	Intimacy	2	0.8	0.157	0.174	0.8	0.157	0.191
	Monotony	3	0.5	0.098	0.049	0.5	0.098	0.135
	Avoidance	4	0.2	0.039	0.043	0.2	0.039	0.019
	Greed	5	0.2	0.039	0.061	0.2	0.039	0.088
	Control	6	0.1	0.020	0.002	0.1	0.020	0.000
Homeostasis	Fatigue	7	0.2	0.039	0.083	0.2	0.039	0.023
	Drowsiness	8	0.2	0.039	0.022	0.2	0.039	0.019
	Hunger	9	0.2	0.039	0.062	0.2	0.039	0.019
Emotion	Happiness	10	0.8	0.157	0.120	0.8	0.157	0.270
	Sadness	11	0.5	0.098	0.070	0.5	0.098	0.051
	Anger	12	0.2	0.039	0.029	0.2	0.039	0.000
	Fear	13	0.2	0.039	0.053	0.2	0.039	0.000
	Neutral	14	0.5	0.098	0.078	0.5	0.098	0.014

Group	State	k	Antagonistic personality					
			Assigned preference	Normalized preference	Optimized genome	Assigned preference	Normalized preference	Optimized genome
Motivation	Curiosity	1	0.2	0.039	0.071	0.2	0.039	0.060
	Intimacy	2	0.2	0.039	0.002	0.2	0.039	0.066
	Monotony	3	0.5	0.098	0.029	0.5	0.098	0.086
	Avoidance	4	0.8	0.157	0.173	0.8	0.157	0.238
	Greed	5	0.8	0.157	0.186	0.8	0.157	0.113
	Control	6	0.7	0.137	0.051	0.7	0.137	0.013
Homeostasis	Fatigue	7	0.2	0.039	0.010	0.2	0.039	0.046
	Drowsiness	8	0.2	0.039	0.000	0.2	0.039	0.073
	Hunger	9	0.2	0.039	0.070	0.2	0.039	0.000
Emotion	Happiness	10	0.2	0.039	0.003	0.2	0.039	0.040
	Sadness	11	0.5	0.098	0.074	0.5	0.098	0.046
	Anger	12	0.8	0.157	0.133	0.8	0.157	0.113
	Fear	13	0.8	0.157	0.134	0.8	0.157	0.106
	Neutral	14	0.2	0.039	0.033	0.2	0.039	0.000

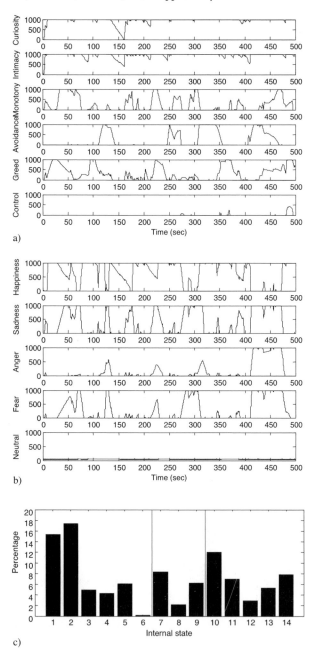

Fig. 9 Internal state responses of agreeable Rity A to the perception scenario 2. **a**) Motivation response. **b**) Emotion response. **c**) Normalized possession ratio histogram

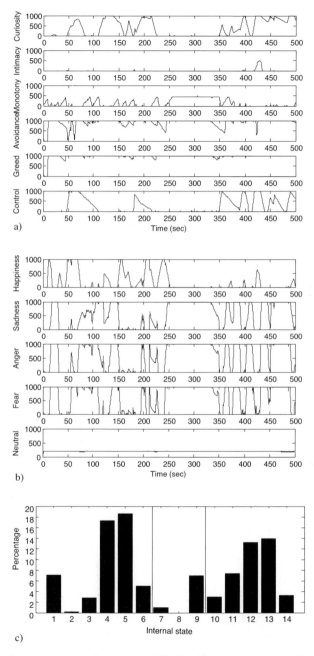

Fig. 10 Internal state responses of antagonistic Rity B to the perception scenario 2. **a**) Motivation response. **b**) Emotion response. **c**) Normalized possession ratio histogram

respectively. Figure 9 shows that states of curiosity and intimacy have wider distribution than those of avoidance, greed, and desire to control in motivation for the perception scenario time of 500 s. Figure 9(b) shows that happiness state has the widest distribution among emotion states. Figure 9(c) shows a histogram of normalized possession ratios calculated in (11) for the perception scenario 2. The horizontal axis represents the index of 14 internal states and the vertical axis represents the normalized possession ratios of internal states. The 1st, 2nd, and 10th internal states have high possession ratios, which indicate strong states of curiosity and intimacy in motivation and of happiness in emotion, while the 4th, 5th, 6th, 12th and 13th internal states have low possession ratios which indicate weak states of avoidance, greed, and desire to control in motivation, and of anger and fear in emotion.

In contrast, Fig. 10(a) shows that states of avoidance and greed have wider distribution than those of curiosity and intimacy in motivation for the same perception scenario 2. Figure 10(b) shows that states of sadness, anger, and fear have wider distribution than happiness state in emotion. In Fig. 10(c), '4-Avoidance', '5-Greed',

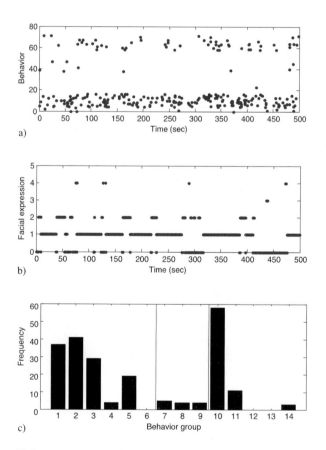

Fig. 11 External behavior responses of agreeable Rity A to the perception scenario 2. **a)** Behavior response. **b)** Facial expression response. **c)** Frequency of behavior groups

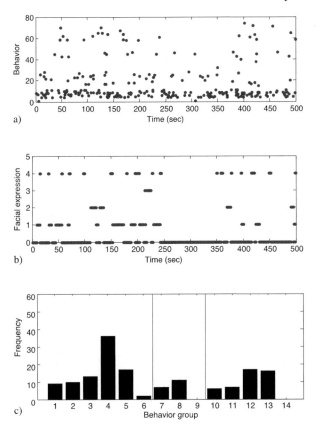

Fig. 12 External behavior responses of antagonistic Rity B to the perception scenario 2. **a**) Behavior response. **b**) Facial expression response. **c**) Frequency of behavior groups

'12-Anger', and '13-Fear' have high possession ratios, while '1-Curiosity', '2-Intimacy', '3-Monotony', '10-Happiness', and '14-Neutral' have low possession ratios.

Verification of behavior responses Figures 11 and 12 show external output responses of agreeable Rity A and antagonistic Rity B, respectively, for the same perception scenario 2 used for Figs. 9 and 10. All the behaviors are indexed sequentially and the indexes of five facial expressions, neutral, happiness, sorrow, anger, and fear are set to 0, 1, 2, 3, and 4. The behavior groups are sequentially indexed based on Table 3. Figures 11(a) and 12(b) show the behavior responses, where the horizontal axis represents the perception scenario time and the vertical axis represents output behavior indexes. In Fig. 11(b), there are many more facial expressions of happiness than other kinds of facial expressions. In contrast, in Fig. 12(b), there are more facial expressions of sorrow and anger than those of happiness. Figure 11(c) shows that the frequencies of the behaviors belonging to the groups such as '1-Curiosity', '2-Intimacy', and '10-Happiness' are high. In contrast, Fig. 12(c) shows that the frequencies of the groups such as '4-Avoidance', '5-Greed', '12-Anger', and '13-Fear' are high.

For both agreeable and antagonistic genomes, plausible artificial creatures, Ritys were observed for all internal states and behaviors simultaneously for the prescribed perception scenario. The genomes obtained defined consistent and distinct personalities for Ritys. These experimental results show the effectiveness of EGPP as an evolutionary gene-generative mechanism for the personality desired by the user. At http://rit.kaist.ac.kr/~ritlab/research/Artificial_Creatures/agreeable_antagonistic. wmv, video clips of two Ritys are available.

6 Conclusions

This chapter has presented an overview of the evolutionary process for generating a genome for an artificial creature with its own personality that resembles its living counterpart. This enabled the creation of interactive and believable Sobots for the ubiquitous robot system. The artificial creature was defined by the genome composed of chromosomes in which genes were devised as the basic building blocks for representing the personality. Using these building blocks, the evolutionary generative process enabled an artificial creature to have a personality desired by the user. The key objective of this process was to generate an artificial creature with a personality that was both complex and feature-rich, but still plausible by human standards as an intelligent life form. This was demonstrably achieved by masking processes and a stochastic voting mechanism for behavior selection. It also outlined evolutionary procedures that allowed for reproduction through artificial means, completing the process of design for a fully functional life form. The technique was utilized to implant the evolved genome into a Sobot artificial creature, Rity and then by confirming its traits in a 3D virtual world. In the fables of the Arabian Nights, a mythical creature, the Genie, emerged from within a magical lamp, and satisfied all of our desires. Future systems based on the ubiquitous robot paradigm, are bringing this dream to fruition through their immense capabilities of context-aware, calm, networked service available at anytime, anyplace and whenever desired. With the limitless possibilities they present coupled with the power of evolutionary computation, the artificial evolution of personality holds great promise to generate believable artificial creatures which can seamlessly interact with humans and provide them with services.

References

1. M. Brady, "Artificial Intelligence and Robotics", in *Artificial Intelligence*, Vol.26, pp. 79–121, 1985
2. Jong-Hwan Kim, "Ubiquitous Robot", in *Computational Intelligence, Theory and Applications* (edited by B. Reusch), in Springer, pp. 451-459, 2004 (Keynote speech paper of the 8th Fuzzy Days International Conference, Dortmund, Germany, Sep. 2004)
3. J.-H. Kim, Y.-D. Kim, and K.-H. Lee, "The 3rd Generation of Robotics: Ubiquitous Robot", in *Proc. of the International Conference on Autonomous Robots and Agents (Keynote Speech Paper)*, Palmerston North, New Zealand, 2004

4. K.-H. Lee and J.-H. Kim, "A Survey of Ubiquitous Space Where Ubiquitous Robots will Exist", in *Proc. of 2004 FIRA Robot World Congress*, 2004

5. Jong-Hwan Kim, "Ubiquitous Robot: Recent Progress and Development", in *SICE-ICASE International Joint Conference 2006 (Keynote Speech Paper)*, Busan, Korea, pp. I-25 – I-30, 2006

6. Jong-Hwan Kim, "Genetic Robot based on Ubiquitous Technologies", in *Proc. of 2nd Asia International Symposium on Mechatronics* (Keynote Speech Paper), Hong Kong, pp. 30–35, Dec. 2006

7. J.-H. Kim, K.-H. Lee, Y.-D. Kim, N.S. Kuppuswamy, and Jun Jo, "Ubiquitous Robot: A new Paradigm for Integrated Services", in *IEEE International Conference on Robotics and Automation* (accepted for publication), Italy, April 2007

8. B.M. Blumberg, *Old Tricks, New Dogs: Ethology and Interactive Creatures*, PhD Dissertation, MIT, 1996

9. S.-Y. Yoon, B.M. Blumberg, and G.E. Schneider, "Motivation driven learning for interactive synthetic characters", in *Proceedings of Autonomous Agents*, pp. 365-372, 2000

10. J. Bates, A.B. Loyall, and W.S. Reilly, "Integrating Reactivity, Goals, and Emotion in a Broad Agent", in *Proc. of 14th Ann. Conf. Cognitive Science Soc.*, Morgan Kaufmann, San Francisco, 1992

11. H. Miwa, T. Umetsu, A. Takanishi, and H. Takanobu, "Robot personality based on the equation of emotion defined in the 3D mental space", in *Proc. of IEEE International Conference on Robotics and Automation*, Vol. 3, pp. 2602–2607, 2001

12. Y.-D. Kim, and J.-H. Kim, "Implementation of Artificial creature based on Interactive Learning", in *Proc. of the FIRA Robot World Congress*, pp. 369–374, 2002

13. Y.-D. Kim, Y.-J. Kim, and J.-H. Kim, "Behavior Selection and Learning for Synthetic Character", in *Proc. of the IEEE Congress on Evolutionary Computation*, pp. 898–903, 2004

14. R. Arkin, M. Fujita, T. Takagi and R. Hasekawa, "An Ethological and Emotional Basis for Human-Robot Interaction", in *Robotics and Autonomous Systems*, Vol. 42, pp. 191–201, 2003

15. L. Cañamero, O. Avila-Garcia and E. Hafner, "First Experiments Relating Behavior Selection Architectures to Environmental Complexity", in *Proc. of the 2002 IEEE/RSJ Int. Conference on Intelligent Robots and Systems*, pp. 3024–3029, 2002

16. D. Cañamero, *Designing Emotions for Activity Selection*, Technical Report DAIMI PB 545, Dept. of Computer Science, University of Aarhus, Denmark, 2000

17. S. Grand, D. Cliff and A. Malhotra, "Creatures: Artificial Life Autonomous Software Agents for Home Entertainment", in *Proc. of the First International Conference on Autonomous Agents*, pp. 22–29, 1997

18. S. Grand and D. Cliff, "Creatures: Entertainment Software Agents with Artificial Life", in *Autonomous Agents and Multi-Agent Systems*, Vol. 1, pp. 39-57, 1998

19. B. D. Bryant and R. Miikkulainen, "Neuroevolution for adaptive teams", in *Proc. of the IEEE Congress on Evolutionary Computation*, Vol. 3, pp. 2194–2201, 2003

20. S. M. Lucas, "Cellz: a simple dynamic game for testing evolutionary algorithms", in *Proc. of the IEEE Congress on Evolutionary Computation*, pp. 1007–1014, 2004

21. K. O. Stanley and R. Miikkulainen, "Competitive coevolution through evolutionary complexification", in *Journal of Artificial Intelligence Resesarch*, Vol. 21, pp. 63–100, 2004

22. K. O. Stanley, B. D. Bryant, and R. Miikkulainen, "Real-time neuroevolution in the NERO video game", in *IEEE Trans. on Evolutionary Computation Special Issue on Evolutionary Computation and Games*, Vol. 9, no. 6, pp. 653–668, 2005

23. B. D. Bryant and R. Miikkulainen, "Evolving Stochastic Controller Networks for Intelligent Game Agents", in *Proc. of the IEEE Congress on Evolutionary Computation*, pp. 1007–1014, 2006

24. D. Fogel, T. Hays, and D. Johnson, "A platform for evolving characters in competitive games", in *Proc. of the IEEE Congress on Evolutionary Computation*, pp. 1420–1425, 2004
25. D. Fogel, T. Hays, and D. Johnson, "A platform for evolving intelligently interactive adversaries", BioSystems, Vol. 85, pp. 72–83, 2006
26. A. Ortony, "Making Believable Emotional Agents Believable", *Emotions in Humans and Artifacts*, MIT Press, pp. 189–211, 2002
27. J.-H. Kim, K.-H. Lee and Y.-D. Kim, "The Origin of Artificial Species: Genetic Robot", in *International Journal of Control, Automation and Systems*, Vol. 3 No. 4, 2005
28. J.-H. Kim, K.-H. Lee, Y.-D. Kim and I.-W. Park, "Genetic Representation for Evolving Artificial Creature", in *Proc. of the IEEE Congress on Evolutionary Computation*, pp. 6838–6843, July 2006
29. R. R. McCrae and P. T. Costa, "Validation of a Five-Factor Model of Personality across Instruments and Observers", in *J. Pers. Soc. Psychol. 52*, pp. 81–90, 1987
30. E. Rolls, "A theory of emotion, its functions, and its adaptive value", *Memory and Emotion*, Eds. P. Calabrese and A. Neugebauer. World Scientific: London, pp. 349–360, 2002
31. C. Breazeal, "Function meets style: Insights from emotion theory applied to HRI", *IEEE Trans. Syst., Man, Cybern. C*, Vol. 34, pp. 187–194, May 2004
32. A. Sloman, "How Many Separately Evolved Emotional Beasties Live Within Us?" *Emotions in Humans and Artifacts*, MIT Press, pp. 35–114, 2002

Some Guidelines for Genetic Algorithm Implementation in MINLP Batch Plant Design Problems

Antonin Ponsich, Catherine Azzaro-Pantel, Serge Domenech, and Lue Pibouleau

Laboratoire de Génie Chimique - INP ENSIACET UMR 5503 CNRS/INP/UPS, 5 rue Paulin Talabot BP 301, 31106 Toulouse Cedex 1, France.
Catherine.AzzaroPantel@ensiacet.fr

Abstract

In recent decades, a novel class of optimization techniques, namely metaheuristics, has been developed and devoted to the solution of highly combinatorial discrete problems. The improvements provided by these methods were extended to the continuous or mixed-integer optimization area. This chapter addresses the problem of adapting a Genetic Algorithm (GA) to a Mixed Integer Non-linear Programming (MINLP) problem. The basis of the work is optimal batch plant design, which is of great interest in the framework of Process Engineering. This study deals with the two main issues for GAs, i.e. the treatment of continuous variables by specific encoding and efficient constraints handling in GA. Various techniques are tested for both topics and numerical results show that the use of a mixed real-discrete encoding and a specific domination-based tournament method is the most appropriate approach.

Key words: Genetic Algorithms, Variable Encoding, Constraint Handling, Batch Plant Design

1 Introduction

A large range of applications drawn from Process Engineering can be expressed as optimization problems. This application range consists of examples formulated as pure continuous problems – for instance the phase equilibrium calculation problem [1], as well as problems involving pure discrete variables – like the discrete job-shop batch plant design [2]. Typically, for the former case, difficulties arise from non-linearities, while they are due to the discontinuous nature of functions and search space, for the latter. Finally, a great variety of models from the Process Engineering area combine both kinds of problems and involve simultaneously (continuous) operation and (discrete) decision variables. Design problems are good examples of this complexity level, such as process network superstructure design problems [3] or multiproduct batch plant design problems [4].

Obviously, significant investigation efforts have been carried out to develop efficient and robust optimization methods, initially especially in the Operational Research and Artificial Intelligence areas, but subsequently within the Process Engineering community. Among the diversity of optimization techniques, two main classes can be distinguished. One consists of deterministic methods, which are based on a rigorous mathematical study (derivability, continuity...) of the objective function and of the constraints to ensure an optimum. However, despite their ability to handle non-linear models, their performances can be strongly affected by non-convexities. This implies great effort to obtain a proper formulation of the model. Grossmann [5] proposes a review of the existing Mixed Integer Non-linear Programming (MINLP) techniques but it is commonly accepted that these methods might be heavily penalized by the NP-hard nature of the problems, and will then be unable to solve large size instances of a problem.

So, increasing effort was applied to the development of methods of the second class, i.e. metaheuristics or stochastic methods. They work by evaluating the objective function at various points of the search space. These points are chosen using a set of heuristics combined with the generation of random numbers. Stochastic techniques do not use any mathematical properties of the functions, so they cannot guarantee to obtain an optimum. Nevertheless, metaheuristics allow the solution of a large range of problems, particularly when the objective function is computed by a simulator embedded in an outer optimization loop [6]. Furthermore, despite their computationally greedy nature, they are quite easily adaptable to highly combinatorial optimization problems. A classification of metaheuristics and a survey of the main techniques is proposed in [7].

This study deals with the treatment of a problem drawn from the Chemical Engineering literature, i.e. optimal design of batch plants usually dedicated to the production of chemicals. The model, involving both real and integer variables, is solved with a Genetic Algorithm. This technique has already shown its efficiency for this problem class, especially when the objective function computation is carried out through the use of Discrete Event Simulators (DES) [6]. Then this chapter studies the adaptation of the Genetic Algorithm to a particular problem and focuses on the two main issues inherent to model formulation: constraint handling and variable encoding. The efficient management of these two GA internal procedures is the key to obtaining good quality results within an acceptable computation time.

This chapter is divided into six sections. The problem formulation and the methodology are presented in Section 2, while Section 3 is devoted to the model development of the Optimal Batch Plant Design problem. Section 4 describes the Genetic Algorithm implemented throughout the study. Some typical results are then analysed in Section 5 and finally, conclusions and perspectives are given in Section 6.

2 Outline of the Problem

2.1 Outline of Metaheuristics

In the last two decades, major advances were made in the optimization area through the use of metaheuristics. These methods are defined as a set of fundamental concepts that lead to design heuristics rules dedicated to the solution of an optimization problem [7]. Basically, they can be divided into two classes: neighbourhood methods and evolutionary algorithms. The former is obviously based on the definition of neighbourhood notion.

Definition 1. *Considering a set X and a string $x = [x_1, x_2, \ldots, x_n] \in X$. Let f be an application that from x leads to $y = [y_1, y_2, \ldots, y_n] \in X$. Then the neighbourhood $Y_x \subset X$ is the set of all possible images y of string x for the application f.*

Then, a neighbourhood method typically proceeds by starting with an initial configuration and iteratively replacing the solution by one of its neighbours according to an appropriate evolution of the objective function. Consequently, neighbourhood methods differ one from another by the application defining the neighbourhood of any configuration and by the strategy used to update the current solution.

A great variety of neighbourhood optimization techniques have been proposed, such as Simulated Annealing (SA, see [8]), Tabu Search (TS, see [9]), threshold algorithms [10] or GRASP methods [11], etc. SA and TS are indeed the most representative examples. Simulated Annealing mimics the physical evolution of a solid to thermal equilibrium, slowly cooling until it reaches its lower energy state. Kirkpatrick et al. [12] studied the analogy between this process and an optimization procedure. A new state, or solution, is accepted if the cost function decreases or if not, according to a probability depending on the cost increase and the current temperature.

Tabu Search tackles a group of neighbours of a configuration s and keeps the best one s' even if it deteriorates the objective function. Then, a tabu list of visited configurations is created and updated to avoid cycles like $s \to s' \to s \ldots$ Furthermore, specific procedures of intensification or diversification allow the search to be concentrated on the most promising zones or to be guided towards unexplored regions.

The second class of metaheuristics consists of evolutionary algorithms. They are based on the principle of natural evolution as stated by Darwin and involve three essential features: (i) a population of solutions to the considered problem; (ii) a technique evaluating each individual adaptation ; (iii) an evolution process made up of operators reproducing elimination of some individuals and creation of new ones (through crossover or mutation). This leads to an increase in the average quality of the solutions in the latest computed generation.

The most widely used techniques are Genetic Algorithms (GAs), Evolutionary Strategies and Evolutionary Programming. Section 4 presents in detail the GA adopted within this investigation. It must be pointed out that a large number of contributions have been published showing how their efficiency can be improved [13]. The second technique, commonly said $(\mu + \lambda) - ES$, generates λ children from μ parents and a selection step reduces the population to μ individuals for the following iteration [14]. Finally, Evolutionary Programming is based on an appropriate coding of the problem to be solved and on an adapted mutation operator [15].

To summarize, as regards metaheuristics performance, their efficiency is generally balanced by two opposite considerations: on the one hand, their general procedures are powerful enough to search for an optimum without much specific information about the problem, i.e. a "black box" context. But the *No Free Lunch* theory shows that no one method can outperform all other mehods on all problems. So, on the other hand, metaheuristics performance can be improved by integrating particular knowledge of the problem studied, but this specialization means, of course, adaptation effort.

2.2 Optimal Batch Plant Design problems

Due to the growing interest in batch operating mode, many studies have dealt with the batch plant design issue. The problem has been modelled in various forms for which assumptions are more or less simplistic. Generally, the objective consists in the minimization of plant investment cost.

Grossmann and Sargent [16] proposed a simple posynomial formulation for multiproduct batch plants. Kocis and Grossmann [17] then used the same approach to validate the good behaviour of a modified version of the Outer Approximation algorithm. This model involved only batch stages and was subjected to a constraint on the total production time. Modi and Karimi [4] modified this MINLP model by taking into account, in addition, semi-continuous stages and intermediate finite storage with fixed location. They solved small size examples (up to two products and eight operating stages) with heuristics. The same model was used again by Patel et al. [18] who treated larger size examples with Simulated Annealing, and by Wang et al. [19] [20] [21], who tackled successively Genetic Algorithms, Tabu Search and an Ants Foraging Method. Nevertheless, Ponsich et al. [22] showed that for this mixed continuous and discrete formulation, and independently from the size of the studied instance, a Branch-and-Bound technique proves to be the most efficient option. This Mathematical Programming (MP) technique is implemented in the SBB solver, which is available in the GAMS modelling environment [23].

The above mentioned formulations were further improved by taking into account continuous process variables [24] or uncertainties in product demand modelled by normal probability distributions [25] or by fuzzy arithmetic concepts, embedded in a multiobjective GA [26]. However, those sophistication levels were not considered in the framework of the present study.

2.3 Methodology

This chapter is dedicated to the treatment of MINLP problems by a Genetic Algorithm. The case study is a typical engineering problem, involving mixed integer variables and constraints. Even though the stochastic technique used is initially devoted to dealing with discrete variables, it was applied also to a large number of either continuous or mixed integer optimization problems. The crucial issue is the necessary adaptation effort to integrate the treatment of real variables and the efficient handling of the constraints.

The basis of this work is several instances of the optimal batch plant design problem and this investigation aims at testing and evaluating various operating modes of the GA. As shown in previous work, (deterministic) MP methods proved to be the most efficient for the model considered. Thus, the results are compared with the optimal solutions provided by the above mentioned SBB solver. The variables encoding issue is studied using three different size examples: the first one is a small size example but quite difficult to solve to global optimality. The two others are larger size instances. The constraint handling problem is analysed by tackling a medium size example in order to force the Genetic Algorithm to cope with a quite complex problem, without being restricted by computation time.

3 Optimal Batch Plant Design Problems

Within the Process Engineering framework, batch processes are of growing industrial importance because of their flexibility and their ability to produce high added-value products in low volumes.

3.1 Problem Presentation

Basically, batch plants are composed of items operating in a discontinuous way. Each batch then visits a fixed number of equipment items, as required by a given synthesis sequence (so-called production recipe). Since a plant is flexible enough to carry out the production of different products, the units must be cleaned after each batch has passed into it. In this study, we will only consider multiproduct plants, which means that all the products follow the same operating steps. Only the operating times may be different from one recipe to another.

The objective of the Optimal Batch Plant Design (OBPD) problem is to minimize the investment cost for all items involved in the plant, by optimizing the number and size of parallel equipment units in each stage. The production requirements of each product and data related to each item (processing times and cost coefficients) are specified, as well as a fixed global production time.

3.2 Assumptions

The model formulation for OBPD problems adopted in this chapter is based on Modi's approach [4]. It considers not only treatment in batch stages, which usually appears in all types of formulation, but also represents semi-continuous units that are part of the whole process (pumps, heat exchangers...). A semi-continuous unit is defined as a continuous unit alternating idle times and normal activity periods.

Besides, this formulation takes into account mid-term intermediate storage tanks. They are just used to divide the whole process into sub-processes in order to store an amount of materials corresponding to the difference of each sub-process productivity. This representation mode confers on the plant better flexibility for numerical resolution: it prevents the whole production process from being paralysed by one limiting stage. So, a batch plant is finally represented by a series of batch stages (B), semi-continuous stages (SC) and storage tanks (T) as shown in Fig. 1.

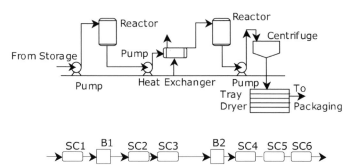

Fig. 1 Typical batch plant and modelling

The model is based on the following assumptions:

(i)   Devices used in the same production line cannot be used again by the same product.
(ii)   Production is achieved through a series of single product campaigns.
(iii)   Units of the same batch or semi-continuous stage have the same type and size.
(iv)   All intermediate tank sizes are finite.
(v)   If a storage tank exists between two stages, the operation mode is "Finite Intermediate Storage". If not, the "Zero-Wait" policy is adopted.
(vi)   There is no limitation for utility.
(vii)   The cleaning time of the batch items is included in the processing time.
(viii)   The size of the items are continuous bounded variables.

3.3 Model Formulation

The model considers the synthesis of I products treated in J batch stages and K semi-continuous stages. Each batch stage consists of m_j out-of-phase parallel items of the same size V_j. Each semi-continuous stage consists of n_k out-of-phase parallel items with the same processing rate R_k (i.e. treatment capacity, measured in volume unit per time unit). The item sizes (continuous variables) and equipment numbers per stage (discrete variables) are bounded. The $S - 1$ storage tanks, with size V_s^*, divide the whole process into S sub-processes.

Following the above mentioned notation, a MINLP problem can be formulated, minimizing the investment cost for all items. This cost is written as an exponential function of the unit size:

$$MinCost = \sum_{j=1}^{J} m_j a_j V_j^{\alpha_j} + \sum_{k=1}^{K} n_k b_k R_k^{\beta_k} + \sum_{s=1}^{S-1} c_s V_s^{*\gamma_s} \qquad (1)$$

where a_j and α_j, b_k and β_k, c_s and γ_s are classical cost coefficients. A complete nomenclature is available in Appendix A. Equation (1) shows that there is no fixed cost coefficient for any item. This may be unrealistic and will not tend towards minimization of the equipment number per stage. Nevertheless, this formulation was kept unchanged in order to compare our results with those found in the literature (see Table 1 in Sect. 4.3).

This problem is subjected to three kinds of constraints:

(i) Variable bounding:

$$\forall j \in \{1,\ldots,J\}, V_{\min} \leq V_j \leq V_{\max} \tag{2}$$

$$\forall k \in \{1,\ldots,K\}, R_{\min} \leq R_k \leq R_{\max} \tag{3}$$

(ii) Time constraint: the total production time for all products must be lower than a given time horizon H:

$$H \geq \sum_{i=1}^{I} H_i = \sum_{i=1}^{I} \frac{Q_i}{Prod_i} \tag{4}$$

where Q_i is the demand for product i.

(iii) Constraint on productivities: the global productivity for product i (of the whole process) is equal to the lowest local productivity (of each sub-process s).

$$\forall i \in \{1,\ldots,I\}, Prod_i = \min_{s \in S}[Prodloc_{is}] \tag{5}$$

These local productivities are calculated from the following equations:

(a) Local productivities for product i in sub-process s:

$$\forall i \in \{1,\ldots,I\}, \forall s \in \{1,\ldots,S\}, Prodloc_{is} = \frac{B_{is}}{T_{is}^L} \tag{6}$$

(b) Limiting cycle time for product i in sub-process s:

$$\forall i \in \{1,\ldots,I\}, \forall s \in \{1,\ldots,S\}, T_{is}^L = \max_{j \in J_s, k \in K_s}[T_{ij}, \theta_{ik}] \tag{7}$$

where J_s and K_s are, respectively, the sets of batch and semi-continuous stages in sub-process s.

(c) Cycle time for product I in batch stage j:

$$\forall i \in \{1,\ldots,I\}, \forall j \in \{1,\ldots,J\}, T_{ij} = \frac{p_{ij} + \theta_{ik} + \theta_{i,k+1}}{m_j} \tag{8}$$

where k and $k+1$ represent the semi-continuous stages before and after batch stage j.

(d) Processing time of product i in batch stage j:

$$\forall i \in \{1,\ldots,I\}, \forall j \in \{1,\ldots,J\}, \forall s \in \{1,\ldots,S\}, p_{ij} = p_{ij}^0 + g_{ij}B_{is}^{d_{ij}} \tag{9}$$

(e) Operating time for product i in semi-continuous stage k:

$$\forall i \in \{1,\ldots,I\}, \forall k \in \{1,\ldots,K\}, \forall s \in \{1,\ldots,S\}, \theta_{ik} = \frac{B_{is}D_{ik}}{R_k n_k} \tag{10}$$

(f) Batch size of product i in sub-process s:

$$\forall i \in \{1,\ldots,I\}, \forall s \in \{1,\ldots,S\}, B_{is} = \min_{j \in J_s}\left[\frac{V_j}{S_{ij}}\right] \tag{11}$$

(g) Finally, the size of intermediate storage tanks is estimated as the greatest size difference between the batches treated in two successive sub-processes:

$$\forall s \in \{1,\ldots,s-1\}, V_s^* = \max_{i \in I}[S_{is}Prod_i(T_{is}^L + T_{i,s+1}^L) - \theta_{ik} - \theta_{i,k+1}] \tag{12}$$

Then, the aim of the OBPD problems is to find the plant structure that respects the production requirements within the time horizon while minimizing the economic criterion. The resulting MINLP problem proves to be non-convex and NP-hard [19].

4 Proposed Genetic Algorithm

The following comments recall the basic principles of the stochastic optimization technique initiated by Holland [27], and then focus on the specific parameters used in this study.

4.1 General Principles

The principles of GAs rely on the analogy between a population of individuals and a set of solutions of any optimization problem. The algorithm makes the solution set evolve towards good quality, or adaptation, and mimics the rules of natural selection stated by Darwin: the weakest individuals will disappear while the best ones will survive and be able to reproduce themselves. By way of genetic inheritance, the features that make these individuals "stronger" will be preserved generation after generation.

The mechanisms implemented in the GAs reproduce this natural behaviour. Good solutions are reached by creating selection rules, that will state whether the individuals are adapted or not to the problem considered. Crossover and mutation operators then contribute to the population evolution in order to obtain, at the end of the run, a population of good quality solutions. This heuristics set is mixed with a strong stochastic feature, leading to a compromise between exploration and intensification in the search space, which contributes to GA efficiency.

The algorithm presented in this study is adapted from a very classical implementation of a GA. A major difficulty for GAs is concerned with parameters tuning. The quality of this tuning depends strongly on the user's experience and problem knowledge. A sensitivity analysis was performed to set parameters such as population size, maximal number of computed generations or survival and mutation rates, to an appropriate value.

As mentioned before, two main features of GA implementation are still a challenge for GA performance: constraint handling and variables encoding. These two points are presented in the following sub-sections.

4.2 Constraint Handling

Since constraints cannot be easily implemented just by additional equations, as in MP techniques, their handling is a key-point of GAs. Indeed, an efficient solution will largely depend on the correct choice of the constraint handling technique, in terms of both result quality and computation time.

In the framework of the studied problem, the constraint on variable bounds is intrinsically considered in the variable encoding while the constraint on productivities is implemented in the model. So the only constraint to be explicitly handled by the GA is the time constraint formulated in Eq. (4), which imposes the condition that I products are to be synthesized before a time horizon H.

The most obvious approach would be to lay down the limits of the feasible space through the elimination of all solutions violating any constraint. That means that only feasible solutions should be generated for the initial population. Then, the more severe the constraints, the more difficult it is to randomly find one feasible solution. So, the effect of this technique on the computation time is strongly penalizing. Furthermore, the search efficiency would greatly benefit from getting information about the infeasible space. Then, allowing some infeasible solutions to survive the selection step should be considered.

This was performed in various alternative constraint handling modes. Thorough reviews are given in [28] and [29]. The most famous technique is the penalization of infeasible individuals, which is typically carried out by adding, in the objective function, a quadratic constraint violation weighted by a penalty factor. This factor can be either static (i.e. set to a fixed value throughout the whole search), dynamic (increasing with the generation number), or set by analogy with simulated annealing (for more details see [29]). The drawback due to the necessity of tuning at least one parameter (the penalty factor or its initial value) can be overcome with self-adaptive penalty approaches [30], but is associated with high computation costs. Some alternative options for constraint handling are based on domination concepts, drawn from multiobjective optimization. They are implemented either within roulette wheel [31] or tournament [32] selection methods.

According to the investigated literature references, the following methods are evaluated in this chapter:

- Elimination as a reference.
- Penalization of the feasible individuals objective function as given by:

$$F = \text{CostFunction} + \rho\left(H - \sum H_i\right) \tag{13}$$

where ρ is a penalization factor. H and H_i are respectively the fixed horizon time and the production time for product i, from Eq. (4). A static penalty factor is used in this study, for implementation simplicity reasons. Obviously, the efficiency of this technique depends strongly on the value of the ρ factor in the added penalization term. Its value was thus the object of a sensitivity analysis.

- Relaxation of the discrete variables range. By setting the discrete upper bounds to a greater value, this technique gives an enlargement of the feasible space: minimization should, anyway, make the variables remain within their initial bounds.
- Tournament based on domination rules. This method, applied in the selection step of the GA, relies on domination rules stated in [32] and [31]. Basically, it is stated that: (i) a feasible individual dominates an infeasible one; (ii) if two individuals are feasible, the one with the best objective function wins; (iii) if two individuals are infeasible, the one with the smallest constraint violation wins. These rules enable the selection of the winners among some randomly chosen competitors. Various combinations of competitors and winners were tested. A special case of this method, namely single tournament (ST), occurs when the number of competitors is equal to the population size, while the number of winners is the survivor number: then, all survivors are determined in one single tournament realization for each selection step.

Specific selection procedures were implemented according to the above mentioned techniques. For the elimination, penalization and relaxation techniques, the selection is performed using Goldberg's roulette wheel [33]. This procedure implies evaluation of the adaptation or strength of each individual. This is computed as being the difference between the highest value of the objective function in the current population and that of the considered individual i: $strength_i = f_{max} - f_i$. This method shows the advantage of being adapted to GA operation, i.e. maximizing the criterion. The last constraint handling method is applied in the selection step itself, which is carried out by tournament.

4.3 Variable Encoding

The way the variables are encoded strongly influences GAs efficiency. In what follows, three encoding techniques, which show increasing adaptation to the continuous context but share some general features, are presented. The array representing the complete set of all variables is called a chromosome. It is composed of genes, each one encoding a variable by means of one or several locus. A difference will be made between genes encoding continuous variables from those encoding discrete ones. Since the formers are bounded, they can be written in reduced form, like a real number α bounded within 0 and 1. Each integer variable is coded directly in a single-locus gene, keeping it unchanged.

The various encoding techniques differ one from another in the way the continuous variables are represented, and by the gene location throughout the chromosome.

Rough Discrete Coding The first encoding method tested consists in discretizing the continuous variables, i.e. the above mentioned α. According to the required precision, a given number of decimals of α is coded. This will logically have an effect on the string length, and consequently on the problem size. The so-called weight box [34] was used in this study (Fig. 2): each decimal is coded by four bits b_1, b_2, b_3, b_4, weighting respectively 1, 2, 3, 3. As an example, the value of decimal d of α is given by the following expression:

$$d = b_1 \cdot 1 + b_2 \cdot 2 + b_3 \cdot 3 + b_4 \cdot 3 \tag{14}$$

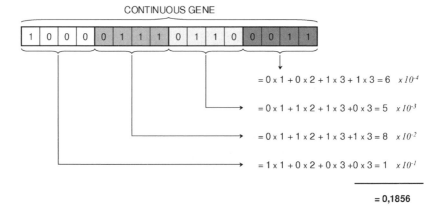

Fig. 2 Coding example with the weight box

This method enables one to prevent bias since the sum of all weights is lower than ten. Also, the duplication of threes in the weighted box means that there exist various ways to encode the same number. This means that the probabilities of selecting a given number are not all equal. Concerning the variable position, all continuous variables are located in the first part of the string, while the discrete genes are positioned at the end. The resulting configuration of a chromosome is shown in Fig. 3, for a small size example.

Obviously, the crossover and mutation operators are to be adapted to the coding configuration. The crossover is implemented by a single cut-point procedure, but two distinct mutation operators must be used: (i) mutation by reversion of the locus value on the continuous part of the string; (ii) mutation by subtraction of one unit of a bit value on the discrete part (when possible). The latter technique is not a symmetric mutation operator, thus it cannot prevent the algorithm from being trapped in some local optimum. However, it proved to lead efficiently towards minimization.

Crossed discrete coding The discretization of the continuous variables is unchanged with regard to the previous case, i.e. with the weight box method. The two methods only differ by the variables location. This change is suggested by the respective size of continuous and discrete parts of the chromosome. Indeed, because of the required precision for continuous variables, the former is much larger than the latter (in our case, the ratio equals 16 to 1). The crossover operator with a single cut-point procedure finds it very difficult to act on the discrete variables, and, consequently, to allow correct exploration of the search space.

In order to deal with this problem, the continuous and discrete variables were mixed up inside the string: since each processing stage induces one continuous and one discrete variable (respectively the size and number of items), the variables are encoded respecting the order of the operating stages in the recipe, as shown in Fig. 4 for the same illustrative example.

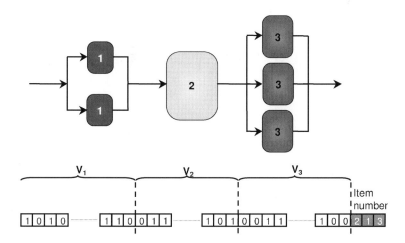

Fig. 3 Chromosome configuration, coding 1

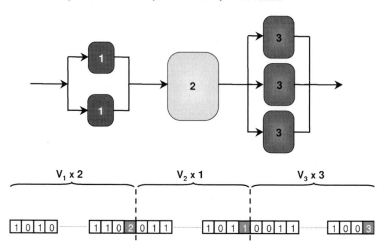

Fig. 4 Chromosome configuration, coding 2

Obviously, this new encoding method strengthens the probability that discrete genes may be involved in the one cut-point crossover process. However, due to the size difference between continuous and integer genes, there is still much more opportunity for the former to be directly involved in crossover.

Mixed real-discrete coding The last coding method seems to be the best suited to the nature of the variables. The reduced form α of continuous variables is coded directly on a real-value locus while, as in previous cases, the discrete variables are still kept unchanged for their coding. Therefore, both continuous and discrete genes have a one-locus size and occupy well-distributed lengths inside the chromosome (Fig. 5). A mixed real-discrete chromosome is obtained, which will require specific genetic operators.

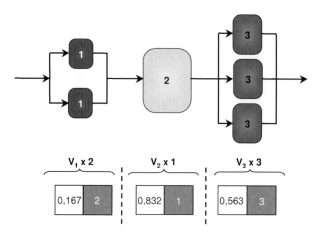

Fig. 5 Chromosome configuration, coding 3

First, the crossover methods applied in real-coded GAs work on each gene and not on the global structure of the chromosome. The simplest method relies on arithmetical combinations of parent genes, such as is presented in the following equations:

$$y_k^{(1)} = \lambda x_k^{(1)} + (1 - \lambda)x_k^{(2)} \qquad (15)$$

$$y_k^{(1)} = (1 - \lambda)x_k^{(1)} + \lambda x_k^{(2)} \qquad (16)$$

where $x_k^{(1)}$ and $x_k^{(2)}$ are genes k of both parents and $y_k^{(1)}$ and $y_k^{(2)}$ are those of the resulting children. λ is a fixed parameter within the range 0 to 1. Then, different procedures are implemented according to the method of determining the λ parameter [35]. The chosen technique is a simulated binary crossover (SBX), proposed in [36]. The method consists in generating a probability distribution around parent solutions to create two offspring.

This probability distribution is chosen in order to mimic single-point crossover behaviour in binary coded GAs, and involves the following two features:

- the mean decoded parameter value of two parents strings is invariant among the resulting children strings;
- if the crossover is applied between two children strings at the same cross site as used to create the children strings, the same parents strings will result.

This feature generates higher probabilities of creating an offspring close to the parents than away from them, as illustrated in Fig. 6. The procedure for the generation of two children solutions from two parents solutions is fully explained in [36]. Note that this crossover procedure is carried out for each locus of the chromosome and as a consequence, the arrangement of the continuous and discrete genes along the string does not matter. However, even though SBX crossover does not induce any problem for real variables, it may lead to real values for the discrete genes of the resulting offspring. So, for the latter case, these real values were truncated in order to keep only their integer part.

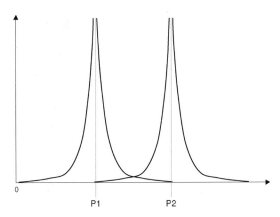

Fig. 6 Probability distribution for the location of an offspring, SBX crossover

With respect to mutation, on the one hand, the method corresponding to discrete variables was kept unchanged from the previous cases. On the other hand, for real-coded genes, an inventory of the variety of techniques is proposed in [35] or [37]. They usually rely on noise added to the initial muted gene value, according to a specific probability distribution. For the technique used in this study, a uniform probability distribution was chosen.

5 Numerical Results and Interpretation

In this section, the two main issues, i.e. variables encoding and constraint handling, are treated on different size instances of Optimal Batch Plant Design problems.

5.1 Variable Encoding

The above mentioned procedures for variable encoding were tested on three examples. Problem 1 is a quite small size example. The plant has to synthesize three products and comprises four batch stages, six semi-continuous stages, and one storage tank. Thus, the problem involves 10 continuous and 10 integer optimization variables. It was previously studied with various techniques as shown in Table 1, which highlights the fact that mathematical programming locates the best solution, for which optimality is proved.

Despite its small size, this example turned out to be difficult to solve to optimality since some of the stochastic techniques mentioned became trapped in a local optimum, showing a set of discrete values different from that of the optimal solution. The optimal values of the discrete variables are actually $m_j = \{1, 2, 2, 1\}$ parallel items for batch stages, while those of the local optimum are $m_j = \{1, 3, 3, 1\}$.

The two other examples are larger size instances of batch plants, both manufacturing three products. Problems 2 and 3 contain, respectively 7 and 18 batch stages, 10 and 24 semi-continuous stages and 3 and 6 sub-processes. They were also solved to optimality by the SBB solver in this study.

The parameters chosen for the Genetic Algorithm were the following: survival (respectively mutation) rate equal to 40% (resp. 30%). The maximum generation number and the population size depend on the complexity of the example. Concerning constraint handling, intuitive choices were adopted: elimination was chosen for problem 1 due to its small size (not expensive in terms of computation time). For the

Table 1 Typical solutions obtained by various methods

Reference	Optimization method	Best solution
Patel et al. [18]	Simulated Annealing	368,883
Wang et al. [19]	GA	362,130
Wang et al. [20]	Tabu Search	362,817
Wang and Xin [21]	Ants Foraging Method	368,858
Ponsich et al. [22]	Math. Programming	356,610

two larger examples, the single tournament method was selected. Table 2 summarizes the main features of each problem.

The results were analysed in terms of quality and computation time. The number of function calls could also be studied, but the time criterion appeared to be more significant when checking the influence of the variable encoding methods. The CPU time was measured on a Compaq Workstation W6000.

Quality is evaluated, of course, by the distance between the best found solution and the optimal value. Since GA is a stochastic method, its results have to be analysed also in terms of repeatability. So, for each test, the GA was run 100 times. The criterion for repeatability evaluation is the dispersion of the runs around the GA best solution F^*_{GA}. The 2%-dispersion or 5%-dispersion are then defined as the percentage of runs providing a result lying, respectively, in the range $[F^*_{GA}, F^*_{GA}+2\%]$ or $[F^*_{GA}, F^*_{GA}+5\%]$.

The results for problem 1 are presented in Table 3. Coding 1, 2 and 3 represent respectively rough discrete, crossed discrete and mixed real-discrete encoding techniques. Clearly, the solution obtained with the mixed real-coding is much better than the others: indeed, the optimal solution previously determined by the SBB solver is almost exactly located. GA with coding 1 stays trapped in a local optimum and was not able to find the set of discrete variables corresponding to the global optimum. Although it finds a slightly lower solution, GA with coding 2 does not show much more efficiency. The 2% and 5%-dispersions seem to be superior for the two first encoding techniques but this trend is due to the high quality of the best solution found with coding 3.

The difference between relative and absolute result quality is highlighted by the previous remark. A run set might show very good 2% and 5%-dispersions, if the best result is far from the optimum, giving poor global quality of the runs. On the other hand, a run set with lower dispersions but a better final result may be better performing. This remark is illustrated in Fig. 7, in which case 1 shows a better relative quality than case 2, which is characterized by a better absolute quality.

Table 2 Characteristics of the problems and solution by GA

	Problem 1	Problem 2	Problem 3
Cont./disc. variables	10/10	17/17	42/42
Optimum	356,610	766,031	1,925,888
GA Parameters			
Population size	200	200	500
Generations number	200	1000	500
Const. Handling	Elim.	Single Tour.	Single Tour.

Table 3 Comparison of variables encoding for Problem 1

	Coding 1	Coding 2	Coding 3
GA best solution	371,957	369,774	356,939
Gap to optimum (%)	4.30	3.69	0.09
2%-dispersion	68	67	1
5%-dispersion	100	98	79
CPU time (s)	3	3	1

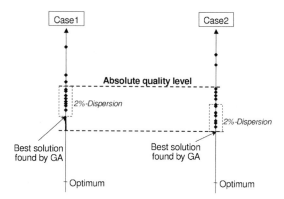

Fig. 7 Example of absolute and relative quality of numerical results

Thus, by considering 5%-dispersion with the optimal solution as a reference (this means calculating an absolute quality), the value is equal to 50%, 48% and 77%, for codings 1, 2 and 3 respectively. This clearly highlights the good global quality of the GA runs performed with coding 3.

The computation time is not significant on such a small example. The number of function evaluations is around 5.5×10^4 for all problems. The population size and generations number are both equal to 200, which would mean 4×10^4 evaluations per run, this shows that more than 25% of the search is spent in randomly looking for an initial population of feasible solutions. This small size but complex problem demonstrates the superiority of mixed real-discrete coding.

The results for problems 2 and 3 are presented in Table 4. It can be observed that the three coding methods find results very close to the optimum. With regard to dispersion evolution, Figure 8 shows the increasing superiority of coding 3 from example 1 to example 3. This behaviour does not seem useful for such simple examples, since the final solutions are quite similar, but it may be interesting for more severely constrained problems, for which the feasible space is reduced: it might be assumed that better solutions could be found, or at least found more easily. This would then require the use of large population sizes and generation numbers and consequently, a lower computation time.

For codings 1 and 2, the quality of all the runs can be related to the percentage of feasible solutions in the last generation and to the number of failures of the GA runs. It is considered that a GA run fails when no feasible solution is found during the whole search. It is, however, clear that this failure number compensates slightly

Table 4 Comparison of variables encoding for Problems 2 and 3

	Problem 2			Problem 3		
	Cod. 1	Cod. 2	Cod. 3	Cod. 1	Cod. 2	Cod. 3
GA best solution	770,837	771,854	767,788	1,986,950	1,997,768	1,975,027
Gap to optimum (%)	0.63	0.76	0.23	3.17	3.73	2.55
% feas. solutions (end search)	65	67	54	41	59	52
% failures	0	0	0	37	10	0
CPU time (s.)	17	17	3	126	126	22

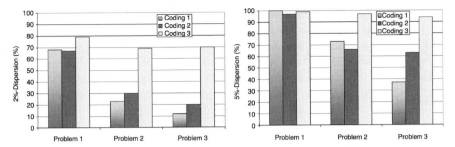

Fig. 8 Dispersion evolution for the three problems with different coding methods

the dispersion fall for encoding 1 and 2, since the percentage of runs finding a result close to the best found solution is based on the total number of runs and not on the number of nonfailed runs.

Note that for examples 1 and 2, coding 1 performs more or less as coding 2. With increasing problem size, the difficulty of acting on discrete variables – which are located at the bottom of the chromosome in coding 1 – increases but there is no great difference between the results provided by the two encoding methods. Only the dispersions seem to slightly favour coding 2 for Problem 3.

Finally, the comparison criterion based on CPU time highlights coding 3 performance, since a run of this GA version is six times quicker than coding 1 or 2. Thus, regarding variable encoding selection, numerical results prove the superiority of the mixed real-discrete encoding method, and this will be used in the following section.

5.2 Constraint Handling

The constraint handling techniques described in Sect. 4 were tested on problem 2.

Elimination of infeasible individuals The results in Table 5 present trends for GA with the elimination technique. These can be seen as a reference in order to evaluate the performance of different constraint handling methods: they show very good quality in terms of both distance to the optimum and dispersion of the runs. However, this technique performs less well in terms of computation time: this could cause a bottleneck in applications that require lengthy computations for the objective function.

Comparison with the behaviour of other techniques that do not need a feasible initial population, like tournament or penalization, highlights the fact that the GA

Table 5 Results for elimination technique

GA best solution	767,182
Gap to optimum (%)	0.15
2%-dispersion	95
5%-dispersion	100
Function evaluation no.	2.346×10^5
CPU time (s)	56

with the elimination method spends most of the computing time randomly searching for feasible solutions. Indeed, for a 1000 generation run, involving a population of 200 individuals, the theoretical function evaluation number is 2×10^5, but it turns out to be approximately 2.35×10^6. So, it can be deduced that less than 10% of the computing time is used for the GA normal sequence while the remainder is devoted to the initial population generation. For all the other techniques, the computation time is around six seconds.

Penalization of Infeasible Individuals This study was carried out for different values of the penalization factor ρ. The results presented in Table 6 underline the logical results of the method. On the one hand, for small values of ρ, priority is assigned to the minimization of the economic term while the time constraint is severely violated. The best solution found is then often infeasible. On the other hand, for higher values of the penalization factor, the result is feasible while the dispersion and optimality gap criteria remain satisfied.

Finally, a compromise solution can be reached with intermediate values of ρ. By giving the same weight to the time constraint and to minimization of the investment cost, the global performances can be improved with solutions slightly exceeding the time horizon. For all cases, the number of feasible solutions remains quite low.

Relaxation of discrete upper bounds In this section, the upper bound on discrete variables, i.e. the maximum number of parallel items per stage, is doubled and set to six. As shown in Table 7, the best result is similar to that previously obtained with the elimination technique. The dispersions fall with regard to the elimination technique but remain acceptable. The ratio of feasible solutions at the end of the search proves the good behaviour of GA, which maintains the discrete variables within their initial

Table 6 Results for penalization technique

ρ factor	100	10	0.1
GA best solution	769,956	766,520	592,032
Gap to optimum (%)	0.51	0.06	−22.71
Constraint violation (%)	0	0.64	37.4
2%-dispersion	44	54	68
5%-dispersion	85	89	96
% feas. solutions (end search)	16	3	0

Table 7 Results for relaxation technique

GA best solution	767,361
Gap to optimum (%)	0.17
2%-dispersion	58
5%-dispersion	95
% feas. solutions (end search)	53
Function evaluation no.	2.196×10^5
CPU time (s.)	7

bounds. Moreover, the CPU time is considerably reduced with regard to the elimination technique, showing the efficiency of relaxation to avoid the wasted time spent in generating the initial solution. Indeed, the function evaluation number is almost equal to the standard number *generation number · population size*.

Thus, the discrete upper bounds relaxation really appears to be a suitable technique for constraint handling. For larger size problems, the issue is how to determine the order of magnitude of the upper bound relaxation. On the one hand, the relaxation should be sufficient to easily create the initial population. On the other hand, too high an increase would cancel the necessary pressure that pushes the individuals towards the initial discrete feasible space, i.e. that leads to minimization of the parallel item number.

Domination Based Tournament This method was applied to various combinations of competitors *Ncomp* and survivors *Nsurv*, referred as (*Ncomp, Nsurv*)-tournaments in the following, except for the single tournament technique (ST). The tested combinations and their corresponding results are shown in Table 8. For all cases, the computation time is equal to 6 seconds, which is almost ten times faster than the elimination technique.

The best results are similar and near-optimal for all kinds of tournaments tested. Due to the low number of feasible solutions obtained at the end of the search, the $(2, 1)$, $(3, 2)$ and $(5, 4)$-tournaments can be discarded. This behaviour can easily be explained by the following assumption: the smaller the difference between *Ncomp* and *Nsurv*, the easier it is to pass the selection step for weak individuals, with a poor objective function. This underlines the efficiency of a more severe selection pressure, which is furthermore confirmed by the number of failures of the $(5, 4)$ version.

To evaluate the other options, the remaining criteria are the evolution of feasible solutions ratio during the search and the dispersions of the runs. The corresponding numerical results are given respectively in Figs. 9 and 10. On the one hand, it can be deduced that the combinations visiting more feasible solutions are single tournament, $(5, 1)$ and $(4, 1)$-tournaments.

On the other hand, Fig. 10 shows that $(4, 2)$ and $(5, 2)$-tournaments as well as single tournament are the best-performing combinations in terms of 2% and 5%-

Table 8 Results for various tournament techniques

Tournament version	$(2, 1)$	$(3, 1)$	$(3, 2)$	$(4, 1)$	$(4, 2)$
GA best solution	767,450	767,334	767,900	768,228	767,587
Gap to optimum (%)	0.19	0.17	0.24	0.29	0.20
% feas. solutions	34	47	23	53	42
% failures	0	0	1	0	0
Tournament version	$(5, 1)$	$(5, 2)$	$(5, 4)$	ST	
GA best solution	768,907	767,981	767,955	767,788	
Gap to optimum (%)	0.38	0.25	0.25	0.25	
% feas. solutions	54	48	10	55	
% failures	0	0	15	0	

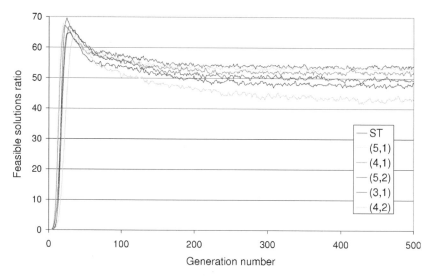

Fig. 9 Evolution of the ratio of feasible solutions (curves in the order of legend)

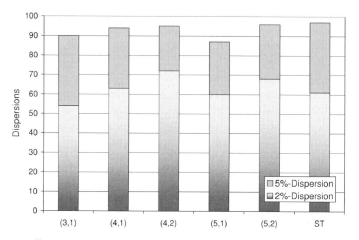

Fig. 10 Dispersions of the run for various tournament versions

dispersion. So, even though several combinations are very close in terms of result quality, the single tournament method proves to be the best compromise, closely followed by the $(4, 1)$ option.

Finally, for this example, the single tournament and relaxation methods are the best-suited constraint handling methods. However, it must be pointed out that the relaxation method needs the relaxed upper bound as a parameter that requires relevant choice, achieved with some knowledge of the problem studied. So, to conclude this study and highlight the general trends, the single domination-based tournament technique appears to be the most efficient constraint handling technique for Genetic Algorithms.

This conclusion is valid for medium and large size problems. Nevertheless, it is obvious that when the computation time is not a limiting factor, preference will be given to the elimination technique.

6 Conclusions

A Genetic Algorithm was adapted to solve Optimal Batch Plant Design problems. This work considered several operating modes of classical GA operators in order to determine the one best suited to the problem studied.

Since the problem involves a MINLP formulation, the first main issue was the evaluation of different variable encoding techniques that deal efficiently with both continuous and integer variables. Computing tests were carried out on three different size benchmark problems, in order to compare the behaviour of rough discrete, crossed discrete and mixed real-discrete coding methods. The result quality was evaluated with regard to the optimum found by a MP technique. The mixed real-discrete method proved to be the best option, since it provided nearly optimal results for all examples with a low computation time. Moreover, it overcame some local optimal difficulties while the two other techniques were trapped in sub-optimal solutions.

The second investigation was devoted to constraint handling techniques; here, the production time constraint. Four constraint handling methods were tested on a medium size problem: elimination, penalization, relaxation of the discrete upper bounds and dominance based tournament. Elimination, which led to very good results in terms of quality but less efficient in terms of computation time, is recommended for small size examples. The use of a penalization technique, depending strongly on the appropriate choice of the penalization factor, does not ensure feasible solutions throughout the whole search. Nevertheless, it could provide compromise solutions that improve the investment criterion, slightly violating the constraint.

Finally, numerical results showed that the relaxation and tournament methods were the most efficient procedures. However, the relaxation technique depends on the physical meaning of the variable and may be viewed as less general since it is a parameter-based technique. Dominance rules implemented in the selection step for the single tournament case are thus revealed to be the best constraint handling technique in the case of severely constrained problems.

This contribution has proposed some guidelines to tackle the two operating mode issues that are limiting factors for GAs. The use of these strategies should enable efficient evolution of the algorithm on different instances of mixed continuous and integer problems.

Appendix A. Nomenclature

a_j: cost factor for batch stage j
b_k: cost factor for semi-continuous stage k
B_{is}: batch size for product i in sub-process s (kg)
c_s: cost factor for intermediate storage tanks
D_{ik}: duty factor for product i in semi-continuous stage k (L/kg)

d_{ij}: power coefficient for processing time of product i in batch stage j
g_{ij}: coefficient for processing time of product i in batch stage j
H: time horizon (h)
H_i: production time for product i (h)
i: product index
I: total number of products
j: batch stage index
J: total number of batch stages
J_s: total number of batch stages in sub-process s
k: semi-continuous stage index
K: total number of semi-continuous stages
K_s: total number of semi-continuous stages in sub-process s
m_j: number of parallel out-of-phase items in batch stage j
n_k: number of parallel out-of-phase items in semi-continuous stage k
p_{ij}: processing time of product i in batch stage j (h)
p_{ij}^0: constant for calculation of processing time of product i in batch stage j
$Prod_i$: global productivity for product i (kg/h)
$Prodloc_{is}$: local productivity for product i in sub-process s (kg/h)
Q_i: demand for product i
R_k: processing rate for semi-continuous stage k (L/h)
R_{max}: maximum feasible processing rate for semi-continuous stage k (L/h)
R_{min}: minimum feasible processing rate for semi-continuous stage k (L/h)
s: sub-process index
S: total number of sub-processes
S_{ij}: size factor of product i in batch stage j (L/kg)
S_{is}: size factor of product i in intermediate storage tanks (L/kg)
T_{ij}: cycling time of product i in batch stage j (h)
T_{is}^L: limiting cycling time of product i in sub-process s (h)
V_j: size of batch stage j (L)
V_{max}: maximum feasible size of batch stage j (L)
V_{min}: minimum feasible size of batch stage j (L)
V_s^*: size of intermediate storage tank (L)
α_j: power cost coefficient for batch stage j
β_k : power cost coefficient for semi-continuous stage k
γ_s : power cost coeficient for intermediate storage
θ_{ik} : operating time of product i in semi-continuous stage k

References

1. Teh YS, Rangaiah GP (2003) Comput Chem Eng 27:1665–1679
2. Dedieu S, Pibouleau L, Azzaro-Pantel C, Domenech S (2003) Comput Chem Eng 27:1723–1740
3. Lee S, Grossmann IE (2000) Comput Chem Eng 24:2125–2141
4. Modi AK, Karimi IA (1989) Comput Chem Eng 13:127–139

5. Grossmann IE (2002) Opt Eng 3:227–252
6. Dietz A, Azzaro-Pantel C, Pibouleau L, Domenech S (2005) Ind Eng Chem Res 44:2191–2206
7. Hao JK, Galinier P, Habib M (1999) Rev Intell Art 1:11-48
8. Triki E, Collette Y, Siarry P (2005) Eur J Op Res 166:77–92
9. Hedar AR, Fukushia M (2006) Eur J Op Res 170:329–349
10. Ducek G, Scheuer T (1990) J Computat Phys 90:161–175
11. Ahmadi S, Osman IH (2005) Eur J Op Res 162:30–44
12. Kirkpatrick S, Gelatt CD Jr, Vecchi MP (1982) Optimization by Simulated Annealing. IBM Research Report RC 9355
13. André J, Siarry P, Dognon P (2001) Adv Eng Softw 32:49–60
14. Beyer HG, Schwefel HP (2002) Nat Comput Int J 1:3–52
15. Yang YW, Xu JF, Soh CK (2006) Eur J Op Res 168:354–369
16. Grossmann IE, Sargent RWH (1979) Ind Eng Chem Proc Des Dev 18:343–348
17. Kocis GR, Grossmann IE (1988) Ind Eng Chem Res 27:1407–1421
18. Patel AN, Mah RSH, Karimi IA (1991) Comput Chem Eng 15:451–469
19. Wang C, Quan H, Xu X (1996) Ind Eng Chem Res 35:3560–3566
20. Wang C, Quan H, Xu X (1999) Comput Chem Eng 23:427–437
21. Wang C, Xin Z (2002) Comput Chem Eng 41:6678–6686
22. Ponsich A, Azzaro-Pantel C, Domenech S, Pibouleau L (2005) About the relevance of Mathematical Programming and stochastic optimisation methods : application to optimal batch plant design problems. In: Espuña and Puigjaner (eds) Proceedings of European Symposium of Computer Aided Process Engineering 15, Barcelona, May 30–June 1, 2005. Elsevier, New York
23. Brooke A, Kendrick D, Meeraus A, Raman R (1998) GAMS User's Guide. GAMS Development Corporation, Washington DC
24. Pinto JM, Montagna JM, Vecchietti AR, Irribarren OA, Asenjo JA (2001) Biotechnol Bioeng 74:451–465
25. Epperly TGW, Ierapetritou MG, Pistikopoulos EN (1997) Comput Chem Eng 21:1411–1431
26. Aguilar Lasserre A, Azzaro-Pantel C, Domenech S, Pibouleau L (2005) Modélisation des imprécisions de la demande en conception optimale multicritère d'ateliers discontinus. Proceedings of SFGP Congress in Toulouse (France), September 20–22, 2005
27. Holland JH (1975) Adaptation in natural and artificial systems. University of Michigan Press, Ann Arbor, MI
28. Coello Coello CA (2002) Comput Meth Appl Mech Eng 191:1245–1287
29. Michalewicz Z, Dasgupta D, Le Riche RG, Schoenauer M (1996) Comput Ind Eng 30:851–870
30. Coello Coello CA (2002b) Comput Ind 41:113–127
31. Deb K (2000) Comput Meth Appl Mech Eng 186:311–338
32. Coello Coello CA, Mezura Montes E (2002) Adv Eng Inf 16:193–203
33. Golberg DE (1989) Genetic Algorithms in Search, Optimization and Machine Learning. Addison-Wesley Publishing Company Inc., MA
34. Montastruc L (2003) Développement d'un pilote automatisé, fiable et sûr pour la dépollution d'effluents aqueux d'origine industrielle. PhD Thesis, INP Toulouse, France
35. Michalewicz Z, Schoenauer M (1996) Evol Comp 4:1–32
36. Deb K, Agrawal RB (1995) Complex Sys 9:115–148
37. Raghuwanshi MM, Kakde OG (2005) Survey on multiobjective evolutionary and real coded genetic algorithms. Proceedings of the 7th International Conference on Adaptive and Natural Computing Algorithms, Coimbra (Portugal), March 21–23, 2005

Coevolutionary Genetic Algorithm to Solve Economic Dispatch

Márcia M. A. Samed[1] and Mauro A. da S. S. Ravagnani[2]

[1] Maringá State University, Computer Science Department, Colombo Avenue, 5790, Bl. 19, 87020-900, Maringá, Paraná, Brazil.
samed@din.uem.br

[2] Maringá State University, Department of Chemical Engineering, Colombo Avenue, 5790, Bl. D-90, 87020-900, Maringá, Paraná, Brazil.
ravag@deq.uem.br

Abstract

Two approaches based on genetic algorithms (GA) to solve economic dispatch (ED) problems are presented. The first approach is based on the hybrid genetic algorithm (HGA). Undesirable premature convergence to local minima can be avoided by means of the mutation operator, which is used to create diversity in the population by penalization or perturbation. Nevertheless, HGA needs to tune parameters before starting a run. A coevolutionary hybrid genetic algorithm (COEHGA) is proposed to improve the performance of the HGA. The COEHGA effectively eliminates the parameter tuning process because the parameters are adjusted while running the algorithm. A case from the literature is studied to demonstrate these approaches.

Key words: Economic Dispatch, Genetic Algorithm, Hybrid Genetic Algorithm, Coevolutionary Algorithm

1 Introduction

One problem of fundamental importance with thermoelectric power generation is optimization of the production cost. The Economic Dispatch (ED) problem has as principal objective, the minimization of the total production cost function by allocation of the power demand among the units available, while satisfying their operational limits. The ED problem has been the subject of a very large number of published papers since its original formulation. Important advances in optimization methods are related to the achievement of solutions to the ED problem.

For a plant with n units, the total production cost function is expressed as:

$$Fe = \sum_{i=1}^{n} Fe_i(P_i) = \sum_{i=1}^{n} a_i P_i^2 + b_i P_i + c_i \tag{1}$$

in which Fe_i represents the cost of the ith units, P_i is the electrical power generated by ith units and a_i, b_i and c_i are the characteristic coefficients of the cost function.

According to [1], the ED problem can be modeled as a constrained optimization problem considering power balance constraints and operational constraints of the form:

$$\min Fe$$

$$\text{subject to} \tag{2}$$

$$P_i^{\min} \leq P_i \leq P_i^{\max}$$

$$\sum_{i=1}^{n} P_i = P_D$$

in which Fe is the objective function; P_i is the electrical power generated by the ith units; P_D is the value of the power demand and $P_i^{\min}$ and $P_i^{\max}$ are, respectively, the lowest and highest operational output limits of the units.

An operationally effective method of optimization for solving dispatch problems should be able to cater for a combination of feasible solutions or, simply, better solutions than those that already exist. All these characteristics are often more desirable than having only one optimal solution. To satisfy the above arguments, two approaches based on GA to solve ED problems were proposed.

The following sections of this chapter are organized as follows. Section 2 presents a review of some methods employed in the solution of ED problems. HGA and CO-EHGA are reported in Sect. 3 and 4, respectively. In Sect. 5, tests and results for a simple case are presented. Section 6 presents some conclusion.

2 Review of the Methods Employed in the Solution of Dispatch Problems

As was reported in [2] and [3], investigation of ED problems started in the early 1920s, when engineers were concerned with the problem of economic allocation of generation or proper division of the load among the generating units available. One of the first methods that yielded economic results was known as the equal incremental method. The use of digital computers to obtain loading schedules was initially investigated in the 1950s and continues today.

Deterministic methods were proposed to solve dispatch problems by means of linear simplifications of the objective function. To do this, there are approaches using linear programming, dynamic programming methods based on linear programming, integer programming, mixed integer programming methods, branch and bound method, among others.

Since the 1990s, due to the advances in computation, approaches to solve ED based on such heuristics as simulated annealing (SA), particle swarm (PS) and evolutionary algorithms (EA) as GA have been proposed. Hybrid techniques that combine GA with local search or other heuristics provide better solutions than the best solutions found so far. SA and GA were combined in [4] to avoid premature convergence in the GA solution process. In [5] techniques were introduced to GA to improve efficiency and precision in the solution of ED. The techniques introduced were prognostic mutation, elitism, approximate interval and penalty factors. A heuristic called stochastic hybrid search was proposed in [6] to solve ED. The proposed method was designed with a genetic operator called blend crossover, which supplied a better search capacity. Two methods based on GA to solve ED by considering discontinuities of the objective function were presented in [7]: generation-apart elitism and atavism. These methods can achieve optimal solutions in cases thath deterministic methods cannot. A micro-genetic algorithm to solve ED subject to merit order loading was developed in [8]. The algorithm worked with a very small population, resulting in a reduced computation time. In [9], an improved GA was proposed to solve ED with a segmented quadratic cost function. In this work, a multi-stage algorithm and a directional crossover were used to improve GA. A hybrid GA consisting of two phases to solve the ED problem with fuel options was presented in [10]. Phase-1 used standard real coded GA while optimization by direct search and systematic reduction of the size of search region method were employed in phase-2. Reference [11] presented a promising technique for solving ED based on artificial-immune-system (AIS) methodology. AIS uses learning, memory and associative retrieval to solve recognition and classification tests. It was reported in [12] that memetic algorithms are capable of accommodating more complicated constraints and giving better quality solutions to solve ED. An approach based on PS with a nonsmooth cost function for finding the global optimum was published in [13]. Different EAs were published in [14] for several kinds of ED problems. The techniques considered different mutation operators using a Gaussian, Cauchy and Gaussian–Cauchy formulation.

The next section describes a hybrid genetic algorithm that was developed to solve ED problem.

3 Hybrid Genetic Algorithm

From the optimization point of view, hybrid methods are advantageous techniques to solve ED problems because they can handle constraints by incorporation of the local search and they do not need rigorous mathematical exigency.

Preliminary studies used binary code. The exclusion of binary code avoids the process coding and decoding and the algorithm becomes faster. Real codes permit the use of deterministic rules, simplifying algorithm control during execution.

The constraints are treated separately from the objective function (fitness function). First, evaluation of the individuals (candidate solutions) occurs in the fitness function and then, the candidate solutions are evaluated within the constraints and classified into feasible or infeasible ones.

Special treatment is given to the infeasible solutions. An "effort" is made to bring them into the feasible domain. A perturbation is applied to the feasible solutions to eliminate premature convergence.

Crossover and mutation operators are formulated from deterministic rules based on [15]. The selection operator uses the Roulette method. The main operator of this algorithm is the mutation operator, which was specially formulated to guarantee the diversity of the candidate solutions. The process of optimization, however, occurs due to the interaction of the operators and, mainly, due to the pressure of the selection.

A self-adaptive control algorithm is added to generate and submit the parameters to the main routine. The control algorithm, in turn, is also a GA, whose operators follow deterministic rules.

The main routine of the algorithm, called HGA, was implemented to solve the optimization problem:

$$\min f(P_i)$$

subject to   (3)

$$g_i(P_i) \leq 0, \quad i = 1, 2, \ldots, m$$
$$h_j(P_i) = 0, \quad j = 1, 2 \ldots, r$$

where f is the objective function, g is the inequality constraint and h is the equality constraint.

The steps of the main routine, HGA, are described below.

3.1 Initial Population

The candidate solutions of the initial population are the power, Pi, represented by real numbers. The initial population is randomly determined according to the form:

$$P_i^{ini} = (P_i^{max} - P_i^{min})RAN$$ (4)

with RAN being a random number in the interval [0,1].

3.2 Evaluation of the Candidate Solutions of the Initial Population

The candidate solutions of the initial population are evaluated in the fitness function and in the constraints. These candidate solutions are classified into $g_i(P_i) \leq 0$ and $g_i(P_i) > 0$. Those that satisfy the condition $g_i(Pi) \leq 0$ (considered feasible solutions) undergo a perturbation and those that satisfy $g_i(P_i) > 0$ (infeasible solutions) are penalized.

3.3 Mutation Operator

This operator gives rise to the "modified" solution, P_i^{mod}.

$$P_i^{mod} = P_i^{ini} + \beta_i d(P_i^{ini})$$ (5)

where $d(P_i^{ini})$ is the gradient (that is evaluated as described in (6)), β_i represents the size of the "step" in the direction of the gradient and is a parameter that may be controlled. According to [16], large mutation steps can be good in the early generations helping exploration of the search space and small mutation steps might be needed in the late generations to help fine tune the suboptimal solution.

3.4 Evaluation of the Gradient

The gradient is evaluated using the function:

$$d(P_i^{ini}) = \delta_0 \nabla f(P_i^{ini}) - \sum_{i=1}^{m} \delta_i \nabla g_i(P_i^{ini}) \tag{6}$$

with δ_0 represents a perturbation factor, δ_i represents a penalty factor, considering that δ_0 and δ_i do not act simultaneously; $\nabla f(P_i^{ini})$ is the gradient of the objective function and $\nabla g_i(P_i^{ini})$ are the gradient of the inequality constraints.

3.5 Perturbation Factor

The perturbation factor, δ_0, is given as:

$$\delta_0 = \begin{cases} 1, \text{ if } g_i(P_i) \leq 0 \\ 0, \text{ if } g_i(P_i) > 0 \end{cases} \tag{7}$$

δ_0 may be considered as an attempt to "escape" from the premature convergence to the undesired local minima.

3.6 Penalty Factor

The penalty factor, δ_i, is only applied to the infeasible candidate solutions and is presented as:

$$\delta_i = \begin{cases} \dfrac{1}{1 - \dfrac{g_i(P_i)}{g_{max}(P_i) + s}}, \text{ if } g_i(P_i) > 0 \\ \qquad\qquad 0, \text{ if } g_i(P_i) \leq 0 \end{cases} \tag{8}$$

where $g_{max}(P_i) = \max\{g_i(P_i), i = 1, 2, \ldots, m\}$, s is a parameter that may be controlled, but in general, is considered a very small number.

3.7 Crossover Operator

The candidate solutions of the initial population and those coming from the mutation are selected using the Roulette method to decide which pairs participate in the Crossover:

$$P_i^{new} = \alpha P_i^{ini} + (1 - \alpha) P_i^{mod} \tag{9}$$

with α randomly selected in the interval [0,1].

3.8 Evaluation and Selection of the New Initial Population

The candidate solutions of the new population are evaluated in the fitness function and selected according to the Roulette method. The best candidate solutions from the initial population and the best candidate solutions from the new population are classified according to the number of candidate solutions from the population to form a new initial population.

3.9 Stopping Criteria

The procedures repeat themselves until the maximum number of generations (iterations) has been reached. However, the number of iterations is a parameter that can be controlled.

4 Coevolutionary Algorithm

The HGA was applied to dispatch problems and the number of candidate solutions, NSC, and the number of iterations parameters, NI, required intensive trials to calibrate the initial estimates. It was noticed that the parameters are not independent and the optimal combination among them can take much time and can make the algorithm impractical for on-line applications. Under these conditions, an accurate result depends on the sensitivity of the programmer and the time used to determine the parameter values, which, when compared to the execution time of the algorithm, is huge.

Therefore, a control algorithm, called coevolutionary (COE), was implemented to avoid the process of manual control of the parameters of HGA.

4.1 Initial Parameters

First, a number of global generations (iterations) may be chosen. This number is considered the general stop criterion. Then, two initial population parameters were randomly chosen, so that, NI is a parameter between $[1, 300]$ and, NSC is a parameter between $[1, 50]$. The choice of these intervals was motivated by the good performance of the parameters when HGA without parameter control was applied to the ED problem. The populations are $Param_1$ and $Param_2$, formed as:

$$Param_1 = (NI_1, NSC_1)$$
$$Param_2 = (NI_2, NSC_2) \tag{10}$$

4.2 Crossover of the Parameters

The parameters $Param_1$ and $Param_2$ are used to generate a new population of parameters:

$$Param_{NEW} = \mu Param_1 + (1 - \mu)Param_2 \tag{11}$$

where μ is a randomly created number in the interval $[0, 1]$.

4.3 Submission of the Parameters to the HGA

The input data for the execution of the HGA are the parameters from the population $Param_{NEW}$. When the NI is achieved in the HGA, the respective parameters are returned to the COE to form the next generations of parameters.

4.4 Mutation of the Parameters

Another population of parameters, $Param_{NEWPOP}$, is inserted to be combined with $Param_{NEW}$ in the mutation operator. $Param_{NEWPOP}$ is randomly chosen with NI in the interval $[1, 300]$ and NSC in the interval $[1, 50]$. The objective of inserting this population is to produce diversity of the parameters.

$$Param_{MOD} = \lambda Param_{NEWPOP} + (1 - \lambda) Param_{NEW} \qquad (12)$$

in which λ is a random number between $[0, 1]$.

$Param_{MOD}$ is submitted to the HGA and the process is repeated until the number of global generations is reached.

5 Results

The algorithm performance analysis considers convergence and mutation operator analysis.

In these analyses, the equality constraint $(P_1 + P_2 + .. + P_n) = P_D$ was rewritten as $P_n = P_D - (P_1 + P_2 + ... + P_{n-1})$ to be incorporated in the inequality constraints.

5.1 Case Studied: Thirteen Generators

An industrial size problem was chosen to show how the accuracy and parameter control are improved by the proposed coevolutionary genetic algorithm. The case studied is an industrial power generation system composed of thirteen generators and it is necessary to allocate 2520 MW of power among these generators. The system characteristics are given in Table 1.

The problem in question is a nonlinear objective function with linear constraints. However, this simple problem was chosen to show how the accuracy and parameter control are improved by the coevolutionary genetic algorithm.

Two approaches, HGA and COEHGA, were used to solve the problem. A solver (IMSL-Fortran) based on the gradient method (GM) for nonlinear optimizations was used as reference. Table 2 shows the results.

As reported in Table 2, HGA provides better results than GM. After a large number of trials running the HGA, the "best" estimates for the population size and the number of iterations were found, respectively, as $NSC = 15$ and $NI = 300$. In the COEHGA the self-adaptive parameters were $NSC = 35$ and $NI = 10$. The process of the automatic control of parameters makes the algorithm more suitable for use

Table 1 System characteristics

Unit	P^{min}	P^{max}	a	b	c
1	0	680	0.00028	8.10	500
2	0	360	0.00056	8.10	309
3	0	360	0.00056	8.10	307
4	60	180	0.00324	7.74	240
5	60	180	0.00324	7.74	230
6	60	180	0.00324	7.74	240
7	60	180	0.00324	7.74	240
8	60	180	0.00324	7.74	240
9	60	180	0.00324	7.74	240
10	40	120	0.00284	8.60	160
11	40	120	0.00284	8.60	160
12	55	120	0.00284	8.60	160
13	55	120	0.00284	8.60	160

Table 2 Results

Results	GM	HGA	COEHGA
P_1 (MW)	643.8791	651.1451	735.6263
P_2 (MW)	330.1394	319.9820	337.4955
P_3 (MW)	309.5107	320.4637	292.6257
P_4 (MW)	124.5300	137.7761	146.7135
P_5 (MW)	145.1631	156.6884	177.3462
P_6 (MW)	150.7257	147.0077	131.5521
P_7 (MW)	160.1543	159.1650	154.1975
P_8 (MW)	172.1578	145.3784	159.5506
P_9 (MW)	176.9499	151.5512	167.3398
P_{10} (MW)	62.5544	82.2596	60.6778
P_{11} (MW)	92.8891	86.3206	74.6819
P_{12} (MW)	62.6343	82.8938	56.5370
P_{13} (MW)	88.7153	79.3682	25.6558
P_{Total} (MW)	2520.0000	2520.0000	2520.0000
Objective Function ($/h)	24703.32	24111.69	24072.03

because, obviously, it avoids the process of trial and error. Moreover, the results were improved. The difference between HGA and COEHGA results shows that a dependency exists among the parameters. This makes the exact determination of the optimal parameters very difficult.

5.2 Convergence Analysis

This analysis was applied to COEHGA with $NSC = 35$ and $NI = 10$. All the candidate solutions were verified in all iterations. This process is carried out to classify the candidate solutions according to the range of values in the objective function.

Table 3 Convergence

Iteration	Range of values of the objective function					
	30,000–29,000	29,000–28,000	28,000–27,000	27,000–26,000	26,000–25,000	25,000–24,000
1	2	2	7	13	9	2
2	0	0	0	10	20	5
3	0	0	0	0	21	14
4	0	0	0	0	6	29
5	0	0	0	0	0	35
6	0	0	0	0	0	35
7	0	0	0	0	0	35
8	0	0	0	0	0	35
9	0	0	0	0	0	35
10	0	0	0	0	0	35

It can be observed that the number of candidate solutions in the lowest values from the objective function increases iteration by iteration. However, all candidate solutions were already in the best range at the 5th iteration and they remain in the best range until the end. In the last iteration all candidate solutions can be considered near optimal solutions or the optimal solution and this process can be considered convergent.

5.3 Mutation Analysis

This analysis was applied to one candidate solution (initially infeasible) that was chosen in the first iteration. The objective of this analysis is to observe the evolution of this candidate solution through its values for the objective function and with a constraint that was not satisfied.

Figure 1 shows the values of the objective function and Fig. 2 shows the constraint values, both through 10 iterations.

In Fig. 2, the process of penalization and perturbation can be observed. In the first iteration, the candidate solution initially infeasible is penalized and conducted to the feasible domain. So, after successive perturbation, the candidate solution was

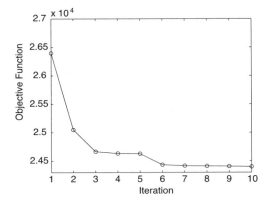

Fig. 1 Analysis of a candidate solution in the objective function

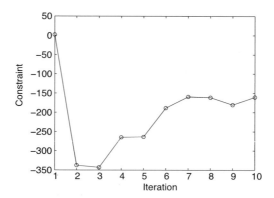

Fig. 2 Analysis of a candidate solution in a constraint

kept in the feasible domain and it improves the results in the objective function, as can be observed in Fig. 1.

6 Conclusions

Economic Dispatch is a simple nonlinear problem that becomes complicated when the operational constraints of the systems are considered.

This chapter presented two evolutionary approaches to solve the ED problem, a HGA and a COEHGA. These approaches were formulated according to hybrid methodologies. The tests proved that the process of GA investigation is not affected by the infeasibility of the initial candidate solutions. In this chapter a local search based on gradient was introduced to treat infeasible solutions.

The convergence process of the algorithms was demonstrated through analysis of the distribution of the candidate solutions in the range of values of the objective function. The efficiency of the mutation operator in treating constraints was confirmed. The accuracy of the results was improved through the COEHGA and refined solutions were found and the time consumed in trial and errors eliminated.

If only the computation running time is taken into account, the HGA approach could be considered the best choice because it can find the optimal solution in less than one minute. The COEHGA approach can find the optimal solution in several minutes in this specific case studied. Therefore, evaluation of the fitness function requires the most computational effort in this specific application.

The results of this coevolutionary algorithm should directly contribute to the development of new computational tools for solving ED problems and, indirectly, should contribute to the better performance of Genetic Algorithms when applied to problems in any other area of optimization.

References

1. Wood, A. J., Wollenberg, B. F. (1884) Power Generation, Operation & Control, John Wiley & Sons, Inc.
2. Chowdhury, B. H., Rahman, S. (1990) A Review of Recent Advances in Economic Dispatch, IEEE Transactions on Power Systems 5:1248–1259
3. Happ, H. H. (1977) Optimal Power Dispatch – A Comprehensive Survey. IEEE Transactions on Power Apparatus and Systems PAS-96:841–854
4. Wong, K. P., Wong, Y. W. (1994) Genetic and Genetic/Simulated Annealing Approaches to Economic Dispatch, IEE Proceedings Gener. Trans. Distrib. 141:507–513
5. Sheblé, G. B., Brittig, K. (1995) Refined Genetic Algorithm – Economic Dispatch Example, IEEE Transactions on Power Systems 10:117–124
6. Bhagwan Das, D., Patvardhan, C. (1999) Solution of Economic Dispatch Using Real Coded Hybrid Stochastic Search, Electric Power and Energy Systems 21:165–170
7. Kim, J. O., Shin, D. J., Park, J. N., Singh, C. (2002) Atavistic Genetic Algorithm for Economic Dispatch with Valve Point Effect, Electric Power Systems Research, 62:201–207
8. Ongsakul, W., Tippayachai, J. (2002) Micro–Genetic Based on Migration and Merit Order Loading Solutions to the Constrained Economic Dispatch Problems. Electric Power and Energy Systems 24:223–231
9. Won, J. R.,Park, Y. M. (2003) Economic Dispatch Solutions with Piecewise Quadratic Function Using Improved Genetic Algorithm. Electric Power and Energy Systems 25:335–361
10. Baskar, S., Subbaraj, P., Rao, M. V. C. (2003) Hybrid Real Coded Genetic Algorithm Solution to Economic Dispatch Problem, Computers and Electrical Engineering 29:407–419
11. Rahman, T. K. A., Yasin, Z. M., Abdullah, W. N. W. (2004) Artificial Imune Based for Solving Economic Dispatch in Power Systems, Proceedings of National Power and Energy Conference, 31–35
12. Yamin, H. Y. (2004) Review on Methods of Generations Scheduling in Electric Power Systems, Electric Power Systems Research, 69:227–248
13. Park, J. B., Lee, J. S., Shi, J. R., Lee, K. Y. (2005) A Particle Swarm Optimization for Economic Dispatch with Nonsmooth Cost Function, IEEE Transactions on Power Systems 20:34–32
14. Jayabarathi, T., Jayaprakashi, K., Jayakumar, D. N., Raghunathan, T. (2005) Evolutionary Programming Techniques for Different Kinds of Economic Dispatch Problems, Electric Power Systems Research 73:169–176
15. Tang, J., Wang, D. (1998) Hybrid Genetic Algorithn for a Type of Nonlinear Programming Problem, Computers Math. Applic. 36:11–21
16. Eiben, A. E., Hinterding, R., Michalewicz, Z. (1999) Parameter Control in Evolutionary Algorithms, IEEE Transactions on Evolutionary Computation 3:124–141

An Evolutionary Approach to Solve a Novel Mechatronic Multiobjective Optimization Problem

Efrén Mezura-Montes[1], Edgar A. Portilla-Flores[2], Carlos A. Coello Coello[3], Jaime Alvarez-Gallegos[4] and Carlos A. Cruz-Villar[4]

[1] Laboratorio Nacional de Informática Avanzada (LANIA A.C.) Rébsamen 80, Centro, Xalapa, Veracruz, 91000 Mexico.
emezura@lania.mx

[2] Autonomous University of Tlaxcala. Engineering and Technology Department, Calz. Apizaquito s/n Km. 15, Apizaco Tlax. 90300 Mexico.
eportilla@ingenieria.uatx.mx

[3] CINVESTAV-IPN (Evolutionary Computation Group), Departamento de Computación, Av. IPN No. 2508, Col. San Pedro Zacatenco, México D.F. 07360, Mexico.
ccoello@cs.cinvestav.mx

[4] CINVESTAV-IPN Departamento de Ingeniería Eléctrica, Sección de Mecatrónica, Av. IPN No. 2508, Col. San Pedro Zacatenco, México D.F. 07360, Mexico.
jalvarez@cinvestav.mx, cacruz@cinvestav.mx

Abstract

In this chapter, we present an evolutionary approach to solve a novel mechatronic design problem of a pinion-rack continuously variable transmission (CVT). This problem is stated as a multiobjective optimization problem, because we concurrently optimize the mechanical structure and the controller performance, in order to produce mechanical, electronic and control flexibility for the designed system. The problem is solved first with a mathematical programming technique called the goal attainment method. Based on some shortcomings found, we propose a differential evolution (DE)-based approach to solve the aforementioned problem. The performance of both approaches (goal attainment and the modified DE) are compared and discussed, based on quality, robustness, computational time and implementation complexity. We also highlight the interpretation of the solutions obtained in the context of the application.

Key words: Parametric Optimal Design, Multiobjective Optimization, Differential Evolution

1 Introduction

The solution of real-world optimization problems poses great challenges, particularly when the problem is relatively unknown, since these uncertainties add an extra complexity layer. Currently, several systems can be considered as mechatronic systems due to the integration of the mechanical and electronic elements in such systems. This is the reason why it is necessary to use new design methodologies that consider integral aspects of the systems.

The traditional approach to the design of mechatronic systems, considers the mechanical behavior and the dynamic performance separately. Therefore, the design of mechanical elements involves kinematic and static behaviors while the design of the control system uses only the dynamic behavior. This design approach from a dynamic point of view cannot produce an optimal system behavior [14, 23]. Recent works on mechatronic systems design propose a concurrent design methodology which considers jointly the mechanical and control performances.

For this concurrent design concept, several approaches have been proposed. However, these concurrent approaches are based on an iterative process. There, the mechanical structure is obtained in a first step and the controller in a second step. If the resulting control structure is very difficult to implement, then the first step must be repeated all over again.

On the other hand, an alternative approach to formulate the system design problem is to consider it as a dynamic optimization problem [2,3]. In order to do this, the parametric optimal design of the mechatronic system needs to be stated as a multi-objective dynamic optimization problem (MDOP). In this approach, both the kinematic and the dynamic models of the mechanical structure and the dynamic model of the controller are considered at the same time, together with system performance criteria. This approach allows us to obtain a set of optimal mechanical and controller parameters in only one step, which could produce a simple system reconfiguration.

In this chapter, we present the parametric optimal design of a pinion-rack continuously variable transmission (CVT). The problem is stated as a multiobjective optimization problem. Two approaches are used to solve it. One is based on a mathematical programming technique called goal attainment [11] and the other is based on an evolutionary algorithm called differential evolution [17]. The remainder of this chapter is organized as follows. In Sect. 2, we detail the transformation of the original problem into a multiobjective optimization problem. In Sect. 3, we present the mathematical programming method, its adaptation to solve the problem and the results obtained. Afterwards, the evolutionary approach is explained and tested in Sect. 4. Later, in Sect. 5, we present a discussion of the behavior of both approaches, based on issues such as quality and robustness of the approach, computation time and implementation complexity. Finally, our conclusions and future paths of research are presented in Sect. 6.

2 Multiobjective Problem

In the concurrent design concept, the mechatronic design problem can be stated as the following general problem:

$$\min \Phi(x, p, t) = [\Phi_1, \Phi_2, \dots, \Phi_n]^T \tag{1}$$

$$\Phi_i = \int_{t_0}^{t_f} L_i(x, p, t)\,dt \qquad i = 1, 2, \dots, n$$

under p and subject to:

$$\dot{x} = f(x, p, t) \tag{2}$$
$$g(x, p, t) \leq 0 \tag{3}$$
$$h(x, p, t) = 0 \tag{4}$$
$$x(0) = x_0$$

In the problem stated by (1) to (4): p is a vector of the design variables from the mechanical and control structure, x is the vector of the state variables and t is the time variable. On the other hand, some performance criteria L must be selected for the mechatronic system. The dynamic model (2) describes the state vector x at time t. Also, the design constraints of the mechatronic system must be developed and proposed, respectively. Therefore, the parameter vector p which is a solution of the previous problem will be an optimal set of structure and controller parameters, which minimize the performance criteria selected for the mechatronic system and subject to the constraints imposed by the dynamic model and the design.

Current research efforts in the field of power transmission of rotational propulsion systems, are dedicated to obtaining low energy consumption with high mechanical efficiency. An alternative solution to this problem is the so called continuously variable transmission (CVT), whose transmission ratio can be continuously changed in an established range. There are many CVT configurations built in industrial systems, especially in the automotive industry, due to the requirements to increase fuel economy without decreasing system performance. The mechanical development of CVTs is well known and there is little to modify regarding its basic operating principles. However, research efforts continue on the controller design and the CVT instrumentation side. Different CVT types have been used in different industrial applications; the Van Doorne belt or V-belt CVT is the most widely studied mechanism [19, 20]. This CVT is built with two variable radii pulleys and a chain or metal-rubber belt. Due to its friction-drive operating principle, the speed and torque losses of rubber V-belts are a disadvantage. The Toroidal Traction-drive CVT uses the high shear strength of viscous fluids to transmit torque between an input torus and an output torus. However, the special fluid characteristic used in this CVT makes the manufacturing process expensive. A pinion-rack CVT is a traction-drive mechanism, presented in [21]. This CVT is built-in with conventional mechanical elements as a gear pinion, one cam and two pairs of racks. The conventional CVT manufacture is advantageous over other existing CVTs. However, in the pinion-rack CVT, it has been determined that the teeth size of the gear pinion is an important factor in the performance of the system.

Because the gear pinion is the main mechanical element of the pinion-rack CVT, determining the optimal teeth size of such a mechanical element to obtain an optimal performance is, by no means, easy. On the other hand, an optimal performance system must consider low energy consumption in the controller. Therefore, in order to obtain an optimal performance of the pinion-rack CVT, it is necessary to propose the parametric optimal design of such system.

The goals of the parametric optimal design of the pinion-rack CVT are to obtain a maximum mechanical efficiency as well as a minimum controller energy. Therefore, a MDOP for the pinion-rack CVT will be proposed in this chapter.

2.1 Description and Dynamic CVT Model

In order to adapt the MDOP to the pinion-rack CVT, it is necessary to develop the dynamic model of such a system. The pinion-rack CVT changes its transmission ratio when the distance between the input and output rotation axes is changed. This distance is called "offset" and will be denoted by "e". As was indicated earlier, this CVT is built-in with conventional mechanical elements such as a gear pinion, one cam and two pairs of racks. An offset mechanism is integrated inside the CVT. This mechanism is built-in with a lead screw attached by a nut to the vertical transport cam. Figure 1 depicts the main mechanical CVT components.

The dynamic model of a pinion-rack CVT is presented in [2]. Ordinary differential equations (5), (6) and (7) describe the CVT dynamic behavior. In Eq. (5): T_m is the input torque , J_1 is the mass moment of inertia of the gear pinion, b_1 is the input shaft coefficient viscous damping, r is the gear pinion pitch circle radius, T_L is the CVT load torque, J_2 is the mass moment of inertia of the rotor, R is the planetary gear pitch circle radius, b_2 is the output shaft coefficient viscous damping and θ is the angular displacement of the rotor. In Eqs. (6) and (7): L, R_m, K_b, K_f and n represent the armature circuit inductance, the circuit resistance, the back electromotive force constant, the motor torque constant and the gearbox gear ratio of the

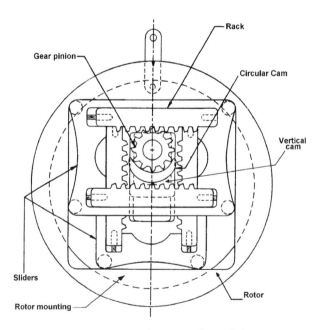

Fig. 1 Main pinion-rack CVT mechanical elements

DC motor, respectively. Parameters r_p, λ_s, b_c and b_l denote the pitch radius, the lead angle, the viscous damping coefficient of the lead screw and the viscous damping coefficient of the offset mechanism, respectively. The control signal $u(t)$ is the input voltage to the DC motor. $J_{eq} = J_{c2} + Mr_p^2 + n^2 J_{c1}$ is the equivalent mass moment of inertia, J_{c1} is the mass moment of inertia of the DC motor shaft, J_{c2} is the mass moment of inertia of the DC motor gearbox and $d = r_p \tan \lambda_s$, is a lead screw function. $\theta_R(t) = \frac{1}{2}\arctan\left[\tan\left(2\Omega t - \frac{\pi}{2}\right)\right]$ is the rack meshing angle. The combined mass to be translated is denoted by M and $P = \frac{T_m}{r_p}\tan\phi\cos\theta_R$ is the load on the gear pinion teeth, where ϕ is the pressure angle.

$$\left(\frac{R}{r}\right)T_m - T_L = \left[J_2 + J_1\left(\frac{R}{r}\right)^2\right]\ddot{\theta} \tag{5}$$

$$-\left[J_1\left(\frac{R}{r}\right)\frac{e}{r}\sin\theta_R\right]\dot{\theta}^2$$

$$+\left[\begin{array}{c}b_2 + b_1\left(\frac{R}{r}\right)^2 \\ +J_1\left(\frac{R}{r}\right)\frac{\dot{e}}{r}\cos\theta_R\end{array}\right]\dot{\theta}$$

$$L\frac{di}{dt} + R_m i = u(t) - \left[\frac{nK_b}{d}\right]\dot{e} \tag{6}$$

$$\left[\frac{nK_f}{d}\right]i - P = \left[M + \frac{J_{eq}}{d^2}\right]\ddot{e} + \left[b_l + \frac{b_c}{r_p d}\right]\dot{e} \tag{7}$$

In order to fulfill the concurrent design concept, the dynamic model of the pinion-rack CVT must be stated with state variables as it is indicated in the general problem stated by (1) to (4). With the state variables $x_1 = \dot{\theta}$, $x_2 = i$, $x_3 = e$, $x_4 = \dot{e}$, the dynamic model given by (5) to (7) can be written as:

$$\dot{x}_1 = \frac{AT_m + \left[J_1 A \frac{2x_3}{p_1 p_2}\sin\theta_R\right]x_1^2 - T_L - \left[b_2 + b_1 A^2 + J_1 A \frac{2x_4}{p_1 p_2}\cos\theta_R\right]x_1}{J_2 + J_1 A^2}$$

$$\dot{x}_2 = \frac{u(t) - \left(\frac{nK_b}{d}\right)x_4 - Rx_2}{L} \tag{8}$$

$$\dot{x}_3 = x_4$$

$$\dot{x}_4 = \frac{\left(\frac{nK_f}{d}\right)x_2 - \left(b_l + \frac{b_c}{r_p d}\right)x_4 - \frac{T_m}{r_p}\tan\phi\cos\theta_R}{M + \frac{J_{eq}}{d^2}}$$

Performance Criteria and Objective Functions

The performance of a system is measured by several criteria. One of the most common is the system efficiency because it reflects the energy loss. In the case of the pinion-rack CVT, the mechanical efficiency criterion of the gear systems is used to state the MDOP. This is because the racks and the gear pinion are the main CVT mechanical elements.

The mathematical equation for mechanical efficiency presented in [22] is used in this work, where μ, N_1, N_2, m, r_1 and r_2 represent the coefficient of sliding friction, the number of gear pinion teeth, the number of spur gear teeth, the gear module, the pitch pinion radius and the pitch spur gear radius, respectively:

$$\eta = 1 - \pi\mu \left(\frac{1}{N_1} + \frac{1}{N_2} \right) = 1 - \frac{\pi\mu}{2m} \left(\frac{1}{r_1} + \frac{1}{r_2} \right) \tag{9}$$

In [2], the speed ratio equation is as below, where ω is the input angular speed and Ω is the output angular speed of the CVT:

$$\frac{\omega}{\Omega} = \frac{R}{r} = 1 + \frac{e}{r} \cos \theta_R \tag{10}$$

Considering $r_1 \equiv r$ and $r_2 \equiv R$, the CVT mechanical efficiency is given by

$$\eta(t) = 1 - \frac{\pi\mu}{N_1} \left(1 + \frac{1}{1 + \frac{e \cos \theta_R}{r}} \right) \tag{11}$$

In order to maximize the mechanical CVT efficiency, $F(\cdot)$, which is given below, must be minimized:

$$F(\cdot) = \frac{1}{N_1} \left(1 + \frac{1}{1 + \frac{e \cos \theta_R}{r}} \right) \tag{12}$$

Equation (12) can be written as follows, and is used to state the MDOP:

$$L_1(\cdot) = \frac{1}{N_1} \left(\frac{2r + e \cos \theta_R}{r + e \cos \theta_R} \right) \tag{13}$$

The second objective function of the MDOP must describe the dynamic behavior. In order to fulfill this, a proportional and integral (PI) controller structure is used in the MDOP. This is because, despite the development of many control strategies, the PI controller structure remains one of the most popular approaches in industrial process control because of its good performance. Then, in order to obtain the minimal controller energy, the objective function for the MDOP, given below, is used:

$$L_2(\cdot) = \frac{1}{2} \left[-K_p(x_{\text{ref}} - x_1) - K_I \int_0^t (x_{\text{ref}} - x_1) dt \right]^2 \tag{14}$$

The objective functions previously established fulfill the concurrent design concept, since structural and dynamic behaviors will be considered at the same time in the MDOP.

Constraint Functions

The design constraints for the CVT optimization problem are proposed according to geometric and strength conditions for the gear pinion of the CVT.

To prevent fracture of the annular portion between the axis bore and the teeth root on the gear pinion, the pitch circle diameter of the pinion gear must be greater than the bore diameter by at least 2.5 times the gear module [16]. Then, in order to avoid fracture, the constraint g_1 must be imposed. To achieve a uniform load distribution on the teeth, the face width must be 6 to 12 times the value of the gear module [14]. This is ensured with constraints g_2 and g_3. To maintain the CVT transmission ratio within the range $[2r, 5r]$ constraints g_4, g_5 are imposed. Constraint g_6 ensures the number of teeth on the gear pinion is equal to or greater than 12 [14]. A practical constraint requires that the gear pinion face width is greater than or equal to 20 mm. In order to ensure this, constraint g_7 is imposed. To constrain the distance between the corner edge in the rotor and the edge rotor, constraint g_8 is imposed. Finally, to ensure a practical design for the pinion gear, the pitch circle radius must be equal to or greater than 25.4 mm. For this, constraint g_9 is imposed.

On the other hand, it can be observed that J_1, J_2 are parameters which are a function of the CVT geometry. For these mechanical elements, the mass moments of inertia are defined by

$$J_1 = \frac{1}{32}\rho\pi m^4 (N + 2)^2 N^2 h \tag{15}$$

$$J_2 = \rho h \left[\frac{3}{4}\pi r_c^4 - \frac{16}{6} (e_{\max} + mN)^4 - \frac{1}{4}\pi r_s^4 \right] \tag{16}$$

where ρ, m, N, h, $e_{\max}$, r_c and r_s are the material density, the module, the number of teeth on the gear pinion, the face width, the highest offset distance between axes, the rotor radius and the bearing radius, respectively.

Design Variables

Because the concurrent design concept considers structural and dynamic behaviors at the same time, the vector of the design variables must describe the mechanical and controller structures. In order to fulfill this, design variables of the mechanical structure related to the standard nomenclature for a gear tooth are used. Moreover, the controller gains K_P and K_I which describe the dynamic CVT behavior, are also used.

Equation (17) establishes a parameter called gear module m for metric gears, where d is the pitch diameter and N is the teeth number.

$$m = \frac{d}{N} = \frac{2r}{N} \tag{17}$$

On the other hand, the face width h, which is the distance measured along the axis of the gear and the highest offset distance between axes $e_{\max}$, are parameters which define the CVT size. Therefore, the vector p^i is proposed in order to establish the MDOP of the pinion-rack CVT:

$$p^i = [p_1^i, p_2^i, p_3^i, p_4^i, p_5^i, p_6^i]^T$$
$$= [N, m, h, e_{\max}, K_P, K_I]^T \tag{18}$$

2.2 Optimization Problem

In order to obtain the mechanical CVT parameter optimal values, we propose a MDOP given by Eqs. (19) to (26), where the control signal $u(t)$ is given by (20). As the objective functions must be normalized to the same scale [15], the corresponding factors $W = [0.4397, 563.3585]^T$ were obtained using the algorithm from Sect. 3 by minimizing each objective function subject to constraints given by Eqs. (8) and (20) to (26).

$$\min_{p \in R^6} \Phi(x, p, t) = [\Phi_1, \Phi_2]^T \tag{19}$$

where

$$\Phi_1 = \frac{1}{W_1} \int_0^{10} \left[\frac{1}{p_1} \left(\frac{p_1 p_2 + x_3 \cos \theta_R}{\frac{p_1 p_2}{2} + x_3 \cos \theta_R} \right) \right] dt$$

$$\Phi_2 = \frac{1}{W_2} \int_0^{10} u^2 dt$$

subject to the dynamic model stated by (8) and subject to:

$$u(t) = -p_5(x_{\text{ref}} - x_1) - p_6 \int_0^t (x_{\text{ref}} - x_1) dt \tag{20}$$

$$J_1 = \frac{1}{32} \rho \pi p_2^4 (p_1 + 2)^2 p_1^2 p_3 \tag{21}$$

$$J_2 = \frac{\rho p_3}{4} \left[3\pi r_c^4 - \frac{32}{3} (p_4 + p_1 p_2)^4 - \pi r_s^4 \right] \tag{22}$$

$$A = 1 + \frac{2x_3}{p_1 p_2} \cos \theta_R \tag{23}$$

$$d = r_p \tan \lambda_s \tag{24}$$

$$\theta_R = \frac{1}{2} \arctan \left[\tan \left(2x_1 t - \frac{\pi}{2} \right) \right] \tag{25}$$

$$g_1 = 0.01 - p_2 (p_1 - 2.5) \le 0$$

$$g_2 = 6 - \frac{p_3}{p_2} \le 0$$

$$g_3 = \frac{p_3}{p_2} - 12 \le 0$$

$$g_4 = p_1 p_2 - p_4 \le 0$$

$$g_5 = p_4 - \frac{5}{2} p_1 p_2 \le 0 \tag{26}$$

$$g_6 = 12 - p_1 \le 0$$

$$g_7 = 0.020 - p_3 \le 0$$

$$g_8 = 0.020 - \left[r_c - \sqrt{2}(p_4 + p_1 p_2) \right] \le 0$$

$$g_9 = 0.0254 - p_1 p_2 \le 0$$

3 Mathematical Programming Optimization

As we can observe, in a general way, a MDOP is composed by continuous functions given by the dynamic model of the system as well as the objective functions of the problem. In order to find the solution of the MDOP, it must be transformed into a Nonlinear Programming Problem (NLP) [4]. Two transformation approaches exist: the sequential and the simultaneous approach. In the sequential approach, only the control variables are discretized. This approach is also known as control vector parameterization. In the simultaneous approach, the state and control variables are discretized resulting in a large-scale NLP problem which requires special algorithms for its solution [1]. Because of the diversity of mathematical programming algorithms already established, transformation of the MDOP into a NLP problem was done adopting the sequential approach.

The NLP problem which is used to approximate the original problem given by (1) to (4) can be stated as:

$$\min_{p} F(p) \tag{27}$$

subject to:

$$c_i \le 0 \tag{28}$$

$$c_e = 0 \tag{29}$$

where p is the vector of the design variables, c_i are the inequality constraints and c_e are the equality constraints. In order to obtain the NLP problem given by (27) to (29), the sequential approach requires the value and the gradient calculation of the objective functions. Moreover, the gradient of the constraints with respect to the design variables must be calculated.

3.1 Gradient Calculation and Sensitivity Equations

The gradient calculation for the objective function uses the following equation:

$$\frac{\partial \Phi_i}{\partial p_j} = \int_{t_0}^{t_f} \left(\frac{\partial L_i}{\partial x} \left[\frac{\partial x}{\partial p_j}(t) \right] + \frac{\partial L_i}{\partial p_j} \right) dt \tag{30}$$

where, it can be seen in the general problem stated by (1) to (4), that L_i is the ith objective function, x is the vector of the state variables, p_j is the jth element of the vector of the design variables and t is the time variable. On the other hand, in order to obtain the partial derivatives $\frac{x}{p_j}$, it is necessary to solve the ordinary differential equations of the sensitivity given by

$$\frac{\partial \dot{x}}{\partial p_j} = \frac{\partial f}{\partial x} \left[\frac{\partial x}{\partial p_j} \right] + \frac{\partial f}{\partial p_j} \tag{31}$$

These sensitivity equations can be obtained by taking the time derivatives with respect to p_j of the dynamic model. Due to the fact that $\dot{x}$ is a function of the time variable t as well as the design variables p_j (we must consider that p_j are independent of t), then:

$$\dot{x} = \frac{dx}{dt} = \frac{\partial x}{\partial t} \tag{32}$$

moreover

$$\frac{d\left(\frac{\partial x}{\partial p_j}\right)}{dt} = \frac{\partial \left(\frac{\partial x}{\partial p_j}\right)}{\partial t} = \frac{\partial \left(\frac{\partial x}{\partial t}\right)}{\partial p_j} = \frac{\partial \left(\frac{dx}{dt}\right)}{\partial p_j} = \frac{\partial \dot{x}}{\partial p_j} \tag{33}$$

Finally, using the equalities (33) and proposing the following variable:

$$y_j = \frac{\partial x}{\partial p_j} \tag{34}$$

the partial derivatives of x with respect to p_j are now given by the following ordinary differential equations:

$$\dot{y}_j = \frac{\partial f}{\partial x} y_j + \frac{\partial f}{\partial p_j} \tag{35}$$

$$y_j(0) = \frac{\partial x_0}{\partial p_j} \tag{36}$$

3.2 Goal Attainment Method

In order to transform the MDOP into a NLP problem, the sequential approach is used. The resulting problem is solved using the Goal Attainment Method [11]. In the remainder of the chapter, we will refer to it as "MPM" (Mathematical Programming Method). In such a technique, a subproblem is obtained as follows:

$$\min_{p,\lambda} G\left(p,\lambda\right) \stackrel{\Delta}{=} \lambda \tag{37}$$

subject to:

$$
\begin{aligned}
g(\boldsymbol{p}) &\leq 0 \\
h(\boldsymbol{p}) &= 0 \\
g_{a1}(\boldsymbol{p}) &= \Phi_1\left(p\right) - \omega_1\lambda - \Phi_1^d \leq 0 \\
g_{a2}(\boldsymbol{p}) &= \Phi_2\left(p\right) - \omega_2\lambda - \Phi_2^d \leq 0
\end{aligned}
\tag{38}
$$

where λ is an artificial variable without sign constraint, and $g(\boldsymbol{p})$ and $h(\boldsymbol{p})$ are the constraints established in the original problem. Moreover, in the last two constraints, ω_1 and ω_2 are the scattering vectors, Φ_1^d and Φ_2^d are the desired goals for each objective function and Φ_1 and Φ_2 are the evaluated functions.

3.3 Numerical Method to Solve the NLP Problem

In order to solve the resulting NLP problem, Eqs. (37) and (38), the Successive Quadratic programming (SQP) method is used. There, a Quadratic Problem (QP) which is a quadratic approximation to the Lagrangian function optimized over a linear approximation to the constraints, is solved. A vector $\boldsymbol{p}^i$ containing the current parameter values is proposed and the NLP problem given by Eqs. (39) and (40) is obtained, where B_i is the Broyden–Fletcher–Goldfarb–Shanno updated (BGFS) positive definite approximation of the Hessian matrix, and the gradient calculation is obtained using sensitivity equations. Hence, if γ solves the subproblem given by (39) and (40) and $\gamma = 0$, then the parameter vector $\boldsymbol{p}^i$ is an original problem optimal solution. Otherwise, we set $\boldsymbol{p}^{i+1} = \boldsymbol{p}^i + \gamma$ and with this new vector the process is repeated all over again.

$$\min_{\gamma} QP(\boldsymbol{p}^i) = G\left(\boldsymbol{p}^i\right) + \nabla G^T\left(\boldsymbol{p}^i\right)\gamma + \frac{1}{2}\gamma^T B_i\gamma \tag{39}$$

subject to

$$
\begin{aligned}
g(\boldsymbol{p}^i) + \nabla g^T\left(\boldsymbol{p}^i\right)\gamma &\leq 0 \\
h(\boldsymbol{p}^i) + \nabla h^T\left(\boldsymbol{p}^i\right)\gamma &= 0 \\
g_{a1}(\boldsymbol{p}^i) + \nabla g_{a1}^T\left(\boldsymbol{p}^i\right)\gamma &\leq 0 \\
g_{a2}(\boldsymbol{p}^i) + \nabla g_{a2}^T\left(\boldsymbol{p}^i\right)\gamma &\leq 0
\end{aligned}
\tag{40}
$$

3.4 Experiments and Results of the Mathematical Programming Method

In order to carry out the parametric optimal design of the pinion-rack CVT, we performed 10 independent runs, all of them using a PC with a 2.8 GHz Pentium IV processor with 1 GB of Memory using Matlab 6.5.0 Release 13. The system parameters used in the numerical simulations were: $b_1 = 1.1$ Nms/rad, $b_2 = 0.05$ Nms/rad, $r = 0.0254$ m, $T_m = 8.789$ Nm, $T_L = 0$ Nm, $\lambda_s = 5.4271$, $\phi = 20$, $M = 10$ Kg, $r_p = 4.188 \times 10^{-03}$ m, $K_f = 63.92 \times 10^{-03}$ Nm/A, $K_b = 63.92 \times 10^{-03}$ Vs/rad, $R = 10$ Ω, $L = 0.01061$ H, $b_l = 0.015$ Ns/m, $b_c = 0.025$ Nms/rad and $n = ((22{\cdot}40{\cdot}33)/(9{\cdot}8{\cdot}9))$. The initial conditions vector was $[x_1(0), x_2(0), x_3(0), x_4(0)]^T = [7.5, 0, 0, 0]^T$ and the output reference was considered to be $x_{\text{ref}} = 3.2$.

Because the goal attainment method requires a goal for each of the objective functions, further calculations were necessary. The goal for Φ_1 was obtained by minimizing this function subject to Eqs. (8) and (20) to (26). The optimal solution vector $\boldsymbol{p}^1$ is shown in Table 1. The goal for Φ_2 was obtained by minimizing this function subject to Eqs. (8) and (20) to (26). The optimal solution vector $\boldsymbol{p}^2$ for this problem is also shown in Table 1.

Varying the scattering vector can produce different nondominated solutions. In Table 1, two cases are presented: $\boldsymbol{p}_A^*$ is obtained with $\omega = [0.5, 0.5]^T$, and $\boldsymbol{p}_B^*$ is obtained with $\omega = [0.4, 0.6]^T$.

As can be seen in the results in Table 3, 80% of the runs diverged. This behavior shows a high sensitivity of the MPM to the starting point (detailed in Table 2) because it must be carefully chosen in order to allow the approach to obtain a good solution. Information about the time required by the MPM per independent run is summarized in Table 3.

Figure 2 shows the mechanical efficiency and the input control of the pinion-rack CVT with both solutions obtained by the MPM ($\boldsymbol{p}^1$, $\boldsymbol{p}^2$ and $\boldsymbol{p}_A^*$). The solution

Table 1 Details of the solutions obtained by the MPM

$[N^*, m^*, h^*, e_{max}^*, K_p^*, K_I^*]$	$\Phi_N(\bullet) = [\Phi_1(\bullet), \Phi_2(\bullet)]$	$\Phi(\bullet) = [\Phi_1(\bullet), \Phi_2(\bullet)]$
$\boldsymbol{p}^1 = [38, 0.0017, 0.02, 0.0636, 10.000, 1.00]$	$\Phi_N(\boldsymbol{p}^1) = [1.0000, 4.7938]$	$\Phi(\boldsymbol{p}^1) = [0.4397, 2700.6279]$
$\boldsymbol{p}^2 = [13.4459, 0.0019, 0.02, 0.0826, 5.000, 0.01]$	$\Phi_N(\boldsymbol{p}^2) = [2.8017, 1.0000]$	$\Phi(\boldsymbol{p}^2) = [1.2319, 563.3585]$
$\boldsymbol{p}_A^* = [26.7805, 0.0017, 0.02, 0.0826, 5.000, 0.01]$	$\Phi_N(\boldsymbol{p}_A^*) = [1.4696, 1.4696]$	$\Phi(\boldsymbol{p}_A^*) = [0.6461, 827.9116]$
$\boldsymbol{p}_B^* = [29.0171, 0.0017, 0.02, 0.0789, 5.000, 0.01]$	$\Phi_N(\boldsymbol{p}_B^*) = [1.3646, 1.5469]$	$\Phi(\boldsymbol{p}_B^*) = [0.6000, 871.4592]$

Table 2 Initial points used for the MPM. Also shown is the corresponding scattering vector

Initial search point	Scattering vector
$[13.4459, 0.0019, 0.02, 0.0826, 5.000, 0.01]$	$[0.5, 0.5]$
$[38, 0.0017, 0.02, 0.0636, 10.000, 1.00]$	$[0.5, 0.5]$
$[38, 0.0017, 0.02, 0.0636, 10.000, 1.00]$	$[0.4, 0.6]$
$[38, 0.0017, 0.02, 0.0636, 10.000, 1.00]$	$[0.6, 0.4]$
$[28.8432, 0.0017, 0.02, 0.0550, 5.024, 0.017]$	$[0.5, 0.5]$
$[13.4459, 0.0019, 0.02, 0.0826, 5.000, 0.01]$	$[0.4, 0.6]$
$[28.8432, 0.0017, 0.02, 0.0550, 5.024, 0.017]$	$[0.4, 0.6]$
$[28.8432, 0.0017, 0.02, 0.0550, 5.024, 0.017]$	$[0.6, 0.4]$
$[30.77, 0.0017, 0.02, 0.0694, 5.121, 0.010]$	$[0.5, 0.5]$
$[30.77, 0.0017, 0.02, 0.0694, 5.121, 0.010]$	$[0.4, 0.6]$

Table 3 Time required by each run of the MPM. Note that only two runs could converge to a solution. The remaining 8 runs could not provide any result

Run	Time required
1	Diverged
2	23.78 Min
3	Diverged
4	Diverged
5	Diverged
6	Diverged
7	Diverged
8	Diverged
9	Diverged
10	48.5 Min
Average	**36.365 Min**

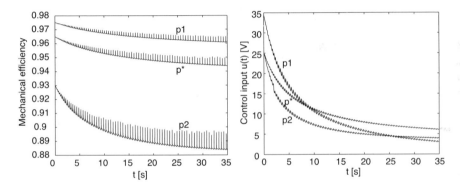

Fig. 2 Mechanical efficiency and control input for the pinion-rack CVT obtained by the MPM

p_A^* was selected because it has the same overachievement of the proposed goal for each objective function.

As can be observed in Fig. 2, when the number of teeth is increased (p_1^*) and their size is decreased (p_2^*), a higher CVT mechanical efficiency is obtained. Also, we can observe perturbations in the mechanical efficiency, which are produced because of tip-to-tip momentary contact prior to full engagement between teeth. With the optimal solution, this tip-to-tip contact is reduced because a better CVT planetary gear is obtained when the tooth size is decreased. Summarizing, the optimal solution implies a lower sensitivity of the mechanical efficiency with respect to reference changes. On the other hand, a more compact CVT size is obtained since (p_3^*) is decreased. Furthermore, a minimal controller energy is obtained when the controller gains (p_5^*) and (p_6^*) are decreased. In Fig. 2, it can be observed that the optimal vector minimizes the initial overshoot of the control input.

Despite the sensitivity of the NLP method, the optimal solutions obtained are good from the mechanical and controller point of view.

4 Evolutionary Optimization

The high sensitivity of the MPM to its initial conditions, and its implementation complexity motivated us to solve the problem using an evolutionary algorithm (EA). This is because one of the main advantages of an EA is that competitive results are obtained regardless of its initial conditions (i.e. a set of solutions is randomly generated). We selected Differential Evolution [17] for several reasons: (1) it is an EA which has provided very competitive results when compared with traditional EAs such as genetic algorithms and evolution strategies in real-world problems [6]; (2) it is very simple to implement [17]; and (3) its parameters for the crossover and mutation operators generally do not require a careful fine-tuning [13].

DE is an evolutionary direct-search algorithm to solve optimization problems. DE shares similarities with traditional EAs, however, it does not use binary encoding as a simple genetic algorithm [8] and it does not use a probability density function to self-adapt its parameters as an Evolution Strategy [18]. Instead, DE performs mutation based on the distribution of the solutions in the current population. In this way, search directions and possible stepsizes depend on the location of the individuals selected to calculate the mutation values.

Several DE variants have been proposed [17]. The most popular is called *"DE/-rand/1/bin"*, where "DE" means Differential Evolution, the word "rand" indicates that the individuals selected to compute the mutation values are chosen at random, "1" is the number of pairs of solutions chosen to calculate the differences for the mutation operator and finally "bin" means that a binomial recombination is used. A detailed pseudocode of this variant is presented in Fig. 3.

Four parameters must be defined in DE: (1) the population size, (2) the number of generations, (3) the factor $F \in [0.0, 1.0]$, which scales the value of the differences computed from randomly selected individuals (typically three, where two are used to compute the difference and the other is only added) from the population (row 11 in Fig. 3). A value of $F = 1.0$ indicates that the complete difference value is used; and finally, (4) the $CR \in [0.0, 1.0]$ parameter, which controls the influence of the parent on its corresponding offspring; a value of $CR = 0.0$ means that the offspring will take its values from its parent instead of taking its values from the mutation values generated by the combination of the differences of the individuals chosen at random (rows 9–15 in Fig. 3).

DE was originally proposed to solve global optimization problems. Moreover, like other EAs, DE lacks a mechanism to handle the constraints of a given optimization problem. Hence, we decided to modify the DE algorithm in order to solve constrained multiobjective optimization problems. It is worth remarking that the goal when performing these modifications was to maintain the simplicity of DE as much as possible.

Three modifications were made to the original DE:

1. The selection criterion between a parent and its corresponding offspring was modified in order to handle multiobjective optimization problems.
2. A constraint-handling technique to guide the approach to the feasible region of the search space was added.

```
1   Begin
2       G=0
3       Create a random initial population X i,G  ∀i, i = 1 ,...,NP
4       Evaluate f (X i,G ) ∀i, i = 1 ,...,NP
5       For G=1 to MAX_GEN Do
6           For i=1 to NP Do
7               Select randomly r₁ ≠ r₂ ≠ r₃ :
8               j rand  = randint(1 ,D )
9               For j=1 to D Do
10                  If (rand j [0, 1) < CR  or j = j rand  ) Then
11                      u ij,G +1 = x r₃ j,G  + F (x r₁ j,G  − x r₂ j,G )
12                  Else
13                      u ij,G +1 = x ij,G
14                  End If
15              End For
16              If (f (U i,G +1 ) ≤ f (X i,G )) Then
17                  X i,G +1 = U i,G +1
18              Else
19                  X i,G +1 = X i,G
20              End If
21          End For
22          G = G + 1
23      End For
24  End
```

Fig. 3 "DE/rand/1/bin" algorithm. randint(min,max) is a function that returns an integer number between min and max. rand[0, 1) is a function that returns a real number between 0 and 1. Both are based on a uniform probability distribution. "NP", "MAX_GEN", "CR" and "F" are user-defined parameters. "D" is the dimensionality of the problem

3. A simple external archive to save the nondominated solutions found during the process was added.

4.1 Selection Criterion

We changed the original criterion to select between parent and offspring (rows 16–20 in Fig. 3) based only on the objective function value. As in multiobjective optimization we are looking for a set of trade-off solutions, we used, as traditionally adopted in Evolutionary Multiobjective Optimization [5], Pareto Dominance as the criterion to select between the parent and its corresponding offspring. The aim is to keep the nondominated solutions from the current population.

A vector $\mathbf{U} = (u_1,\ldots,u_k)$ is said to dominate $\mathbf{V} = (v_1,\ldots,v_k)$ (denoted by $\mathbf{U} \preceq \mathbf{V}$) if and only if $\mathbf{U}$ is partially less than $\mathbf{V}$, i.e. $\forall i \in \{1,\ldots,k\}, u_i \leq v_i \wedge \exists i \in \{1,\ldots,k\}: u_i < v_i$. If we denote the feasible region of the search space as $\mathcal{F}$, the evolutionary multiobjective algorithm will look for the Pareto optimal set ($\mathcal{P}^*$) defined as:

$$\mathcal{P}^* := \{x \in \mathcal{F} \mid \neg \exists\, x' \in \mathcal{F}\ \mathbf{F}(x') \preceq \mathbf{F}(x)\}. \tag{41}$$

In our case, $k = 2$, as we are optimizing two objectives.

4.2 Constraint Handling

The most popular approach to incorporate the feasibility information into the fitness function of an EA is the use of a penalty function. The aim is to decrease the fitness value of the infeasible individuals (i.e., those that do not satisfy the constraints of the problem). In this way, feasible solutions will have a higher probability of being selected and the EA will eventually reach the feasible region of the search space. However, the main drawback of penalty functions is that they require the definition of penalty factors. These factors determine the severity of the penalty. If the penalty value is very high, the feasible region will be approached mostly at random and the feasible global optimum will be hard to find. On the other hand, if the penalty is too low, the probability of not reaching the feasible region will be high. Based on the aforementioned disadvantage, we decided to avoid the use of a penalty function. Instead, we incorporated a set of criteria based on feasibility, originally proposed by Deb [7] and further extended by other researchers [9,10,12]:

- Between two feasible solutions, the one which dominates the other wins.
- If one solution is feasible and the other one is infeasible, the feasible solution wins.
- If both solutions are infeasible, the one with the lowest sum of constraint violation is preferred.

We combine Pareto dominance and the set of feasibility rules into one selection criterion, which substitutes rows 16–20 in Fig. 3 as presented in Fig. 4.

4.3 External Archive

One of the features that distinguishes a modern evolutionary multiobjective optimization algorithm is the concept of elitism [5]. In our modified DE, we added an external archive, which stores the set of nondominated solutions found during the evolutionary process. This archive is updated at each generation in such a way that all nondominated solutions from the population will be included in the archive. After that, nondominance checking is performed with respect to all the solutions (the newcomers and also the solutions in the archive). The solutions that are nondominated with respect to everybody else will remain in the archive. When the search ends, the set of nondominated solutions in the archive will be reported as the final set of solutions obtained by the approach.

If (U^i_{G+1} is better than X^i_G (based on the three *selection criteria*)) **Then**
  $X^i_{G+1} = U^i_{G+1}$
Else
  $X^i_{G+1} = X^i_G$
End If

Fig. 4 Modified selection mechanism added to the DE algorithm in order to solve the multiobjective optimization problem

4.4 Results of the EA Approach

In our experiments, we performed 10 independent runs. A fixed set of values for the parameters was used in all runs and they were defined as follows: Population size NP = 200, $MAX\_GENERATIONS$ = 100; the parameters F and CR were randomly generated within an interval. The parameter F was generated per generation in the range $[0.3, 0.9]$ (the differences can be scaled in different proportions without affecting the performance of the approach) and CR was generated per run in the range $[0.8, 1.0]$ (greater influence of the mutation operator instead of having such influence from the parent when generating the offspring). These values were empirically derived. This way of defining the values for F and CR makes it evident that they do not require to be fine-tuned. We will refer to the evolutionary approach as "EA" (Evolutionary Algorithm).

The experiments were performed on the same platform on which the goal attainment experiments were carried out. This was done to have a common point of comparison to measure the computational time required by each approach.

In Table 4 we present the number of nondominated solutions and also the time required per run.

The 10 different Pareto fronts obtained are presented in Fig. 5.

In order to help the decision maker, we filtered the 10 different set of solutions in order to obtain the final set of nondominated solutions. The final Pareto front obtained from the 10 runs contains 28 nondominated points and is presented in Fig. 6. Finally, the details of the 28 solutions are presented in Table 5.

Figure 7 shows the mechanical efficiency and the input control of the pinion-rack CVT with the optimal solution obtained with the MPM and the solution ([30.185435, 0.017, 0.020075, 0.059569, 5.133269,0.019914]) in the middle of the filtered Pareto front obtained with the EA (Fig. 6). We can observe that the mechanical efficiency found by the EA is better than that of the MPM solution. We can also see a smooth behavior of

Table 4 Time required and number of nondominated solutions found at each independent run by the EA

Run	Time required	Nondominated solutions
1	18.53 Hrs.	17
2	20.54 Hrs.	15
3	18.52 Hrs.	25
4	18.63 Hrs.	16
5	18.55 Hrs.	17
6	17.57 Hrs.	19
7	18.15 Hrs.	18
8	18.47 Hrs.	24
9	18.67 Hrs.	16
10	20.24 Hrs.	18
Average	**18.78 Hrs**	**18.5 solutions**

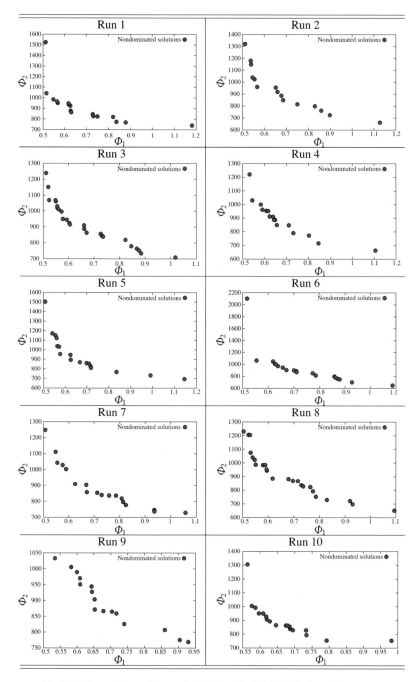

Fig. 5 Different Pareto fronts obtained by the EA in 10 independent runs

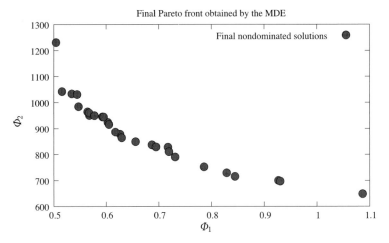

Fig. 6 Final set of solutions obtained by the EA in 10 independent runs

Table 5 Details of the trade-off solutions found by the EA. All solutions are feasible

$[N^*, m^*, h^*, e^*_{max}, K_P^*, K_I^*]$	$[\Phi_1(\bullet), \Phi_2(\bullet)]$
$[32.949617, 0.001780, 0.020413, 0.063497, 5.131464, 0.022851]$	$[0.534496, 1033.243548]$
$[25.022005, 0.001699, 0.020103, 0.052385, 5.087026, 0.024991]$	$[0.687214, 837.167059]$
$[24.764331, 0.001723, 0.020662, 0.048119, 5.104801, 0.011072]$	$[0.694969, 828.856396]$
$[32.203853, 0.001793, 0.021356, 0.066703, 5.033164, 0.012833]$	$[0.547385, 984.149814]$
$[30.774167, 0.001710, 0.020092, 0.069459, 5.129618, 0.010260]$	$[0.568131, 950.480089]$
$[34.231339, 0.001756, 0.020974, 0.065426, 5.104461, 0.023469]$	$[0.515604, 1042.009590]$
$[31.072336, 0.001760, 0.020295, 0.072332, 5.018621, 0.024963]$	$[0.564775, 964.310541]$
$[27.647589, 0.001685, 0.020151, 0.069264, 5.001687, 0.031805]$	$[0.627021, 877.670407]$
$[27.548056, 0.001696, 0.020083, 0.067970, 5.006868, 0.017859]$	$[0.629913, 864.206663]$
$[30.866972, 0.001735, 0.020305, 0.058766, 5.002777, 0.032694]$	$[0.567519, 960.120458]$
$[28.913492, 0.001747, 0.020478, 0.058322, 5.021887, 0.027174]$	$[0.603222, 923.771423]$
$[28.843277, 0.001764, 0.020282, 0.055027, 5.024443, 0.017157]$	$[0.605340, 915.753294]$
$[30.185435, 0.001700, 0.020075, 0.059569, 5.133269, 0.019914]$	$[0.577733, 949.842309]$
$[29.448640, 0.001755, 0.020601, 0.063276, 5.019318, 0.033931]$	$[0.593085, 944.906551]$
$[20.002905, 0.001697, 0.020098, 0.053235, 5.114809, 0.018447]$	$[0.844657, 715.605541]$
$[26.373053, 0.001718, 0.020176, 0.068410, 5.031773, 0.014986]$	$[0.656264, 849.215816]$
$[32.227085, 0.001764, 0.020567, 0.070369, 5.178989, 0.026127]$	$[0.544721, 1030.722785]$
$[23.476167, 0.001731, 0.020618, 0.057264, 5.050345, 0.010533]$	$[0.730990, 790.412654]$
$[23.853314, 0.001696, 0.020054, 0.063646, 5.097374, 0.040464]$	$[0.717403, 827.978369]$
$[23.936736, 0.001767, 0.020179, 0.054081, 5.026456, 0.013965]$	$[0.719347, 810.685134]$
$[18.094865, 0.001754, 0.020097, 0.033930, 5.263513, 0.012051]$	$[0.926890, 700.251032]$
$[15.287561, 0.001836, 0.020539, 0.065247, 5.001634, 0.077960]$	$[1.086582, 648.563140]$
$[20.410186, 0.001689, 0.020082, 0.067889, 5.005502, 0.046545]$	$[0.828891, 729.481066]$
$[29.319668, 0.001754, 0.020557, 0.057790, 5.140154, 0.012875]$	$[0.595073, 944.511281]$
$[28.165197, 0.001722, 0.020449, 0.069922, 5.035457, 0.013965]$	$[0.617721, 886.468167]$
$[34.733111, 0.001738, 0.020849, 0.064827, 5.470063, 0.078838]$	$[0.504179, 1230.655492]$
$[18.028162, 0.001753, 0.021026, 0.075356, 5.185506, 0.027797]$	$[0.930299, 697.362827]$
$[21.642511, 0.001694, 0.020196, 0.061009, 5.040619, 0.029378]$	$[0.785859, 752.464167]$

the mechanical efficiency for the EA, maintaining a more compact CVT size for the EA solution. However, the initial overshoot of the input control is greater than that of the MPM solution. These behaviors are observed with all the solutions lying on the middle of the Pareto front, because a higher number of teeth and a corresponding smaller size are obtained (p_1^* was increased and p_2^* was decreased) whereas the input

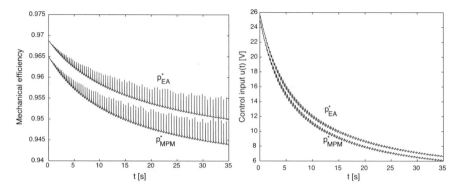

Fig. 7 Mechanical efficiency and Control input for the pinion-rack CVT. obtained by the EA approach

energy controller is greater (p_5^* and p_6^* were increased) in these optimal solutions. In conclusion, from a mechanical point of view, the solutions in the middle of the Pareto front, offer many possible system reconfigurations of the pinion-rack CVT.

5 Advantages and Disadvantages of Both Approaches

5.1 Quality and Robustness

As we can see, the results provided by the EA were as good as those obtained by the MPM method because the latter solutions were also nondominated with respect to those found by the EA. However, the EA was not sensitive to the initial conditions (a randomly generated set of solutions was adopted at all times). The EA approach provided a more robust behavior than that shown by the MPM. Despite the fact that the results obtained by both approaches are considered similar (from a mechanical and from a control point of view), as the EA obtains several solutions from a single run, it gives the designer the chance to select from them, the best choice based on his preferences.

5.2 Computation Cost

It is clear, based on the results shown in Tables 3 and 4 for the MPM and the EA approaches respectively, that the EA is the most expensive, computationally speaking. However, as pointed out in Table 4, the EA obtains a set of nondominated solutions per single run. In contrast, the MPM always returns a single solution on each run. Therefore, based on the average time (18.78 h) and the average number of solutions obtained (18.5 solutions), approximately, one solution per hour is obtained. On the other hand, the MPM obtained a solution after approximately 36 minutes computation time. Therefore, we can conclude that the EA requires, roughly, twice the time used by the MPM to find a single solution.

5.3 Implementation Issues

As mentioned in Sect. 3.3, in order to solve the multiobjective optimization problem using the MPM, a sequential quadratic programming method was used. There, a quadratic programming problem, which is an approximation to the original CVT problem, was solved, and some difficulties detected:

- This method requires gradient calculation, sensitivity equations and gradient equations of the constraints. In general, the number of sensitivity equations is the product of the number of state variables and the number of design variables. Gradient equations are related to the number of design variables. Summarizing, we must calculate two objective functions equations, 24 sensitivity equations, six gradient equations and 54 constraint gradient equations. On the other hand, with the EA only two objective functions equations must be calculated. Therefore, reconfiguration of the EA is simple.
- Due to the fact that the QP problem is an approximation to the original problem and that the constraints are a linear approximation, this problem might be unbounded or infeasible, whereas the original problem is not. With the EA, the original problem is solved. Therefore, the search for the optimal solution is performed in the feasible region of the search space, directly. In this way, in the case of the EA, new structural parameters can be obtained when additional mechanical constraints to the design problem are added. These mechanical constraints could be considered directly in the constraint-handling mechanism of the algorithm without the need for any further changes.

It is worth recalling that another additional step related to the use of the MPM is that it requires minimizing each objective function considered, separately. This is because the goal attainment method requires a goal for each function to be optimized. This step is not required by the EA. Finally, the EA showed no significant sensitivity to its parameters.

5.4 Goal Attainment to Refine Solutions

It is important to mention that we carried out a set of runs of the MPM using a non-dominated solution obtained by the EA, as a starting point. However, the approach was unable to improve the solution in all cases.

6 Conclusions and Future Work

We have presented the multiobjective optimization of a pinion-rack continuously variable transmission (CVT). The aim is to maximize the mechanical efficiency and to minimize the corresponding control. The problem is subject to geometric and strength conditions for the gear pinion of the CVT. Two different approaches were used to solve the problem: A mathematical programming method called Goal Attainment and also an evolutionary algorithm. The first one was very sensitive to the initial start point of the search (the point must be given by the user and must be carefully selected), but the computation time required was of about 30 minutes to obtain a solution. On the other hand, the evolutionary algorithm, which in our case was differential evolution, showed no sensitivity to the initial conditions, i.e. a set of randomly generated solutions was used in all experiments. Also, the approach did not show any sensitivity to the values of the parameters related to the crossover and mutation operators. Furthermore, the EA returned a set of solutions on each single run, which gave the designer more options to select the best solutions, based on his preferences. The computational time required for the EA was about 60 minutes to find a solution. The results obtained with the two approaches were similar based on quality, but the EA was more robust (in each single run it obtained feasible results). Finally, the EA was easier to implement, which is one clear advantage of the approach.

Future work will include designing a preferences-handling mechanism in order to let the EA concentrate the search on those regions of the Pareto front where the most convenient solutions are located. Furthermore, we plan to solve other mechatronic problems using the proposed approach.

Acknowledgement. The first author gratefully acknowledges support from CONACyT project number 52048-Y. The third author also acknowledges support from CONACyT project number 42435-Y.

References

1. A. Cervantes and L.T. Biegler. Large-scale DAE optimization using a simultaneous NLP formulation. *AIChe Journal*, 44:1038–1050, 1998
2. J. Alvarez-Gallegos, C. Cruz-Villar, and E. Portilla-Flores. Parametric optimal design of a pinion-rack based continuously variable transmission. In *Proceedings of the 2005 IEEE/ASME International Conference on Advanced Intelligent Mechatronics*, pages 899–904, Monterey California, 2005
3. J. Alvarez-Gallegos, C. Alberto Cruz Villar, and E. Alfredo Portilla Flores. Evolutionary Dynamic Optimization of a Continuously Variable Transmission for Mechanical Efficiency Maximization. In A. Gelbukh, Álvaro de Albornoz and H. Terashima-Marín, editors, *MICAI 2005: Advances in Artificial Intelligence*, pages 1093–1102, Monterrey, México, November 2005. Springer. Lecture Notes in Artificial Intelligence Vol. 3789
4. J.T. Betts. *Practical Methods for Optimal Control using Nonlinear Programming.* SIAM, Philadelphia, USA, 2001
5. C. A. Coello Coello, D. A. Van Veldhuizen, and G. B. Lamont. *Evolutionary Algorithms for Solving Multi-Objective Problems.* Kluwer Academic Publishers, New York, June 2002. ISBN 0-3064-6762-3
6. I.L. Lopez Cruz, L.G. Van Willigenburg, and G. Van Straten. Parameter Control Strategy in Differential Evolution Algorithm for Optimal Control. In M.H. Hamza, editor, *Proceedings of the IASTED International Conference Artificial Intelligence and Soft Computing (ASC 2001)*, pages 211–216. Cancun, México, ACTA Press, May 2001. ISBN 0-88986-283-4
7. K. Deb. An efficient constraint handling method for genetic algorithms. *Computer Methods in Applied Mechanics and Engineering*, 186(2/4):311–338, 2000

8. D. E. Goldberg. *Genetic Algorithms in Search, Optimization and Machine Learning*. Addison-Wesley Publishing Co., Reading, Massachusetts, 1989

9. F. Jiménez and J. L. Verdegay. Evolutionary Techniques for Constrained Optimization Problems. In H.-J. Zimmermann, editor, *7th European Congress on Intelligent Techniques and Soft Computing (EUFIT'99)*, Aachen, Germany, 1999. Verlag Mainz. ISBN 3-89653-808-X

10. J. Lampinen. A Constraint Handling Approach for the Differential Evolution Algorithm. In *Proceedings of the Congress on Evolutionary Computation 2002 (CEC'2002)*, volume 2, pages 1468–1473, Piscataway, New Jersey, May 2002. IEEE Service Center

11. G. P Liu, J. Yang, and J. Whidborne. *Multiobjective Optimisation and Control*. Research Studies Press, 2003

12. E. Mezura-Montes and C. A. Coello Coello. A simple multimembered evolution strategy to solve constrained optimization problems. *IEEE Transactions on Evolutionary Computation*, 9(1):1–17, February 2005

13. E. Mezura-Montes, J. Velázquez-Reyes, and C. A. Coello Coello. Promising Infeasibility and Multiple Offspring Incorporated to Differential Evolution for Constrained Optimization. In H.-G. Beyer and et al., editors, *Proceedings of the Genetic and Evolutionary Computation Conference (GECCO'2005)*, volume 1, pages 225–232, New York, June 2005. ACM Press

14. R. Norton. *Machine Design. An Integrated Approach*. Prentice-Hall Inc., 1996

15. A. Osyczka. *Multicriterion Optimization in Engineering*. John Wiley and Sons, 1984

16. P. Papalambros and D. Wilde. *Principles of Optimal Design. Modelling and Computation*. Cambridge University Press, 2000

17. K. V. Price. An Introduction to Differential Evolution. In David Corne, Marco Dorigo, and Fred Glover, editors, *New Ideas in Optimization*, pages 79–108. Mc Graw-Hill, UK, 1999

18. H.-P. Schwefel, editor. *Evolution and Optimization Seeking*. John Wiley & Sons, New York, 1995

19. P. Setlur, J. Wagner, D. Dawson, and B. Samuels. Nonlinear control of a continuously variable transmission (cvt). In *Transactions on Control Systems Technology, Vol 11*, pp. 101–108. IEEE, 2003

20. E. Shafai, M. Simons, U. Neff, and H. Geering. Model of a continuously variable transmission. In *First IFAC Workshop on Advances in Automotive Control*, pp. 575–593, 1995

21. C. De Silva, M. Schultz, and E. Dolejsi. Kinematic analysis and design of a continuously variable transmission. *Mechanism and Machine Theory*, 29(1):149–167, 1994

22. M. Spotts. *Mechanical Design Analysis*. Prentice Hall Inc, 1964

23. H. van Brussel, Németh I. Sas P, P.D. De Fonseca, and P. van den Braembussche. Towards a mechatronic compiler. *IEEE/ASME Transactions on Mechatronics*, 6(1):90–104, 2001

Optimizing Stochastic Functions Using a Genetic Algorithm: an Aeronautic Military Application

Hélcio Vieira Junior

Instituto Tecnológico de Aeronáutica, São José dos Campos - SP, Brazil
junior_hv@yahoo.com.br

Abstract

Air defense systems based on MAN Portable Air Defense Systems (MANPADS) have demonstrated exceptional effectiveness against aircraft. To counter these systems, an artefact named flare was developed.

The purpose of this chapter is to suggest a methodology to determine an optimum sequence of flare launch, which aims to maximize the survival probability of an aircraft against a MANPADS missile.

The proposed method consists in simulating the missile/aircraft/flare engagement to obtain the flare launch program success' probability distribution function. In addition to utilizing this simulation, the use of a Genetic Algorithm is suggested to optimize the flare launch program.

Employment of the proposed methodology increased the aircraft success probability by 51% under the same conditions of generic parameters for the missile, aircraft and flare.

Key words: Metaheuristics, Combinatorial Optimization, Military

1 Introduction

The air defense system based on MANPADS (MAN Portable Air Defense Systems), besides having low cost and great mobility, has demonstrated, throughout history, great lethality: estimates suggest that 90% of worldwide combat aircraft losses between 1984 and 2001 were due to this kind of missile [6].

The majority of these weapon systems are based on an infrared sensor, i.e., they follow the radiation emitted by the aircraft's heat.

To counter these weapons, an artefact named flare was developed. This object, after being launched by the target aircraft, begins to burn itself, in this way producing much more infrared radiation than the target aircraft. The missile, which follows the biggest source of radiation inside its field of view, will change its trajectory to intercept the flare, causing no more threat to the aircraft.

The target aircraft has three options for flare launch:

- the first is an automatic flare launch by a device that senses the approaching missile;
- another option is a manual flare launch by the pilot when he sees the approaching missile;
- the third option is an automatic launch of a predetermined sequence of flares by the pilot when he enters into an area with great probability of having a MANPADS based air defense system.

Despite the existence of devices that sense the approaching missile (first option), they are not totally reliable and/or are not very widespread. In the absence of these equipments, the pilot has the choice to detect the threat himself (second option) or to use a preemptive flare launch sequence (third option).

This work is limited to the third option, its purpose being to suggest a methodology to determine an optimum sequence of flare launch, which aims to maximize the survival probability of an aircraft against a MANPADS missile.

The proposed method consists of a simulation of the missile/aircraft/flare engagement to obtain the flare launch program success probability distribution function. Besides utilizing this simulation, a Genetic Algorithm is suggested to optimize the flare launch program.

The utilization of simulations to infer the effectiveness of missiles against aircraft has been discussed by several researchers [4, 5, 7–9] and demonstrated to be a very efficient tool. The difference between this and previous work, is the addition of a third entity into the simulation (the flare), and the employment of an optimization method (Genetic Algorithm).

This chapter is organized as follows: in Sect. 2, the problem is defined. The heuristic methodology (Genetic Algorithm) is reviewed in Sect. 3. Section 4 describes the Monte Carlo simulation and Sect. 5 concludes.

2 Problem Definition

The aircraft's flare launch system has the following programming parameters: the number of flares (n_f) by salvo[1], the number of salvos (n_s), the time interval between successive flares (δ_f) and the time interval between successive salvos (δ_s), where the first two variables are discrete and the last two are continuous. These parameters are illustrated in Fig. 1, where $n_f = 4$, $n_s = 3$ and the symbol $\wedge$ represents the launch of one flare.

The main problem is how to define optimal parameters that produce the aircraft's greatest survival probability. Due to the stochastic behavior of the problem (the unknown distance and bearing of the missile in relation to the aircraft and the time of missile launch) and as field tests required to obtain these parameters are not a viable option, Monte Carlo Simulation was selected as a tool to measure the efficiency of a specific flare launch sequence.

---

[1] A simultaneous discharge of weapons or projectiles from an aircraft.

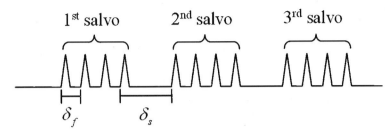

Fig. 1 Parameters of the flare launch system

"The most commonly used optimization procedures – linear programming, non-linear programming and (mixed) integer programming – require an explicit mathematical formulation. Such a formulation is generally impossible for problems where simulation is relevant, which are characteristically the types of problems that arise in practical applications" [1].

As the objective function is not known (either mathematical formulae or properties), the use of conventional optimization algorithms is not possible. We chose the Genetic Algorithm as the solution for this problem.

3 Genetic Algorithm

Genetic Algorithms (GA) are search and optimization algorithms based on the natural selection theory of Charles Darwin. These algorithms were initially proposed by John Holland [3] and they follow Darwin's assertion: the better an individual adapts to the environment, the better his chances of survival and generating descendants.

The GA starts by randomly selecting a number of individuals inside the search space. These individuals are then evaluated and assigned a fitness value. According to the fitness value, a new population is created through stochastic operators of selection, recombination (crossover) and mutation.

A generic GA can be described formally as ($P(g)$ is the population of the generation g):

Algorithm 1: Genetic Algorithm

$g = 0$;
Start $P(g)$;
Evaluate $P(g)$;
Repeat
    $g = g + 1$;
    Select $P(g)$ from $P(g - 1)$;
    Do crossover in $P(g)$;
    Do mutation in $P(g)$;
    Evaluate $P(g)$;
Until the stop criterion is reached.

3.1 Encoding and Generation Gaps of the Implemented GA

Each component of one solution, in genetics terms, is named a gene. Each gene represents, in our work, one parameter of the flare launch program. The ways the genes may be represented are by binary (canonical) and real encodings. We opted for real-value encoding, i.e., the flare launch program parameters are represented as vectors in the n-dimensional Euclidean space $\mathbb{R}^n$. For example, the codification for launching 8 salvos, with 2 flares each, time interval between successive flares of 1.23 seconds and time interval between successive salvos of 2.76 seconds is:

$$p_i = (8.00 \quad 2.00 \quad 1.23 \quad 2.76)$$

We utilized the generational replacement with elitism as the implemented Generation Gap[2]. In this procedure, the whole population, but the best individual, is replaced in each generation.

3.2 Operators of the Implemented GA

Evaluate
To evaluate is to measure the performance with respect to an objective function.

In our problem, we utilized the statistic *mean* obtained by Monte Carlo Simulation as the value of the objective function. Further details of this simulation are provided in Sect. 4.

Selection
The selection operator is used to elect the individuals more adapted to the environment with the objective of allowing them to reproduce.

We used, as selection operator, linear ranking. In this option, the fitness value of an individual is not the value of its objective function, but is re-defined as:

$$fitness(i) = Min + \frac{(Max - Min).(n - i)}{n - 1} \tag{1}$$

where: $fitness(i)$ is the fitness value of the individual i, Max is the maximum value of the fitness, Min is the minimum value of the fitness, $1 \le Max \le 2$, $Max + Min = 2$, n is the number of individuals of the population and i is the index of the individual on the population arranged in descending order by objective function value.

After that, the selection operator will choose, with probability proportional to the fitness value, an intermediate population (mating pool). This choice was implemented through the method known as roulette wheel.

Algorithm 2: Roulette Wheel

$Total\_fitness = \sum_{i=1}^{n} fitness(i);$
$Random = U(0, Total\_fitness);$
$Partial\_fitness = 0;$
$i = 0;$

---

[2] The proportion of individuals in the population which are replaced in each generation.

Repeat
 $i = i + 1$;
 $Partial\_fitness = Partial\_fitness + fitness(i)$;
Until $Partial\_fitness \leq Random$;
Return individual p_i.

---

p_i is the ith individual of population $P(g)$ and $U(a, b)$ is a uniform distribution in the interval $[a, b]$.

Crossover

Two individuals k and l are chosen from the mating pool and, with a certain probability (normally between 60% and 90%), the crossover operator will be employed to generate two new individuals (offspring).
 We chose the BLX-α crossover operator for our GA:

$$offspring_i = parent_k + \beta(parent_l - parent_k) \tag{2}$$

where: $offspring_i$ is the ith individual of population g, $parent_j$ is the jth individual of the mating pool (extracted from population $g - 1$) and $\beta \in U(-\alpha, 1 + \alpha)$.

Mutation

The mutation operator improves the individual diversity of the population, but, as it also destroys the individual's data, it must be used with low probability rates (normally between 0.1% and 5%).
 Uniform mutation was the type implemented in our GA:

$$offspring^i = \begin{cases} U(a_i, b_i), & \text{if } i = j \\ parent^i, & \text{if } i \neq j \end{cases} \tag{3}$$

where: $offspring^i$ is the ith gene of individual $offspring$, a_i and b_i are the viable interval limits for the gene $offspring^i$, j is the selected gene for mutation on the individual $parent$ and $parent^i$ is the ith gene of individual $parent$.
 Additionally, we took another procedure when dealing with the two first flare programming parameters (number of flares by salvo and number of salvos): as these parameters are not continuous, after the use of the GA operators crossover and mutation, we took as result the integer value furnished by the operators. For example: the crossover operator applied to the individuals $p_1 = \left(2\ 5\ x_1\ x_2 \right)$ and $p_2 = \left(6\ 1\ x_3\ x_4 \right)$ provided: $p_3 = \left(3.12\ 1.87\ x_5\ x_6 \right)$. Therefore, the resultant individual will be: $p_3 = \left(3\ 2\ x_5\ x_6 \right)$.
 For a detailed review of GA, readers can refer to [2].

4 Monte Carlo Simulation

The Monte Carlo Simulation (MCS) is based on the simulation of random variables with the objective of solving a problem. Its main characteristic is that the outcomes of a process are determined, even those resulting from random causes. As it is a very simple and flexible method, it can be used on problems of different complexity levels. The biggest inconvenience of this methodology is the great number of runs necessary to reduce the error estimate of the solution (normally values around 10,000), making the simulation a very slow process.

The MCS of the flare program fitness calculation randomly generates scenarios for the random variables of the entity missile: distance, bearing and launch time. After this initial step, simulation of the missile, aircraft and flare path determines the flare program's success or failure in decoying the missile. Repeating this procedure n times, we will obtain the flare program success' probability distribution function. The fitness value of a specific flare program will be the mean of the respective probability distribution function. In other words: for each individual of each generation of the GA, the MCS will generate lots of random scenarios for the entity missile and the mean of the outcomes will be the value that the MCS will return to the GA.

Observe that the result of one simulation of the MCS is a success or a failure in decoying the missile, i.e., a zero or one, and, consequently, the probability distribution function is a binomial one. We named the MCS outcome (mean of a specific flare program probability distribution function) as the *effectiveness*.

The implemented MCS uses the following parameters:

- Aircraft parameters: the strike aircraft begins its path by heading for its target (it is assumed that the aircraft is attacking a ground target) at distance d_1 from the target, altitude alt_{nav} and velocity V_{acft}. At distance d_2 from the target, the aircraft turns $\widehat{a_1}$ degrees. At distance d_3, the aircraft climbs at $\widehat{a_2}$ degrees of attitude until altitude alt_{high}, when it heads for its target again and begins to descend with $\widehat{a_3}$ degrees of attitude. At alt_{bomb} altitude, the aircraft releases its weapons, turns $\widehat{a_4}$ degrees and levels at alt_{nav} altitude. This procedure, called pop-up at-

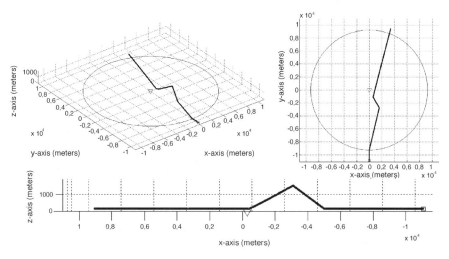

Fig. 2 Perspective (*up/left*), top (*up/right*) and side (*down*) views of the aircraft's path

tack, is illustrated in Figure 2, where the continuous line represents the aircraft's path, the dotted line represents the coordinates system's reference and the triangle represents the aircraft's target.

- Missile parameters: the MCS uses, as data for the missile, its mean velocity, turn rate, sensor movement maximum limit, maximum distance of proximity fuse, maximum and minimum distance of employment, sensor's field of view, time constant and proportional navigation constant. The interception law implemented was proportional navigation. It is assumed that the missile is pointed toward the aircraft at the moment of its launch. All the random variables of this MCS are associated with the entity missile (bearing and distance in relation to the target aircraft and launch time). As these variables behave completely randomly in the real world, we chose the uniform distribution inside the viable intervals to represent their probability distribution function:

$$Bearing \in U(0, 2\pi) \tag{4}$$

$$Distance \in U(\text{minimum distance, maximum distance}) \tag{5}$$

$$Launch\ Time \in U(0, \text{simulation's maximum time}) \tag{6}$$

- Flare parameters: the flare characteristics that are significant in the MCS are the available number of flares, burn time, time to reach the effective radiation (rise time), ejection velocity, deceleration and ejection angle in relation to the aircraft body. The flare–aircraft separation's path has great influence in decoying or not decoying the missile, and it is illustrated in Fig. 3, where the bold line represents the aircraft path, the thin line represents the flare path and the asterisk represents the point where the flare reaches its effective radiation (rise time point).

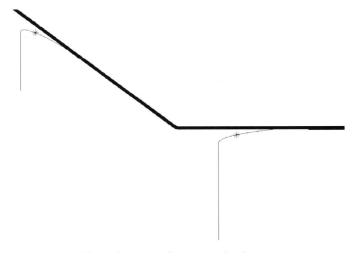

Fig. 3 Flare-Aircraft separation's side view

As an example of the simulation's geometry, one MCS history (named engagement ε), in which the missile was decoyed by a flare, is shown in Figs. 4 and 5, where the bold line represents the aircraft path, the thin line represents the missile path and the dotted line represents the coordinates system's reference.

The moments of the engagement ε marked by the symbols ◯ and ⋆, in the Figures 4 and 5, represent the positions of the aircraft and the missile, respectively, at missile

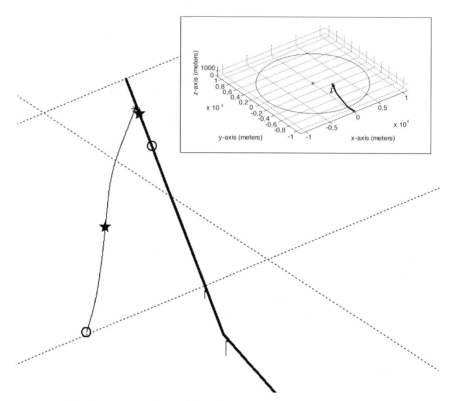

Fig. 4 Perspective (*up/right*) and close-up views of the engagement ε

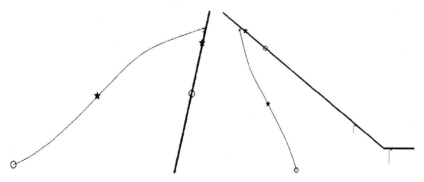

Fig. 5 Top (*left*) and side (*right*) views of the engagement ε

launch time and flare launch time. Observe that the missile had already fixed a bearing to intercept the aircraft and then, after the moment ⋆, it changed its trajectory to intercept the flare.

As parameters to validate the MCS, we used generic values for the missile, the flare and the aircraft, not representing any simulation of a real missile/aircraft/flare.

We ran five MCS, with 30,000 histories each, with the objective of confirming model convergence. These simulations showed that, for a precision of 1.50%, the simulation of 2000 histories would be sufficient. The graphics of these simulations can be checked in Figs. 6 and 7, where the y-axis measures the effectiveness differ-

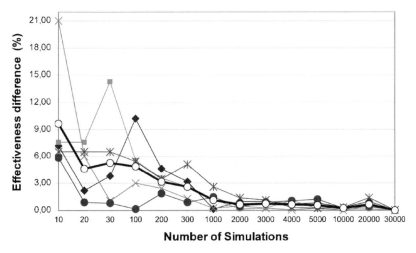

Fig. 6 Convergence graphic of the simulations

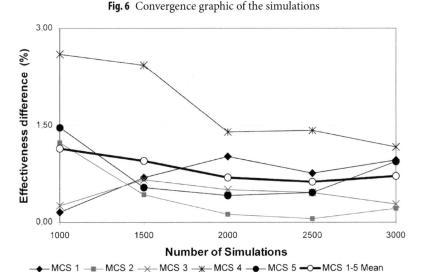

Fig. 7 Detail of the convergence graphic shown in Fig. 6

ence between 30,000 simulations and the x-axis is the number of simulations. For example: MCS 3 had, after 10 histories, an effectiveness of μ_{10} and, after 30,000 histories, an effectiveness of $\mu_{30,000}$. In Fig. 6, the value of $\mu_{30,000} - \mu_{10} = 21.00\%$.

After determining the number of histories (2000) required for each MCS, we ran the optimization model. The results of the first 10 generations can be seen in Table 1, where the numbers in brackets are the 95% confidence interval estimates.

Observe that the relation between the best efficiency of the tenth generation and the worst of the first generation was approximately 1.51, i.e., an improvement of 51%.

We ran only 10 generations with only 20 individuals because the computation time required to perform the 2000 MCS associated with each individual was very long, e.g., in a Pentium 4, 3.2 GHz and 1 GB RAM, these 10 generations required 8 days to run. As our purpose was to infer if this methodology would work and not to optimize a real-world problem, we stopped the simulation after these 10 generations.

5 Conclusions

This chapter proposed a methodology to determine the optimum sequence of flare launching, which aims to maximize the survival probability of an aircraft against a MANPADS missile. This methodology utilizes the heuristic Genetic Algorithm to optimize a stochastic objective function. As the engagement missile/flare/aircraft is a stochastic problem, we applied Monte Carlo Simulation to determine its probability distribution function, which was then used as objective function value for the GA.

In Sect. 2, the problem was detailed for a better understanding of the proposed method; the heuristic Genetic Algorithm and the implemented stochastic operators were reviewed in Sect. 3; the Monte Carlo Simulation and the significant parameters of the problem were described in Sect. 4.

In the implemented MCS, we used generic parameters for the aircraft, the flare and the missile, not representing any simulation of a real missile/aircraft/flare. The worst success probability found by the first generation of the GA was 53.85% (95%

Table 1 Optimization's GA: 10 first generations' efficiency

Individual	Effectiveness (%)				
	Generation 1	Generation 2	...	Generation 9	Generation 10
p_1	78.55	80.75	...	81.75	81.75
	[76.68, 80.33]	[78.95, 82.45]		[79.98, 83.42]	[79.98, 83.42]
p_2	78.35	78.70	...	81.65	81.65
	[76.47, 80.13]	[76.83, 80.47]		[79.88, 83.32]	[79.88, 83.32]
⋮	⋮	⋮	⋱	⋮	⋮
p_{20}	53.85	65.10	...	75.00	76.25
	[51.63, 56.05]	[62.96, 67.19]		[73.04, 76.88]	[74.32, 78.10]

CI: [51.63, 56.05]). After 10 generations, it was possible to reach a success probability of 81.75% (95% CI: [79.98, 83.42]), showing the coherency and efficiency of the proposed methodology.

As future work, we will use the implemented method with real aircraft, flare and missile parameters, and also compare the use of GA against other simulation optimization techniques, such as the sequential response surface methodology procedure.

References

1. Fu, MC (2002) Optimization for Simulation: Theory vs. Practice. INFORMS Journal on Computing, v. 14, n. 3:192–215
2. Goldberg, DE (1989) Genetic Algorithms in Search, Optimization and Machine Learning. Addison-Wesley, Reading, MA.
3. Holland, JH (1975) Adaptation in Natural and Artificial Systems. University of Michigan Press, Michigan, USA
4. Lukenbill, FC (1990) A target/missile engagement scenario using classical proportional navigation. MA Thesis, Naval Postgraduate School, Monterey, California, USA
5. Moore, FW, Garcia, ON (1997) A Genetic Programming Approach to Strategy Optimization in the Extended Two-Dimensional Pursuer/Evader Problem. In: Koza et al (eds) Genetic Programming 1997: Proceedings of the Second Annual Conference: 249–254, Morgan Kaufmann Publishers, Inc.
6. Puttre, M (2001) Facing the shoulder-fired threat. Journal of Electronic Defense, April
7. Shannon, RE (1968) Systems Simulation: The Art and Science. Prentice-Hall, New Jersey
8. Sherif, YS, Svestka, JA (1984) Tactical air war: A SIMSCRIPT model. Microelectronics and Reliability, v. 24, n. 1:65–71
9. Vieira, WJ, Prati, A, Destro, JPB (2004) Simulação Monte Carlo de Mísseis Antiáereos. In: Oliveira, JEB (eds) Proceedings of the 2004–VI Simpósio de Guerra Eletrônica. Instituto Tecnológico de Aeronáutica, São José dos Campos, Brazil

Learning Structure Illuminates Black Boxes – An Introduction to Estimation of Distribution Algorithms

Jörn Grahl[1], Stefan Minner[1], and Peter A.N. Bosman[2]

[1] Department of Logistics, University of Mannheim, Schloss, 68131 Mannheim.
joern.grahl@bwl.uni-mannheim.de, minner@bwl.uni-mannheim.de

[2] Centre for Mathematics and Computer Science, P.O. Box 94079, 1090 GB Amsterdam.
Peter.Bosman@cwi.nl

Abstract

This chapter serves as an introduction to estimation of distribution algorithms (EDAs). Estimation of distribution algorithms are a new paradigm in evolutionary computation. They combine statistical learning with population-based search in order to automatically identify and exploit certain structural properties of optimization problems. State-of-the-art EDAs consistently outperform classical genetic algorithms on a broad range of hard optimization problems. We review fundamental terms, concepts, and algorithms which facilitate the understanding of EDA research. The focus is on EDAs for combinatorial and continuous non-linear optimization and the major differences between the two fields are discussed.

Key words: Black Box Optimization, Probabilistic Models, Estimation of Distributions

1 Introduction

In this chapter, we give an introduction to estimation of distribution algorithms (EDA, see [60]). Estimation of distribution algorithms is a novel paradigm in evolutionary computation (EC). The EDA principle is still being labelled differently in the literature: estimation of distribution algorithms, probabilistic model building genetic algorithms (PMBGA), iterated density estimation evolutionary algorithms (IDEA) or optimization by building and using probabilistic models (OBUPM). For the sake of brevity we call this class of algorithms EDAs.

EDAs have emerged in evolutionary computation from research into the dynamics of the simple genetic algorithm (sGA, see [45] and [32]). It has been found in this research that using standard variation operators, e.g., two-parent recombination or mutation operators, easily leads to exponentially scaled-up behavior of the sGA. This means that the required time measured by the number of fitness evaluations to reliably solve certain optimization problems grows exponentially with the size of the problem. Loosely speaking, the use of fixed variation operators can easily cause sGA behavior that moves towards enumeration of the search space, admittedly in an elegant manner.

The failure of the sGA is systematic on certain problems. This has triggered research that replaces the traditional variation steps in a GA. Briefly stated, what differentiates EDAs from simple GAs and other evolutionary and non-evolutionary optimizers is that the main variation in EDAs comes from applying statistical learning concepts as follows:

1. The joint probability density of the selected individuals' genotypes is estimated.
2. This density is sampled to generate new candidate solutions.

As we will illustrate in this chapter, estimating a density from the genotypes of selected solutions and subsequently sampling from it to generate new candidate solutions is a powerful tool to make variation more flexible. This is because density estimation can be regarded as a *learning* process. EDAs try to learn the structure of problems. They attempt to adapt their search bias to the structure of the problem at hand by applying statistical- and machine-learning techniques on a population of solutions.

The learning capabilities of EDAs render them especially suitable for black-box-optimization (BBO). In BBO one seeks to find the extremum of a fitness function, without having a formal representation of the latter. One is solely given candidate solutions and their fitness values. The fitness function itself is unknown, it is encapsulated in a so-called black box. BBO often appears in practice, if little knowledge on the formal structure of the fitness function is available, and solutions are evaluated with a complex virtual or physical simulation model.

EDAs have successfully been developed for combinatorial optimization. State-of-the-art EDAs systematically outperform standard Genetic Algorithms with fixed variation operators on a broad range of GA-hard problems, such as deceptive problems, MAXSAT, or Ising Spins. Many problems that are intractable for standard GAs can reliably be solved to optimality by EDAs within a low-order polynomial number of fitness evaluations depending on the problem size.

Because of this success in the discrete domain, the EDA principle has been adapted for the continuous domain. Continuous EDAs are intended to solve non-linear optimization problems in continuous spaces that cannot be handled by analytical or classical numerical techniques. It must be noted though, that a straightforward adaptation of the lessons learned from the discrete domain to the continuous domain does not necessarily lead to efficient EDAs. We will discuss additional necessary steps for the continuous domain that have led to optimizers that solve hard non-linear problems reliably.

This chapter is structured as follows. In Sect. 2, we briefly review the simple genetic algorithm and present a general framework for estimation of distribution algorithms. We focus on discrete EDAs for combinatorial optimization in Sect. 3. Fundamental relationships between problem decompositions and search distributions are explained in Sect. 3.1. Standard test functions for discrete EDAs are presented in Sect. 3.2. Subsequently, a literature review on discrete EDAs for combinatorial optimization is given. Section 4 focuses on EDAs for non-linear optimization in continuous spaces. A literature review for continuous EDAs is presented in Sects. 4.1 and 4.2. Consequently, we present material on convergence defects of EDAs for continuous optimization in Sect. 4.3 and discuss recent approaches to enhance the efficiency of continuous EDAs. The chapter ends with concluding remarks.

2 Preliminaries and Notation

2.1 The Simple Genetic Algorithm

The simple genetic algorithm is a cornerstone in GA theory. It is a stochastic search strategy that maintains a set of solutions $\mathcal{P}$ of size $|\mathcal{P}| = n$, called the population, throughout the search. The population undergoes generational changes. We denote a population in generation t by $\mathcal{P}^t$. A single solution is referred to as an individual. Each individual has an associated fitness value that measures its quality. The goal of the sGA is to find the individual that has the highest quality. An individual consists of a phenotype and a genotype. The phenotype is its physical appearance (the actual solution to the problem at hand) whereas the genotype is the genetic encoding of the individual. The sGA processes genotypes that are binary (bit) strings of a fixed length l. A single bit string is also referred to as a chromosome. A single bit at position $i, i = 1, 2, \dots, l$ in the chromosome is also referred to as an allele. The genotype–phenotype mapping is called the representation of the problem and is an important ingredient of GAs, see [75]. Whereas the fitness of an individual is computed with its phenotype, new solutions are built on the basis of the genotype. For an illustration of the genotype–phenotype mapping, see Fig. 1.

The sGA processes binary strings of fixed length as follows. The first population of individuals is filled with a random sample from the complete solution space. All solutions are drawn with equal likelihood, and all genotypes are assumed to be feasible. The fitness of all individuals is evaluated and the better solutions are selected

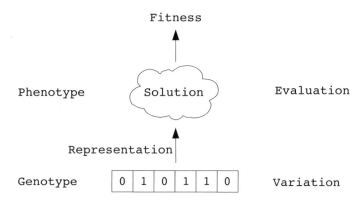

Fig. 1 Genotype–phenotype mapping

for variation using fitness proportionate selection. In fitness proportionate selection, each individual's probability of being selected is proportional to its quality. Selection intends to push the population into promising parts of the search space. For a comparison of different selection schemes [9]. The set of selected individuals from generation t is called the mating-pool and is denoted by $\mathcal{S}^t$.

New candidate solutions are generated by applying variation operators on elements of $\mathcal{S}^t$. The variation operators of a sGA are recombination operators of the two-parent crossover-type (Fig. 2) and bit-flip mutation. For recombination, one-point crossover and uniform crossover are used. One-point crossover combines parts of two randomly selected individuals (the so-called parents) from $\mathcal{S}^t$ by cutting them into two pieces at a randomly selected locus. New candidate solutions, the so-called offspring $\mathcal{O}^t$, are generated by exchanging partial solutions between the parents. Uniform crossover produces offspring by exchanging every single bit between two randomly chosen parents with a predefined probability. Bit-flip mutation modifies single solutions by inverting each bit of the string with a usually small probability.

The offspring $\mathcal{O}^t$ replaces the parents and the next iteration $t+1$ of the sGA begins with an evaluation of the newly generated population $\mathcal{P}^{t+1}$. The process of evaluation, selection, variation, and replacement is repeated until a predefined stopping criterion is met, e.g., the optimal solution has been found, the best found solution cannot be improved further, or a maximal running time has been reached. See Fig. 3 for pseudo-code of the sGA.

2.2 A General Framework for Estimation of Distribution Algorithms

Similar to the sGA, EDAs are stochastic, population-based search algorithms. What differentiates EDAs from sGA and other evolutionary and non-evolutionary optimizers is that the primary source of variation in EDAs is not driven by the application of

a) One-point crossover   b) Uniform crossover

Fig. 2 Recombination in the sGA

1. Set generation counter $t = 1$
2. Fill $\mathcal{P}^1$ with uniformly distributed solutions
3. Evaluate $\mathcal{P}^t$
4. Select $\mathcal{S}^t$ from $\mathcal{P}^t$
5. Apply variation operators on $\mathcal{S}^t$, fill $\mathcal{O}^t$ with newly generated solutions
6. Replace parents with $\mathcal{O}^t$
7. If termination criterion is met, then **end**,
        else $t = t + 1$, go to step 3.

Fig. 3 Pseudo-code for simple genetic algorithm

variation operators to subsets of solutions. Instead, it comes from estimating a probability distribution from all selected individuals $\mathcal{S}^t$ and consequently sampling from this probability distribution to generate offspring $\mathcal{O}^t$. In the following explanation, assume maximization of an objective function.

To illustrate the EDA principle, we introduce a random variable $\mathcal{Z}$ that has an associated probability distribution $P(\mathcal{Z})$ that covers all possible genotypes. To be more concrete, $P(\mathcal{Z})$ denotes a joint probability distribution of all l alleles in the chromosome. A single genotype is denoted with Z, its probability is denoted by $P(Z)$. The random variable associated with a single, ith allele Z_i is denoted by Z_i. The probability distribution of a single allele is denoted by $P(Z_i)$. We further write $P^{\mathfrak{T}}(\mathcal{Z})$ for a probability distribution over all genotypes that has a uniform distribution for all genotypes with an associated fitness value larger than $\mathfrak{T}$ and is equal to 0 otherwise. If the probability distribution $P^{\mathfrak{T}^*}(\mathcal{Z})$ of the optimal solution $\mathfrak{T}^*$ was known, we would simply have to sample from it to obtain the optimum because all solutions worse than $\mathfrak{T}^*$ have a chance of zero to be drawn. In practical optimization we do not know $P^{\mathfrak{T}^*}(\mathcal{Z})$. EDAs try to approximate it iteratively.

The first population $\mathcal{P}^1$ of n individuals is usually generated uniformly from all feasible solutions. All individuals are evaluated and the selection step yields a mating-pool $\mathcal{S}^t$ with solutions of higher quality. Now, a probability distribution is estimated from the genotypes of the solutions in $\mathcal{S}^t$. This is achieved by learning a probabilistic model $\mathcal{M} = (\varsigma, \theta)$ from $\mathcal{S}^t$ that is composed of a structure ς and a set of parameters θ. The structure defines (in)dependence between random variables and the associated alleles. Learning the structure ς can be a complex task that is related to learning (in)dependence relationships between alleles. The (in)dependence assumptions of the model are chosen such that they match those of the sample $\mathcal{S}^t$ as close as possible. This can be an optimization problem itself. The parameters θ of the model are mostly probabilities and conditional probabilities. They are estimated after the structure that fits best has been found. The model $\mathcal{M}$ represents a probability distribution that approximates the true distribution of the selected solutions' genotypes in $\mathcal{S}^t$. Let $\mathfrak{T}$ denote the worst fitness from the selected individuals. The model $\mathcal{M}$ approximates the true distribution of $P^{\mathfrak{T}}(\mathcal{Z})$, that is the distribution of all individuals that have a better quality than $\mathfrak{T}$. The estimated probability distribution represented by $\mathcal{M}$ is now randomly sampled from to generate offspring $\mathcal{O}^t$. $\mathcal{O}^t$ replaces the worst solutions in the old population $\mathcal{P}^t$, and the population advances to population $\mathcal{P}^{t+1}$. The replacement step uses elitism – it is assured that the best found solution is not replaced by a possibly worse solution. As a result the quality of the best found solution does not decrease over time.

The process of evaluation, selection, model building, model sampling, and replacement is iteratively performed until a predefined convergence criterion is met. Pseudo-code for the general EDA framework can be found in Fig. 4.

Note that in other metaheuristics one is often interested in building efficient variation operators that are applied on single or subsets of solutions for a specific optimization problem. In EDAs the focus shifts from single solutions to the statistical distribution of sets of high-quality solutions in the search space. Loosely speaking, EDAs approximate a density function that tells the decision maker where high-quality so-

1. Set generation counter $t = 1$
2. Fill $\mathcal{P}^1$ with uniformly distributed solutions
3. Evaluate $\mathcal{P}^t$
4. Select $\mathcal{S}^t$ from $\mathcal{P}^t$
5. Learn probabilistic model $\mathcal{M} = (\varsigma, \theta)$ from $\mathcal{S}^t$
6. Sample offspring $\mathcal{O}^t$ from $\mathcal{M}$
7. Replace the worst solutions in $\mathcal{P}^t$ with $\mathcal{O}^t$
8. If termination criterion is met, then **end**,
   else $t = t + 1$, go to step 3.

Fig. 4 Pseudo-code for EDA framework.

lutions can be found, what probabilities they have in good populations, and which decision variables are (in)dependent from each other.

In this chapter, we focus on EDAs that operate on fixed-length strings. A discussion of variable length EDAs for Genetic Programming, such as Probabilistic Incremental Program Evolution (PIPE, [77]), Extended Compact Genetic Programming (eCGP, [78]), or grammar learning approaches, [12], is beyond the scope of this chapter.

3 Binary Estimation of Distribution Algorithms

3.1 Problem Decomposition and Factorized Search Distributions

In this section, we focus on the special case that the genotype is a binary string of fixed length l. Although the major results apply to higher alphabets as well, and the proposed algorithms are extendable into this direction, the main stream of research covers the binary case. This means that single alleles have either the value 1 or 0. A simple and straightforward way to implement an EDA that follows the general EDA framework of Sect. 2.2 for binary genotypes of length l would be to use a frequency table of size 2^l as the probabilistic model. The frequency table holds a probability $P(Z)$ for each solution Z. The parameters of this model are the 2^l probabilities of the solutions. These probabilities can be estimated by the relative frequency $\hat{P}(Z)$ of single solutions in the set of selected solutions $\mathcal{S}$ as

$$\hat{P}(Z) = \frac{\alpha}{|\mathcal{S}|}, \tag{1}$$

where α denotes the number of solutions in $\mathcal{S}$ that equal Z. Generating new solutions from this probabilistic model can be done in a straightforward manner by setting the probability to sample Z to $\hat{P}(Z)$ and sample the offspring individual by individual. The structure of this model implicitly assumes that all alleles depend on each other. Using a frequency table of size 2^l exploits no independence assumptions between alleles.

Note that if we let the population size (and henceforth $|\mathcal{S}|$) tend to infinity, the estimated density expressed by the frequency table converges towards the true probability distribution $P(\mathcal{Z})$. An iterative procedure of selection, estimation, sampling, and replacement would steadily increase $\mathfrak{T}$ until $P^{\mathfrak{T}}(\mathcal{Z})$ only has positive probability for optimal solutions and has zero probability for all other solutions. However, this approach is generally intractable because the size of the frequency table and thus the effort to estimate its parameters grows exponentially with the size l of the problem. Also, population sizes are finite in practice.

To overcome the drawback of exponentially growing frequency tables, we can allow for the estimation of factorized probability models. Factorizations of joint densities of several random variables are products of marginal densities defined over subsets of the random variables. Factorizations result from a joint density by assuming statistical independence between random variables. The structure of a probabilistic model relates to the (in)dependence relationships between the random variables. The use of factorizations reduces the number of parameters that have to be estimated. Estimating the parameters of factorized probability models is relatively easy, as the parameters can independently be estimated for each factor [54].

A simple example: we assume that all l distributions of the alleles are independent from each other. Then, the joint distribution of the chromosomes can be expressed as a univariate factorization, see Sect. 3.3. The l-dimensional density is decomposed into a product of l one-dimensional densities. The univariate factorization is defined as

$$P(\mathcal{Z}) = \prod_{i=1}^{l} P(Z_i). \tag{2}$$

The *structure* ς of this probabilistic model is fixed. The alleles are statistically independent from each other. The parameters θ of this model are the l probabilities of each allele being 0 or 1 (in the binary case). Factorizing the joint probability table thus results in a reduction of dimensionality and, henceforth, probability tables that can be estimated more efficiently without a reduction in precision.

Different types of factorizations have been used in EDAs, see Sects. 3.3–3.5. Not surprisingly, depending on the type of factorization that is used, the corresponding EDA exploits different structures of the optimization problem at hand and exhibits a different type of search bias. In general, however, the EDA approach of building a model with respect to a certain factorization-type and sampling it to generate offspring is especially suited when it comes to solving *additively decomposable problems*.

According to [57] the fitness function $f(Z)$ is additively decomposable if it can be formulated as

$$f(Z) = \sum_{i=1}^{m} f_i(Z_{s_i}). \tag{3}$$

$f(Z)$ is additively defined over m subset of the alleles. The $s_1, s_2, \dots, s_m$ are index sets, $s_i \subseteq \{1, 2, \dots, l\}$. The f_i are sub-problems that are only defined on the alleles Z_j with $j \in s_i$. The subproblems can be non-linear. The Z_{s_i} are subsets of all alleles. These subsets can overlap.

Equation (3) exhibits a modular structure. It consists of m components that can, but may not, be coupled. If the s_i are disjunct, $s_i \cap s_j = \varnothing \ \forall \ i \neq j$, the functions do not overlap and the overall problem is called *separable*. Separable problems can be solved by solving the m subproblems f_i and summing the results. Depending on the size of the subproblems, separation reduces the dimensionality of the search space significantly. Assuming that $|s_i| = k \ \forall \ i$ the dimensionality is reduced from l to k and the size of the solution space is reduced from 2^l to $m2^k$. Problem (3) is called *decomposable* if some sets s_i, s_j exist for which $s_i \cap s_j \neq \varnothing$. In this case, a strict separation of the sub-functions is no longer possible because a single decision variable influences more than one sub-function.

What makes decomposable problems hard to solve? This is a non-trivial question and several answers can be found in the literature. Most obviously the hardness of the sub-functions directly contributes to the overall complexity of the problem. Deceptive problems (see Sect. 3.2) are hard to solve for GA and EDA and are often assumed as sub-functions for testing purposes. Deceptive functions are typically harder to solve for GA and EDA than non-deceptive functions. Further, subproblems can contribute to the overall fitness on a similar scale, or the scaling of the sub-functions can differ greatly. In the first case, all sub-functions of equal importance and convergence towards the partial solutions will happen simultaneously. If the sub-functions are exponentially scaled however, the most salient of them will converge first. The other sub-functions may converge later and some instantiations might already be lost at that time. Additively decomposable functions with exponentially scaled sub-functions are harder to solve for GA and EDA – they require a higher population size [86]. [47] discusses whether the size $|s_i|$ of the sets influences the hardness of a problem. This can be the case, if for solving $f_i(s_i)$ all associated variables must be regarded simultaneously. It may not be the case, however, if interactions are not very strong and only some of the dependencies are important. The size of the sets can thus be a source for the hardness of a problem but the degree of connectivity and importance of the dependencies appears to be a more important source for the GA- or EDA-complexity of a function.

High-quality configurations of alleles that belong to the sets s_i are referred to as building blocks (BBs, [45], [33]). It is commonly assumed in GA and EDA literature that BBs are not further decomposable. This means that to solve a subproblem defined by f_i, all associated alleles Z_{s_i} have to be considered simultaneously. In experiments, this can be achieved by using deceptive functions as subproblems, see Sect. 3.2.

The building block structure of an ADF is called *problem decomposition*. A problem decomposition indicates which alleles depend on each other and which are independent from each other. Information on the problem decomposition is also referred to as linkage information [40], [38], [41] and [33]. Tight linkage is a feature of a representation that encodes alleles belonging to the same sets closely to each other. Loose linkage characterizes a solution representation that spreads alleles belonging to the same set widely over the chromosome.

The relationship between a problem decomposition and the factorization of a search distribution is important. Assume a given population $\mathcal{P}$ that contains high-quality solutions for a decomposable problem. The necessity of simultaneous appearance of certain configurations of alleles within a sub-function will cause a statistical dependency between these alleles in $\mathcal{P}$. Alleles from different sub-functions can be set separately from each other to optimize (3), and in general they will be statistically independent from each other in $\mathcal{P}$ except for noise and finite population effects. A central element of factorized probability distributions is the possibility to assume independence between random variables. If these assumptions are exactly in accordance with the decomposition of the problem, then the joint probability of dimensionality l is factorized into several marginal densities of possibly smaller dimensionality – each modeling the distribution of alleles in a sub-function. In this case, the factorization is called exact.

Sampling from an exactly factorized distribution is a powerful tool to solve combinatorial optimization problems [58]. However, efficient sampling is not always possible. In the following paragraphs, we refer to work that has been developed elsewhere [59], [57] to illustrate for which decompositions efficient sampling is possible. Assume that a fitness function of type (3) is given and one tries to solve the optimization problem $Z^* = \arg\max f(Z)$ by sampling solutions Z from a search distribution. A candidate for the search distribution is the Boltzmann distribution which is given as [56]

$$P_\beta(Z) = \frac{e^{\beta f(Z)}}{\sum_y e^{\beta f(y)}},$$

(4)

where $\beta \geq 0$ and y denotes the set of all solutions. The Boltzmann distribution has the appealing property, that for increasing β it focuses on global optima of $f(\mathcal{Z})$. For $\beta \to \infty$, only global optima have positive probabilities. Unfortunately, sampling from the Boltzmann distribution needs exponential effort because the denominator is defined over all possible solutions. This is not a tractable search strategy.

If the fitness function is additively decomposable, the sampling effort can sometimes be reduced by sampling from a factorization of the Boltzmann distribution. If it can be shown that for a given fitness function $f(Z)$ the Boltzmann distribution can be decomposed into boundedly smaller marginal distributions, sampling candidate solutions from it can potentially be a promising search strategy.

To analyze whether this is the case, we define the sets $d_i, b_i,$ and c_i for the index sets s_i for $i = 1, 2, \ldots, m$ as follows:

$$d_i = \bigcup_{j=1}^{i} s_j, \qquad b_i = s_i \setminus d_{i-1}, \qquad c_i = s_i \cap d_{i-1}.$$

If the following Factorization Theorem [59], [57], [58] holds for a given decomposable function, the Boltzmann distribution can be factorized exactly into some marginal distributions.

Factorization Theorem: Let the fitness function $f(Z) = \sum_{i=1}^{m} f_i(Z_{s_i})$ be an additive decomposition. If

$$b_i \neq 0 \quad \forall\ i = 1, 2, \ldots, m$$

(5)

and

$$\forall\ i \geq 2 \ \exists\ j < i \text{ such that } c_i \subseteq s_j,$$

(6)

then

$$q_\beta(Z) = \prod_{i=1}^{m} P_\beta(Z_{b_i}|Z_{c_i}) = P_\beta(Z). \tag{7}$$

Condition (6) is called the running intersection property (RIP). If conditions (5) and (6) hold, then the Boltzmann distribution can be obtained by an exact factorization of marginal distributions. But, it is only reasonable to sample new solutions from (7) in order to solve (3), if sampling new solutions from (7) is computationally easier than solving (3) directly. This is not the case if the marginal distributions are of arbitrary dimensionality, because the sampling effort could then grow exponentially with the problem size l. It is indeed the case, if the size of the sets b_i and c_i is bounded by a constant that is independent of the bit-string length l. Then, the factorization is called polynomially bounded.

The effort of sampling a polynomially bounded factorization is much smaller than sampling the unfactorized distribution. Exactly factorizing a search distribution with respect to a problem decomposition can lead to a significant reduction in dimensionality of the size of the problems that one attempts to solve.

A major result of EDA theory is that if the factorization of the Boltzmann distribution for a combinatorial optimization problem is polynomially bounded, new solutions can efficiently be generated and an EDA can theoretically solve the problem to optimality with a polynomial number of fitness evaluations [58]. This is an important theoretical result that holds for infinite population sizes and if the exact problem decomposition is known. In Sects. 3.3–3.5, we will describe how different EDAs attempt to transfer this theoretical result into scalable optimizers.

3.2 Decomposable Test Functions

The dynamics of the simple GA and EDAs are commonly investigated on decomposable test functions, especially the One-Max function and deceptive trap functions described below. Both functions are defined on binary strings of fixed length l and are separable. The One-Max function $f_{\text{One-Max}}$ is defined as

$$f_{\text{One-Max}}(Z) = \sum_{i=1}^{l} Z_i. \tag{8}$$

$f_{\text{One-Max}}$ simply counts the number of ones in a binary string Z. From a decomposition point of view, the One-Max function is decomposable into l sub-problems that are independent from each other as the fitness contribution of a single bit does not depend on any other bit. To solve One-Max, each bit can be set independently from each other, its building blocks each consist of a single bit and are of size one.

Deceptive trap functions [1], [24] of order k are defined as

$$f_{\text{Trap}}(Z, u, k, d) = \begin{cases} k - u(Z) - d & \text{if } u < k \\ k & \text{if } u = k, \end{cases} \tag{9}$$

where $u(Z)$ denotes the number of ones in the binary string Z of length k. The deceptive trap function of order $k = 5$ is defined as

$$f_{\text{Trap5}}(Z, u) = \begin{cases} 4 - u & \text{otherwise} \\ 5 & \text{if } u = 5 \end{cases} \tag{10}$$

where d has been set to $d = 1$. The trap function of order 5 is illustrated in Fig. 5. Deceptive functions are designed such that all k bits have to be considered simultaneously in order to maximize them. The average fitness of partial solutions of sizes that are $< k$ that have all values of 0 is higher than the average fitness of the same partial solutions that have all values of 1. Decision making about allele values that is not based on the joint consideration of all k bits in a trap function will therefore systematically mislead genetic algorithms and EDAs into choosing the local maximum of $Z = (0, 0, \dots, 0)$ instead of the globally optimal chromosome $Z = (1, 1, \dots, 1)$.

From a decomposition point of view, trap functions are artificial worst case functions that are not further decomposable. k bits that belong to a single trap represent bits of a single building block of size k. Trap functions are often used as sub-functions in problems with more than one BB, like in composed trap functions.

Composed trap functions of order k are generated by defining m trap functions on a binary string whose length l is $k \cdot m$. This is done by partitioning the chromosome into $m = \frac{l}{k}$ sets of alleles s_i, $i = 1, 2, \dots, m$, $|s_i| = k$ and defining a trap function on each of the m sets of alleles. The resulting function is defined as

$$f_{\text{Comp-Trap}}(Z) = \sum_{i=1}^{m} f_{\text{Trap}}(Z_{s_i}, u_i, k), \text{ with } u_i = \sum_{j=1}^{k} Z_{(s_i)_j}, \tag{11}$$

where $(s_i)_j$ denotes the jth bit in the ith trap function. If $s_i \cap s_j = \varnothing$, then alleles are only assigned to a single trap function and $f_{\text{Comp-Trap}}$ is separable. In this case, the BBs are non-overlapping. If on the other hand $s_i \cap s_j \neq \varnothing$, then alleles are assigned to several trap functions. In this case, the BBs are overlapping.

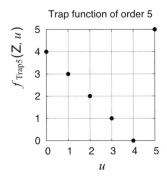

Fig. 5 Trap function of order $k = 5$

Composed trap functions with tight linkage have the bits Z_{s_i} arranged next to each other on the string. Composed trap functions with loose linkage have the bits distributed all over the chromosome, e.g., in a random fashion.

Composed trap function with loose linkage are artificial test functions that represent hard BBO problems. Composed traps assume that the overall problem can be decomposed in sub-problems, potentially yielding a reduction of the overall problem dimensionality and making the problem much easier to solve. To solve composed traps, we need to evaluate at most $m \cdot 2^k$ solutions instead of 2^{mk} if we *know* the assignment of bits to traps. In BBO, we do not know the structure of an optimization problem. Thus, we do *not know* the assignments of the bits to the trap functions and cannot properly decompose the problem. Instead, we are forced to learn the problem decomposition before exploiting it. This is exactly what EDAs do.

3.3 No Interactions

Historically, the first EDA approaches in binary search spaces used univariate factorizations of search distributions. A univariately factorized probability distribution is a product of l univariate probability distributions $\prod_{i=1}^{l} P_{\theta^i}(Z_i)$, where $P_{\theta^i}(Z_i)$ models the distribution of a single allele Z_i and θ^i is the parameter vector that has to be estimated for the ith allele. In the binary domain, θ^i simply holds the probability of the ith allele being 1 or 0. Note, that univariately factorized probability distributions have a fixed structure. All alleles are assumed to be independent from each other. Univariate factorizations are special cases of general Bayesian factorizations described in Sect. 3.4 with l independent variables.

Estimating a model on the basis of univariate factorizations reduces to parameter estimation because the structure of the density in terms of (in)dependence assumptions is fixed. Estimating parameters for univariate factorizations has a computational complexity of $\mathcal{O}(n|\theta|)$ where n is the population size and $|\theta|$ is the number of parameters that has to be estimated. As the number of parameters that has to be estimated for a single univariate density $P_{\theta^i}(Z_i)$ is usually a small constant, $|\theta| = \mathcal{O}(l)$.

Different EDAs based on univariate factorizations of frequency tables can be found in the literature. The bit-based simulated crossover operator [83] uses a binary string of fixed length l. In the model building process, it computes for each allele i, $i = 1, 2, \ldots, l$ the univariate probability of a 0 and a 1 at bit position i. Therefore, the relative frequencies of single alleles are weighted by taking into account a fitness measure from high quality solutions. New solutions are sampled bit by bit. The sampling process uses the probabilities that were obtained for each bit independently from each other.

Population-based Incremental Learning (PBIL) [4] uses the same vector of univariate probabilities for each bit as [83]. However, instead of re-estimating the complete probability vector from the selected individuals, the probability vector is adapted in each generation from elitist individuals using an update rule. Starting with an initial setting where the probability of each bit being 1 is set to exactly 0.5, the probabilities for each bit are shifted towards 0 or 1. The directions for the shifts are obtained from elitist samples. Offspring are generated bit by bit. The alleles are sampled independently from each other using a univariate probability for each bit.

The compact Genetic Algorithm (cGA) [42] is very similar to PBIL. The cGA also works on a probability vector of length l but does not maintain a complete population of solutions. Instead, it updates entries in the probability vectors from two sampled individuals only. The updates are made with an updating rule that is similar to that of PBIL.

In a similar fashion, the Univariate Marginal Distribution Algorithm (UMDA, [60]) estimates the probability of being 1 or zero for each bit position from its relative frequency in the population. n new individuals are sampled from the distribution to replace the population as a whole. In contrast to previous work, UMDA does not alter the probabilities after estimating them. New solutions are generated bit by bit from the estimated univariate factorization. The UMDA algorithm approximates the behavior of a simple GA with uniform crossover.

Univariate factorizations have been used [87] to solve permutation problems. Binary random variables are introduced for each pairwise combination of permutation elements. The probability associated with each random variable relates to the probability that the associated permutation elements are positioned next to each other in good solutions.

All algorithms discussed in this section are similar to each other and the sGA. They do *not* respect linkage information. This means that subsolutions can be cut into several parts by the crossover operators and can get lost. The sGA potentially disrupts building blocks in its crossover step. The EDAs based on univariate factorization set each bit independently from each other. They do not take into account, that for solving decomposable problems with BBs of size > 1, the joint appearance of configurations of several alleles has to be accounted for.

This is a valid approach, if the problem is decomposable into subproblems of order $k = 1$ like the One-Max function. The simple genetic algorithm with uniform crossover converges on One-Max in $\mathcal{O}(l \log l)$ evaluations [61], [39]. However, if the size k of the sub-problems grows, the performance of the sGA can easily deteriorate. It is well known that the sGA scales exponentially with the problem size, if the alleles of a composed deceptive trap function are arbitrarily distributed over the chromosome and the order k of the traps is $k > 3$. In this case, the scalability of the simple GA drops from $\mathcal{O}(l \log l)$ to $\mathcal{O}(2^l)$ [85]. This clearly demonstrates the boundaries of genetic algorithms with fixed operators. Loosely speaking, the behavior of exponentially scaling GAs moves towards that of complete enumeration of the search space. Note that the effect of generating offspring from a univariately factorized probability distribution is similar to using uniform crossover [63]. Thus, the EDAs discussed in this section are expected to scale similarly to the sGA. They are relatively efficient on decomposable problems with sub-problems of smallest sizes and scale badly on problems where the BBs are of higher order.

For real-world optimization, it is not realistic to assume that the BBs of a problem are always of size 1. Linkage information is often not known *a priori*. This is especially true for black box optimization. Using simple GA with fixed recombination operators or univariate EDAs can thus easily result in an exponential scale-up behavior of the algorithm. Note that this can also happen with other fixed recombination operators, and is not related to using one-point crossover or uniform crossover [86].

3.4 Bivariate Interactions

Exponential scalability of univariate EDAs on problems with BBs of higher order has motivated the development of EDAs based on more involved probabilistic models. This subsection reviews EDAs that are capable of capturing bivariate dependencies between alleles. In contrast to EDAs presented in Sect. 3.3, the structure of the probabilistic model of the EDAs in this section is no longer fixed. Instead, the structure of the model is flexible to a certain degree, allowing for an adjustment of the factorization that is built in every generation. Flexible probabilistic models allow for an adaptive bias of the algorithm. To be more specific: EDAs discussed in this section attempt to convert linkage information into statistical (in)dependence relationships correctly. Learning probabilistic models as done in many EDAs corresponds to minimizing the discrepancy between the distribution expressed by the model and the empirical distribution of the alleles in the set of selected individuals. Inside the boundaries imposed by a model type, EDAs search for that model that fits the distribution of the genotypes best.

In the binary problem domain, Bayesian factorizations of search distributions based on frequency counts are commonly used as probabilistic models. A Bayesian factorization of a joint density is a product of conditional densities $P(Z_i|Z_{\pi_i})$ for each random variable Z_i. π_i denotes a vector of so-called parents of Z_i that indicates on which variables the univariate density Z_i is conditioned. A Bayesian factorization of a joint density $P(\mathcal{Z})$ is defined as

$$P(\mathcal{Z}) = \prod_{i=1}^{l} P_{\theta^i}(Z_i|Z_{\pi_i}), \tag{12}$$

where θ^i denotes the vector of parameters that has to be estimated for each density $P_{\theta^i}(Z_i|Z_{\pi_i})$. The matrix $\pi = (\pi_1, \pi_2, \dots, \pi_l)$ represents the structure of a Bayesian factorization. π has a representation as a directed acyclic graph with l nodes where each node corresponds to a single Z_i, $i = 1, 2, \dots, l$ and an arc from node j to node i indicates that Z_i is conditioned on Z_j. Importantly, the Bayesian factorization is only a valid density if and only if the factorization graph is acyclic. These probabilistic models are also referred to as graphical models or Bayesian networks, see [44], [54] and [46] for details.

EDA that are capable of capturing bivariate dependencies between variables restrict the general form of a Bayesian network given in (12) such that each variable can at most depend on a single other variable. This means, that $|\pi_i| = 1 \ \forall \ i = 1, 2, \dots, l$.

Estimating the parameters of a given Bayesian factorization is straightforward. The probabilities of bit i of being 0 or 1 is estimated by the relative frequency of this bit being 0 or 1 given all possible configurations of the parents π_i. This means, that the probability of this bit of being 1 or 0 is held in a frequency table of size $2^{|\pi_i|}$.

The frequency table lists all possible configurations of the values of the $|\pi_i|$ parent variables and for each configuration the corresponding relative frequency of the ith bit being 1 or 0 in the mating-pool.

Sampling from Bayesian factorizations is done in two consecutive steps. First, the nodes are sorted in topological order. If nodes are visited in topological order, a node i is visited after all of its parent nodes π_i have been visited already. After the nodes have been sorted, sampling starts at a root node and then visits nodes in topological order. The probabilities of a bit i of being 0 or 1 is set to the corresponding relative frequency that has been calculated in the parameter estimation process.

The Mutual Information Maximization Input Clustering algorithm (MIMIC, [23]) uses a factorization graph that has the structure of a chain. In a chain, each random variable has exactly one parent and conditions exactly one random variable (except for starting and ending nodes in the chain). To select the chain that models the distribution of the selected individuals as well as possible, MIMIC minimizes the difference between the distribution expressed by the factorization and the joint distribution of all alleles in the population. Therefore, the Kullback–Leibler divergence between both distributions is taken as a distance measure that is minimized with a greedy chain constructing approach. The computational complexity of the model building process is $\mathcal{O}(l|\theta|n)$. As $|\theta| = \mathcal{O}(l)$ in a chain, the overall model building complexity in MIMIC is quadratic in l.

The Combining Optimizers with Mutual Information Trees algorithm (COMIT, [5]) uses a factorization graph that has the structure of a tree. If the factorization structure is a tree, then every random variable $\mathcal{Z}_i$ is conditioned on exactly one parent. In contrast to the chain model of MIMIC, several variables can be conditioned on the same variable. To find a dependency tree that models the distribution of the genotypes of the selected individuals as well as possible, the COMIT algorithm uses a learning technique from [20] that results in a maximum-likelihood factorization with tree structure. The computational complexity of the model building process of COMIT is similar to that of MIMIC.

The Bivariate Marginal Distribution Algorithm (BMDA, [69]) uses a set of independent trees as its factorization graph. In this model, each random variable has at most a single parent. The model building process starts with an empty factorization graph. The factorization is constructed greedily by adding edges on basis of a χ^2 dependency test between pairs of random variables, where the edges are chosen in descending order of the related χ^2 statistic. The computational complexity of the model building process is $\mathcal{O}(l^2|\theta|n)$. As $|\theta| = \mathcal{O}(l)$, the model building complexity is cubic in l.

In contrast to EDAs that use univariate factorizations of the search distribution, EDAs that allow bivariate dependencies between alleles are more powerful. However, this comes at the price of a more complex model-building process. Moreover, EDAs that can model bivariate dependencies between alleles are still not able to solve general k-decomposable problems efficiently. As reported in [10], the number of fitness evaluations that an EDA based on tree-like factorization of the joint density requires to successfully solve this problem grows exponentially with the size of the problem l.

This is caused by inexact factorizations with respect to the problem decomposition that allows mixing solutions of different subproblems instead of sampling partial solutions to the subproblems independently from each other.

3.5 Multivariate Interactions

Probabilistic models that are able to capture multivariate interactions between alleles were proposed in order to advance the model flexibility further. Therefore, marginal product factorizations and Bayesian networks without restriction on the network structure were used.

A marginal-product factorization is a decomposition of a joint density into a product of multiple multivariate densities, where the multivariate densities are defined over mutually exclusive subsets of all considered random variables. This means that each random variable appears in a single factor of the product only. We call the set of variables that forms the ith multivariate density a node-vector $\mathbf{v}_i$. The node partition vector $\mathbf{v} = (\mathbf{v}_1, \mathbf{v}_2, \dots, \mathbf{v}_m)$ represents the structure of the marginal product factorization. A marginal product factorization is defined as

$$P(\mathcal{Z}) = \prod_{i=1}^{m} P_{\theta^{\mathbf{v}_i}}(Z_{\mathbf{v}_i}). \tag{13}$$

Marginal product factorizations decompose a multidimensional density into a product of possibly multidimensional densities. In contrast to Bayesian factorizations, marginal product factorizations do *not* use conditional densities. They simply assume that all variables that appear in a factor $\mathbf{v}_i$ are jointly distributed. Thus, the size of the frequency tables associated with a node vector $\mathbf{v}_i$ grows exponentially with $|\mathbf{v}_i|$, whereas in Bayesian factorizations, it grows exponentially with the number of parent variables $|\pi_i|$. Since a single allele cannot appear in several factors in the marginal product factorization, this model is suited to modeling problem decompositions with non-overlapping building blocks, that is separable ADFs.

Bayesian factorizations are a more general class of distributions that characterize marginal product models. In Bayesian factorizations, single alleles can be, but do not necessarily have to be, associated with a single marginal density. Thus, Bayesian factorizations are suitable for problem decomposition in which building blocks are non-overlapping *and* in which they are overlapping.

Marginal product factorizations where the multivariate densities $P_{\theta^{\mathbf{v}_i}}(Z_{\mathbf{v}_i})$ are represented by frequency tables were proposed in the Extended Compact Genetic Algorithm (ECGA, [39]). The ECGA builds a marginal product model, starting with a univariate factorization of the joint density of the binary genotype. A greedy heuristic is used to iteratively join marginal densities such that the improvement in a minimum description length (MDL) metric is maximized. The greedy heuristic stops, when no potential merging of marginal densities improves the MDL metric further. The total number of fitness evaluations required by the ECGA to solve additively decomposable problems of order k grows with $\mathcal{O}(2^k m^{1.5} \log m)$ where m is the number of BBs in the problem. Note that the ECGA is able to properly decompose a search density based on frequency tables with respect to a problem decomposition. The time required to solve additively decomposable problems measured by the number of fitness evaluations grows sub-quadratically with the size of the problem if the size of the problems k is bounded independent from l.

To solve permutation problems, marginal product factorizations were used by [16] and [17]. The factorizations were built on permutation random variables that allow for direct representation of permutations in combination with the greedy model building process of the ECGA. Both the Akaike Information Criterion (AIC) metric and the Bayesian Information Criterion (BIC) metric are used. The results indicate a sub-quadratic growth of the minimally required population size with respect to the problem size on decomposable deceptive permutation problems. Simple GAs scale exponentially on these problems with respect to the problem size.

Acyclic Bayesian networks without further restrictions on the structure of the graph were independently proposed for use in the Bayesian Optimization Algorithm (BOA, [67], [66], [63]), the Estimation of Bayesian Network Algorithm (EBNA, [28]) and the Learning Factorized Distribution Algorithm (LFDA, [58]).

In these algorithms, the model building starts with a univariate factorization. In univariate factorizations, all variables are assumed to be independent from each other. Iteratively arcs are added to the probabilistic model in a greedy fashion. This relates to assuming dependence between alleles. A first version of the BOA uses a Bayesian–Dirichlet metric [43] to measure the fit between the distribution expressed by the model and the empirical distribution of the alleles in the mating-pool. Greedily, arcs that maximize the Bayesian–Dirichlet metric are added to the graph, where arcs that cause cycles are skipped. If the maximum number of parent variables π_i is bounded from above by κ, then this greedy approach to model building has a computational complexity of $\mathcal{O}(\kappa l^3 + \kappa l^2 |\theta| n)$.

Later versions of the BOA [68] and the EBNA and LFDA use penalization metrics similar to the Bayesian Information Criterion (BIC) metric. Using a greedy way to estimate a Bayesian network from selected individuals with the BIC metric has led to sub-quadratic scale-up behavior of these algorithms on additively decomposable problems [58], [63]. Scalability results for the BOA that also suits the LFDA nd EBNA can be found in [63] and [72]. According to BOA scalability theory, the number of fitness evaluations that the BOA requires to reliably solve additively decomposable problems grows between $\mathcal{O}(l^{1.55})$ for uniformly scaled subproblems and $\mathcal{O}(l^2)$ for exponentially scaled subproblems independent of k.

Unconstrained Bayesian factorizations with local structures represented by decision graphs were used by [64] in the Hierarchical BOA (hBOA) to solve hierarchically decomposable problems. Hierarchically decomposable problems are decomposable and introduce additional dependencies on several levels of interpretation. The large number of dependencies in these problems can hardly be expressed in a straightforward manner different from decision graphs. In hBOA, the decision graphs are combined with a niching scheme. Niching localizes competition between solutions and ensures that only similar solutions compete with each other. hBOA uses restricted tournament replacement. For each newly generated solution, a set of currently available solutions is picked randomly. The most similar solution is replaced by the new solution, if the fitness of the new solution is higher. The combination of Bayesian factorization with local structures and restricted tournament replacement has led to sub-quadratic scale-up behavior on hierarchically decomposable problems [70]. hBOA was successfully applied to Ising spin-glass problems and MAXSAT [65].

The BOA and the hBOA have both been used to solve multi-objective problems [49], [73]. The BOA was able to solve deceptive multi-objective problems that other evolutionary algorithms could not solve. In a practical application, no significant superiority to GAs that use classical crossover operators could be found [53]. The hBOA is found to scale up well on multi-objective decomposable deceptive problems. To obtain good scale-up however, clustering techniques in the objective space are needed. On the test problems considered, multi-objective variants of the simple GA and the UMDA scale badly and are incapable of solving problem instances of medium sizes.

As we have seen in this chapter, the use of flexible probabilistic models allows for a flexible bias of the EDA. State-of-the-art EDAs that are built on multivariate probabilistic models solve decomposable problems in sub-quadratic time. The increase of the model flexibility has led to more involved structure learning processes. At the same time, the class of problems that can reliably be solved was enlarged such that the resulting EDAs consistently outperform standard genetic algorithms with fixed operators on hard decomposable optimization problems.

3.6 Summary

Discrete EDAs use probabilistic models to guide their search for high quality solutions. State-of-the-art EDAs incorporate statistical learning techniques for building a probabilistic model during the optimization and thereby are able to adapt their bias to the structure of the problem at hand. For the important class of additively decomposable problems, the use of Bayesian factorizations or multivariate factorizations has led to scalable optimizers that reliably solve additively decomposable problems within a sub-quadratic number of fitness evaluations.

4 Continuous Estimation of Distribution Algorithms

The considerable success of discrete EDAs has motivated researchers to adapt the general EDA principle to the continuous problem domain. Continuous EDAs are proposed for function optimization in continuous spaces. Their application can be promising if classical numerical methods like gradient-based methods fail or are not applicable because derivatives are not available or due to outliers or noise. Continuous EDAs are used for what is often referred to as global optimization. For continuous optimization with the EDAs of this section a genotype is a vector of size l of real values if not stated otherwise.

Continuous EDAs mostly use variants of the normal probability density function (pdf) as the basis of their probabilistic model because the normal pdf is a commonly-used and computationally tractable approach to represent probability distributions in continuous spaces. The normal pdf $P^{\mathcal{N}}_{(\mu,\Sigma)}$ for l-dimensional random variables $\mathcal{Z}$ is parameterized by a vector $\mu' = (\mu_1, \mu_2, \ldots, \mu_l)$ of means and a symmetric covariance matrix Σ and is defined by

$$P^{\mathcal{N}}_{(\mu,\Sigma)}(\mathcal{Z} = Z) = \frac{(2\pi)^{-\frac{l}{2}}}{(\det \Sigma)^{\frac{1}{2}}} e^{-\frac{1}{2}(Z-\mu)^T (\Sigma)^{-1}(Z-\mu)}. \tag{14}$$

The number of parameters to be estimated from data to fit the normal distribution to selected individuals equals $\frac{1}{2}l^2 + \frac{3}{2}l$. A maximum likelihood estimation for the normal pdf is obtained from a vector $\mathcal{S}$ of samples if the parameters are estimated by the sample average and the sample covariance matrix [2], [84]:

$$\hat{\mu} = \frac{1}{|\mathcal{S}|} \sum_{j=0}^{|\mathcal{S}|-1} \mathcal{S}_j, \quad \hat{\Sigma} = \frac{1}{|\mathcal{S}|} \sum_{j=0}^{|\mathcal{S}|-1} (\mathcal{S}_j - \hat{\mu})(\mathcal{S}_j - \hat{\mu})^T \tag{15}$$

On the basis of the normal pdf, different probabilistic models can be estimated from the selected individuals. The resulting EDAs are discussed in the following sections.

4.1 No Interactions

A univariate factorization of a multidimensional normal pdf is a product of several independent univariate normal pdfs. This means that all covariances between two normally distributed random variables $Z_i, Z_j, i \neq j$ are 0. The variance of the ith allele random variable is denoted by σ_i^2. Univariate factorizations of normal distributions are defined as

$$P(\mathcal{Z}) = \prod_{i=1}^{l} \frac{1}{\sqrt{2\pi}\sigma_i} e^{-2(\frac{Z_i-\mu_i}{\sigma_i})^2}. \tag{16}$$

The first continuous EDA based on the normal pdf was proposed in [76]. The algorithm is an adaptation of the binary PBIL algorithm ([4]) to continuous domains. The distribution of each allele is modeled by a univariate normal distribution. In the initial phase of the optimization run, the variances are set to high values to stimulate exploration of the search space. They are adapted in further generations with a geometrically decaying schedule to enforce convergence. The means are adjusted with a learning rule similar to the original discrete PBIL algorithm. New solutions are generated allele by allele by sampling the values for each allele from the univariate Gaussian distributions independently from each other.

A second adaptation of the binary PBIL algorithm is presented in [81]. In this algorithm, a lower and an upper bound for each variable is given that defines a range where values for the variables lie. A simple histogram model with two bins is maintained. The first bin corresponds to the lower half of the value range, the second bin corresponds to the upper half of the value range. The binary random variables of the PBIL algorithm correspond to the probability that a solution lies in the upper half of the range. For example $P(Z_4 == 1)$ denotes the probability that the fourth allele has a value in the upper half of the initialization range. If the values converge towards a single bin, then the bin sizes are adaptively resized to the half of that range, similar to bisection. Sampling is again done allele by allele.

Another adaptation of PBIL to continuous spaces is proposed in [80]. A univariate normal pdf is used for modeling the distribution of each allele independently from each other as in [76]. In contrast to the latter, the variance and the mean are updated with the same learning rule.

In [31], the parameters of the univariate factorization are estimated from a set of selected individuals with the standard maximum-likelihood estimator for mean and variance. The approach of [31] is more similar to the EDA principle than are previous approaches. Similarly, the Univariate Marginal Distribution Algorithm in the Continuous Domain (UMDA$_c$) was proposed in [50]. UMDA$_c$ is an adaptation of the UMDA algorithm [60] to continuous domains. UMDA$_c$ estimates a single normal pdf for the distribution of each of the l alleles using maximum-likelihood estimators for mean and variance. New solutions are sampled allele by allele from the univariate normal distributions.

4.2 Multivariate Interactions

To increase the modeling capabilities of the probabilistic model, research into more involved probabilistic models on the basis of the normal pdf was conducted. To be more specific, similar to the discrete domain, the model structure is no longer assumed to be fixed, but is allowed to be flexible to a certain degree. This is achieved by allowing for modeling multivariate interactions between continuous alleles. Consequently, the probabilistic models comprehend multivariate densities. For real-valued optimization, marginal product factorization, Bayesian factorizations, and mixture-based factorizations have been proposed.

Marginal product factorizations of the normal pdf are products of possibly multivariate normal densities where each allele random variable belongs to exactly one factor. Similar to multivariate factorizations in the discrete problem domain, they are defined for the continuous domain as:

$$P(\mathcal{Z}) = \prod_{i=1}^{m} P^{\mathcal{N}}_{(\mu^{v_i}, \Sigma^{v_i})}, \tag{17}$$

where μ^{v_i} denotes the $|v_i|$-dimensional mean vector and Σ^{v_i} the covariance matrix of the ith partition. Multivariate factorizations of normals have first been proposed in [16].

Bayesian factorizations based on the normal pdf are referred to as Gaussian networks [43]. For real-valued optimization, the use of Gaussian networks has independently been proposed by [14] in the iterative density estimation evolutionary algorithm (IDEA)-framework and in [50]. The latter uses a variant of the MIMIC algorithm based on the normal pdf, called MIMIC$_c$, and unrestricted Gaussian networks in the Estimation of Gaussian Network Algorithm (EGNA).

As a first approach to learning the Gaussian network, [14] used an algorithm by [27] to build a factorization of the search distribution with minimum entropy.

In this factorization, each variable is allowed to depend on at most one other variable. In [15], unrestricted Gaussian networks are used. A greedy factorization learning scheme in combination with a Bayesian Information Criterion metric is used to learn the structure of the Gaussian network. In [50], the model building process starts with a complete factorization graph. Arcs are greedily removed from the factorization based on a likelihood-ratio test.

Mixture distributions are weighted sums of $M > 1$ pdfs. The probabilistic model defined by a mixture distribution is a collection ς of M (simpler) probabilistic model structures ς_m and a collection θ of M parameter vectors where $m = 1, 2, \ldots, M$. A mixture distribution is then defined as

$$P_{(\varsigma,\theta)}(\mathcal{Z}) = \sum_{m=1}^{M} \beta_m P_{(\varsigma_m, \theta_m)}, \text{where} \tag{18}$$

$$\beta_m \geq 0, \quad \text{and} \quad \sum_{m=1}^{M} \beta_m = 1. \tag{19}$$

The factors of this product are called mixture components, the weights β_m are called mixing coefficients. The interesting feature of mixture-based probabilistic models is that they allow to model the distribution of solutions independently on different peaks, potentially allowing a population to concentrate on more than a single peak in the search space. This is of special importance, if the search function is multimodal and several basins of attraction should be investigated simultaneously. To achieve this, mixture-based probabilistic models for continuous optimization have been proposed by [15]. Clustering techniques like k-means clustering are used to divide the population into M subpopulations from which the parameters for each of the mixing components are estimated. Maximum likelihood estimates for (18) cannot be obtained analytically. Instead, an iterative procedure defined in [25] is used to obtain the estimates for the mixing coefficients, the mean vector and the covariance matrix.

Real-valued mixture probability distributions have been used for multi-objective optimization in [18] and [19]. The results indicate that a mixture-based model can effectively help to maintain diversity along the Pareto-frontier of a multi-objective problem. Similar observations were made in [22], where pdfs based on Parzen-windows, which are similar to mixture distributions, are used.

Continuous EDAs can readily be used to solve permutation problems if a real-valued representation of permutations is chosen [16], [52], [74]. These approaches are, however, not very effective, as the commonly used "random-key" representation is highly redundant [16].

4.3 Enhancing the Efficiency of Continuous Estimation of Distribution Algorithms

Convergence Defects

For discrete search spaces, the use of Bayesian factorization based on frequency counts has led to scalable evolutionary optimizers that outperform classical genetic algorithms on a broad range of problems. It has been noted however, that simply estimating and sampling probabilistic models on the basis of the normal pdf as illustrated in Sects. 4.1 and 4.2 does not automatically lead to efficient EDAs for the continuous problem domain.

Analytical results on the convergence properties of the univariate marginal distribution algorithm in the continuous domain are available in [34] and [36]. The analysis of UMDAc revealed important peculiarities of continuous EDAs based on the normal pdf.

To be more precise, the performance of UMDAc depends heavily on the structure of the area of the fitness function that UMDAc is currently exploring. We can regard continuous search spaces as arrangements of two elemental structures: peaks (i.e., local and global optima) and slopes. At the beginning of the search, the EDA will in general be approaching a local or global optimum by exploring a region that has a slope-like function. Eventually the search focuses around an optimum (either local or global) in its final phases, i.e., the region to be explored is shaped like a peak.

It has been shown that UMDAc can only reach the optimum if the set of search points is close to the optimum [34], [36]. The reason for this is that the mean of the normal distribution that is estimated by UMDAc can only move a limited distance before converging due to shrinking of the estimated variance. This means that on slope-parts of the search space, UMDAc will perform extremely poor whereas on peak-parts UMDAc will perform nicely. Both studies assume that UMDAc uses the estimated normal density to generate new candidate solutions with no modification.

The bad performance of UMDAc when traversing a slope has caused poor experimental results. In fact, continuous EDAs that estimate and sample variants for the normal distribution fail on some test problems of numerical optimization where other evolutionary algorithms or classical gradient-based methods do not fail. This was first noticed by [15] and confirmed by [48] and [88].

On the Choice of Search Distributions

To account for the inability of the normal distribution to successfully exploit the structure of some continuous fitness functions, [13] introduced the notion of adequacy and competency of a search distribution in EDAs.

Estimating the contours of the fitness landscapes on the basis of a probability distribution, as done in any EDA, results in a probabilistic representation of the true fitness landscape. The induced bias of an EDA is based on this internal probabilistic representation. The restrictions of the model used and the estimation procedure however cause the representation to be only an *approximation* of the optimal distribution; the latter being a close representation of the contours of the fitness landscape.

A class of probability distributions is considered *adequate* with respect to a given optimization problem, if it is able to closely model the contours of the fitness function of that problem with arbitrary exactness. If a probability distribution is inadequate, then the estimated probabilistic representation of the fitness landscape can be (but not necessarily has to be) misleading, because it is different from the true structure of the problem. It should therefore be carefully assessed whether this density can be seen as a reliable source of information for guiding the search.

The density estimation procedure is considered *competent* if it is actually able to obtain an estimate for the probability distribution that closely models the fitness landscape and properly generalizes the sample set. This means that the probabilistic representation of the true landscape is correct. Additional requirements to this end are that the density estimation procedure is tractable with respect to the computation time and the population-sizing requirement.

For solving discrete additively decomposable functions, probabilistic models based on Bayesian networks are adequate. Moreover, for these problems the commonly adopted greedy way to build the Bayesian network is competent. For the continuous problem domain, we now assess briefly whether the normal pdf is competent and adequate for peaks and slopes; the two basic structures of continuous fitness landscapes.

The normal pdf can match contour-lines of a single peak nicely as it always concentrates the search around its mean and therefore can contract around a single peak with arbitrary exactness. If the search is initialized near the peak, selection can shift the mean of the pdf onto the peak. Thus, the normal pdf is adequate for search on a single peak. An estimation procedure based on the standard maximum-likelihood estimates is competent, because by using the maximum-likelihood estimates for the normal pdf, a properly generalizing estimate can be constructed from data in computationally tractable time. As a result, the UMDAc algorithm is able to converge on a peak, if it is initialized near to it. This agrees with initial results on research into continuous EDAs [10], [51].

This becomes different for slope-like regions of the search space. Contour-lines of slopes can not be matched with the normal pdf. The true structure is misrepresented using a maximum-likelihood estimation as the normal kernel introduces an additional basin of attraction around its mean. Estimates from the normal pdf are thus a much less reliable source of information for guiding the search compared to exploring a single peak. Relying the search on maximum-likelihood estimates of the normal pdf potentially misleads the EDA and can cause premature convergence on slope-like regions of the search space.

Recent Approaches

In order to solve the problem of premature convergence, a class of more involved probability distributions could theoretically be introduced for use as a search distribution in continuous EDAs. However, contours of continuous fitness landscapes can be of virtually any shape. As universal approximation in arbitrary exactness is computationally intractable, recently developed EDAs still stick to the use of the normal pdf, but emphasize a more sensible tuning of the parameters of the normal distribution.

A self-adaptation approach adopted from evolution strategies is used in [62] to scale the variance after the distribution estimation. The results indicate that the performance of the resulting algorithm is comparable to that of the evolution strategy with covariance adaptation (CMA-ES, [37]), a state-of-the-art evolution strategy, on separable univariate functions.

A scheme that tries to approximate the Boltzmann distribution (see Sect. 3.1) in continuous spaces using the normal pdf is proposed in [30]. Note that the Factorization Theorem and the related theoretical work is valid for discrete and continuous domains. The results from [30] indicate that indeed a more sensible tuning of the parameters of the normal pdf that is not limited to using maximum-likelihood estimates results in more efficient continuous EDAs.

The estimation scheme is modified in [88] such that it maintains diversity in the population by restricting the variances to values greater than 1. This reduces the risk of premature convergence, because it directly enlarges the area that the algorithm is exploring.

Graham et al. in [35] propose to adaptively scale the covariance matrix after the estimation process. The scaling of the variance is triggered on slope-like regions of the search space and is disabled when the currently investigated region is shaped like a peak. To identify which structure currently dominates the investigated region, the use of ranked correlation estimates between density of the normal pdf and the fitness values is proposed. The results on non-linear problems that can not be solved by simple hill-climbing algorithms (e.g. Rosenbrock function in high dimensions) show that the proposed algorithm scales with a low order polynomial depending on the problem size. Computational results are very close to that of the CMA-ES [37], one of the leading algorithms in continuous evolutionary optimization.

All of the above approaches advance the class of problems that continuous EDA are able to solve reliably. A more sensible adaptation of the EDA principle to the continuous domain seems to be required to develop efficient continuous EDAs. Recently obtained results are promising and indicate the potential of continuous EDA when it comes to solving continuous optimization problems. Adaptation of the normal density in the continuous domain appears to be a central element of study for EDAs to come.

4.4 Summary

The success of discrete EDAs has motivated researchers to adapt the EDA principle to the continuous domain. A direct adaptation of the principle has however proven not to be effective. In contrast to the discrete field, recent work shows that the model structure is not as important as the right treatment of the model parameters. A more sensible adjustment or alteration of the density estimation or sampling process for continuous EDAs is required to advance the performance of continuous EDAs. Recent approaches have proven to be successful and reliable for hard continuous non-linear optimization problems.

5 Conclusions and Outlook

This chapter intended to serve as an introduction to estimation of distribution algorithms. We discussed the major terms, concepts, and algorithms for the discrete and the continuous problem domain and pointed out the difference between the two fields. State-of-the-art EDAs systematically outperform standard GAs with fixed recombination operators on hard additively decomposable problems. To be more specific, the total number of fitness evaluations that is required to reliably solve additively decomposable problems to optimality is often found to grow subquadratically with respect to the problem size. In the continuous domain, the use of probabilistic models has not directly led to effective optimizers. In contrast to the discrete problem domain, a more sensible adjustment of the estimated parameters is necessary to avoid premature convergence and boost performance. State-of-the-art EDAs for the continuous domain are mature, powerful optimizers for non-linear unconstrained problems.

In future, we expect for the continuous domain, efficiency enhancements that are guided by formal models. Most of the available results for the continuous domain are still experimental. Developing formal models that help us to understand the dynamics of continuous EDAs is a formidable, but promising task. Also, it will be important to work out links between continuous EDAs and evolution strategies on the basis of these theoretical results.

Currently, there is still a lack of research in constraint handling for both discrete and continuous EDAs. Almost all real-world problems include (non)linear (in)equality-constraints. It is still an open question, how constraints can be integrated best into the EDA principle and whether and how available constraint handling literature (see [21] for an overview) can readily be adopted or not. Also, it would be interesting to see whether the probabilistic model itself can serve as a starting point to develop constraint handling techniques. As an example, it would be nice to build probabilistic models that have zero probabilities for infeasible solutions. From these models, only feasible solutions can be sampled.

Most of the publications on EDAs come from the computer-science, EC-theory and machine-learning community. This research has led to effective optimizers and theoretical insights. In industry, however, the sGA and its variants are still predominantly used, although EDA research has clearly shown us the boundaries of these approaches. This is partly due to the fact that the EDA principle is relatively new and still unknown in industry. In addition to existing applications of EDAs [8], [82], [51], [6], [26], [65], [7], [71], more applications of EDAs to problems of industrial relevance are needed that illustrate to practitioners the effectiveness of EDAs and the drawback of classical approaches. Research on EDA applications will drive diffusion of the EDA principle and establish EDA in the long run. A promising field for EDA applications is that of simulation optimization, where virtual or physical simulation models are coupled with optimizers [29], [3], [55], [11]. The simulation model is used as a black box that estimates the fitness of solutions. Simulation is a widely adopted tool in industry, e.g., when it comes to evaluating cost and time in complex manufacturing systems or supply networks. To this end it might prove helpful that the simulation and the EDA community speak a similar language. Finally, EDAs that are coupled with simulation models could readily benefit from the growing field of efficiency enhancement techniques [79].

The term *problem decomposition* and how it is understood in GA and EDA research is still largely unknown in operations research. Consequently, the vast majority of GA applications neglect an analysis of dependencies between decision variables. A decomposition analysis that utilizes the Factorization Theorem and related techniques can answer questions that are important for every optimization practitioner, e.g.:

1. Is the overall optimization problem decomposable or separable?
2. How do the decision variables interact?
3. Which features at the phenotype level cause dependencies between alleles at the genotype level?
4. Which dependencies need to be considered in order to solve the problem reliably?
5. Does tight linkage pay off? Is there a natural way to obtain such a coding?
6. What is the structure (size, overlap) of dependency sets in existing benchmark instances? Is there a relationship between the observed hardness of a benchmark instance and these (or other) characteristics?

As a concluding remark, we would like to stress the fact that the application of an EDA to your problem of choice is easy. Most of the algorithms that we have discussed in this chapter are available in standard scripting languages and can be downloaded from the homepages of the respective authors. The implementations typically provide documented interfaces for the integration of custom fitness functions.

References

1. D. H. Ackley. *A Connectionist Machine for Genetic Hillclimbing*. Kluwer Academic, Boston, 1987
2. T. W. Anderson. *An Introduction to Multivariate Statistical Analysis*. John Wiley & Sons Inc., New York, 1958
3. S. Andradottir. *Handbook of Simulation: Principles, Methodology, Advances, Applications, and Practice*, Chapter 9, Simulation Optimization. John Wiley, New York, 1998
4. S. Baluja. Population-based incremental learning: A method for integrating genetic search based function optimization and competitive learning. Technical Report CMU-CS-94-163, Carnegie Mellon University, 1994
5. S. Baluja and S. Davies. Using optimal dependency-trees for combinatorial optimization: Learning the structure of the search space. In D.H. Fisher, editor, *Proceedings of the 1997 International Conference on Machine Learning*, pages 30–38, Madison, Wisconsin, 1997 Morgan Kaufmann.
6. E. Bengoetxea, P. Larrañaga, I. Bloch, A. Perchant, and C. Boeres. Learning and simulation of Bayesian networks applied to inexact graph matching. *Pattern Recognition*, 35(12):2867–2880, 2002
7. R. Blanco, I. Inza, and P. Larrañaga. Learning Bayesian networks in the space of structures by estimation of distribution algorithms. *International Journal of Intelligent Systems*, 18:205–220, 2003
8. R. Blanco, P. Larrañaga, I. Inza, and B. Sierra. Selection of highly accurate genes for cancer classification by estimation of distribution algorithms. In P. Lucas, editor, *Proceedings of the Bayesian Models in Medicine Workshop at the 8th Artificial Intelligence in Medicine in Europe AIME-2001*, pages 29–34, 2001

9. T. Blickle and L. Thiele. A comparison of selection schemes used in evolutionary algorithms. *Evolutionary Computation*, 4(4):361–394, 1996

10. P.A.N. Bosman. *Design and Application of Iterated Density-Estimation Evolutionary Algorithms*. PhD Thesis, University of Utrecht, Institute of Information and Computer Science, 2003

11. P.A.N. Bosman and T. Alderliesten. Evolutionary algorithms for medical simulations – a case study in minimally-invasive vascular interventions. In S.L. Smith and S. Cagnoni, editors, *Proceedings of the Medical Applications of Genetic and Evolutionary Computation MedGEC Workshop at the Genetic and Evolutionary Computation Conference GECCO–2005*, pages 125–132, New York, 2005. ACM Press

12. P.A.N. Bosman and E.D. De Jong. Learning probabilistic tree grammars for genetic programming. In X. Yao, E. Burke, J.A. Lozano, J. Smith, J.J. Merelo-Guervós, J.A. Bullinaria, J. Rowe, P.T.A. Kabán, and H.P. Schwefel, editors, *Parallel Problem Solving From Nature-PPSN VII*, pages 192–201, Springer, Berlin, 2004

13. P.A.N. Bosman and J. Grahl. Matching inductive search bias and problem structure in continuous estimation-of-distribution algorithms. *European Journal of Operational Research*, Forthcoming

14. P.A.N. Bosman and D. Thierens. Expanding from discrete to continuous estimation of distribution algorithms: The IDEA. In M. Schoenauer, K. Deb, G. Rudolph, X. Yao, E. Lutton, J.J. Merelo, and H.-P. Schwefel, editors, *Parallel Problem Solving from Nature – PPSN VI*, pages 767–776, Springer, Berlin, 2000

15. P.A.N. Bosman and D. Thierens. Advancing continuous IDEAs with mixture distributions and factorization selection metrics. In M. Pelikan and K. Sastry, editors, *Proceedings of the Optimization by Building and Using Probabilistic Models OBUPM Workshop at the Genetic and Evolutionary Computation Conference GECCO–2001*, pages 208–212, San Francisco, California, 2001. Morgan Kaufmann

16. P.A.N. Bosman and D. Thierens. Crossing the road to efficient IDEAs for permutation problems. In L. Spector, E.D. Goodman, A. Wu, W.B. Langdon, H.-M. Voigt, M. Gen, S. Sen, M. Dorigo, M.H. Garzon S. Pezeshk, and E. Burke, editors, *Proceedings of the Genetic and Evolutionary Computation Conference – GECCO–2001*, pages 219–226, San Francisco, California, 2001. Morgan Kaufmann

17. P.A.N. Bosman and D. Thierens. New IDEAs and more ICE by learning and using unconditional permutation factorizations. In *Late-breaking Papers of the Genetic and Evolutionary Computation Conference GECCO–2001*, pages 16–23, 2001

18. P.A.N. Bosman and D. Thierens. Multi-objective optimization with diversity preserving mixture-based iterated density estimation evolutionary algorithms. *International Journal of Approximate Reasoning*, 31:259–289, 2002

19. P.A.N. Bosman and D. Thierens. The balance between proximity and diversity in multi-objective evolutionary algorithms. *IEEE Transactions on Evolutionary Computation*, 7:174–188, 2003

20. C.K. Chow and C. N. Liu. Approximating discrete probability distributions with dependence trees. *IEEE Transactions on Information Theory*, 14:462–467, 1968

21. C.A. Coello Coello. Theoretical and numerical constraint-handling techniques used with evolutionary algorithms: A survey of the state of the art. *Computer Methods in Applied Mechanics and Engineering*, 191(11-12):1245–1287, 2002

22. M. Costa and E. Minisci. MOPED: A multi-objective parzen-based estimation of distribution for continuous problems. In C.M. Fonseca, P.J. Fleming, E. Zitzler, K. Deb, and L. Thiele, editors, *Evolutionary Multi-Criterion Optimization Second International Conference – EMO 2003*, pages 282–294, Springer, Berlin, 2003.

23. S.J. De Bonet, C.L. Isbell, and P. Viola. MIMIC: Finding optima by estimating probability densities. In M.C. Mozer, M.I. Jordan, and T. Petsche, editors, *Advances in Neural Information Processing Systems*, Volume 9, page 424. The MIT Press, 1997

24. K. Deb and D. E. Goldberg. Analysing deception in trap functions. In L.D. Whitley, editor, *Foundations of Genetic Algorithms 2*, pages 93–108. Morgan Kaufmann, 1993

25. A.P. Dempster, N.M. Laird, and D. B. Rubin. Maximum likelihood from incomplete data via the EM algorithm. *Journal of the Royal Statistic Society*, Series B 39:1–38, 1977

26. E.I. Ducheyne, R. R. De Wulf, and B. De Baets. Using linkage learning for forest management planning. In *Late-Breaking Papers of the Genetic and Evolutionary Computation Conference GECCO–2002*, pages 109–114, 2002

27. J. Edmonds. Optimum branchings. *Journal of Research of the National Bureau of Standards*, 71b:233–240, 1976

28. R. Exteberria and P. Larrañaga. Global optimization using bayesian networks. In *Proceedings of the Second Symposium on Artificial Intelligence CIMAF-1999*, pages 332–339, 1999

29. M. C. Fu. Optimization via simulation: A review. *Annals of Operations Research*, 53:199–248, 1994

30. M. Gallagher and M. Frean. Population-based continuous optimization, probabilistic modeling and the mean shift. *Evolutionary Computation*, 13(1):29–42, 2005

31. M. Gallagher, M. Frean, and T. Downs. Real-valued evolutionary optimization using a flexible probability density estimator. In W. Banzhaf et al., editors, *Proceedings of the Genetic and Evolutionary Computation Conference GECCO–1999*, pages 840–846, San Francisco, California, 1999. Morgan Kaufmann

32. D. E. Goldberg. *Genetic Algorithms in Search, Optimization, and Machine Learning*. Addison-Wesley, Reading, MA, 1989

33. D. E. Goldberg. *The Design of Innovation: Lessons from and for Competent Genetic Algorithms*, volume 7 of *Genetic Algorithms and Evolutionary Computation*. Kluwer Academic, 2002

34. C. González, J.A. Lozano, and P. Larrañaga. Mathematical modelling of UMDAc algorithm with tournament selection. Behaviour on linear and quadratic functions. *International Journal of Approximate Reasoning*, 31(3):313–340, 2002

35. J. Grahl, P.A.N. Bosman, and F. Rothlauf. The correlation-triggered adaptive variance scaling IDEA (CT-AVS-IDEA). In M. Keijzer et. al., editor, *Proceedings of the Genetic and Evolutionary Computation Conference GECCO–2006*, pages 397–404. ACM Press, 2006

36. J. Grahl, S. Minner, and F. Rothlauf. Behavior of UMDAc with truncation selection on monotonous functions. In *Proceedings of the 2005 IEEE Congress on Evolutionary Computation*, volume 2, pages 2553–2559, Scotland, 2005. IEEE Press

37. N. Hansen and A. Ostermeier. Completely derandomized self-adaptation in evolution strategies. *Evolutionary Computation*, 9(2):159–195, 2001

38. G. Harik. *Learning Gene Linkage to Efficiently Solve Problems of Bounded Difficulty Using Genetic Algorithms*. PhD Thesis, University of Michigan, Ann Arbor, Michigan, 1997

39. G. Harik. Linkage learning via probabilistic modeling in the ECGA. Technical Report 99010, IlliGAL, University of Illinois, Urbana, Illinois, 1999

40. G. Harik and D. E. Goldberg. Learning linkage. In R.K. Belew and M.D. Vose, editors, *Foundations of Genetic Algorithms 4*, pages 247–262, San Francisco, California, 1997. Morgan Kaufmann

41. G. Harik and D. E. Goldberg. Linkage learning through probabilistic expression. *Computer Methods in Applied Mechanics and Engineering*, 186:295–310, 2000

42. G. Harik, F. Lobo, and D. E. Goldberg. The compact genetic algorithm. In *Proceedings of the 1998 IEEE International Conference on Evolutionary Computation*, pages 523–528, IEEE Press, Piscataway, NJ, 1998

43. D. Heckerman and D. Geiger. Learning Bayesian networks: A unification for discrete and Gaussian domains. In P. Besnard and S. Hanks, editors, *Proceedings of the Eleventh Conference on Uncertainty in Artificial Intelligence UAI–1995*, pages 274–284, Morgan Kaufmann, San Mateo, CA, 1995

44. D. Heckerman, D. Geiger, and D. M. Chickering. Learning Bayesian networks: The combination of knowledge and statistical data. In R. Lopez de Mantaras and D. Poole, editors, *Proceedings of the Tenth Conference on Uncertainty in Artificial Intelligence UAI–1994*, pages 293–301, Morgan Kaufmann, San Mateo, CA, 1994

45. J. H. Holland. *Adaptation in Natural and Artificial Systems*. University of Michigan Press, Ann Arbor, Michigan, 1975

46. M. I. Jordan. *Learning in Graphical Models*. MIT Press, Cambridge, MA, 1999

47. L. Kallel, B. Naudts, and C. R. Reeves. Properties of fitness functions and search landscapes. In B. Naudts L. Kallel and A. Rogers, editors, *Theoretical Aspects of Evolutionary Computing*, pages 175–206. Springer, Berlin, 2001

48. S. Kern, S.D. Müller, N. Hansen, D. Büche, J. Ocenasek, and P. Koumoutsakos. Learning probability distributions in continuous evolutionary algorithms - a comparative review. *Natural Computing*, 3(1):77–112, 2004

49. N. Khan, D.E. Goldberg, and M. Pelikan. Multi-objective Bayesian optimization algorithm. Technical Report 2002009, IlliGAL, University of Illinois, Urbana, Illinois, 2002

50. P. Larrañaga, R. Etxeberria, J.A. Lozano, and J.M. Peña. Optimization in continuous domains by learning and simulation of Gaussian networks. In A.S. Wu, editor, *Proceedings of the 2000 Genetic and Evolutionary Computation Conference Workshop Program*, pages 201–204, 2000

51. P. Larrañaga and J. Lozano, editors. *Estimation of Distribution Algorithms: A New Tool for Evolutionary Computation*, volume 2 of *Genetic Algorithms and Evolutionary Computation*. Kluwer Academic Publishers, 2001

52. P. Larrañaga, J.A. Lozano, V. Robles, A. Mendiburu, and P. de Miguel. Searching for the best permutation with estimation of distribution algorithms. In H.H. Hoos and T. Stuetzle, editors, *Proceedings of the Workshop on Stochastic Search Algorithms at the IJCAI–2001*, pages 7–14, San Francisco, California, 2002. Morgan Kaufmann

53. M. Laumanns and J. Ocenasek. Bayesian optimization algorithms for multi-objective optimization. In J.J. Merelo, P. Adamidis, H.G. Beyer, J. L.F.V. Martin, and H.P. Schwefel, editors, *Parallel Problem Solving from Nature – PPSN VII*, pages 298–307, Berlin, 2002. Springer

54. S. L. Lauritzen. *Graphical Models*. Clarendon Press, Oxford, 1996

55. A.M. Law and W. D. Kelton. *Simulation Modeling and Analysis*, 3rd edn. McGraw-Hill, New York, 2000

56. F. Mandl. *Statistical Physics*, 2nd edn. The Manchester Physics Series. Wiley, 1988

57. H. Mühlenbein and R. Höns. The estimation of distributions and the mimimum relative entropy principle. *Evolutionary Computation*, 13(1):1–27, 2005

58. H. Mühlenbein and T. Mahnig. FDA – a scalable evolutionary algorithm for the optimization of additively decomposed functions. *Evolutionary Computation*, 7(4):353–376, 1999

59. H. Mühlenbein, T. Mahnig, and A. O. Rodriguez. Schemata, distributions and graphical models in evolutionary optimization. *Journal of Heuristics*, 5:215–247, 1999

60. H. Mühlenbein and G. Paaß. From recombination of genes to the estimation of distributions I. Binary parameters. In *Lecture Notes in Computer Science 1411: Parallel Problem Solving from Nature - PPSN IV*, pages 178–187, Berlin, 1996. Springer
61. H. Mühlenbein and D. Schlierkamp-Voosen. Predictive models for the breeder genetic algorithm. *Evolutionary Computation*, 1(1):25–49, 1993
62. J. Ocenasek, S. Kern, N. Hansen, and P. Koumoutsakos. A mixed Bayesian optimization algorithm with variance adaptation. In J.A. Lozano J. Smith J.J. Merelo Guervós J.A. Bullinaria J. Rowe P. Tino A. Kaban X. Yao, E. Burke and H.P. Schwefel, editors, *Parallel Problem Solving from Nature – PPSN VIII*, pages 352–361, Springer, Berlin, 2004
63. M. Pelikan. *Bayesian Optimization Algorithm: From Single Level to Hierarchy*. PhD Thesis, University of Illinois at Urbana-Champaign, Dept. of Computer Science, Urbana, IL, 2002
64. M. Pelikan and D. E. Goldberg. Escaping hierarchical traps with competent genetic algorithms. In L. Spector, E.D. Goodman, A. Wu, W.B. Langdon, H.M. Voigt, M. Gen, S. Sen, M. Dorigo, S. Pezeshk, M.H. Garzon, and E. Burke, editors, *Proceedings of the GECCO–2001 Genetic and Evolutionary Computation Conference*
65. M. Pelikan and D. E. Goldberg. Hierarchical BOA solves ising spin glasses and MAXSAT. In E. Cantú-Paz, J.A. Foster, K. Deb, L. Davis, R. Roy, U.-M. O'Reilly, H.-G. Beyer, R. Standish, G. Kendall, S. Wilson, M. Harman, J. Wegener, D. Dasgupta, M.A. Potter, A.C. Schultz, K. Dowsland, N. Jonoska, and J. Miller, editors, *Proceedings of the GECCO–2003 Genetic and Evolutionary Computation Conference*, pages 1271–1282, Springer, Berlin, 2003
66. M. Pelikan, D.E. Goldberg, and E. Cantú-Paz. Bayesian optimization algorithm, population sizing, and time to convergence. In *Proceedings of the Genetic and Evolutionary Computation Conference GECCO–2000*
67. M. Pelikan, D.E. Goldberg, and E. Cantú-Paz. BOA: The Bayesian optimization algorithm. 1999
68. M. Pelikan, D.E. Goldberg, and K. Sastry. Bayesian optimization algorithm, decision graphs and occam's razor. In L. Spector, E.D. Goodman, A. Wu, W.B. Langdon, H.M. Voigt, M. Gen, S. Sen, M. Dorigo, S. Pezeshk, M.H. Garzon, and E. Burke, editors, *Proceedings of the GECCO–2001 Genetic and Evolutionary Computation Conference*
69. M. Pelikan and H. Mühlenbein. The bivariate marginal distribution algorithm. In R. Roy, T. Furuhashi, and P.K. Chawdhry, editors, *Advances in Soft Computing – Engineering Design and Manufacturing*, pages 521–535, Springer, London, 1999
70. M. Pelikan, K. Sastry, M.V. Butz, and D. E. Goldberg. Hierarchical BOA on random decomposable problems. IlliGAL Report 2006002, Illinois Genetic Algorithms Laboratory, University of Illinois at Urbana-Champaign, IL, 2006
71. M. Pelikan, K. Sastry, and E. Cantú-Paz, editors. *Scalable optimization via probabilistic modeling: From algorithms to applications*. Springer, 2006
72. M. Pelikan, K. Sastry, and D. E. Goldberg. Scalability of the bayesian optimization algorithm. *International Journal of Approximate Reasoning*, 31(3):221–258, 2003
73. M. Pelikan, K. Sastry, and D. E. Goldberg. Multiobjective hBOA, clustering, and scalability. In *Proceedings of the Genetic and Evolutionary Computation Conference GECCO–2005*, Volume 1, pages 663–670, ACM Press, New York, 2005
74. V. Robles, P. de Miguel, and P. Larrañaga. Solving the traveling salesman problem with EDAs. In P. Larrañaga and J.A. Lozano, editors, *Estimation of Distribution Algorithms. A New Tool for Evolutionary Computation*. Kluwer Academic, London, 2001
75. F. Rothlauf. *Representations for Genetic and Evolutionary Algorithms*. Number 104 in Studies on Fuzziness and Soft Computing. Springer, Berlin, 2002

76. S. Rudlof and M. Köppen. Stochastic hill climbing with learning by vectors of normal distributions. In T. Furuhashi, editor, *Proceedings of the First Online Workshop on Soft Computing (WSC1)*, pages 60–70, Nagoya, Japan, 1996. Nagoya University

77. R.P. Salustowicz and J. Schmidhuber. Probabilistic incremental program evolution. *Evolutionary Computation*, 5(2):123–141, 1997

78. K. Sastry and D. E. Goldberg. *Genetic Programming Theory and Practice*, Chapter Probabilistic model building and competent genetic programming, pages 205–220. Kluwer Academic, Boston, MA, 2003

79. K. Sastry, M. Pelikan, and D. E. Goldberg. Efficiency enhancement of probabilistic model building algorithms. In *Proceedings of the Optimization by Building and Using Probabilistic Models Workshop at the Genetic and Evolutionary Computation Conference*, 2004

80. M. Sebag and A. Ducoulombier. Extending population-based incremental learning to continuous search spaces. In A.E. Eiben, T. Bäck, M. Schoenauer, and H.P. Schwefel, editors, *Parallel Problem Solving from Nature – PPSN V*

81. I. Servet, L. Trave-Massuyes, and D. Stern. Telephone network traffic overloading diagnosis and evolutionary computation technique. In J.K. Hao, E. Lutton, E.M.A. Ronald, M. Schoenauer, and D. Snyers, editors, *Proceedings of Artificial Evolution '97*, pages 137–144, Springer, Berlin, 1997

82. B. Sierra, E. Lazkano, I. Inza, M. Merino, P. Larrañaga, and J. Quiroga. Prototype selection and feature subset selection by estimation of distribution algorithms. a case study in the survival of cirrhotic patients treated with tips. In A.L. Rector et al., editors, *Proceedings of the 8th Artificial Intelligence in Medicine in Europe AIME-2001*, pages 20–29, Springer, Berlin, 2001

83. G. Syswerda. Simulated crossover in genetic algorithms. In L.D. Whitley, editor, *Proceedings of the Second Workshop on Foundations of Genetic Algorithms*, pages 239–255, Morgan Kaufmann, San Mateo, CA, 1993

84. M. M. Tatsuoka. *Multivariate Analysis: Techniques for Educational and Psychological Research*. John Wiley & Sons Inc., New York, 1971

85. D. Thierens. *Analysis and Design of Genetic Algorithms*. PhD Thesis, Leuven, Belgium: Katholieke Universiteit Leuven, 1995

86. D. Thierens. Scalability problems of simple genetic algorithms. *Evolutionary Computation*, 7(4):331–352, 1999

87. S. Tsutsui. Probabilistic model-building genetic algorithms in permutation representation domain using edge histogram. In M. Schoenauer, K. Deb, G. Rudolph, X. Yao, E. Lutton, J.J. Merelo, and H.P. Schwefel, editors, *Parallel Problem Solving from Nature – PPSN VII*, pages 224–233, Springer, Berlin, 2002

88. B. Yuan and M. Gallagher. On the importance of diversity maintenance in estimation of distribution algorithms. In H.G. Beyer, editor, *Proceedings of the Genetic and Evolutionary Computation Conference GECCO-2005*, Volume 1, pages 719–726, ACM Press, Washington DC, 2005

Making a Difference to Differential Evolution

Zhenyu Yang[1], Jingsong He[1] and Xin Yao[1,2]

[1] Nature Inspired Computation and Applications Laboratory (NICAL),
University of Science and Technology of China, Hefei, Anhui, China.
zhyuyang@mail.ustc.edu.cn, hjss@ustc.edu.cn

[2] School of Computer Science, University of Birmingham, Edgbaston,
Birmingham B15 2TT, UK.
x.yao@cs.bham.ac.uk

Abstract

Differential evolution (DE) and evolutionary programming (EP) are two major algorithms in evolutionary computation. They have been applied with success to many real-world numerical optimization problems. Neighborhood search (NS) is a main strategy underpinning EP. There have been analyses of different NS operators' characteristics. Although DE might be similar to the evolutionary process in EP, it lacks the relevant concept of neighborhood search. In this chapter, DE with neighborhood search (NSDE) is proposed based on the generalization of NS strategy. The advantages of NS strategy in DE are analyzed theoretically. These analyses mainly focus on the change of search step size and population diversity after using neighborhood search. Experimental results have shown that DE with neighborhood search has significant advantages over other existing algorithms on a broad range of different benchmark functions. NSDE's scalability is also evaluated on a number of benchmark problems, whose dimension ranges from 50 to 200.

Key words: Differential Evolution, Global Optimization, Evolutionary Algorithms, Neighbourhood Search, Hybrid Algorithms

1 Introduction

Differential evolution (DE) is a simple yet effective global optimization algorithm with superior performance in several real-world applications [1,2]. Several variations of DE have been proposed to improve its performance. A self-adaptive strategy has also been investigated to adapt between different variants of DE [3]. Although there are only three strategy parameters (population size NP, crossover rate CR and mutation scaling factor F), it was found that the performance of DE is very sensitive to the setting of these control parameters. Zaharie [4] analyzed how these control parameters influence the population diversity of DE, while [5] used extensive experiments to study how the performance of DEs is affected by these factors. Although empirical rules can be found for choosing these control parameters, they are not general and therefore not suitable for practical applications.

There are two main steps in DE: mutation and crossover. The mutation will create a trial vector for each individual, and then crossover will recombine each pair of trial vectors and individuals to produce offspring. Since DE only uses discrete recombination for crossover, mutation supplies the major power to make progress during evolution. As another mutation-based strategy, evolutionary programming (EP) is also a major branch of evolutionary computation. In EP, new offspring are obtained by giving a perturbation to the original individuals. That means all offspring for the next generation are generated in the neighborhood of current solutions. Thus EP uses a neighborhood search (NS) strategy to improve the quality of solutions. The characteristics of different NS operators have been analyzed in [6,7]. EP's evolutionary behavior has shown that NS is an efficient operator for such a generate-and-test method. However there is no definite neighborhood search concept in DE, and little work has been done to investigate how this NS strategy will affect DE's evolutionary behaviors. Based on the success of the NS strategy in EP, the idea of DE with a similar strategy deserves more investigation.

In this chapter, the common features of DE and EP are generalized into a uniform framework, and based on this understanding, DE with neighborhood search (NSDE) is proposed to improve its neighborhood search ability. Then the advantages of DE with neighborhood search are analyzed theoretically. These analyses mainly focus on the change of search step size and population diversity after using a neighborhood search (NS) strategy. Experimental evidence is also given to regarding the evolutionary behavior of DE with neighborhood search on widely used benchmark functions.

2 Preliminaries

2.1 Evolutionary Programming

Optimization by evolutionary programming (EP) can be summarized into two major steps: first mutate the solutions in the current population, and then select the next generation from the mutated and the current solutions [8].

$$x_i'(j) = x_i(j) + \eta_i(j)N_j(0,1) \tag{1}$$

$$\eta_i'(j) = \eta_i(j)\exp(\tau'N(0,1) + \tau N_j(0,1)) \tag{2}$$

where $i \in \{1,\dots,\mu\}$, μ is the population size, $j \in \{1,\dots,n\}$, n is the dimension of object parameters, $N(0,1)$ denotes a normally distributed one-dimensional random number with mean zero and standard deviation one, and $N_j(0,1)$ indicates that the random number is generated anew for each value of j. Usually, evolutionary programming using Eqs. (1) and (2) is called classical evolutionary programming (CEP).

The Cauchy operator is introduced into Eq. (1) to substitute Gaussian mutation in [6], i.e. the update equation is replaced by:

$$x_i'(j) = x_i(j) + \eta_i(j)\delta_j \tag{3}$$

where δ_j is a Cauchy random variable with the scale parameter $t = 1$ and is generated anew for each j. EP using Cauchy mutations is called fast evolutionary programming (FEP). IFEP (Improved FEP) based on mixing different mutation operators, and LEP based on mutations with the Lévy probability distribution were also proposed in [6] and [7], respectively.

By Eq. (1) or Eq. (3), we can find that new offspring are obtained by giving a perturbation to the original individual. This means all offspring for the next generation are generated in the neighborhood of current solutions. Thus EP improves the quality of solutions through a neighborhood search (NS) strategy. Operators such as Gaussian $N_j(0, 1)$ and Cauchy δ_j are used to control the size and shape of the neighborhood: η_j is a self-adaptive scaling factor for the neighborhood size. The characteristics of different NS operators have been analyzed in [6,7]. How these NS operators are used will strongly affect EP's performance. Based on the importance of the NS strategy in EP, we intend to investigate whether such an NS strategy can be generalized and used in other evolutionary algorithms.

2.2 Differential Evolution

Individuals are represented by D-dimensional vectors $x_i, \forall i \in \{1, \ldots, NP\}$ in DE, where D is the number of optimization parameters and NP is the population size. According to the description by Storn and Price [1], the mutation and crossover of original DE can be summarized as follows:

$$y_i = x_{i_1} + (x_{i_2} - x_{i_3}) \cdot F \tag{4}$$

$$z_i(j) = \begin{cases} y_i(j), & \text{if} \quad U_j(0,1) < CR \\ x_i(j), & \text{otherwise} \end{cases} \tag{5}$$

with $i, i_1, i_2, i_3 \in [1, NP]$ are integers and mutually different. $F > 0$ is a real constant factor to control the differential variation $d_i = x_{i_2} - x_{i_3}$, and $U_j(0, 1)$ denotes a uniform random number between 0 and 1. To represent crossover result z_i more formally, we can define a Boolean mask $M = (M(1), M(2), \ldots, M(D))$ as follows:

$$M(j) = \begin{cases} 1, & \text{if} \quad U_j(0,1) < CR \\ 0, & \text{otherwise.} \end{cases}$$

and then z_i can be represented as:

$$\begin{aligned} z_i &= y_i \cdot M + x_i \cdot \overline{M} &= y_i \cdot M + x_i \cdot (I - M) \\ &= y_i \cdot M + x_i - x_i \cdot M &= x_i - (x_i - y_i) \cdot M \end{aligned}$$

where $\mathbf{I}$ is an all-'1' vector $\mathbf{I} = (1, \dots , 1)_D$. With these equations, we know that $\mathbf{M}$ is a componential mask of vector $(\mathbf{x}_i - \mathbf{y}_i)$. For each offspring, $\mathbf{z}_i(j) = \mathbf{x}_i(j)$ if $\mathbf{M}(j) = 0$, otherwise $\mathbf{z}_i(j) = \mathbf{y}_i(j)$. This means crossover in DE can be regarded as a selection process on the mutated components. After mutation DE performs a subspace selection process through crossover, and then it will search for better solutions in the subspace.

Now, similarity can be found between the evolutionary processes of DE and EP. Although there is no concept of neighborhood in DE, it has been carried out with a scaling factor F that has some relation with DE's search step size [5]. A large F value is expected to increase the probability of escaping from local optima. However, it also increases the perturbation of mutation, which will decrease DE's convergence speed. This is similar to the characteristics of neighborhood search operators used in EP, except there F was restricted to a constant number. So within the selected subspace after crossover, DE will have similar evolutionary behaviour to EP. To make use of the successful neighborhood search (NS) operators in EP, the idea of DE with similar NS strategy deserves more investigation.

3 DE with Neighborhood Search

3.1 Analyses of DE's Search Characteristics

In Eq. (4), note that the smaller the difference between parameters $\mathbf{x}_{i_2}$ and $\mathbf{x}_{i_3}$, the smaller the difference vector $\mathbf{d}_i = \mathbf{x}_{i_2} - \mathbf{x}_{i_3}$ and therefore the perturbation. That means if the population becomes close to the optimum, the step length is automatically decreased. This is similar to the self-adaptive step size control found in evolutionary programming. Based on this understanding, we can use an uniform mutation equation for DE and EP as follows:

$$x'_i(j) = x_i(j) + \xi_i(j)\, \psi \tag{6}$$

where $\xi_i(j)$ means $\mathbf{d}_i(j)$ ($\mathbf{d}_i$ is the difference vector $\mathbf{x}_{i_2} - \mathbf{x}_{i_3}$), and is $\boldsymbol{\eta}_i(j)$ in EP. ψ is a constant number F in DE, a Gaussian random number $N_j(0, 1)$ in CEP , and a Cauchy random number δ_j in FEP .

Generalizing the analysis method for the mean search step size in [6], the expected length of ψ jumps in the universal equation can be calculated as follows:

$$E_\psi = \int_{-\infty}^{+\infty} x\, \Psi(x)\, \mathrm{d}x$$

where $\Psi(x)$ is the distribution density function for generating the number ψ. When $\Psi(x)$ takes a Gaussian function and a Cauchy function , the expected length of Gaussian and Cauchy jumps are 0.8 and $+\infty$, respectively. Obviously, $\Psi(x)$ takes an impulse function $\delta(x - F)$ in DE and the expected jump will be:

$$E_{DE} = \int_{-\infty}^{+\infty} x\, \delta(x - F)\, \mathrm{d}x = F$$

Up to now we have shown the relation between scaling factor F and search step size theoretically. After setting a value for F, the search step size will be determined directly. Since different optimization problems or even different stages of the same evolution may demand different kinds of search step size, it is easy to understand why the empirical value $F = 0.5$ is not always suitable, and DE with more universal NS operators will have greater potential.

3.2 Mixing Search Biases of Different NS Operators

In Sect. 3.1, we have given a uniform equation for DE and EP, and within the frame we can generalize Eq. (4) to:

$$y_i = x_{i_1} + d_i \cdot \psi \tag{7}$$

where ψ denotes a NS operator, which can be substituted with any kind of neighborhood search operator. The NS operator is used to control the size and shape of neighborhood during the evolutionary process. Assume the density function of ψ is $f_\psi(x)$, then the probability of generating jumps l smaller than a specified step L and larger than L will be:

$$P(l < L) = \int_{-L}^{+L} f_\psi(x) \, \mathrm{d}x$$

$$P(l > L) = \int_{-\infty}^{-L} f_\psi(x) \, \mathrm{d}x + \int_{+L}^{+\infty} f_\psi(x) \, \mathrm{d}x$$

With these equations, the probability of generating different jumps by the NS operator can be calculated, and thus we can achieve some basic search biases for different NS operators. To show the expected advantages of changing the constant number F to a NS operator, two widely used NS operator candidates will be analyzed here. They are Gaussian random $N(0, 1)$, and Cauchy random $C(t = 1)$. The probabilities of them generating different jumps are given in Fig. 1.

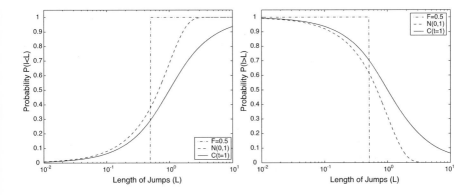

Fig. 1 The probability of generating different jumps for the given NS operators.

First, from Fig. 1 we can observe that all of the NS operators are more flexible than the constant setting for F. Constant F can only produce fixed length jump steps, while the two NS operators have the ability to produce many kinds of jump steps with different probabilities. For a common optimization problem, the probability is very low that the fixed jump steps produced by F are just right for evolution. So NS operators will be more universal than a constant setting for F. Deeper observation shows that Gaussian random $N(0,1)$ is very localized. It is more likely to produce small jumps. In contrast, Cauchy random $C(t = 1)$ is more expandable. It has a greater probability of producing long jumps. Small jumps will be beneficial when the current search point is near the global optimum. However, as analyzed in [6], convergence of the evolution process will be very slow and will risk being trapped in some of the local optima. Long jumps have the ability to escape from poor local optima and locate a good near-global optimum. But this will be beneficial only when the global is sufficiently far away from the current search point. Generally, the Cauchy operator will perform better when far away from the global optimum, while the Gaussian operator is better at finding a local optimum in a good region. It will therefore be beneficial to mix different search methods but to also bias them differently. Here we change DE's mutation to:

$$y_i = x_{i_1} + \begin{cases} d_i \cdot N(0.5, 0.5), & \text{if} \quad U(0,1) < 0.5 \\ d_i \cdot \delta, & \text{otherwise.} \end{cases} \quad (8)$$

where $N(0.5, 0.5)$ denotes a normally distributed one-dimensional random number with mean 0.5 and standard deviation 0.5, and δ is a Cauchy random variable with scale parameter $t = 1$. The parameters $(0.5, 0.5)$ for the Gaussian operator are taken after determining an empirical value for F in DE. Equation (8) introduces a neighborhood search strategy that can produce many more kinds of step sizes, and this is expected to be beneficial when dealing with real-world optimization problems. Here the neighborhood search inspired DE is denoted NSDE .

Population diversity can be used to analyze the expected advantages of DE with neighborhood search. It has been found that if $\psi \sim F \cdot N(0,1)$, the expected population variance after mutation and crossover becomes:

$$E(\text{Var}(Y)) = \left(2F^2 + \frac{m-1}{m}\right)\text{Var}(x) \quad (9)$$

$$E(\text{Var}(Z)) = \left(2F^2 p - \frac{2p}{m} + \frac{p^2}{m} + 1\right)\text{Var}(x) \quad (10)$$

where m is the dimension of object parameters, and p is the value of crossover rate CR. These equations are proved in [4] in detail.

Similar analysis can be carried out when ψ is restricted to a constant number F. For $E((F \cdot N(0,1))^2) = E(F^2) = F^2$, the result for $E(Var(Y))$ and $E(Var(Z))$ will remain the same as Eqs. (9) and (10), i.e. the population diversity stays the same after using the Gaussian NS operator. In Eq. (4), the mutation will be determined by d_i directly. For each mutation, the object parameters can only change with step d_i. After introducing the Gaussian NS operator, DE's search step size is determined not only by difference vector d_i, but also by the NS operator $N(0,1)$. The operator can scale d_i into many kinds of search step sizes, while the original constant number can only produce a fixed step size $\boldsymbol{d}_i \cdot F$. This will be an advantage when DE is searching for the global optimum in an unknown environment, where no prior knowledge exists about what search step size is prefered.

However, a localized Gaussian operator will only be beneficial when near the small neighborhood of the global optimum. In contrast, the Cauchy operator can overcome this limitation. For the Cauchy variable with scale parameter $t = 1$, we know that

$$\text{Var}(\delta) = \int_{-\infty}^{+\infty} x^2 \frac{1}{\pi(1 + x^2)} \, dx = +\infty$$

$$\text{E}(\delta^2) = Var(\delta) + (\text{E}(\delta))^2 = +\infty$$

So after changing ψ in Eq. (7) to a Cauchy random variable, it is easy to carry out the expected population variance after mutation, and crossover that becomes:

$$\text{E}(\text{Var}(Y)) = E(\text{Var}(Z)) = +\infty$$

Obviously, the Cauchy operator is much more global than the Gaussian, so DE with a Cauchy NS operator will have superior ability to escape from local optima. Equation (8) has considered not only a Gaussian operator , but also the Cauchy operator's search biases, so NSDE is expected to be more powerful when searching in an environment without prior knowledge. In the next section, experiments will be provided to test the performance of NSDE on a set of widely used benchmark functions.

4 Experimental Studies

Here experimental evidence is provided to study how neighborhood search operators influence DE's performance. We evaluate the performance of the proposed NSDE algorithm on both classical test functions and the new set of functions provided by CEC2005 special session. NSDE follows Eq. (8) to replace F with NS operators, and CR is set to U(0,1) to enhance DE's subspace search ability, here U(0,1) stands for a uniform random between 0 and 1. The algorithms used for comparison are classical DE with empirical parameter setting [5,9] and other widely used algorithms such as FEP, CMAES .

4.1 NSDE on classical benchmark functions

First we test NSDE's performance on a set of 23 classical functions for numeric optimization. Functions $f_1 - f_{13}$ are high-dimensional problems. Functions $f_1 - f_5$ are unimodal. Function f_6 is the step function which has one minimum and is discontinuous. Function f_7 is a noisy quartic function where $random[0,1)$ is a uniformly distributed random variable in $[0, 1)$. Functions $f_8 - f_{13}$ are multimodal functions where the number of local minima increases exponentially with the problem dimension [6,7]. Functions $f_{14} - f_{23}$ are low-dimensional functions which have only a few local minima [6,7]. Details of these functions can be found in the appendix to [6]. The average experimental results of 50 independent runs are summarized in Tables 1–3 (the results for FEP are taken from [6]).

For unimodal functions $f_1 - f_7$, it is apparent that the proposed NSDE performs better than both DE and FEP. In particular, NSDE's results on $f_1 - f_4$ are significantly better than DE and FEP. It seems a large neighborhood search is efficient in speeding up the evolutionary convergence on unimodal functions. However, NSDE performs worse than DE on f_5. This is a generalized Rosenbrock's function, and there are correlations between each pair of neighboring object parameters. It can be inferred that the one-dimension based neighborhood search strategy has difficulty in optimizing such functions. No strong conclusion can be drawn for f_6 and f_7. There is no significant difference among all the algorithms.

Table 2 shows the experimental results for functions $f_8 - f_{13}$. These functions are multimodal functions with many local optima. Their landscapes appear to be very "rugged" [6], and are often regarded as being difficult to optimize. Figure 3 shows the evolutionary processes of NSDE and DE for these functions. It can be observed that DE stagnates rather early in the search and makes little progress thereafter, while NSDE keeps making improvements throughout the evolution. It appears that DE is trapped in one of the local optima and is unable to get out. NSDE on the other hand, has the ability to produce many kinds of search step sizes with neighborhood search operators and thus has a higher probability of escaping from a local optimum when trapped. A good near-global optimum is more likely to be found by NSDE. In terms of detailed results in Table 2, NSDE performs significantly better than DE and FEP on almost all these functions. It can be concluded that DE with neighborhood search is more effective and efficient when searching in mutimodal functions with many optima.

For multimodal functions with only a few local optima, such as $f_{14} - f_{23}$, both NSDE and DE have better performance than FEP, except that the three algorithms performed exactly the same on f_{16}, f_{17} and f_{19}. NSDE have a similar performance to classical DE. They performed exactly the same on six (i.e. $f_{14} - f_{19}$) out of ten functions. For the rest of the functions, NSDE performed better on f_{20}, but was outperformed by DE on $f_{21} - f_{23}$. To trace why NSDE is inefficient on these 3 functions, further experiments are conducted to observe its evolutionary behaviors. In these experiments we give more computational effort to NSDE, and the results of 50 independent runs are summarized in Table 4.

With a few more generations, NSDE can find the global optima of these functions. Figure 4 shows the evolutionary processes for $f_{21} - f_{23}$. It is clear that NSDE's

Fig. 2 Evolution process of the mean best values found for unimodal functions $f_1 - f_7$. The results were averaged over 50 runs. The *vertical axis* is the function value and the *horizontal axis* is the number of generations

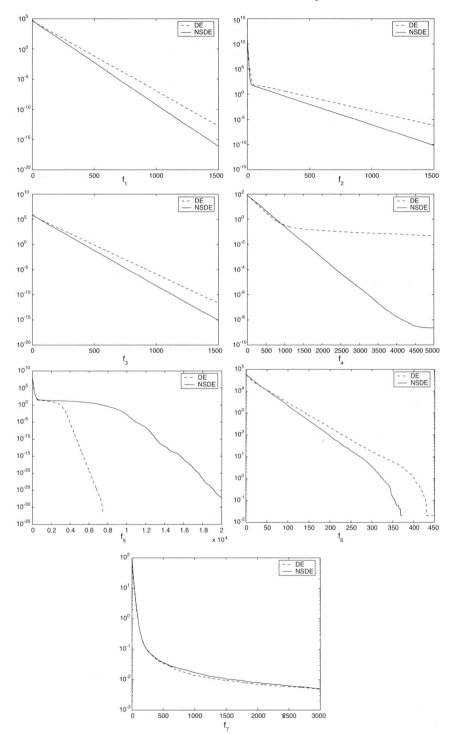

Table 1 Comparison between NSDE, DE and FEP on $f_1 - f_7$. All results have been averaged over 50 independent runs

Test Func	# of Gen's	NSDE Mean	DE Mean	# of Gen's	FEP Mean	vs DE t-test	vs FEP t-test
f_1	1500	7.10×10^{-17}	1.81×10^{-13}	1500	5.70×10^{-04}	$-10.84^\dagger$	$-31.00^\dagger$
f_2	1500	6.49×10^{-11}	6.43×10^{-07}	2000	8.10×10^{-03}	$-17.22^\dagger$	$-74.38^\dagger$
f_3	1500	7.86×10^{-16}	2.12×10^{-12}	5000	1.60×10^{-02}	$-10.86^\dagger$	$-8.08^\dagger$
f_4	5000	2.27×10^{-09}	4.61×10^{-02}	5000	0.3	$-2.05^\dagger$	$-4.24^\dagger$
f_5	20,000	5.90×10^{-28}	0	20,000	5.06	1.03	$-6.10^\dagger$
f_6	1500	0	0	1500	0	0	0
f_7	3000	4.97×10^{-03}	4.84×10^{-03}	3000	7.60×10^{-03}	0.52	$-6.48^\dagger$

† The value of t with 49 degrees of freedom is significant at $\alpha = 0.05$ by a two-tailed test

Table 2 Comparison between NSDE, DE and FEP on $f_8 - f_{13}$. All results have been averaged over 50 independent runs

Test Func	# of Gen's	NSDE Mean	DE Mean	# of Gen's	FEP Mean	vs DE t-test	vs FEP t-test
f_8	1500	$-12,569.5$	$-11,362.1$	9000	$-12,554.5$	$-5.08^\dagger$	$-2.02^\dagger$
f_9	3000	3.98×10^{-02}	138.70	5000	4.60×10^{-02}	$-39.63^\dagger$	-0.22
f_{10}	1500	1.69×10^{-09}	1.20×10^{-07}	1500	1.80×10^{-02}	$-19.86^\dagger$	$-60.61^\dagger$
f_{11}	1500	5.80×10^{-16}	1.97×10^{-04}	2000	1.60×10^{-02}	-1.00	$-5.14^\dagger$
f_{12}	1500	5.37×10^{-18}	1.98×10^{-14}	1500	9.20×10^{-06}	$-6.79^\dagger$	$-18.07^\dagger$
f_{13}	1500	6.37×10^{-17}	1.16×10^{-13}	1500	1.60×10^{-04}	$-7.39^\dagger$	$-15.50^\dagger$

† The value of t with 49 degrees of freedom is significant at $\alpha = 0.05$ by a two-tailed test.

Table 3 Comparison between NSDE, DE and FEP on $f_{14} - f_{23}$. All results have been averaged over 50 independent runs

Test Func	# of Gen's	NSDE Mean	DE Mean	# of Gen's	FEP Mean	vs DE t-test	vs FEP t-test
f_{14}	100	0.998	0.998	100	1.22	0	$-2.80^\dagger$
f_{15}	4000	3.07×10^{-04}	3.07×10^{-04}	4000	5.00×10^{-04}	0	$-4.26^\dagger$
f_{16}	100	-1.03	-1.03	100	-1.03	0	0
f_{17}	100	0.398	0.398	100	0.398	0	0
f_{18}	100	3.00	3.00	100	3.02	0	-1.29
f_{19}	100	-3.86	-3.86	100	-3.86	0	0
f_{20}	200	-3.32	-3.28	200	-3.27	$-4.97^\dagger$	$-5.99^\dagger$
f_{21}	100	-9.68	-10.15	100	-5.52	$4.58^\dagger$	$-16.83^\dagger$
f_{22}	100	-10.33	-10.40	100	-5.52	1.49	$-15.85^\dagger$
f_{23}	100	-10.48	-10.54	100	-6.57	$2.91^\dagger$	$-8.80^\dagger$

† The value of t with 49 degrees of freedom is significant at $\alpha = 0.05$ by a two-tailed test.

Table 4 Further experiments for NSDE on $f_{21} - f_{23}$. All results have been averaged over 50 independent runs

Test Func	# of Gen's	NSDE Mean	# of Gen's	DE Mean	FEP Mean	vs DE t-test	vs FEP t-test
f_{21}	200	−10.15	100	−10.15	−5.52	0	−20.59[†]
f_{22}	150	−10.40	100	−10.40	−5.52	0	−16.28[†]
f_{23}	150	−10.54	100	−10.54	−6.57	0	−8.94[†]

† The value of t with 49 degrees of freedom is significant at $\alpha = 0.05$ by a two-tailed test

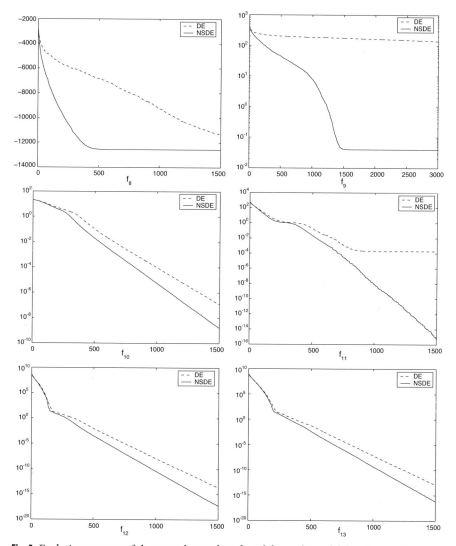

Fig. 3 Evolution process of the mean best values found for multimodal functions with many local optima, i.e. $f_8 - f_{13}$. The results were averaged over 50 runs. The *vertical axis* is the function value and the *horizontal axis* is the number of generations

convergence is a little slower than DE on these functions. The major difference between functions $f_8 - f_{13}$ and $f_{14} - f_{23}$ is that $f_{14} - f_{23}$ appears to be simpler than $f_8 - f_{13}$ due to their low dimensionalities and a smaller number of local optima [6]. But NSDE still spends some computational effort blindly to avoid being trapped in local optima when optimizing these functions. This will weaken the search strength in the direction towards the optimum. NSDE's performance indicates that the advantages of introducing neighborhood search become insignificant on this class of problems.

Scalability of an algorithm is also an important measurement of how good and how applicable the algorithm is [10]. So further experiments were conducted to evaluate NSDE's scalability against the growth of problem dimensions. We selected 9 scalable functions from the 23 benchmark functions. The dimensions D of them were set to 50, 100, 150 and 200, respectively. The computation times used by algorithms were set to grow in the order of $O(D)$ [10]. For example, for function f_1, 2500 generations were set for $D = 50$, 5000 for $D = 100$, 7500 for $D = 150$, and 10000 for $D = 200$. The average results of 50 independent runs are summarized in Tables 5 and 6.

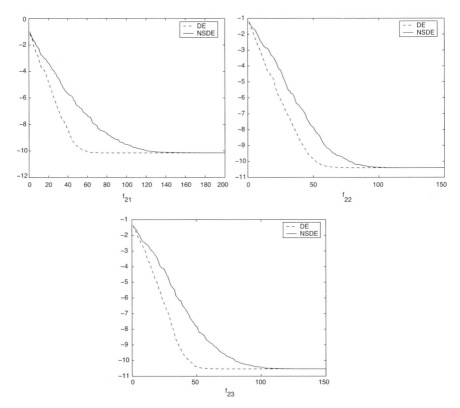

Fig. 4 The evolution process of the mean best values found for $f_{21} - f_{23}$ in further experiments. The results were averaged over 50 runs. The *vertical axis* is the function value and the *horizontal axis* is the number of generations

For unimodal functions, NSDE outperformed DE on $f_1 - f_3$ and f_6. The classical DE has only advantages on f_4 for dimensions 150 and 200. For functions $f_1 - f_3$, NSDE gained better and better results from 50-D to 200-D, with the computational time growth in the order of $O(D)$. But such good scalability was not observed for classical DE. Although DE appeared to achieve better results on the 100-D problem than the 50-D, its performance became poorer for problems of 150-D and 200-D. For function f_6, NSDE found the optimum from 50-D to 200-D, while DE only found the optimum for 50-D. For function f_4, both NSDE and DE performed worse and worse with the growth of dimensions. NSDE's performance decreased faster than DE. Function f_4's fitness value is determined by the maximum component of the D-dimensional vector (see the definition of f_4 [6]). Global evolutionary operator is needed to make progress when optimizing this function. Maybe the neighborhood search strategy in NSDE is too local to create better offspring for large scale f_4.

Table 6 shows the results for multimodal functions with many local optima. As mentioned in Sect. 4.1, this class of functions are often regarded as being difficult to optimize because the number of local optima increase exponentially as their dimension increases. It was encouraging to find NSDE outperformed DE on all of these functions, from 50-D to 200-D. For function f_8, NSDE's results were not only closer

Table 5 Comparison between NSDE and DE on $f_1 - f_6$, with dimension $D = 50, 100, 150$ and 200, respectively. All results have been averaged over 50 independent runs

Test Func	# of Dim's	# of Gen's	NSDE Mean	NSDE Std	DE Mean	DE Std	vs DE t-test
f_1	50	2500	4.28×10^{-18}	7.86×10^{-18}	5.48×10^{-15}	4.05×10^{-15}	$-9.56^{\dagger}$
	100	5000	2.07×10^{-22}	2.72×10^{-22}	2.91×10^{-17}	1.84×10^{-17}	$-11.18^{\dagger}$
	150	7500	1.58×10^{-25}	2.48×10^{-25}	4.82×10^{-17}	4.70×10^{-17}	$-7.25^{\dagger}$
	200	10,000	1.10×10^{-27}	1.47×10^{-27}	2.77×10^{-15}	5.26×10^{-15}	$-3.72^{\dagger}$
f_2	50	2500	2.02×10^{-11}	1.49×10^{-11}	1.05×10^{-07}	3.44×10^{-08}	$-21.58^{\dagger}$
	100	5000	1.58×10^{-13}	1.30×10^{-13}	7.45×10^{-10}	4.63×10^{-10}	$-11.38^{\dagger}$
	150	7500	2.27×10^{-15}	1.69×10^{-15}	2.13×10^{-09}	8.22×10^{-09}	$-1.83^{\dagger}$
	200	10,000	4.88×10^{-17}	3.49×10^{-17}	1.43×10^{-07}	6.50×10^{-07}	-1.56
f_3	50	2500	1.35×10^{-16}	2.44×10^{-16}	1.15×10^{-13}	8.34×10^{-14}	$-9.74^{\dagger}$
	100	5000	9.98×10^{-21}	1.55×10^{-20}	8.35×10^{-16}	6.79×10^{-16}	$-8.70^{\dagger}$
	150	7500	1.12×10^{-23}	1.34×10^{-23}	4.57×10^{-15}	4.58×10^{-15}	$-7.06^{\dagger}$
	200	10,000	1.08×10^{-25}	1.44×10^{-25}	1.89×10^{-13}	5.35×10^{-13}	$-2.50^{\dagger}$
f_4	50	2500	$1.12 \times 10^{+00}$	$1.85 \times 10^{+00}$	$4.86 \times 10^{+00}$	$2.26 \times 10^{+00}$	$-9.05^{\dagger}$
	100	5000	$2.00 \times 10^{+01}$	$6.10 \times 10^{+00}$	$2.26 \times 10^{+01}$	$4.44 \times 10^{+00}$	$-2.43^{\dagger}$
	150	7500	$3.34 \times 10^{+01}$	$4.83 \times 10^{+00}$	$3.09 \times 10^{+01}$	$3.95 \times 10^{+00}$	$2.81^{\dagger}$
	200	10,000	$4.31 \times 10^{+01}$	$6.20 \times 10^{+00}$	$3.41 \times 10^{+01}$	$3.32 \times 10^{+00}$	$8.95^{\dagger}$
f_6	50	2500	$0.00 \times 10^{+00}$	$0.00 \times 10^{+00}$	$0.00 \times 10^{+00}$	$0.00 \times 10^{+00}$	0.00
	100	5000	$0.00 \times 10^{+00}$	$0.00 \times 10^{+00}$	2.00×10^{-02}	1.41×10^{-01}	-1.00
	150	7500	$0.00 \times 10^{+00}$	$0.00 \times 10^{+00}$	4.00×10^{-02}	1.98×10^{-01}	-1.43
	200	10,000	$0.00 \times 10^{+00}$	$0.00 \times 10^{+00}$	$4.02 \times 10^{+00}$	$1.34 \times 10^{+01}$	$-2.12^{\dagger}$

† The value of t with 49 degrees of freedom is significant at $\alpha = 0.05$ by a two-tailed test

Table 6 Comparison between NSDE and DE on $f_8 - f_{11}$, with dimension $D = 50, 100, 150$ and 200, respectively. All results have been averaged over 50 independent runs

Test Func	# of Dim's	# of Gen's	NSDE Mean	NSDE Std	DE Mean	DE Std	vs DE t-test
f_8	50	2500	$-20{,}946.80$	16.70	$-15{,}928.00$	3731.52	$-9.51^\dagger$
	100	5000	$-41{,}860.40$	69.50	$-31{,}729.30$	7533.67	$-9.51^\dagger$
	150	7500	$-62{,}568.70$	216.56	$-54{,}961.90$	5736.55	$-9.37^\dagger$
	200	10,000	$-83{,}044.50$	348.88	$-73{,}964.00$	2124.57	$-29.82^\dagger$
f_9	50	2500	6.57×10^{-01}	8.67×10^{-01}	$3.23 \times 10^{+02}$	$2.84 \times 10^{+01}$	$-80.31^\dagger$
	100	5000	$8.78 \times 10^{+00}$	$3.11 \times 10^{+00}$	$5.56 \times 10^{+02}$	$1.04 \times 10^{+02}$	$-37.18^\dagger$
	150	7500	$2.69 \times 10^{+01}$	$6.12 \times 10^{+00}$	$3.46 \times 10^{+02}$	$3.23 \times 10^{+02}$	$-6.99^\dagger$
	200	10,000	$5.30 \times 10^{+01}$	$1.05 \times 10^{+01}$	$1.61 \times 10^{+02}$	$1.69 \times 10^{+01}$	$-38.37^\dagger$
f_{10}	50	2500	3.74×10^{-10}	4.44×10^{-10}	1.64×10^{-08}	6.13×10^{-09}	$-18.44^\dagger$
	100	5000	1.91×10^{-12}	1.02×10^{-12}	8.59×10^{-10}	4.41×10^{-10}	$-13.74^\dagger$
	150	7500	4.81×10^{-14}	2.29×10^{-14}	3.09×10^{-01}	4.80×10^{-01}	$-4.55^\dagger$
	200	10,000	2.04×10^{-14}	3.53×10^{-15}	$1.48 \times 10^{+00}$	3.66×10^{-01}	$-28.59^\dagger$
f_{11}	50	2500	$0.00 \times 10^{+00}$	$0.00 \times 10^{+00}$	1.97×10^{-04}	1.39×10^{-03}	-1.00
	100	5000	$0.00 \times 10^{+00}$	$0.00 \times 10^{+00}$	5.92×10^{-04}	2.37×10^{-03}	$-1.77^\dagger$
	150	7500	$0.00 \times 10^{+00}$	$0.00 \times 10^{+00}$	1.13×10^{-03}	3.16×10^{-03}	$-2.53^\dagger$
	200	10,000	2.46×10^{-04}	1.74×10^{-03}	3.05×10^{-03}	8.14×10^{-03}	$-2.38^\dagger$

† The value of t with 49 degrees of freedom is significant at $\alpha = 0.05$ by a two-tailed test.

to optimum, but also more stable than classical DE (see the values of Std. Dev and t-test). For function f_9, NSDE's performance decreased a little as the dimensions increased, but was still much better than DE. For function f_{10}, NSDE showed similar good scalability as for unimodal functions, i.e., its results became better and better from 50-D to 200-D problems. For function f_{11}, NSDE found the optimum for 50-D, 100-D and 150-D problems, while DE never found the optimum for them. Although NSDE failed to find the optimum for 200-D f_{11}, it still outperformed DE significantly.

4.2 NSDE on CEC2005's Functions

To evaluate NSDE further, a new set of benchmark functions were used, including 25 functions with different complexity [11]. Many of them are the shifted, rotated, expanded or combined variants of classical functions. Functions $f_1 - f_5$ are unimodal while the remaining 20 functions are multimodal. The experimental results will compare with not only classical DE, but also another widely used algorithm, CMAES. DE and CMAES's experimental results were provided in [12,13]. To be consistent with their experimental setting, experiments are conducted on all 25 $30 - D$ problems, and we chose the function evaluations (FEs) to be $3.0 \times 10^{+05}$. Error value, i.e. the difference between current fitness value and optimum value, is used to compare algorithm's performance. The average error values of 25 independent runs are summarized in Tables 7–9.

For unimodal functions $f_1 - f_5$, the three algorithms have comparable performance. NSDE and DE have exactly the same results on the Shifted Sphere function f_1. NSDE performed better on $f_2 - f_4$, but was outperformed by DE on f_5. It is remarkable that CMAES performed far better than NSDE on f_3 and f_5, but far worse on f_4. One possible reason is that CMAES and DE-based algorithms have very different search biases on these functions. Later we will trace the reason through characteristics of different functions.

For basic and expanded multimodal functions $f_6 - f_{14}$, the advantages of introducing neighborhood search are much more significant. In terms of experimental results, NSDE outperformed DE on almost all functions except f_7 and f_{12}. Although CMAES performed better on $f_6 - f_8$, and was outperformed by NSDE on the other six functions $f_9 - f_{14}$. NSDE is superior on these multimodal functions, which is consistent with the conclusions given for the classical functions.

After analyzing the characteristics of these functions, i.e. the first column of Tables 7 and 8, it is found that DE has rather poor performance on rotated or non-

Table 7 Comparison between NSDE, DE and CMAES on $f_1 - f_5$. All results have been averaged over 25 independent runs (S means the function is Shifted, R means Rotated, and N means Non-separable)

CEC'05 Func	NSDE Mean	DE Mean	CMAES Mean	vs DE t-test	vs CMAES t-test
f_1(S/-/-)	$0.00 \times 10^{+00}$	$0.00 \times 10^{+00}$	5.28×10^{-09}	0	$-26.9^\dagger$
f_2(S/-/N)	5.62×10^{-08}	3.33×10^{-02}	6.93×10^{-09}	$-3.40^\dagger$	$3.78^\dagger$
f_3(S/R/N)	$6.40 \times 10^{+05}$	$6.92 \times 10^{+05}$	5.18×10^{-09}	$-74.9^\dagger$	$11.4^\dagger$
f_4(S/-/N)	$9.02 \times 10^{+00}$	$1.52 \times 10^{+01}$	$9.26 \times 10^{+07}$	-1.41	$-2.76^\dagger$
f_5(-/-/N)	$1.56 \times 10^{+03}$	$1.70 \times 10^{+02}$	8.30×10^{-09}	$15.3^\dagger$	$18.8^\dagger$

† The value of t with 24 degrees of freedom is significant at $\alpha = 0.05$ by a two-tailed test

Table 8 Comparison between NSDE, DE and CMAES on $f_6 - f_{14}$. All results have been averaged over 25 independent runs (S means the function is Shifted, R means Rotated, and N means Non-separable)

CEC'05 Func	NSDE Mean	DE Mean	CMAES Mean	vs DE t-test	vs CMAES t-test
f_6(S/-/N)	$2.45 \times 10^{+01}$	$2.51 \times 10^{+01}$	6.31×10^{-09}	$-7.60^\dagger$	$4.57^\dagger$
f_7(S/R/N)	1.18×10^{-02}	2.96×10^{-03}	6.48×10^{-09}	$3.41^\dagger$	$5.04^\dagger$
f_8(S/R/N)	$2.09 \times 10^{+01}$	$2.10 \times 10^{+01}$	$2.00 \times 10^{+01}$	$-6.42^\dagger$	$76.7^\dagger$
f_9(S/-/-)	7.96×10^{-02}	$1.85 \times 10^{+01}$	$2.91 \times 10^{+02}$	$-1.77^\dagger$	$-41.1^\dagger$
f_{10}(S/R/N)	$4.29 \times 10^{+01}$	$9.69 \times 10^{+01}$	$5.63 \times 10^{+02}$	$-3.22^\dagger$	$-10.5^\dagger$
f_{11}(S/R/N)	$1.41 \times 10^{+01}$	$3.42 \times 10^{+01}$	$1.52 \times 10^{+01}$	$-7.28^\dagger$	-0.56
f_{12}(S/-/N)	$6.59 \times 10^{+03}$	$2.75 \times 10^{+03}$	$1.32 \times 10^{+04}$	$2.76^\dagger$	$-2.53^\dagger$
f_{13}(S/-/N)	$1.62 \times 10^{+00}$	$3.23 \times 10^{+00}$	$2.32 \times 10^{+00}$	$-9.25^\dagger$	$-7.84^\dagger$
f_{14}(S/R/N)	$1.32 \times 10^{+01}$	$1.34 \times 10^{+01}$	$1.40 \times 10^{+01}$	$-5.14^\dagger$	$-9.40^\dagger$

† The value of t with 24 degrees of freedom is significant at $\alpha = 0.05$ by a two-tailed test

Table 9 Comparison between NSDE, DE and CMAES on $f_{15} - f_{25}$. All results have been averaged over 25 independent runs

CEC'05 Func	NSDE Mean	DE Mean	CMAES Mean	vs DE t-test	vs CMAES t-test
f_{15}	$3.64 \times 10^{+02}$	$3.60 \times 10^{+02}$	$2.16 \times 10^{+02}$	0.15	$6.59^\dagger$
f_{16}	$6.90 \times 10^{+01}$	$2.12 \times 10^{+02}$	$5.84 \times 10^{+01}$	$-6.26^\dagger$	1.68
f_{17}	$1.01 \times 10^{+02}$	$2.37 \times 10^{+02}$	$1.07 \times 10^{+03}$	$-5.15^\dagger$	$-9.40^\dagger$
f_{18}	$9.04 \times 10^{+02}$	$9.04 \times 10^{+02}$	$8.90 \times 10^{+02}$	0	1.52
f_{19}	$9.04 \times 10^{+02}$	$9.04 \times 10^{+02}$	$9.03 \times 10^{+02}$	0	0.61
f_{20}	$9.04 \times 10^{+02}$	$9.04 \times 10^{+02}$	$8.89 \times 10^{+02}$	0	1.65
f_{21}	$5.00 \times 10^{+02}$	$5.00 \times 10^{+02}$	$4.85 \times 10^{+02}$	0	$2.21^\dagger$
f_{22}	$8.89 \times 10^{+02}$	$8.97 \times 10^{+02}$	$8.71 \times 10^{+02}$	$-2.00^\dagger$	$3.43^\dagger$
f_{23}	$5.34 \times 10^{+02}$	$5.34 \times 10^{+02}$	$5.35 \times 10^{+02}$	0	$-3.27^\dagger$
f_{24}	$2.00 \times 10^{+02}$	$2.00 \times 10^{+02}$	$1.41 \times 10^{+03}$	0	$-11.1^\dagger$
f_{25}	$2.00 \times 10^{+02}$	$7.30 \times 10^{+02}$	$6.91 \times 10^{+02}$	$-7.09 \times 10^{+03\dagger}$	$-3.12^\dagger$

† The value of t with 24 degrees of freedom is significant at $\alpha = 0.05$ by a two-tailed test.

separable functions. CMAES has superior performance on f_2, f_3 and $f_5 - f_7$, but still gained poor results on the remaining functions. As an improved version of DE, although NSDE outperformed DE on some of these rotated or non-separable functions, the results of f_3, $f_5 - f_8$ and f_{12} are still unsatisfactory. Despite success in expanding the neighborhood search ability, NSDE's performance is still limited by the inherited framework of original DE.

For hybrid composition functions $f_{15} - f_{25}$, the results in Table 9 show that all algorithms not only failed to locate the optimum, but also become trapped in local optima that are far from optimum. These functions are much more difficult, and no effective algorithms have yet been found to slve them [14]. A more detaile investigation of the results shows that NSDE still outperformed DE on f_{17}, f_{22}, f_{25}, and outperformed CMAES on f_{17}, $f_{23} - f_{25}$. These functions are the hybrid composition of basic rotated or non-separable functions. The reason why NSDE is inefficient on some of these functions is as found in the analysis of $f_1 - f_{14}$, i.e. NSDE's performance on non-separable functions is limited by the inherited framework of DE. It is interesting to note that NSDE's results on f_{17} and f_{25} are much closer to the optimum than those of DE and CMAES. The strategy of introducing large jumps to escape from local optima in NSDE is still useful even on some of these composition multimodal functions.

5 Conclusions

This chapter proposes NSDE , an improved variation of classical DE, which is inspired by EP's neighborhood search (NS) strategy. Gaussian and Cauchy NS operators are introduced into NSDE. The advantages of DE with neighborhood search are analyzed theoretically. It has been shown that NS operators will improve the diversity of DE's search step size and population, which will be beneficial to escape from local optima when searching in environments without prior knowledge of what search step size is prefered.

Experimental evidence is also provided showing how the neighborhood search (NS) strategy affects DE's evolutionary behavior. A total of 48 widely used benchmark problems were employed to test NSDE's performance. Our experimental results show that DE with neighborhood search has significant advantages over classical DE.

Acknowledgement. The authors are grateful to Prof. P.N. Suganthan and Dr. Tang Ke for their constructive comments on this chapter. This work is partially supported by the National Science Foundation of China (Grant No. 60428202 and 60573170).

References

1. R. Storn, K. Price (1997) Differential Evolution – A Simple and Efficient Heuristic Strategy for Global Optimization over Continuous Spaces. Journal of Global Optimization, 11:341–359

2. R. Thomsen (2003) Flexible Ligand Docking using Differential Evolution. Proc. of the 2003 Congress on Evolutionary Computation, 4:2354–2361

3. A.K. Qin, P.N. Suganthan (2005) Self-adaptive Differential Evolution Algorithm for Numerical Optimization. Proc. of the 2005 Congress on Evolutionary Computation, 2:1785–1791

4. D. Zaharie (2002) Critical Values for the Control Parameters of Differential Evolution Algorithms. Proc. of Mendel 2002, 8th International Conference on Soft Computing, 62–67

5. R. Gämperle, S. D. Müller, P. Koumoutsakos (2002) A Parameter Study for Differential Evolution. Advances in Intelligent Systems, Fuzzy Systems, Evolutionary Computation, 293–298

6. X. Yao, Y. Liu, G. Lin (1999) Evolutionary Programming Made Faster. IEEE Transactions on Evolutionary Computation, 3:2:82–102

7. C. Lee, X. Yao (2004) Evolutionary Programming Using Mutations Based on the Lévy Probability Distribution. IEEE Transactions on Evolutionary Computation, 8:1:1–13

8. T. Bäck, H. P. Schwefel (1993) An Overview of Evolutionary Algorithms for Parameter Optimization. Evolutionary Computation, 1:1–23

9. J. Vesterstrom, R. Thomsen (2004) A Comparative Study of Differential Evolution, Particle Swarm Optimization, and Evolutionary Algorithms on Numerical Benchmark Problems. Evolutionary Computation, 2:1980–1987

10. Y. Liu, X. Yao, Q. Zhao, T. Higuchi (2001) Scaling Up Fast Evolutionary Programming with Cooperative Coevolution. Proc. of the 2001 Congress on Evolutionary Computation, 1:1101–1108

11. P. N. Suganthan et al. (2005) Problem Definitions and Evaluation Criteria for the CEC 2005 Special Session on Real-Parameter Optimization. http://www.ntu.edu.sg/home/EPNSugan

12. J. Rönkkönen, S. Kukkonen, K. V. Price (2005) Real-Parameter Optimization with Differential Evolution. Proc. of the 2005 Congress on Evolutionary Computation, 1:567–574

13. A. Auger, S. Kern, N. Hansen (2005) Performance Evoluation of an Advanced Local Search Evolutionary Algorithm. Proc. of the 2005 Congress on Evolutionary Computation, 2:1777–1784

14. H. Nikolaus (2005) Compilation of Results on the 2005 CEC Benchmark Function Set. http://www.ntu.edu.sg/home/EPNSugan

Hidden Markov Models Training
Using Population-based Metaheuristics

Sébastien Aupetit, Nicolas Monmarché, and Mohamed Slimane

Université François Rabelais de Tours, Laboratoire d'Informatique, Polytech'Tours, Département Informatique, 64 Avenue Jean Portalis, 37200 Tours, France. {sebastien.aupetit, nicolas.monmarche, mohamed.slimane}@univ-tours.fr

Abstract

In this chapter, we consider the issue of Hidden Markov Model (HMM) training. First, HMMs are introduced and then we focus on the particular HMM training problem. We emphasize the difficulty of this problem and present various criteria that can be considered. Many different adaptations of metaheuristics have been used but, until now, few extensive comparisons have been performed for this problem. We propose to compare three population-based metaheuristics (genetic algorithm, ant algorithm and particle swarm optimization) with and without the help of a local optimizer. These algorithms make use of solutions that can be explored in three different kinds of search space (a constrained space, a discrete space and a vector space). We study these algorithms from both a theoretical and an experimental perspective: parameter settings are fully studied on a reduced set of data and the performances of algorithms are compared on different sets of real data.

Key words: Hidden Markov Model, Likelihood Maximization, API Algorithm, Genetic Algorithm, Particle Swarm Optimization

1 Introduction

Hidden Markov models (HMMs) are statistical tools allowing one to model stochastic phenomena. HMMs are in use in lots of domains [1], such as speech recognition or synthesis, biology, scheduling, information retrieval, image recognition or time series prediction. Using HMMs in an efficient manner implies tackling an often neglected but important issue: training of HMMs. In this chapter, this is what we are concerned with, and we show how the training can be improved using metaheuristics.

This chapter is structured in three parts. We begin with an introduction to hidden Markov models, and associated notation and algorithms. We follow with an inventory of metaheuristics used for HMMs training. Finally, we close the chapter by tuning and comparing six adaptations of three population-based metaheuristics for HMMs training.

2 Hidden Markov Models (HMMs)

Hidden Markov models have a long history behind them. In 1913, A.A. Markov introduced the Markov chain theory [2], but it is only since the 1960s that the main principles and algorithms of HMMs have appeared [3–6]. Over time, many models [7–15] have been derived from the original HMM in order to tackle specificities of particular applications. In this chapter, we discuss a particular kind of HMM for which we give the notation below. However, the techniques described below can easily be adapted to other kinds of HMM.

2.1 Definition

A discrete hidden Markov model allows one to model a time series using two stochastic processes: one is a Markov chain and is called the hidden process, the other is a process depending on states of the hidden process and is called the observed process.

Definition 1. *Let $\mathbb{S} = \{s_1, \ldots, s_N\}$ be the set of the N hidden states of the system and let $S = (S_1, \ldots, S_T)$ be a Tuple of random variables defined on $\mathbb{S}$. Let $\mathbb{V} = \{v_1, \ldots, v_M\}$ be the set of M symbols that can be emitted by the system and let $V = (V_1, \ldots, V_T)$ be a Tuple of random variables defined on $\mathbb{V}$. A first-order discrete hidden Markov model is then defined by:*

- $P(S_1 = s_i)$: *the probability of being in the hidden state s_i at the beginning of the series,*
- $P(S_t = s_j | S_{t-1} = s_i)$: *the probability of transiting from the hidden state s_i to the hidden state s_j between times $t-1$ and t,*
- $P(V_t = v_j | S_t = s_i)$ *the probability of emitting the symbol v_j at time t for the hidden state s_i.*

If the hidden Markov model is a stationary one then transition and emission probabilities are time independent. Consequently, we can define, for all $t > 1$:

- $A = (a_{i,j})_{1 \leq i,j \leq N}$ with $a_{i,j} = P(S_t = s_j | S_{t-1} = s_i)$,
- $B = (b_i(j))_{1 \leq i \leq N, 1 \leq j \leq M}$ with $b_i(j) = P(V_t = v_j | S_t = s_i)$ and
- $\Pi = (\pi_1, \ldots, \pi_N)'$ with $\pi_i = P(S_1 = s_i)$.

A first-order stationary hidden Markov model denoted λ is then completely defined by the triple (A, B, Π). In the following, we denote $\lambda = (A, B, \Pi)$, the triple, and use the term hidden Markov model (HMM) to design a first order stationary hidden Markov model.

Let $Q = (q_1, \ldots, q_T) \in \mathbb{S}^T$ be a sequence of hidden states of a HMM λ and $O = (o_1, \ldots, o_T) \in \mathbb{V}^T$ a sequence of observed symbols of the same HMM. The probability [1] of the realization of the sequences Q and O for a given HMM λ is:

---

[1] Formally, the HMM λ would be considered as the realization of a random variable $P(V = O, S = Q | l = \lambda)$. However, to simplify formulas, the random variable l will be omitted in the following.

$P(V = O, S = Q|A, B, \Pi) = P(V = O, S = Q|\lambda)$. Dependencies between random variables give:

$$P(V = O, S = Q|\lambda) = P(V = O|S = Q, \lambda)P(S = Q|\lambda)$$

$$= \left(\prod_{t=1}^{T} b_{q_t}(o_t) \right) \cdot \left(\pi_{q_1} \prod_{t=1}^{T-1} a_{q_t, q_{t+1}} \right) .$$

When the hidden state sequence is unknown, the likelihood of an observed sequence O of symbols for a HMM λ can be computed. The likelihood is the probability that an observed sequence O was generated from the model and is given by:

$$P(V = O|\lambda) = \sum_{Q \in \mathbb{S}^T} P(V = O, S = Q|\lambda) .$$

For the classical use of HMMs, three main issues need to be solved:

- computing the likelihood, $P(V = O|\lambda)$, of an observed sequence O of symbols for a HMM λ. This is efficiently done by the Forward algorithm or by the Backward algorithm with a complexity of $O(N^2 T)$ [9];
- computing the hidden state sequence Q^* that was the most likely followed to generate the observed sequence O of symbols for the HMM λ. The sequence Q^* is defined by:

$$Q^* = \arg\max_{Q \in \mathbb{S}^T} P(V = O, S = Q|\lambda) ,$$

  and is efficiently determined by the Viterbi algorithm of a complexity $O(N^2 T)$ [4]);
- learning/adjusting/training one or many HMMs from one or many observed sequences of symbols when the number of hidden states is known. Learning by HMMs can be viewed as a constrained maximization of some criterion for which constraints are due to the fact that models are parametrized by stochastic matrices. Lots of criteria can be used to train HMMs, and a short review of the most widely used criteria follows.

2.2 Training Criteria

Let O be the sequence of observed symbols to learn. Let Λ be the set of all HMMs for a fixed number of hidden states N and a fixed number of symbols M. Five main kinds of training criteria can be found in the literature:

- maximum likelihood criteria need, in their most simple forms, to find a HMM λ^* such that

$$\lambda^* = \arg\max_{\lambda \in \Lambda} P(V = O|\lambda) .$$

Up to now, no method has been found that always finds such an optimal model but there exist two algorithms (the Baum–Welch algorithm [3] and the gradient descent [16,17]) that can improve a model. Iterative utilization of such algorithms on an initial model allows one to find a local optimum for the criteria.

- Maximization of the *a posteriori* probability requires finding a HMM λ^* such that

$$\lambda^* = \arg\max_{\lambda \in \Lambda} P(\lambda|V = O)$$

  This criterion is linked to Bayesian decision theory [18] and it can be rewritten in many forms depending on the hypothesis. For simple forms, it can maximize many independent likelihoods, but for more complex forms, it is not possible to consider one or many likelihood maximizations. In such cases, one of the only available ways to find an optimal model is to consider gradient descent. This approach has the disadvantage that only local optima can be found.

- Maximization of the mutual information allows the simultaneous training of many HMMs in order to maximize their discriminative power. Such criteria have been used many times [9,19–21]. Again, one of the only ways to tackle such criteria is to consider gradient descent.

- Minimization of error rate of classification aims at reducing misclassification of observed sequences in a class. Such an approach has been used many times [17,22,23]. These criteria have two particular properties: no gradient can be computed and they are not continuous. In order to tackle them, they are approximated by functions for which a gradient can be computed. Then gradient descent is used to find a model.

- The segmental k-means criterion [24,25] consists in finding a HMM λ^* such that:

$$\lambda^* = \arg\max_{\lambda \in \Lambda} P(V = O, S = Q_\lambda^*|\lambda)$$

  with Q_λ^* the hidden state sequence obtained by the Viterbi algorithm [4] for the model λ. For this criterion, we try to find a model that maximizes the joint probability of the observed sequence and the hidden state sequence that generates the observed sequence. This criterion is neither continuous nor derivable. The approximation of the criterion by a derivable function is difficult. However, there exists the segmental k-means algorithm [24] that allows one to improve the model. This algorithm is similar in its behaviors to the Baum–Welch algorithm because it converges towards a local optimum of the criterion.

As we can see, many optimization criteria can be considered to train HMMs. Previous criteria are not the only ones but are the most widely used. For most criteria, there does not exist an explicit way to find an optimal model, however, there exist algorithms able to improve a model and there exists a way to find a local optimum. In practical use, the local optima are sometimes sufficient but sometimes they are not. It is necessary to find optimal models or at least to find models that are as good as possible for the given criterion.

The difficulty remains in the fact that criteria are difficult to optimize. Let us take one of the most simple criteria: the maximization of the likelihood. The criterion is given by:

$$P(V = O|\lambda) = \sum_{Q \in \mathbb{S}^T} \left[\pi_{q_1} \cdot \left(\prod_{t=1}^{T-1} a_{q_t, q_{t+1}} \right) \cdot \left(\prod_{t=1}^{T} b_{q_t}(o_t) \right) \right].$$

It is a polynomial of degree $2T$ using $N(N + M + 1)$ continuous variables that are constrained. When considering more complex criteria such as maximization of the *a posteriori* probability or maximization of the mutual information, the criterion becomes a function of many polynomials. Note that the degree of the polynomial depends on the length of the data so the more data, the more difficult it is to find optimal models. One way to find such models or at least to find the best possible models is to adapt metaheuristics [26] to explore the hidden Markov model space. This is the aim of this chapter.

3 Training Hidden Markov Models with Metaheuristics

In this section, we introduce adaptations of metaheuristics that have been used to train HMMs. To do so, we begin by presenting the three kinds of search spaces that can be used, and follow with a brief review of the adaptations.

3.1 Search Spaces for Hidden Markov Models Training

To train HMMs, three kinds of search spaces can be considered [10]: Λ, $\mathbb{S}^T$ and Ω. To describe them, we consider HMMs with N hidden states and M symbols. The observed sequence of T symbols is named O.

The search space Λ

The search space Λ is the most commonly used search space. It corresponds to the triple (A, B, Π) of stochastic matrices defining a HMM. Λ is then isomorph to the Cartesian product $\mathbb{G}_N \times (\mathbb{G}_N)^N \times (\mathbb{G}_M)^N$ with $\mathbb{G}_K$ the set of stochastic vectors of dimension K. $\mathbb{G}_K$ is convex so Λ is also convex.

The search space $\mathbb{S}^T$

The search space $\mathbb{S}^T$ corresponds to the set of all sequences of hidden states of length T. This space is discrete, finite and of size N^T. A HMM is defined by a triple of stochastic matrices so it is not possible to directly utilize elements of $\mathbb{S}^T$. In place, the labeled training algorithm[2] can be used to transform any point in $\mathbb{S}^T$ into a point in Λ. Let be $\gamma(Q)$ the model given by the labeled training algorithm from the hidden state sequence $Q \in \mathbb{S}^T$ and the observed sequence O of symbols. The set $\gamma(\mathbb{S}^T) = \{\gamma(Q)|Q \in \mathbb{S}^T\}$ is a finite subset of Λ. It must be noted that γ is neither injective nor surjective. As a consequence, it is not possible to use the Baum–Welch algorithm with this space without losing consistency. It must be noted that $\gamma(\mathbb{S}^T)$ may not contain an optimal HMM for the criterion.

The search space Ω

The search space Ω was defined in order to furnish a vector space structure to train HMMs [10]. Let $\mathbb{G}_K$ be the set of stochastic vectors of dimension K. Let $\mathbb{G}_K^*$ be the set

---

[2] The labeled training algorithm does a statistical estimation of probabilities from the hidden state sequence and the observed sequence of symbols.

of stochastic vectors of dimension K for which none of the coordinates is null that is to say $\mathbb{G}_K^* = \{\boldsymbol{x} \in \mathbb{G}_K | \forall\ i = 1..K,\ \boldsymbol{x}_i > 0\}$. We define $r_K : \mathbb{R}^K \mapsto \mathbb{R}^K$ a regularization function on $\mathbb{R}^K$. If we denote $r_K(\boldsymbol{x})_i$ the ith coordinate of the vector $r_K(\boldsymbol{x})$, we have:

$$r_K(\boldsymbol{x})_i = \boldsymbol{x}_i - \max_{j=1..K} \boldsymbol{x}_j$$

We define the set Ω_K by $\Omega_K = r_K(\mathbb{R}^K) = \{\boldsymbol{x} \in \mathbb{R}^K | r_K(\boldsymbol{x}) = \boldsymbol{x}\}$. Let $\oplus_K : \Omega_K \times \Omega_K \mapsto \Omega_K$, $\odot_K : \mathbb{R} \times \Omega_K \mapsto \Omega_K$ and $\ominus_K : \Omega_K \times \Omega_K \mapsto \Omega_K$ be three symmetric operators such that, for all $(\boldsymbol{x}, \boldsymbol{y}) \in (\Omega_K)^2$ and $c \in \mathbb{R}$, we have:

$$x \oplus_K y = y \oplus_K x = r_K(x + y)$$
$$c \odot_K x = x \odot_K c = r_K(c \cdot x)$$
$$x \ominus_K y = y \ominus_K x = r_K(x - y) = x \oplus (-1 \odot_K y)$$

$(\Omega_K, \oplus_K, \odot_K)$ is then a vector space. We define $\psi_K : \mathbb{G}_K^* \mapsto \Omega_K$ and $\phi_K : \Omega_K \mapsto \mathbb{G}_K^*$ two operators allowing to transform any point from $\mathbb{G}_K^*$ into a point in Ω_K and reciprocally using the following equations:

$$\text{for all } \boldsymbol{x} \in \mathbb{G}_K^*,\quad \psi_K(\boldsymbol{x})_i = \ln \boldsymbol{x}_i - \max_{j=1..K} \ln \boldsymbol{x}_j\ ,$$

$$\text{for all } \boldsymbol{y} \in \Omega_K,\quad \phi_K(\boldsymbol{y})_i = \frac{\exp \boldsymbol{y}_i}{\sum_{j=1}^{K} \exp \boldsymbol{y}_j}\ .$$

Let $\Omega = \Omega_N \times (\Omega_N)^N \times (\Omega_M)^N$ and $\Lambda^* = \mathbb{G}_N^* \times (\mathbb{G}_N^*)^N \times (\mathbb{G}_M^*)^N$. By generalization of the operators $\oplus_K, \odot_K, \ominus_K, \psi_K$ and ϕ_K to the Cartesian products Ω and Λ^* (removing the index K), it can be shown that $(\Omega, \oplus, \odot)$ is a vector space and that $\psi(\Lambda^*) = \Omega$ and $\phi(\Omega) = \Lambda^*$. It is important to note that $\Lambda^* \subset \Lambda$. Figure 1 synthesizes these transformations and relationships between spaces.

3.2 Metaheuristics for Hidden Markov Models Training

Seven main kinds of generic metaheuristics have been adapted to tackle HMMs training: simulated annealing (SA) [27], tabu search (TS) [28–32], genetic algorithms (GA) [33, 34], population-based incremental learning (PBIL) [35, 36], the API algorithm [37, 38] and particle swarm optimization (PSO) [39, 40]. As can be seen, many

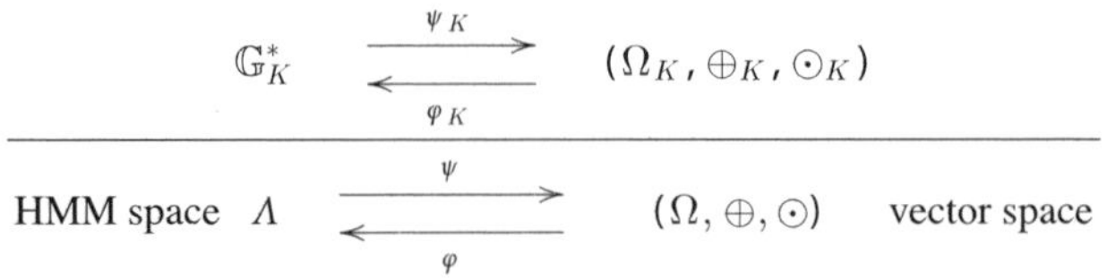

$$\mathbb{G}_K^* \quad \xrightarrow[\ \varphi_K\ ]{\ \psi_K\ } \quad (\Omega_K, \oplus_K, \odot_K)$$

$$\text{HMM space } \Lambda \quad \xrightarrow[\ \varphi\ ]{\ \psi\ } \quad (\Omega, \oplus, \odot) \quad \text{vector space}$$

Fig. 1 Transformations and relationships between the HMM space and the vector space

metaheuristics have been adapted. For some metaheuristics, many different adaptations have been tried (see Table 1). These adaptations do not all try to maximize the same criterion, but the maximum likelihood criterion is nevertheless the criterion most often used. Moreover, they do not explore the same search space, as can be seen in Table 1. When considering these adaptations, we do not know which adaptation is the most efficient. Indeed, they have not been extensively compared and no standard dataset is available to allow cross-experiments. Moreover, even if a standard dataset was available, we would be frustrated by the parametrization of algorithms, which are in many cases an empirical choice. The choices do not guarantee efficiency. In order to reduce overcome this problem, we propose in the next section, to tune and to compare six adaptations of three metaheuristics.

4 Description, Tuning and Comparison of Six Adaptations of Three Metaheuristics for HMMs Training

In the following, we consider as criterion the maximization of the likelihood. The six adaptations that we propose to study are obtained from three population-based metaheuristics: genetic algorithms, the API algorithm and particle swarm optimization.

4.1 Genetic Algorithms

The form of the genetic algorithms [33, 34] that we consider, is described in [10] and synthesized in Fig. 2. This algorithm is the result of extension of the GHOSP algorithm [44], to which an optional step of mutation and/or optimization of parents by the Baum–Welch algorithm was added. We consider two adaptations of this algorithm.

The adaptations

The first adaptation of the GA, which we will call GA-A, explores the space Λ. The chromosome associated with a HMM $\lambda = (A, B, \Pi)$ is the matrix $(\Pi\,A\,B)$ computed by concatenating the three stochastic matrices of the model. The selection operator is an elitist one. The crossover operator is a one point crossover (1X) consisting in

Table 1 Adaptation of metaheuristics to train HMMs against the search space used

Algorithm \ Search space	Λ	$\mathbb{S}^T$	$(\Omega, \oplus, \odot)$
Simulated annealing	[41]	[42]	
Tabu search	[43]		
Genetic algorithms	[44, 45]	[10]	
Population-based incremental learning	[46]		
API (artificial ants)	[37]	[10]	[47]
Particle swarm optimization	[48]		[49]

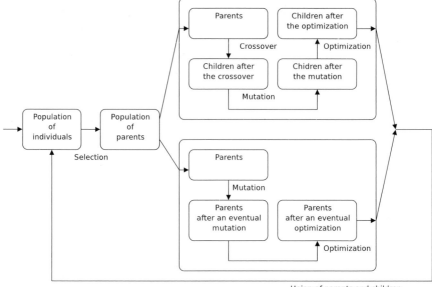

Fig. 2 Principle of the genetic algorithm for HMMs training

randomly choosing a horizontal cut point and exchanging corresponding parts (see Fig. 3). The mutation operator is applied to each coefficient of the chromosome according to the probability p_{mut}. When applied to a coefficient h, the operator replaces h by the value $(1 - \theta)h$ and adds the quantity θh to another coefficient (chosen uniformly) of the same stochastic constraint. For each mutation, the coefficient θ is uniformly chosen in $[0; 1]$. The optimization operator consists in doing $\mathcal{N}_{\text{BW}}$ iterations of the Baum–Welch algorithm on the individual.

The second adaptation, named GA-B, explores the HMMs space using $\mathbb{S}^T$. A chromosome is a sequence of hidden states labeling the observed sequence of symbols to learn. The associated model is computed using the labeled training algorithm. The

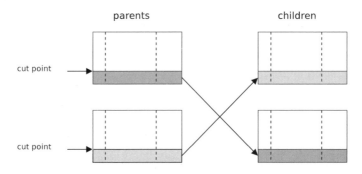

Fig. 3 Principle of the one point crossover operator for GA-A

selection operator is an elitist one. The crossover operator is the classical one point crossover operator used for binary GA. The mutation operator modifies each state of the sequence with a probability of p_{mut}. When a modification is done, the new state is uniformly chosen in the set $\mathbb{S}$. The optimization of parents or children is not considered because of the non-bijectivity of the labeled training algorithm.

Parametrization

The parameters of both algorithms are: $\mathcal{N}$ (the size of the population), p_{mut} (the mutation probability), *MutateParents* (is the mutation operator applied to the parents?), *OptimizeParents* (is the optimization operator applied to the parents?) and $\mathcal{N}_{BW}$ (the number of iterations of the Baum-Welch algorithm to use). The size of the parent population is arbitrarily fixed to half of the total population.

4.2 The API Algorithm

The API algorithm [37, 38] is a metaheuristic inspired from the foraging behavior of primitive ants: the *Pachycondyla apicalis*. When a prey is captured, the ant memorizes the hunting site of the capture and goes back to the nest. The next time the ant leaves the nest, it goes to the last successful hunting site and searches for prey. After many unsuccessful trials for a hunting site, the ant forgets the site and searches for another one. The number of trials before abandoning a site is called the local patience of an ant. Periodically, the nest of the colony decays and it is moved to a new site. These priciples are built into the API algorithm (see Algorithm 1). Experiments conducted in [10] have shown that the size of the colony and the size of the memory of each ant are statistically anti-correlated parameters. Consequently, without reduction of generality, we can suppose that the size of the memory of each ant is 1.

We consider three adaptations of the API algorithm. To present them, it is only necessary to define the initialization operator (used to define the initial nest position) and to define the exploration operator.

The Adaptations

The first adaptation [37], named API-A, searches a model in Λ. The initial position of the nest is obtained by uniformly choosing a model in the space Λ. The exploration operator depends on a parameter called the amplitude. If we denote by $\mathcal{A}$ the amplitude, then application of the operator consists in applying the function AM_A to each coefficient x of the model:

$$AM_A(x) = \begin{cases} -v & \text{if } v < 0 \\ 2 - v & \text{if } v > 1 \\ v & \text{otherwise} \end{cases} \quad \text{and} \quad v = x + \mathcal{A} \cdot (2\mathcal{U}([0,1[) - 1) \,.$$

$\mathcal{U}(X)$ represents an element uniformly chosen in the set X. Coefficients are normalized by dividing each of them by the sum of coefficients of the associated stochastic constraint. Finally, $\mathcal{N}_{BW}$ iterations of the Baum–Welch algorithm can eventually be applied to solutions.

The second adaptation [10], named API-B, explores the space $\mathbb{S}^T$. The labeled training algorithm is used as for algorithm GA-B. The initial position of the nest is defined by uniformly choosing T states in the set $\mathbb{S}$. The exploration operator for a solution $x \in \mathbb{S}^T$ and for an amplitude $A \in [0; 1]$ consists in modifying L states of the sequence x. The number of states L is computed as $L = \min\{A \cdot T \cdot \mathcal{U}([0; 1]), 1\}$. States to modify are uniformly chosen in the sequence. The modification of a state is made by generating a new one uniformly in the set $\mathbb{S}$. Solutions are not optimized because of the non-bijectivity of the labeled training algorithm.

The third adaptation [10], named API-C, explores the space Ω. The initial position of the nest is obtained by uniformly choosing a model in the space Λ^*. The exploration operator around a solution $x \in \Omega$ for an amplitude A consists in choosing a solution $y = \psi(\mathcal{U}(\Lambda^*))$ and in computing the position:

$$x \oplus \left(\frac{-\ln \mathcal{U}(]A; 1[)}{\|y\|_{\max}} \odot y \right)$$

denoting by $\| \cdot \|_{\max}$ the classical max norm ($\max_i |x_i|$).

Algorithm 1 The API algorithm

---

1. Randomly choose a position for the nest
2. Empty the memory of each ant
3. **While** all iterations are not all done **do**
4.   **For** each ant **do**
5.   **If** the ant has not already all its hunting sites **Then**
6.   The ant chooses a new hunting site
7.   **Else**
8.   **If** the last exploration is a failure **Then**
9.   Uniformly choose a hunting site in memorized hunting sites
10.   **Else**
11.   Choose the last explored hunting site
12.   **End If**
13.   Explore a new solution around the hunting site
14.   **If** the new solution is better than the hunting site **Then**
15.   Replace the hunting site by the new solution in the memory
16.   **Else**
17.   **If** there are too many failures **Then**
18.   Forget the hunting site
19.   **End If**
20.   **End If**
21.   **End If**
22.   **End For**
23.   **If** it's time to move the nest **Then**
24.   Move the nest to the best ever found solution
25.   Empty the memory of each ant
26.   **End If**
27. **End While**

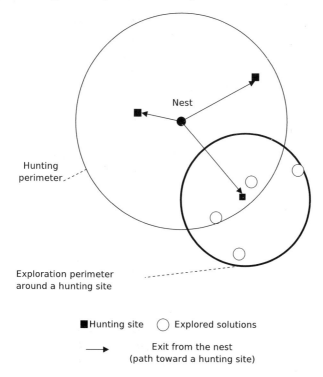

Fig. 4 Exploration principles for an ant in the API algorithm

Parametrization

The parameters of the algorithms are: $\mathcal{N}$ (the size of the colony), $\mathcal{A}^i_{\text{site}}$ (the amplitude for the exploration operator to choose a new hunting site), $\mathcal{A}^i_{\text{local}}$ (the amplitude for the exploration operator to choose a solution around a hunting site), $\mathcal{T}_{\text{Move}}$ (the number of iterations between two nest displacements), $e_{\max}$ (the local patience of a hunting site) and $\mathcal{N}_{\text{BW}}$ (the number of iterations of the Baum–Welch algorithm applied to each explored solution).

Two kinds of parameter settings of amplitudes are considered: the first one is called homogeneous and the second one is called heterogeneous. For the homogeneous parameters, the amplitudes are the same for all ants. For the heterogeneous parameters, amplitudes vary with the ant. If we suppose that ants are numbered from 1 to $\mathcal{N}$ then we have for all $i = 1 \ldots \mathcal{N}$:

$$\mathcal{A}^i_{\text{site}} = 0.01 \left(\frac{1}{0.01} \right)^{i/\mathcal{N}} \quad \text{and} \quad \mathcal{A}^i_{\text{local}} = \mathcal{A}^i_{\text{site}}/10 \,.$$

4.3 Particle Swarm Optimization

Particle swarm optimization (PSO) [39, 40] is a technique that consists in moving $\mathcal{N}$ particles in a search space. Each particle has, at time t, a position denoted $\boldsymbol{x}_i(t)$ and a velocity denoted $v_i(t)$. Let be $x_i^+(t)$ the best position ever found by the particle i until time t. Let be $V_i(t)$ the set of particles of the neighborhood of the particle i at time t and $\hat{x}_i(t)$ the best position ever seen by particles in the neighborhood $V_i(t)$ until time t. We have when maximizing the criterion f:

$$\hat{x}_i(t) = \arg \max_{x_j \in V_i(t)} f(\hat{x}_j(t-1))$$

The particular PSO that we consider is controlled by:

- three parameters ω, c_1 and c_2: ω controls the inertia of the velocity vector, c_1 the cognitive component and c_2 the social component;
- the equations:

$$x_i(t) = x_i(t-1) + x_i(t-1)$$

$$\begin{aligned} v_i(t) = {}& \omega \cdot v_i(t-1) \\ & + \left[c_1 \cdot \mathcal{U}([0,1]) \right] \cdot \left(x_i^+(t) - x_i(t) \right) \\ & + \left[c_2 \cdot \mathcal{U}([0,1]) \right] \cdot \left(\hat{x}_i(t) - x_i(t) \right) \end{aligned}$$

The Adaptation
Adaptation of particle swarm optimization to hidden Markov models training [49] is done in the search space Ω. Algorithm 2 shows the PSO Algorithm.

Parametrization
The parameters of the algorithm are $\mathcal{N}$ (the number of particles), ω (the inertia parameter), c_1 (the cognitive parameter), c_2 (the social parameter), V (the size of the neighborhood of particles) and $\mathcal{N}_{BW}$ (the number of iterations of the Baum–Welch algorithm). The neighborhood is a circular social neighborhood. When the size of the neighborhood is V, the neighborhood $V_i(t)$ of the particle i at time t is constant and equal to V_i. V_i is composed of the $V/2$ particles preceding the particle on the circle and of the $V/2$ particles succeeding the particle on the circle. Figure 5 shows the composition of the neighborhood of size 2 for the particles 1 and 5.

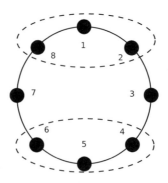

Fig. 5 Example of circular neighborhood of size 2 for the PSO

Algorithm 2 The PSO algorithm

---

1. **For** each particle i **do**
2. $x_i(0) = \mathcal{U}(\psi(\Lambda))$
3. $v_i(0) = \mathcal{U}(\psi(\Lambda))$
4. $x_i^+(0) = x_i(0)$
5. **End For**
6. **While** all iterations are not done **do**
7.   **For** each particle **do**
8. $x' = x_i(t-1) \oplus v_i(t-1)$ // move
9. $x_i(t) = \psi(BW(\phi(x')))$ // optimization
10. $v_i(t) = x_i(t) \ominus x_i(t-1)$ // computation of the effective move
11.   **If** $P(V = O|\phi(x_i(t))) > P(V = O|\phi(x_i^+(t-1)))$ **Then**
12. $x_i^+(t) = x_i(t)$
13.   **Else**
14. $x_i^+(t) = x_i^+(t-1)$
15.   **End If**
16.   **End For**
17.   **For** each particle **do**
18.   Compute the neighborhood $\mathcal{V}_i(t)$ at time t
19. $\hat{x}_i(t) = x_j(t)^+$ with $j = \arg\max_{k \in \mathcal{V}_i(t)} P(V = O|\phi(x_k^+(t)))$
20. $v_i(t) = \omega \odot v_i(t-1)$
                $\oplus [c_1 \cdot \mathcal{U}([0,1])] \odot (x_i^+(t) \ominus x_i(t))$
                $\oplus [c_2 \cdot \mathcal{U}([0,1])] \odot (\hat{x}_i(t) \ominus x_i(t))$
21.   **End For**
22. **End While**

4.4 Analysis of Metaheuristics

The above metaheuristics are very similar since they use a population of agents or solutions but differences in the HMM search spaces considered are important. Table 2 summarizes the following discussion about metaheuristics main characteristics.

Table 2 Main properties of the algorithms

	GA-HMM	API-HMM	PSO-HMM
How do agents interact?	crossover	nest move	velocity vector update
When do agents interact?	at each iteration	every $\mathcal{T}_{\text{Move}}$ iterations	at each iteration
Who interacts?	best ones from elitist selection	the best one ever found	the best one in a geo-graphical neighborhood
How to explore new solutions?	mutation	local exploration	random contribution to velocity vector
How to exploit solutions?	statistical rein-forcement due to the selection	use of nest and sites to concentrate exploitation	density of particles in geographical locations
How to forget unattractive zone?		memory limit, local patience, nest move	

Interactions Between Solutions

In GA-HMM, interactions take place with crossover between best solutions by means of genes exchanges at each iteration. Consequently, interactions directly transfer good properties from parents to children. In PSO-HMM, particle interactions are made at each iteration through the velocity vector update step. Direct properties of particles, such as their position, are not transmitted but directions toward good solutions of the neighborhood are spread between particles. in contrast, interactions in API-HMM are made only every $\mathcal{T}_{\text{Move}}$ iterations when the nest is moved. Interactions at each iteration, as for GA-HMM and PSO-HMM, can make all particles/solutions/agents move similarly in the search space and consequently reduce diversity. Moreover, if many interesting search locations are found by particles, it is possible that search effort is spread over all these locations, and because solutions can be situated at opposite sides of the search space, particles would oscillate between these locations. In some cases this distribution of effort may be useful to guarantee convergence towards a near optimum. On the contrary, rare update, as in API-HMM, can lead to a poorly guided search. However, the small number of interactions can reduce the time spent taking decisions between many sites: a decision is made between many interesting sites when the nest moves. This can be bad for some problems because too strict decisions are made too rapidly. In both cases, the point in time at which interactions take place can be a pro and a con for each algorithm. Sometimes, it is important to hesitate in exploring many sites in parallel as for GA-HMM and PSO-HMM; and sometimes hesitation reduces the effectiveness of the search.

Exploration of Solutions

Exploration of new solutions is made in the three algorithms by a guided random search followed in some cases by a local optimization heuristic. For GA-HMM, the exploration consists in mutation. For API-HMM a solution is generated in the neighborhood of the particle. In a certain manner, local exploration in API-HMM plays a similar role to the mutation operator in GA-HMM. In PSO-HMM, exploration is obtained from velocity update where two components contribute to the new speed vector with random coefficients. The local optimization consists in the application of $\mathcal{N}_{\text{BW}}$ iterations of the Baum–Welch algorithm.

Exploitation of Solutions

With GA-HMM, the exploitation of solutions is made by statistical reinforcement. This reinforcement is due to the elitist selection and each individual represents a sampling of the best solution locations. With PSO-HMM, the exploitation is made by the geographical density of particles present in interesting zones. With API-HMM, the exploitation is made through two mechanisms: the nest position and foraging sites. Each mechanism can be viewed as hierarchical reinforcement. The nest determines the focal point around which ants perform searches. One ant's site determines the search area for this ant. The API-HMM has a particular capability that both the other algorithms do not have: it can forget uninteresting search locations through two mechanisms: patience on sites and nest moves. This can be very useful when the search is not profitable in order to rapidly search other locations or concentrate around a better one.

4.5 Tuning of Algorithms

To experiment with the six adaptations for HMMS training, we need to proceed in two steps. We begin by determining robust parameter settings on a reduced set of data, and we follow this by evaluating and comparing the performance of various adaptations on a bigger data set.

Estimated Probability Distribution to Compare Algorithm Parameterizations

Let $f_{A,X}$ be the probability distribution of the random variable measuring the performance of the algorithm A for the parameter configuration X. Let $\Delta_{v=x} = (*, \ldots, *, v = x, *, \ldots, *)$ be the set of parameter configurations for which the parameter v has the value x. We can define the probability distribution $f_{A,\Delta_{v=x}}$ by:

$$f_{A,\Delta_{v=x}} = \frac{1}{|\Delta_{v=x}|} \sum_{X \in \Delta_{v=x}} f_{A,X}$$

A parameter configuration $X = (x_1, \ldots, x_K)$ for a stochastic algorithm A will be said to be robust if for all values x_i, the probability distribution $f_{A,\Delta_{v_i=x_i}}$ has a high mean and a low standard deviation.

Let $EG = \{(e_1, g_1), \ldots, (e_L, g_L)\}$ be a set of realizations such that e_i is a parameter configuration of $\Delta_{v=x}$ and g_i is the associated measured performance (i.e. the result of one run of A). The probability distribution $f_{A,\Delta_{v=x}}$ can then be approximated by the probability distribution:

$$\frac{1}{|EG|} \sum_{(e,g) \in EG} \mathcal{N}(g,\sigma)$$

with $\mathcal{N}(m,s)$ the normal law of mean m and of standard deviation s. The standard deviation is fixed to:

$$\sigma = 0.1 \cdot \left(\max_{(e,g) \in EG} g - \min_{(e,g) \in EG} g \right)$$

in order to smooth the probability distribution and to reduce the need to consider a huge EG set. Figure 6 presents the five probability distributions obtained for five different mutation rates for the algorithm GA-B on the first image.

Protocol

In order to determine robust parameter configurations, it is sufficient to compare probability distributions of performances. To do a fair comparison of algorithms, we consider parameter configurations offering a similar chance to each algorithm. Without the Baum–Welch algorithm, algorithms evaluate approximately 30000 HMMs

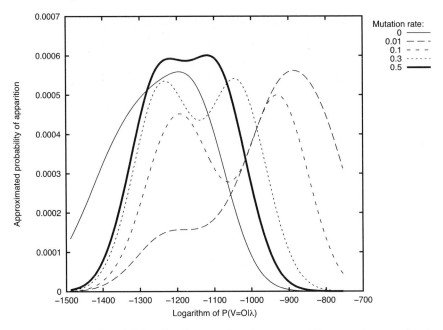

Fig. 6 Approximated probability distribution of performances of five mutation rates for the algorithm GA-B

(i.e. 30000 runs of the Forward algorithm are made). With the Baum–Welch algorithm, about 1000 iterations of the Baum–Welch algorithm are made (2 or 5 iterations are made for each explored solution). Observation sequences are computed from images of the ORL faces database [50]. The four faces of Fig. 8 have been used. These images are coded using 256 grey levels. We recoded them using 32 levels of grey and we linearized them using principles described in Fig. 7. Many parameter configurations

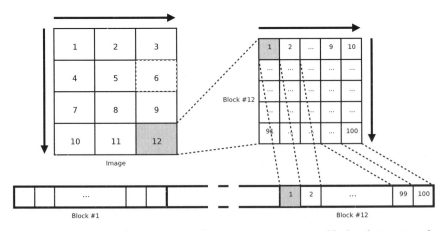

Fig. 7 Linearization of an image into an observation sequence using blocks of 10 × 10 pixels

The use of metaheuristics for HMMs training has been reported in other works and has produced promising results. In the third part of this chapter, we described a comparative study of three population-based bio-inspired metaheuristics for HMMs training. The first metaheuristic is a genetic algorithm that was originally designed for discrete search spaces. The second is the API algorithm, designed to tackle any kind of search spaces that possesses a neighborhood structure. The third metaheuristic is particle swarm optimization, which was originally designed to search in vector spaces. The six adaptations resulting from the three metaheuristics and the three search spaces allow exploitation of the various properties of the three search spaces. We concentrate our comparison on a critical aspect: algorithm tuning. To obtain an objective answer to this issue, we tuned a genetic algorithm, an ant-based method and a particle swarm-based approach. Of course, we need to keep in mind that the conclusions depend on the particular application domain that we have chosen: training HMMs using images, and using the criterion maximization of the likelihood.

Results from our experiments lead to the following conclusions. The search space $\mathbb{S}^T$ is quite useless for HMMs training. Hybridization with the Baum–Welch algorithm improves high performance models. However, when we consider criteria than likelihood maximization that do not have a local optimizer or a too expensive local optimizer, metaheuristics might be useful. Experiments show that the algorithms furnishing the best models are not necessarily those converging the most rapidly. Moreover, it has been shown that the search space Λ is more propitious for fast convergence than other spaces.

Conclusions drawn from experiment results suggest a number of further investigations. Are the results generalizable to other criteria, to other application domains, to other kinds of HMMs and even to other metaheuristics? Even if some properties seem to be quite generalizable, inevitably, they are not. Studies need to confirm them and this study may be considered as a model for future comparative studies.

References

1. O. Cappé, Ten years of hmms, http://www.tsi.enst.fr/ cappe/docs/hmmbib.html (March 2001)
2. A. A. Markov, An example of statistical investigation in the text of "Eugene onyegin" illustrating coupling of "tests" in chains, in: Proceedings of Academic Scienctific St. Petersburg, VI, 1913:153–162
3. L. E. Baum, J. A. Eagon, An inequality with applications to statistical estimation for probabilistic functions of markov processes to a model for ecology, Bull American Mathematical Society 73 (1967):360–363
4. A. J. Viterbi, Error bounds for convolutionnal codes and asymptotically optimum decoding algorithm, IEEE Transactions on Information Theory 13 (1967):260–269
5. L. E. Baum, An inequality and associated maximisation technique in statistical estimation for probabilistic functions of markov processes, Inequalities 3 (1972):1–8
6. G. D. Forney Jr., The Viterbi algorithm, in: Proceedings of IEEE, Vol. 61, 1973:268–278
7. T. Brouard, M. Slimane, J. P. Asselin de Beauville, Modélisation des processus simultanés indépendants par chaînes de Markov cachées multidimensionnelles (CMC-MD/I), Tech.

Rep. 200, Laboratoire d'Informatique, Université François-Rabelais de Tours (December 1997)

8. S. Fine, Y. Singer, N. Tishby, The hierarchical hidden markov model: Analysis and applications, Machine Learning 32 (1) (1998):41–62 citeseer.ist.psu.edu/fine98hierarchical.html

9. L. R. Rabiner, A tutorial on hidden Markov models and selected applications in speech recognition, Proceedings of the IEEE 77 (2) (1989):257–286

10. S. Aupetit, Contributions aux modèles de Markov cachés: métaheuristiques d'apprentissage, nouveaux modèles et visualisation de dissimilarité, Thèse de doctorat, Laboratoire d'Informatique de l'Université François-Rabelais de Tours, Tours, France (30 November 2005)

11. H. Bourland, C. Wellekens, Links between Markov models and multiplayer perceptrons, IEEE Transactions on Pattern Analysis and Machine Inteligence 12 (10) (1990):1–4

12. A. Berchtold, The double chain Markov model, Communications in Statistics: Theory and Methods 28 (11) (1999):2569–2589.

13. S. R. Eddy, Profile hidden Markov models, Bioinformatics 14 (9) (1998):755–63

14. Y. Bengio, Markovian models for sequential data, Neural Computing Surveys 2 (1999):129–162

15. W. Pieczynski, Arbres de Markov Triplet et fusion de Dempster-Shafer, Comptes Rendus de l'Académie des Sciences – Mathématique 336 (10) (2003):869–872

16. S. Kapadia, Discriminative training of hidden Markov models, Ph.D. Thesis, Downing College, University of Cambridge (18 March 1998)

17. A. Ganapathiraju, Discriminative techniques in hidden Markov models, Course paper (1999) http://www.isip.msstate.edu/publications/courses/ece_7000_speech/lectures/1999/lecture_11/

18. M. Berthold, D. J. Hand (Eds), Intelligent Data Analysis: An Introduction, Springer-Verlag, 1998

19. R. Schluter, W. Macherey, S. Kanthak, H. Ney, L. Welling, Comparison of optimization methods for discriminative training criteria, in: EUROSPEECH '97, 5th European Conference on Speech Communication and Technology, Rhodes, Greece, 1997:15–18

20. M. Giurgiu, Maximization of mutual information for training hidden markov models in speech recognition, in: 3rd COST #276 Workshop, Budapest, Hungary, 2002:96–101

21. K. Vertanen, An overview of discriminative training for speech recognition, Tech. Rep., University of Cambridge (2004)

22. A. Ljolje, Y. Ephraim, L. R. Rabiner, Estimation of hidden Markov model parameters by minimizing empirical error rate, in: IEEE International Conference on Acoustic, Speech, Signal Processing, Albuquerque, 1990. 709–712 http://ece.gmu.edu/~yephraim/Papers/icassp 1990 mcehmm.pdf

23. L. Saul, M. Rahim, Maximum likelihood and minimum classification error factor analysis for automatic speech recognition, IEEE Transactions on Speech and Audio Processing 8 (2) (2000):115–125

24. B.-H. Juang, L. R. Rabiner, The segmental k-means algorithm for estimating parameters of hidden Markov models, IEEE Transactions on Acoustics, Speech and Signal Processing 38 (9) (1990):1639–1641

25. R. Dugad, U. B. Desai, A tutorial on hidden Markov models, Tech. Rep. SPANN-96.1, Indian Institute of Technology, Bombay, India (May 1996)

26. J. Dreo, A. Petrowski, P. Siarry, E. Taillard, Métaheuristiques pour l'optimisation difficile, Eyrolles, Paris, 2003

27. S. Kirkpatrick, C. D. Gelatt, M. P. Vecchi, Optimizing by simulated annealing, Science 220 (4598) (1983):671–680

28. F. Glover, Future paths for integer programming and links to artificial intelligence, Computers and Operations Research 13 (1986):533–549
29. F. Glover, Tabu search – part I, ORSA Journal on Computing 1 (3) (1989):190–206
30. F. Glover, Tabu search – part II, ORSA Journal on Computing 2 (1) (1989):4–32
31. A. Hertz, E. Taillard, D. de Werra, A Tutorial on tabu search, in: Proceedings of Giornate di Lavoro AIRO'95 (Enterprise Systems: Management of Technological and Organizational Changes, 1992:13–24
32. F. Glover, M. Laguna, Tabu Search, Kluwer Academic Publishers, 1997
33. J. H. Holland, Adaptation in Natural and Artificial Systems, University of Michigan Press: Ann Arbor, MI, 1975
34. D. E. Goldberg, Genetic Algorithms in Search, Optimization and Machine Learning, Addison-Wesley, 1989
35. S. Bulaja, Population-based incremental learning: a method for integrating genetic search based function optimization and competitive learning, Tech. Rep. CMU-CS-94-163, Carnegie Mellon University (1994)
36. S. Bulaja, R. Caruana, Removing the genetics from the standard genetic algorithm, in: A. Prieditis, S. Russel (Eds.), The International Conference on Machine Learning (ML'95), Morgan Kaufman Publishers, San Mateo, CA, 1995:38–46
37. N. Monmarché, Algorithmes de fourmis artificielles: applications à la classification et à l'optimisation, Thèse de doctorat, Laboratoire d'Informatique de l'Université François-Rabelais de Tours (20 December 2000)
38. N. Monmarché, G. Venturini, M. Slimane, On how *Pachycondyla apicalis* ants suggest a new search algorithm, Future Generation Computer Systems 16 (8) (2000):937–946
39. J. Kennedy, R. Eberhart, Particle swarm optimization, in: Proceedings of the IEEE International Joint Conference on Neural Networks, Vol. 4, IEEE, 1995:1942–1948
40. M. Clerc, L'optimisation par essaims particulaires : versions paramétriques et adaptatives, Hermes Science - Lavoisier, Paris, 2005
41. D. B. Paul, Training of HMM recognizers by simulated annealing, in: Proceedings of IEEE International Conference on Acoustics, Speech and Signal Processing, 1985:13–16
42. Y. Hamam, T. Al Ani, Simulated annealing approach for hidden Markov models, in: 4th WG-7.6 Working Conference on Optimization-Based Computer-Aided Modeling and Design, ESIEE, France, 1996
43. T.-Y. Chen, X.-D. Mei, J.-S. Pan, S.-H. Sun, Optimization of hmm by the tabu search algorithm., Journal Information Science and Engineering 20 (5) (2004):949–957
44. M. Slimane, T. Brouard, G. Venturini, J.-P. Asselin de Beauville, Apprentissage non-supervisé d'images par hybridation génétique d'une chaîne de Markov cachée, Traitement du signal 16 (6) (1999):461–475
45. R. Thomsen, Evolving the topology of hidden Markov models using evolutionary algorithms, in: Proceedings of Parallel Problem Solving from Nature VII (PPSN-2002), 2002:861–870
46. B. Maxwell, S. Anderson, Training hidden markov models using population-based learning, in: W. Banzhaf, J. Daida, A. E. Eiben, M. H. Garzon, V. Honavar, M. Jakiela, R. E. Smith (Eds.), Proceedings of the Genetic and Evolutionary Computation Conference (GECCO'99), Vol. 1, Morgan Kaufmann, Orlando, Florida, USA, 1999, p. 944. http://citeseer.ist.psu.edu/maxwell99training.html
47. S. Aupetit, N. Monmarché, M. Slimane, S. Liardet, An exponential representation in the API algorithm for hidden Markov models training, in: Proceedings of the 7th International Conference on Artificial Evolution (EA'05), Lille, France, 2005, cD-Rom

48. T. K. Rasmussen, T. Krink, Improved hidden Markov model training for multiple sequence alignment by a particle swarm optimization – evolutionary algorithm hybrid, BioSystems 72 (2003):5–17

49. S. Aupetit, N. Monmarché, M. Slimane, Apprentissage de modèles de Markov cachés par essaim particulaire, in: J.-C. Billaut, C. Esswein (Eds.), ROADEF'05 : 6ème congrès de la Société Française de Recherche Opérationnelle et d'Aide à la Décision, Vol. 1, Presses Universitaire François-Rabelais, Tours, France, 2005:375–391

50. F. Samaria, A. Harter, Parameterisation of a stochastic model for human face identification, in: IEEE workshop on Applications of Computer Vision, Florida, 1994

51. T.C. Design, Free Background Textures, Flowers, http://www.tcdesign.net/ free_textures_flowers.htm , accessed junuary 2006 (January 2006)

52. Textures Unlimited, Black & white textures, http://www.geocities.com/texturesunlimited/ blackwhite.html, accessed junuary 2006 (January 2006)

53. S. Agarwal, A. Awan, D. Roth, Learning to detect objects in images via a sparse, part-based representation, IEEE Transactions on Pattern Analysis and Machine Intelligence 26 (11) (2004):1475–1490

Inequalities and Target Objectives for Metaheuristic Search – Part I: Mixed Binary Optimization

Fred Glover

University of Colorado, Boulder, CO 80309-0419, USA.
Fred.Glover@Colorado.EDU

Abstract

Recent adaptive memory and evolutionary metaheuristics for mixed integer programming have included proposals for introducing inequalities and target objectives to guide the search. These guidance approaches are useful in intensification and diversification strategies related to fixing subsets of variables at particular values, and in strategies that use linear programming to generate trial solutions whose variables are induced to receive integer values. We show how to improve such approaches by new inequalities that dominate those previously proposed and by associated target objectives that underlie the creation of both inequalities and trial solutions.

We also propose supplementary linear programming models that exploit the new inequalities for intensification and diversification, and introduce additional inequalities from sets of elite solutions that enlarge the scope of these models. Part I (the present chapter) focuses on 0–1 mixed integer programming, and Part II covers the extension to more general mixed integer programming problems. Our methods can also be used for problems that lack convenient mixed integer programming formulations, by generating associated linear programs that encode part of the solution space in mixed binary or general integer variables

Key words: Zero–one Mixed Integer Programming, Adaptive Search, Valid Inequalities, Parametric Tabu Search

1 Notation and Problem Formulation

We represent the mixed integer programming problem in the form

$$\text{(MIP) Minimize } x_0 = fx + gy$$
$$\text{subject to}$$
$$(x, y) \in Z = \{(x, y) : Ax + Dy \geq b\}$$
$$x \text{ integer}$$

We assume that $Ax + Dy \geq b$ includes the inequalities $U_j \geq x_j \geq 0, j \in N = \{1, \ldots, n\}$, where some components of U_j may be infinite. The linear programming relaxation of

(MIP) that results by dropping the integer requirement on x is denoted by (LP). We further assume $Ax + Dy \geq b$ includes an objective function constraint $x_0 \leq U_0$, where the bound U_0 is manipulated as part of a search strategy for solving (MIP), subject to maintaining $U_0 < x_0^*$, where x_0^* is the x_0 value for the currently best known solution x^* to (MIP).

The current chapter focuses on the zero–one version of (MIP) denoted by (MIP:0–1), in which $U_j = 1$ for all $j \in N$. We refer to the LP relaxation of (MIP:0–1) likewise as (LP), since the identity of (LP) will be clear from the context.

Several recent papers have appeared that evidence a sudden rekindling of interest in metaheuristic methods for pure and mixed integer programming problems, and especially problems in zero–one variables. The issue of identifying feasible integer solutions is addressed in Fischetti, Glover and Lodi [4] and Patel and Chinneck [18], and the challenge of solving Boolean optimization problems, which embrace a broad range of classical zero–one problems, is addressed in Davoine, Hammer and Vizvári [3], and Hvattum, Løkketangen and Glover [15]. Metaheuristics for general zero–one problems are examined in Pedroso [19] and in Nediak and Eckstein [16]. The current chapter focuses on metaheuristic approaches from a perspective that complements (and contrasts with) the one introduced in Glover [9].

In the following we make reference to two types of search strategies: those that fix subsets of variables to particular values within approaches for exploiting strongly determined and consistent variables, and those that make use of solution targeting procedures. As developed here, the latter solve a linear programming problem LP(x', c')[1] that includes the constraints of (LP) (and additional bounding constraints in the general (MIP) case) while replacing the objective function x_0 by a linear function $v_0 = c'x$. The vector x' is called a *target solution*, and the vector c' consists of integer coefficients c_j' that seek to induce assignments $x_j = x_j'$ for different variables with varying degrees of emphasis.

We adopt the convention that each instance of LP(x', c') implicitly includes the (LP) objective of minimizing the function $x_0 = fx + gy$ as a secondary objective, dominated by the objective of minimizing $v_0 = c'x$, so that the true objective function consists of minimizing $\omega_0 = Mv_0 + x_0$, where M is a large positive number. As an alternative to working with ω_0 in the form specified, it can be advantageous to solve LP(x', c') in two stages. The first stage minimizes $v_0 = c'x$ to yield an optimal solution $x = x''$ (with objective function value $v_0'' = c'x''$), and the second stage enforces $v_0 = v_0''$ to solve the residual problem of minimizing $x_0 = fx + gy$.[2]

---

[1] The vector c' depends on x'. As will be seen, we define several different linear programs that are treated as described here in reference to the problem LP(x', c').

[2] An effective way to enforce $v_0 = v_0''$ is to fix all non-basic variables having non-zero reduced costs to compel these variables to receive their optimal first stage values throughout the second stage. This can be implemented by masking the columns for these variables in the optimal first stage basis, and then to continue the second stage from this starting basis while ignoring the masked variables and their columns. (The masked non-basic variables may incorporate components of both x and y, and will generally include slack variables for some of the inequalities embodied in $Ax + Dy \geq b$.) The resulting residual problem for the second

A second convention involves an interpretation of the problem constraints. Selected instances of inequalities generated by approaches of the following sections will be understood to be included among the constraints $Ax + Dy \geq b$ of (LP). In our definition of LP(x', c') and other linear programs related to (LP), we take the liberty of representing the currently updated form of the constraints $Ax + Dy \geq b$ by the compact representation $x \in X = \{x : (x, y) \in Z\}$, recognizing that this involves a slight distortion in view of the fact that we implicitly minimize a function of y as well as x in these linear programs.[3]

To launch our investigation of the problem (MIP:0–1) we first review previous ideas for generating guiding inequalities for this problem in Sect. 2 and associated target objective strategies Sect. 3. We then present new inequalities in Sect. 4 that improve on those previously proposed. Section 5 describes models that can take advantage of these new inequalities to achieve intensification and diversification of the search process. The fundamental issue of creating the target objectives that can be used to generate the new inequalities and that lead to trial solutions for (MIP: 0–1) is addressed in Sect. 6. Section 7 shows how to generate additional inequalities by "mining" reference sets of elite solutions to extract characteristics these solutions exhibit in common. Supplemental strategic considerations are identified in Sect. 8 and concluding remarks are given in Sect. 9.

2 Inequalities and Sub-optimization for Guiding Intensification and Diversification Phases for (MIP:0–1)

Let x' denote an arbitrary binary solution, and define the two associated index sets $N'(0) = \{j \in N : x'_j = 0\}$ and $N'(1) = \{j \in N : x'_j = 1\}$. Then it is evident that the inequality

$$\sum_{j \in N'(0)} x_j + \sum_{j \in N'(1)} (1 - x_j) \geq 1 \tag{1}$$

or equivalently

$$\sum_{j \in N'(0)} x_j - \sum_{j \in N'(1)} x_j \geq 1 - |N'(1)| \tag{2}$$

eliminates the assignment $x = x'$ as a feasible solution, but admits all other binary x vectors. The inequality (2) has been used, for example, to produce 0–1 "short hot starts" for branch and bound by Spielberg and Guignard [22] and Guignard and Spielberg [13].

---

stage can be significantly smaller than the first stage problem, allowing the problem for the second stage to be solved very efficiently.

[3] In some problem settings, the inclusion of the secondary objective x_0 in $v_{00} = Mv_0 + x_0$ is unimportant, and in these cases our notation is accurate in referring to the explicit minimization of $v_0 = c'x$.

Remark 1. Let x denote an arbitrary binary solution, and define the norm L1 of x as

$$\|x\| = ex, \quad \text{where} \quad e = (1,\ldots,1)$$

Note that the Hamming distance from the binary vectors x and x' can be expressed as

$$d(x,x') = \|x - x'\| = (e - x')x + (e - x)x'.$$

Hence the constraint (2) can be written in the following form:

$$d(x,x') = (e - x')x + (e - x)x' \geq 1.$$

Remark 2. The constraint (2) is called *canonical cut* on the unit hypercube by Balas and Jeroslow [1]. The constraint (2) has been used also by Soyster, Lev and Slivka [21], Hanafi and Wilbaut [14] and Wilbaut and Hanafi [24].

To simplify the notation, we find it convenient to give (2) an alternative representation. Let e' denote the vector given by

$$e'_j = 1 - 2x'_j, \quad j \in N$$

or equivalently

$$e' = 1 - 2x',$$

hence

$$e'_j = 1 \quad \text{if} \quad x'_j = 0 \quad \text{and} \quad e'_j = -1 \quad \text{if} \quad x'_j = 1.$$

Then, letting $n'(1) = |N'(1)|$, we can also write (2) in the form

$$e'x \geq 1 - n'(1). \tag{3}$$

More generally, for any positive integer e'_0 satisfying $n \geq e'_0 \geq 1$, the binary vectors x that lie at least a Hamming distance e'_0 from x' are precisely those that satisfy the inequality

$$e'x \geq e'_0 - n'(1). \tag{4}$$

The inequality (4) has been introduced within the context of adaptive memory search strategies (Glover [6] to compel new solutions x to be separated from a given solution x' by a desired distance. In particular, upon identifying a reference set $R = \{x^r, r \in R\}$, which consists of elite and diverse solutions generated during prior search, the approach consists of launching a diversification strategy that requires new solutions x to satisfy the associated set of inequalities

$$e^r x \geq e^r_0 - n^r(1), \quad r \in R. \tag{5}$$

This system also gives a mechanism for implementing a proposal of Shylo [20][4] to separate new binary solutions by a minimum specified Hamming distance from a set of solutions previously encountered.

---

[4] See also Pardalos and Shylo [17] and Ursulenko [23].

The inequalities of (5) constitute a form of *model embedded memory* for adaptive memory search methods where they are introduced for two purposes: (a) to generate new starting solutions and (b) to restrict a search process to visiting solutions that remain at specified distances from previous solutions. A diversification phase that employs the strategy (b) operates by eventually reducing the e_0^r values to 1, in order to transition from diversification to intensification. One approach for doing this is to use tabu penalties to discourage moves that lead to solutions violating (5). We discuss another approach in the next section.

A more limiting variant of (5) arises in the context of exploiting strongly determined and consistent variables, and in associated adaptive memory *projection* strategies that iteratively select various subsets of variable to hold fixed at specific values, or to be constrained to lie within specific bounds (Glover [6]). This variant occurs by identifying sub-vectors $x^{r_1}, x^{r_2}, \ldots,$ of the solutions x^r (thus giving rise to associated sub-vectors $e^{r_1}, e^{r_2}, \ldots,$ of e^r) to produce the inequalities

$$e^{r_h} \geq e_0^{r_h} - n^{r_h}(1), \quad r \in R, \quad h = 1, 2, \ldots \tag{6}$$

The inequalities of (6) are evidently more restrictive than those of (5), if the values $e_0^{r_h}$ are chosen to have the same size as the values e_0^r (i.e., if $e_0^{r_h} \geq e_0^r$ for each r and h).

The inequalities (6) find application within two main contexts. The first occurs within a diversification segment of alternating intensification and diversification phases, where each intensification phase holds certain variables fixed and the ensuing diversification divides each x^r into two sub-vectors x^{r_1} and x^{r_2} that respectively contain the components of x^r held fixed and the components permitted to be free during the preceding intensification phase.

The second area of application occurs in conjunction with frequency memory by choosing three sub-vectors x^{r_1}, x^{r_2} and x^{r_3} (for example) to consist of components of solution x^r that have received particular values with high, middle and low frequencies, relative to a specified set of previously visited solutions. (The same frequency vector, and hence the same way of sub-dividing the x^r vectors, may be relevant for all x^r solutions generated during a given phase of search.)[5] Our following ideas can be implemented to enhance these adaptive memory projection strategies as well as the other strategies previously described.

3 Exploiting Inequalities in Target Solution Strategies

We begin by returning to the simple inequality (3) given by

$$e'x \geq 1 - n'(1)$$

and show how to exploit it in a somewhat different manner. The resulting framework also makes it possible to exploit the inequalities of (5) and (6) more effectively.

---

[5] The formulas of Glover [6] apply more generally to arbitrary integer solution vectors.

We make use of solutions such as x' by assigning them the role of *target solutions*. In this approach, instead of imposing the inequality (3) we adopt the strategy of first seeing how close we can get to satisfying $x = x'$ by solving the LP problem[6]

$$LP(x') : \min_{x \in X} u_0 = e'x$$

where as earlier, $X = \{x : (x, y) \in Z\}$. We call x' the target solution for this problem. Let x'' denote an optimal solution to $LP(x')$, and let u_0'' denote the corresponding value of u_0, i.e., $u_0'' = e'x''$. If the target solution x' is feasible for $LP(x')$ then it is also uniquely optimal for $LP(x')$ and hence $x'' = x'$, yielding $u_0'' = -n'(1)$. In such a case, upon testing x' for feasibility in (MIP:0–1) we can impose the inequality (3) as indicated earlier in order to avoid examining the solution again. However, in the case where x' is not feasible for $LP(x')$, an optimal solution x'' will yield $u_0'' > -n'(1)$ and we may impose the valid inequality[7]

$$e'x \geq \lceil u_0'' \rceil. \tag{7}$$

The fact that $u_0'' > -n'(1)$ discloses that (7) is at least as strong as (3). In addition, if the solution x'' is a binary vector that differs from x', we can also test x'' for feasibility in (MIP:0–1) and then redefine $x' = x''$, to additionally append the constraint (3) for this new x'. Consequently, regardless of whether x'' is binary, we eliminate x'' from the collection of feasible solutions as well as obtaining an inequality (7) that dominates the original inequality (3).

Upon generating the inequality (7) (and an associated new form of (3) if x'' is binary), we continue to follow the policy of incorporating newly generated inequalities among the constraints defining X, and hence those defining Z of (MIP:0–1). Consequently, we assure that X excludes both the original x' and the solution x''. This allows the problem $LP(x')$ to be re-solved, either for x' as initially defined or for a new target vector (which can also be x'' if the latter is binary), to obtain another solution x'' and a new (7).

It is worthwhile to use simple forms of tabu search memory based on recency and frequency in such processes to decide when to drop previously introduced inequalities, in order to prevent the collection of constraints from becoming unduly large. Such approaches can be organized in a natural fashion to encourage the removal of older constraints and to discourage the removal of constraints that more recently or frequently have been binding in the solutions to the $LP(x')$ problems produced (Glover and Laguna [11]). Older constraints can also be replaced by one or several surrogate constraints.

The strategy for generating a succession of target vectors x' plays a critical role in exploiting such a process. The feasibility pump approach of Fischetti, Glover and Lodi [4] applies a randomized variant of nearest neighbor rounding to each non-binary solution x'' to generate the next x', but does not make use of associated in-

---

[6] This strategy is utilized in the parametric branch and bound approach of Glover [5] and in the feasibility pump approach of Fischetti, Glover and Lodi [4].

[7] For any real number z, $\lceil z \rceil$ and $\lfloor z \rfloor$ respectively identify the least integer $\geq z$ and the greatest integer $\leq z$.

equalities such as (3) and (7). In subsequent sections we show how to identify more effective inequalities and associated target objectives to help drive such processes.

3.1 Generalization to Include Partial Vectors and More General Target Objectives

We extend the preceding ideas in two ways, drawing on ideas of parametric branch and bound and parametric tabu search (Glover [5, 7]). First we consider *partial x vectors* that may not have all components x_j determined, in the sense of being fixed by assignment or by the imposition of bounds. Such vectors are relevant in approaches where some variables are compelled or induced to receive particular values, while others remain free or are subject to imposed bounds that are not binding.

Relative to a given vector x' that may contain both assigned and unassigned (free) components, define $N'(0) = \{j \in N : x'_j = 0\}$, $N'(1) = \{j \in N : x'_j = 1\}$ and $N'(\Phi) = \{j \in N : x'_j = \Phi\}$, where $x'_j = \Phi$ signifies that x'_j is not assigned a value (i.e., is not subject to a binding constraint or target affecting its value). Accompanying the vector x' an associated target objective $c'x$ where c' is an integer vector satisfying the condition

$$
\begin{aligned}
c'_j > 0 \quad & \text{if } j \in N'(0), \\
c'_j < 0 \quad & \text{if } j \in N'(1), \\
c'_j = 0 \quad & \text{if } j \in N'(\Phi).
\end{aligned}
$$

The vector e', given by $e'_j = 1$ for $j \in N'(0)$ and $e'_j = -1$ for $j \in N'(1)$, evidently constitutes a special case. We couple the target solution x' with the associated vector c' to yield the problem

$$
LP(x', c') : \min_{x \in X} v_0 = c'x.
$$

An optimal solution to $LP(x', c')$, as a generalization of $LP(x')$, will likewise be denoted by x'', and we denote the corresponding optimum v_0 value by v''_0 $(= c'x'')$. Finally, we define $c'_0 = \lceil v''_0 \rceil$ to obtain the inequality

$$
c'x \geq c'_0. \tag{8}
$$

By an analysis similar to the derivation of (7), we observe that (8) is a valid inequality, i.e., it is satisfied by all binary vectors that are feasible for (MIP:0–1) (and more specifically by all such vectors that are feasible for $LP(x', c')$), with the exception of those ruled out by previous examination. We address the crucial issue of how to generate the target objectives and associated target solutions x' to produce such inequalities that aid in guiding the search after first showing how to strengthen the inequalities of (8).

4 Stronger Inequalities and Additional Valid Inequalities from Basic Feasible LP Solutions

Our approach to generate inequalities that dominate those of (8) is also able to produce additional valid inequalities from related basic feasible solution to the LP problem $LP(x', c')$, expanding the range of solution strategies for exploiting the use of target solutions. We refer specifically to the class of basic feasible solutions that may be called *y-optimal* solutions, which are dual feasible in the continuous variables y (including in y any continuous slack variables that may be added to the formulation), disregarding dual feasibility relative to the x variables. Such y-optimal solutions can easily be generated in the vicinity of an optimal LP solution by pivoting to bring one or more non-basic x variables into the basis, and then applying a restricted version of the primal simplex method that re-optimizes (if necessary) to establish dual feasibility relative only to the continuous variables, ignoring pivots that would bring x variables into the basis. By this means, instead of generating a single valid inequality from a given LP formulation such as $LP(x', c')$, we can generate a collection of such inequalities from a series of basic feasible y-optimal solutions produced by a series of pivots to visit some number of such solutions in the vicinity of an optimal solution.

As a foundation for these results, we assume x'' (or more precisely, (x'', y'')) has been obtained as a y-optimal basic feasible solution to $LP(x', c')$ by the bounded variable simplex method (Dantzig [2]). By reference to the linear programming basis that produces x'', which we will call the x'' basis, define $B = \{j \in N : x_j \text{ is basic}\}$ and $NB = \{j \in N : x_j \text{ is non-basic}\}$. We subdivide NB to identify the two subsets $NB(0) = \{j \in NB : x_j'' = 0\}$, $NB(1) = \{j \in NB : x_j'' = 1\}$. These sets have no necessary relation to the sets $N'(0)$ and $N'(1)$, though in the case where x'' is an optimal basic solution[8] to $LP(x', c')$, we would normally expect from the definition of c' in relation to the target vector x' that there would be some overlap between $NB(0)$ and $N'(0)$ and similarly between $NB(1)$ and $N'(1)$.

The new inequality that dominates (8) results by taking account of the reduced costs derived from the x'' basis. Letting rc_j denote the reduced cost for the variable x_j, the rc_j values for the basic variables satisfy

$$rc_j \quad \text{for} \quad j \in B$$

and the rc_j values for the non-basic variables assure optimality for x'' under the condition that they satisfy

$$rc_j \geq 0 \quad \text{for} \quad j \in NB(0)$$
$$rc_j \leq 0 \quad \text{for} \quad j \in NB(1).$$

Associated with $NB(0)$ and $NB(1)$, define

$$\Delta_j(0) = \lfloor rc_j \rfloor \quad \text{for} \quad j \in NB(0)$$
$$\Delta_j(0) = \lfloor -rc_j \rfloor \quad \text{for} \quad j \in NB(1)$$

Finally, to identify the coefficients of the new inequality, define the vector d' and the scalar d_0' by

$$\begin{aligned} d_j' &= c_j' & \text{for} \quad j \in B \\ d_j' &= c_j' - \Delta_j(1) & \text{for} \quad j \in NB(1) \\ d_j' &= c_j' + \Delta_j(1) & \text{for} \quad j \in NB(1) \\ d_0' &= c_0' + \sum_{j \in NB(1)} \Delta_j(1). \end{aligned}$$

---

[8] We continue to apply the convention of referring to just the x-component x'' of a solution (x'', y''), understanding the y component to be implicit.

We then express the inequality as

$$d' \geq d'_0. \tag{9}$$

We first show that (9) is valid when generated from an arbitrary y-optimal basic feasible solution, and then demonstrate in addition that it dominates (8) in the case where (8) is a valid inequality (i.e., where (8) is derived from an optimal basic feasible solution). By our previously stated convention, it is understood that X (and (MIP:0–1)) may be modified by incorporating previously generated inequalities that exclude some binary solutions originally admitted as feasible.

Our results concerning (9) are based on identifying properties of basic solutions in reference to the problem

$$LP(x', d') : \min_{x \in X} z_0 = d'x.$$

Proposition 1. *The inequality (9) derived from an arbitrary y-optimal basic feasible solution x'' for $LP(x', c')$ is satisfied by all binary vectors $x \in X$, and excludes the solution $x = x''$ when v''_0 is fractional.*

Proof. We first show that the basic solution x'' for $LP(x', c')$ is an optimal solution to $LP(x', d')$. Let rd_j denote the reduced cost for x_j when the objective function $z_0 = d'x$ for $LP(x', d')$ is priced out relative to the x'' basis, thus yielding $rd_j = 0$ for $j \in B$. From the definitions of the coefficients d'_j, and in particular from $d'_j = c'_j$ for $j \in B$, it follows that the relation between the reduced costs rd'_j and rc'_j for the non-basic variables is the same as that between the coefficients d'_j and c'_j; i.e.,

$$rd'_j = rc'_j - \Delta_j(0) \quad \text{for} \quad j \in NB(0)$$
$$rd'_j = rc'_j + \Delta_j(1) \quad \text{for} \quad j \in NB(1).$$

The definitions $\Delta_j(0) = \lfloor rc_j \rfloor$ for $j \in NB(0)$ and $\Delta_j(1) = \lfloor -rc_j \rfloor$ for $j \in NB(1)$ thus imply

$$rd_j \geq 0 \quad \text{for} \quad j \in NB(0)$$
$$rd_j \leq 0 \quad \text{for} \quad j \in NB(1).$$

This establishes the optimality of x'' for $LP(x', d')$. Since the d'_j coefficients are all integers, we therefore obtain the valid inequality

$$d'x \geq \lceil z''_0 \rceil.$$

The definition of d' yields

$$d'x'' = c'x'' + \sum_{j \in NB(1)} \Delta_j(1)$$

and hence

$$z''_0 = v''_0 + \sum_{j \in NB(1)} \Delta_j(1).$$

Since the $\Delta_j(1)$ values are integers, z_0'' is fractional if and only if v_0'' is fractional, and we also have

$$\lceil z_0'' \rceil = \lceil v_0'' \rceil + \sum_{j \in NB(1)} \Delta_j(1).$$

The proposition then follows from the definitions of c_0' and d_0'. $\square$

Proposition 1 has the following novel consequence.

Corollary 1. *The inequality* (9) *is independent of the* c_j' *values for the non-basic* x *variables. In particular, for any y-feasible basic solution and specified values c_j' for $j \in B$, the coefficients d_0' and d_j' of d' are identical for every choice of the integer coefficients c_j', $j \in NB$.*

Proof. The Corollary follows from the fact that any change in the value of c_j' for a non-basic variable x_j (which must be an integer change) produces an identical change in the value of the reduced cost rc_j and hence also in the values $\Delta_j(0)$ and $-\Delta_j(1)$. The argument of the Proof of Proposition 1 thus shows that these changes cancel out, to produce the same final d_0' and d' after implementing the changes that existed before the changes. $\square$

In effect, since Corollary 1 applies to the situation where $c_j' = 0$ for $j \in NB$, it also allows each d_j' coefficient for $j \in NB$ to be identified by reference to the quantity that results by multiplying the vector of optimal dual values by the corresponding column A_j of the matrix A defining the constraints of (MIP), excluding rows of A corresponding to the inequalities $1 \geq x_j \geq 0$. (We continue to assume this matrix is enlarged by reference to additional inequalities such as (8) or (9) that may currently be included in defining $x \in X$.)

Now we establish the result that (9) is at least as strong as (8).

Proposition 2. *If the basic solution x'' for LP(x', c') is optimal, and thus yields a valid inequality* (8), *then the inequality* (9) *dominates* (8).

Proof. We use the fact that x'' is optimal for LP(x', d') as established by Proposition 1. When x'' is also optimal for LP(x', c'), i.e., the x'' is dual feasible for the x variables as well as being y-optimal, the reduced costs rc_j satisfy $rc_j \geq 0$ for $j \in NB(0)$ and $rc_j \leq 0$ for $j \in NB(1)$. The definitions of $\Delta_j(0)$ and $\Delta_j(1)$ thereby imply that these two quantities are both non-negative. From the definitions of d_j' and d_0' we can write the inequality $d'x \geq d_0'$ as

$$\sum_{j \in B} c_j'x_j + \sum_{j \in NB(0)} (c_j' - \Delta_j(0))x_j + \sum_{j \in NB(1)} (c_j' + \Delta_j(1))x_j \geq c_0' + \sum_{j \in NB(1)} \Delta_j(1). \quad (10)$$

From $\Delta_j(0), \Delta_j(1) \geq 0$, and from $1 \geq x_j \geq 0$, we obtain the inequalities $\Delta_j(0)x_j \geq 0$ and $-\Delta_j(1)x_j \geq -\Delta_j(1)$. Hence

$$\sum_{j \in NB(0)} \Delta_j(0)x_j + \sum_{j \in NB(1)} -\Delta_j(1)x_j \geq \sum_{j \in NB(1)} -\Delta_j(1). \quad (11)$$

Adding the left and right sides of (11) to the corresponding sides of (10) and clearing terms gives

$$\sum_{j \in N} c'_j x_j \geq c'_0$$

Consequently, this establishes that (9) implies (8). $\square$

As in the use of the inequality (7), if a basic solution x'' that generates (9) is a binary vector that differs from x', then we can also test x'' for feasibility in (MIP:0–1) and then redefine $x' = x''$, to additionally append the constraint (3) for this new x'.

The combined arguments of the proofs of Propositions 1 and 2 lead to a still stronger conclusion. Consider a linear program $LP(x', h')$ given by

$$LP(x', h') : \min_{x \in X} h_0 = h',$$

where the coefficients $h'_j = d'_j$ (and hence $= c'_j$) for $j \in B$ and, as before, B is defined relative to a given y-optimal basic feasible solution x''. Subject to this condition, the only restriction on the h'_j coefficients for $j \in NB$ is that they be integers. Then we can state the following result.

Corollary 2. *The x'' basis is an optimal LP basis for $LP(x', h')$ if and only if*

$$h'_j \geq d'_j \quad for \quad j \in NB(0)$$
$$h'_j \leq d'_j \quad for \quad j \in NB(1)$$

and the inequality (9) *dominates the corresponding inequality derived by reference to* $LP(x', h')$.

Proof. Immediate from the proofs of Propositions 1 and 2. $\square$

The importance of Corollary 2 is the demonstration that (9) is the strongest possible valid inequality from those that can be generated by reference to a given y-optimal basic solution x'' and an objective function that shares the same coefficients for the basic variables.

It is to be noted that if (MIP:0–1) contains an integer valued slack variable s_i upon converting the associated inequality $A_i x + D_i y \geq b_i$ of the system $Ax + Dy \geq b$ into an equation – hence if A_i and b_i consist only of integers and D_i is the 0 vector – then s_i may be treated as one of the components of the vector x in deriving (9), and this inclusion serves to sharpen the resulting inequality. In the special case where all slack variables have this form, i.e., where (MIP:0–1) is a pure integer problem having no continuous variables and all data are integers, then it can be shown that the inclusion of the slack variables within x yields an instance of (9) that is equivalent to a fractional Gomory cut, and a stronger inequality can be derived by means of the foundation-penalty cuts of Glover and Sherali [12]. Consequently, the primary relevance of (9) comes from the fact that it applies to mixed integer as well as pure integer problems, and more particularly provides a useful means for enhancing target objective strategies for these problems. As an instance of this, we now examine methods that take advantage of (9) in additional ways by extension of ideas proposed with parametric tabu search.

5 Intensification and Diversification Based on Strategic Inequalities

5.1 An Intensification Procedure

Consider an indexed collection of inequalities of the form of (9) given by

$$d^p x \geq d_0^p, \quad p \in P. \tag{12}$$

We introduce an intensification procedure that makes use of (12) by basing the inequalities indexed by P on a collection of high quality binary target solutions x'. Such solutions can be obtained from past search history or from approaches for rounding an optimal solution to a linear programming relaxation (LP) of (MIP:0–1), using penalties to account for infeasibility in ranking the quality of such solutions. The solutions x' do not have to be feasible to be used as target solutions or to generate inequalities. In Section 6 we give specific approaches for creating such target solutions and the associated target objectives $c'x$ that serve as a foundation for producing the underlying inequalities.

Our goal from an intensification perspective is to find a new solution that is close to those in the collection of high quality solutions that give rise to (12). We introduce slack variables $s_p, p \in P$, to permit the system (12) to be expressed equivalently as

$$d^p x - s_p = d_0^p, \quad s_p \geq 0, \ p \in P. \tag{13}$$

Then, assuming the set X includes reference to the constraints (13), we create an *Intensified LP Relaxation*

$$\min_{x \in X} s_0 = \sum_{p \in P} w_p s_p$$

where the weights w_p for the variables s_p are selected to be positive integers.

An important variation is to seek a solution that minimizes the maximum deviation of x from solutions giving rise to (12). This can be accomplished by introducing the inequalities

$$s_0 \geq d_0^p - d^p x, \quad p \in P. \tag{14}$$

Assuming these inequalities are likewise incorporated into X^9, the Min(Max) goal is achieved by solving the problem

$$\min_{x \in X} s_0.$$

An optimal solution to either of these two indicated objectives can then be used as a starting point for an intensified solution pass, performing all-at-once or successive rounding to replace its fractional components by integers. [10]

---

[9] The inclusion of (13) and (14) is solely for the purpose of solving the associated linear programs, and these temporarily accessed constraints do not have to be incorporated among those defining Z.

[10] Successive rounding normally updates the LP solution after rounding each variable in order to determine the effects on other variables and thereby take advantage of modified rounding options.

5.2 A Diversification Analog

To create a diversification procedure for generating new starting solutions, we seek an objective function to drive the search to lie as far as possible from solutions in the region defined by (12). For this purpose we introduce the variables s_p as in (13), but utilize a maximization objective rather than a minimization objective to produce the problem

$$\max_{x \in X} s_0 = \sum_{p \in P} w_p s_p.$$

The weights w_p are once again chosen to be positive.

A principal alternative in this case consists of maximizing the minimum deviation of x from solutions giving rise to (12). For this, we additionally include the inequalities

$$s_0 \le d_0^p - d^p x, \quad p \in P \tag{15}$$

giving rise to the problem

$$\max_{x \in X} s_0.$$

The variable s_0 introduced in (15) differs from its counterpart in (14). In the case where the degree of diversification provided by this approach is excessive, by driving solutions too far away from solutions expected to be good, control can be exerted through bounding X with other constraints, and in particular by manipulating the bound U_0 identified in Sect. 1.

6 Generating Target Objectives and Solutions

We now examine the issue of creating the target solution x' and associated target objective $c'x$ that underlies the inequalities of the preceding sections. This is a key determinant of the effectiveness of targeting strategies, since it determines how quickly and effectively such a strategy can lead to new integer feasible solutions.

Our approach consists of two phases for generating the vector c' of the target objective. The first phase is relatively simple and the second phase is more advanced.

6.1 Phase 1 – Exploiting Proximity

The Phase 1 procedure for generating target solutions x' and associated target objectives $c'x$ begins by solving the initial problem (LP), and then solves a succession of problems LP(x', c') by progressively modifying x' and c'. Beginning from the linear programming solution x'' to (LP) (and subsequently to LP(x', c')), the new target solution x' is derived from x'' simply by setting $x'_j = \langle x''_j \rangle, j \in N$, where $\langle v \rangle$ denotes the nearest integer neighbor of v. (The value $\langle .5 \rangle$ can be either 0 or 1, by employing an arbitrary tie-breaking rule.)

Since the resulting vector x' of nearest integer neighbors is unlikely to be feasible for (MIP:0–1), the critical element is to generate the target objective $c'x$ so that the solutions x'' to successively generated problems $LP(x', c')$ will become progressively closer to satisfying integer feasibility. If one or more integer feasible solutions is obtained during this Phase 1 approach, each such solution qualifies as a new best solution x^*, due to the incorporation of the objective function constraint $x_0 = U_0 < x_0^*$.

The criterion of Phase 1 that selects the target solution x' as a nearest integer neighbor of x'' is evidently myopic. Consequently, the Phase 1 procedure is intended to be executed for only a limited number of iterations. However, the possibility exists that for some problems the target objectives of Phase 1 may quickly lead to new integer solutions without invoking more advanced rules. To accommodate this eventuality, we include the option of allowing Phase 1 to continue its execution as long as it finds progressively improved solutions.

Phase 1 is based on the principle that some variables x_j should be more strongly induced to receive their nearest neighbors target values x'_j than other variables. In the absence of other information, we may tentatively suppose that a variable whose LP solution value x''_j is already an integer or is close to being an integer is more likely to receive that integer value in a feasible integer solution. Consequently, we are motivated to choose a target objective $c'x$ that will more strongly encourage such a variable to receive its associated value x'_j. However, the relevance of being close to an integer value needs to be considered from more than one perspective.

The targeting of $x_j = x'_j$ for variables whose values x''_j already equal or almost equal x'_j does not exert a great deal of influence on the solution of the new $LP(x', c')$, in the sense that such a targeting does not drive this solution to differ substantially from the solution to the previous $LP(x', c')$. A more influential targeting occurs by emphasizing the variables x_j whose x''_j values are more "highly fractional", and hence which differ from their integer neighbors x'_j by a greater amount. There are evidently trade-offs to be considered in the pursuit of influence, since a variable whose x''_j value lies close to .5, and hence whose integer target may be more influential, has the deficiency that the likelihood of this integer target being the "right" target is less certain. A compromise targeting criterion is therefore to give greater emphasis to driving x_j to an integer value if x''_j lies "moderately" (but not exceedingly) close to an integer value. Such a criterion affords an improved chance that the targeted value will be appropriate, without abandoning the quest to identify targets that exert a useful degree of influence. Consequently, we select values λ_0 and $\lambda_1 = 1 - \lambda_0$ that lie moderately (but not exceedingly) close to 0 and 1, such as $\lambda_0 = 1/5$ and $\lambda_1 = 4/5$, or $\lambda_0 = 1/4$ and $\lambda_1 = 3/4$, and generate c'_j coefficients that give greater emphasis to driving variables to 0 and 1 whose x''_j values lie close to λ_0 and λ_1.

The following rule creates a target objective $c'x$ based on this compromise criterion, arbitrarily choosing a range of 1 to 21 for the coefficient c'_j. (From the standpoint of solving the problem $LP(x', c')$, this range is equivalent to any other range over positive values from v to $21v$, except for the necessity to round the c'_j coefficients to integers.)

Finally, replace the specified value of c'_j by its nearest integer neighbor $\langle c'_j \rangle$.

The absolute values of c'_j coefficients produced by the preceding rule describe what may be called a *batwing* function – a piecewise linear function resembling the

Algorithm 3 – Phase 1 Rule for Generating c_j'

Choose λ_0 from the range $.1 \le \lambda_0 \le .4$, and let $\lambda_1 = 1 - \lambda_0$.
if $x_j' = 0$ (hence $x_j'' \le .5$) **then**
  if $x_j'' \le \lambda_0$ **then**
    $c_j' = 1 + 20x_j''/\lambda_0$
  else if $x_j'' > \lambda_0$ **then**
    $c_j' = 1 + 20(.5 - x_j'')/(.5 - \lambda_0)$
  end if
else if $x_j' = 1$ (hence $x_j'' \ge .5$) **then**
  if $x_j'' \le \lambda_1$ **then**
    $c_j' = -(1 + 20(x_j'' - .5)/(\lambda_1 - .5))$
  else if $x_j'' > \lambda_1$ **then**
    $c_j' = -(1 + 20(1 - x_j'')/(1 - \lambda_1))$
  end if
end if

wings of a bat, with shoulders at $x_j'' = 0.5$, wing tips at $x_j'' = 0$ and $x_j'' = 1$, and the angular joints of the wings at $x_j'' = \lambda_0$ and $x_j'' = \lambda_1$. Over the x_j'' domain from the left wing tip at 0 to the first joint at λ_0, the function ranges from 1 to 21, and then from this joint to the left shoulder at 0.5 the function ranges from 21 back to 1. Similarly, from the right shoulder, also at 0.5, to the second joint at λ_1, the function ranges from 1 to 21, and then from this joint to the right wing tip at 1 the function ranges likewise from 21 to 1. (The coefficient c_j' takes the negative of these absolute values from the right shoulder to the right wing tip.).

In general, if we let Tip, $Joint$ and $Shoulder$ denote the $|c_j'|$ values to be assigned at these junctures (where typically $Joint > Tip, Shoulder$), then the generic form of a batwing function results by replacing the four successive c_j' values in the preceding method by

$$c_j' = Tip + (Joint - Tip)x_j''/\lambda_0,$$
$$c_j' = Shoulder + (Joint - Shoulder)(.5 - x_j'')/(.5 - \lambda_0),$$
$$c_j' = -(Shoulder + (Joint - Shoulder)(x_j'' - .5)/(\lambda_1 - .5)),$$
$$c_j' = -(Tip + (Joint - Tip)(1 - x_j'')/(1 - \lambda_1)).$$

The image of such a function more nearly resembles a bat in flight as the value of Tip is increased in relation to the value of $Shoulder$, and more nearly resembles a bat at rest in the opposite case. The function can be turned into a piecewise convex function that more strongly targets the values λ_0 and λ_1 by raising the absolute value of c_j' to a power $p > 1$ (affixing a negative sign to yield c_j' over the range from the right shoulder to the right wing tip). Such a function (e.g., a quadratic function) more strongly resembles a bat wing than the linear function.[11]

---

[11] Calibration to determine a batwing structure, either piecewise linear or nonlinear, that proves more effective than other alternatives within Phase 1 would provide an interesting study.

Design of the Phase 1 Procedure

We allow the Phase 1 procedure that incorporates the foregoing rule for generating c'_j the option of choosing a single fixed λ_0 value, or of choosing different values from the specified interval to generate a greater variety of outcomes. A subinterval for λ_0 centered around 0.2 or 0.25 is anticipated to lead to the best outcomes, but it can be useful to periodically choose values outside this range for diversification purposes.

We employ a stopping criterion for Phase 1 that limits the total number of iterations or the number of iterations since finding the last feasible integer solution. In each instance where a feasible integer solution is obtained, the method re-solves the problem (LP), which is updated to incorporate both the objective function constraint $x_0 \leq U_0 < x_0^*$ and inequalities such as (9) that are generated in the course of solving various problems LP(x', c'). The instruction "Update the Problem Inequalities" is included within Phase 1 to refer to this process of adding inequalities to LP(x', c') and (LP), and to the associated process of dropping inequalities by criteria indicated in Sect. 3.

Algorithm 4 – Phase 1

---

1. Solve (LP). (If the solution x'' to the first instance of (LP) is integer feasible, the method stops with an optimal solution for (MIP:0–1).)
2. Apply the Rule for Generating c'_j, to each $j \in N$, to produce a vector c'.
3. Solve LP(x', c'), yielding the solution x''. Update the Problem Inequalities.
4. If x'' is integer feasible: update the best solution $(x^*, y^*) = (x'', y'')$, update $U_0 < x_0^*$, and return to Step 1. Otherwise, return to Step 2.

A preferred variant of Phase 1 does not change all the components of c' each time a new target objective is produced, but changes only a subset consisting of k of these components, for a value k somewhat smaller than n. For example, a reasonable default value for k is given by $k = 5$. Alternatively, the procedure may begin with $k = n$ and gradually reduce k to its default value. Within Phase 2, as subsequently noted, it can be appropriate to reduce k all the way to 1.

This variant of Phase 1 results by the following modification. Let c^0 identify the form of c' produced by the Rule for Generating c'_j, as applied in Step 2 of the Phase 1 Procedure. Re-index the x_j variables so that $|c_1^0| \geq |c_2^0| \geq \ldots \geq |c_n^0|$, and let $N(k) = \{1, \ldots, k\}$, thus identifying the variables $x_j, j \in N(k)$, as those having the k largest $|c_j^0|$ values. Then Phase 1 is amended by setting $c' = 0$ in Step 1 and then setting $c'_j = c_j^0$ for $j \in N(k)$ in Step 2, without modifying the c'_j values for $j \in N - N(k)$. Relevant issues for research involve the determination of whether it is better to better to begin with k restricted or to gradually reduce it throughout the search, or to allow it to oscillate around a preferred value. Different classes of problems will undoubtedly afford different answers to such questions, and may be susceptible to exploitation by different forms of the batwing function (allowing different magnitudes for the *Tip*, *Joint* and *Shoulder*, and possibly allowing the location of the shoulders to be different than the 0.5 midpoint, with the locations of the joints likewise asymmetric).

6.2 Phase 2 – Exploiting Reaction and Resistance

Phase 2 is based on exploiting the mutually reinforcing notions of *reaction* and *resistance*. The term "reaction" refers to the change in the value of a variable as a result of creating a target objective $c'x$ and solving the resulting problem $LP(x', c')$. The term "resistance" refers to the degree to which a variable fails to react to a non-zero c'_j coefficient by receiving a fractional value rather than being driven to 0 or 1.

To develop the basic ideas, let NF identify the set of variables that receive fractional values in the solution x'' to the problem $LP(x', c)$, given by $NF = \{j \in N : 0 < x''_j < 1\}$, and let $N'(0, 1)$ identify the set of variables that have been assigned target values x'_j, given by $N'(0, 1) = N'(0) \cup N'(1)$ (or equivalently, $N'(0, 1) = N' - N'(\Phi)$). Corresponding to the partition of N' into the sets $N'(\Phi)$ and $N'(0, 1)$, the set NF of fractional variables is partitioned into the sets $NF(\Phi) = NF \cap N'(\Phi)$ and $NF(0, 1) = NF \cap N'(0, 1)$.

We identify two different sets of circumstances that are relevant to defining reaction, the first arising where none of the fractional variables x_j is assigned a target x'_j, hence $NF = NF'(\Phi)$, and the second arising in the complementary case where at least one fractional variable is assigned a target, hence $NF(0, 1) \neq \emptyset$. We start by examining the meaning of reaction in the somewhat simpler first case.

Reaction When No Fractional Variables Have Targets

Our initial goal is to create a measure of reaction for the situation where $NF = NF'(\Phi)$, i.e., where all of the fractional variables are unassigned (hence, none of these variables have targets). In this context we define reaction to be measured by the change in the value x''_j of a fractional variable x_j relative to the value x^0_j received by x_j in an optimal solution x^0 to (LP), as given by [12]

$$\Delta_j = x^0_j - x''_j.$$

We observe there is some ambiguity in this Δ_j definition since (LP) changes as a result of introducing new inequalities and updating the value U_0 of the inequality $x_0 \le U_0$. Consequently, we understand the definition of Δ_j to refer to the solution x^0 obtained by the most recent effort to solve (LP), though this (LP) may be to some extent out of date, since additional inequalities may have been introduced since it was solved. For reasons that will become clear in the context of resistance, we also allow the alternative of designating x^0 to be the solution to the most recent problem $LP(x', c')$ preceding the current one; i.e., the problem solved before creating the latest target vector c'.

The reaction measure Δ_j is used to determine the new target objective by reindexing the variables $x_j, j \in NF = NF'(\Phi)$, so that the absolute values $|\Delta_j|$ are in descending order, thus yielding $|\Delta_1| \ge |\Delta_2| \ge \ldots$. We then identify the k-element subset $N(k) = \{1, 2, \ldots, k\}$ of NF that references the k largest $|\Delta_j|$ values, where $k =$

---

[12] These Δ_j values are not to be confused with the $\Delta_j(0)$ and $\Delta_j(1)$ of Sect. 4.

$\min(|NF|, k_{\max})$. We suggest the parameter $k_{\max}$ be chosen at most 5 and gradually decreased to 1 as the method progresses.

The c'_j coefficients are then determined for the variables $x_j, j \in N(k)$, by the following rule. (The constant 20 is the same one used to generate c'_j values in the Phase 1 procedure, and $\langle v \rangle$ again denotes the nearest integer neighbor of v.)

$NF'(\Phi)$ *Rule for Generating* c'_j *and* $x'_j, j \in N(k)$ *(for* $N(k) \subset NF = NF'(\Phi)$):

$$\begin{aligned}
&\text{If} \quad \Delta_j \geq 0, \quad \text{set} \quad c'_j = 1 + \langle 20\Delta_j/|\Delta_1| \rangle \quad \text{and} \quad x'_j = 0 \\
&\text{If} \quad \Delta_j \leq 0, \quad \text{set} \quad c'_j = -1 + \langle 20\Delta_j/|\Delta_1| \rangle \quad \text{and} \quad x'_j = 1.
\end{aligned}$$

When $\Delta_j = 0$, a tie-breaking rule can be used to determine which of the two options should apply, and in the special case where $\Delta_1 = 0$ (hence all $\Delta_j = 0$), the c'_j assignment is taken to be 1 or -1 for all $j \in N(k)$.

To determine a measure of reaction for the complementary case $NF(0,1) \neq \emptyset$, we first introduce the notion of resistance.

Resistance

A *resisting variable* (or *resistor*) x_j is one that is assigned a target value x'_j but fails to satisfy $x_j = x'_j$ in the solution x'' to LP(x', c'). Accordingly the index set for resisting variables may be represented by $NR = \{j \in N'(0,1) : x''_j \neq x''_j\}$. If x''_j is fractional and $j \in N'(0,1)$ then clearly $j \in NR$ (i.e., $NF(0,1) \subset NR$). Consequently, the situation $NF(0,1) \neq \emptyset$ that was previously identified as complementary to $NF = NF(\Phi)$ corresponds to the presence of at least one fractional resistor.

If a resistor x_j is not fractional, i.e., if the value x''_j is the integer $1 - x'_j$, we say that x_j *blatantly resists* its targeted value x'_j. Blatant resistors x_j are automatically removed from NR and placed in the unassigned set $N'(\Phi)$, setting $c' = 0$. (Alternatively, a blatant resistor may be placed in $N'(1 - x'_j)$ by setting $c'_j = -c'_j$ and $x'_j = 1 - x'_j$.) After executing this operation, we are left with $NR = NF(0,1)$, and hence the condition $NF(0,1) \neq \emptyset$ (which complements the condition $NF = NF'(\Phi)$) becomes equivalent to $NR \neq \emptyset$.

Let V_j identify the amount by which the LP solution value $x_j = x''_j$ violates the target assignment $x_j = x'_j$; i.e, $V_j = x''_j$ if $x'_j = 0$ (hence if $c'_j > 0$) and $V_j = 1 - x''_j$ if $x'_j = 1$ (hence if $c'_j < 0$). We use the quantity V_j to define a *resistance measure* RM_j for each resisting variable $x_j, j \in NR$, that identifies how strongly x_j resists its targeted value x'_j. Two simple measures are given by $RM_j = V_j$, and $RM_j = |c_j|V_j$.

The resistance measure RM_j is used in two ways: (a) to select specific variables x_j that will receive new x'_j and c'_j values in creating the next target objective; (b) to determine the relative magnitudes of the resulting c'_j values. For this purpose, it is necessary to extend the notion of resistance by making reference to *potentially resisting* variables (or *potential resistors*) $x_j, j \in N'(0,1) - NR$, i.e., the variables that have been assigned target values x'_j and hence non-zero objective function coefficients c'_j, but which yield $x''_j = x'_j$ in the solution x'' to LP(x', c'). We identify a resistance measure RM_j^0 for potential resistors by reference to their reduced cost values rc_j (as identified in Sect. 4):

$$RM_j^0 = -rc_j \quad \text{for} \quad j \in N'(0) - NR \quad \text{and} \quad RM_j^0 = rc_j \quad \text{for} \quad j \in N'(1) - NR.$$

We note that this definition implies $RM_j^0 \le 0$ for potentially resisting variables. (Otherwise, x_j would be a non-basic variable yielding $x_j'' = 1$ in the case where $j \in N'(0)$, or yielding $x_j'' = 0$ in the case where $j \in N'(1)$, thus qualifying as a blatant resistor and hence implying $j \in NR$.) The closer that RM_j^0 is to 0, the closer x_j is to qualifying to enter the basis and potentially to escape the influence of the coefficient c_j' that seeks to drive it to the value 0 or 1. Thus larger values of RM_j^0 indicate greater potential resistance. Since the resistance measures RM_j are positive for resisting variables x_j, we see that there is an automatic ordering whereby $RM_p > RM_q^0$ for a resisting variable x_p and a potentially resisting variable x_q.

Combining Measures of Resistance and Reaction

The notion of reaction is relevant for variables x_j assigned target values x_j ($j \in N'(0,1)$) as well as for those not assigned such values ($j \in N'(\Phi)$). In the case of variables having explicit targets (hence that qualify either as resistors or potential resistors) we combine measures of resistance and reaction to determine which of these variables should receive new targets x_j' and new coefficients c_j'.

Let x^0 refer to the solution x'' to the instance of the problem $LP(x', c')$ that was solved immediately before the current instance;[13] hence the difference between x_j^0 and x_j'' identifies the reaction of x_j to the most recent assignment of c_j' values. In particular, we define this reaction for resistors and potential resistors by

$$\delta_j = x_j'' - x_j^0 \quad \text{for } x_j' = 0 \ (j \in N'(0))$$
$$\delta_j = x_j^0 - x_j'' \quad \text{for } x_j' = 1 \ (j \in N'(1)).$$

If we use the measure of resistance $RM_j = V_j$, which identifies how far x_j lies from its target value, a positive δ_j implies that the resistance of x_j has decreased as a result of this assignment. Just as the resistance measure RM_j is defined to be either V_j or $V_j|c_j'|$, the corresponding reaction measure $R\delta_j$ can be defined by either $R\delta_j = \delta_j$ or $R\delta_j = \delta_j|c_j'|$. Based on this we define a composite resistance–reaction measure RR_j for resisting variables as a convex combination of RM_j and $R\delta_j$; i.e., for a chosen value of $\lambda \in [0, 1]$:

$$RR_j = \lambda RM_j + (1 - \lambda)R\delta_j, \quad j \in NR.$$

Similarly, for implicitly resisting variables, we define a corresponding composite measure RR_j^0 by

$$RR_j^0 = \lambda RM_j^0 + (1 - \lambda)R\delta_j, \quad j \in N'(0,1) - NR.$$

---

[13] This is the "alternative definition" of x^0 indicated earlier.

In order to make the interpretation of λ more consistent, it is appropriate first to scale the values of RM_j, RM_j^0 and $R\delta_j$. If v_j takes the role of each of these three values in turn, then v_j may be replaced by the scaled value $v_j = v_j/|Mean(v_j)|$ (bypassing the scaling in the situation where $|Mean(v_j)| = 0$).

To give an effective rule for determining RR_j and RR_j^0, a few simple tests can be performed to determine a working value for λ, as by limiting λ to a small number of default values (e.g., the three values 0, 1 and 0.5, or the five values that include 0.25 and 0.75). More advanced methods for handling these issues are described in Sect. 8, where a linear programming post-optimization process for generating stronger evaluations is given in Sect. 8.4, and a target analysis approach for calibrating parameters and combining choice rules more effectively is given in Sect. 8.5.

Including Reference to a Tabu List

A key feature in using both RR_j and RR_j^0 to determine new target objectives is to make use of a simple tabu list T to avoid cycling and insure a useful degree of variation in the process. We specify in the next section a procedure for creating and updating T, which we treat both as an ordered list and as a set. (We sometimes speak of a variable x_j as belonging to T, with the evident interpretation that $j \in T$.) It suffices at present to stipulate that we always refer to non-tabu elements of $N'(0,1)$, and hence we restrict attention to values RR_j and RR_j^0 for which $j \in N'(0,1) - T$. The rules for generating new target objectives make use of these values in the following manner.

Because RR_j and RR_j^0 in general are not assured to be either positive or negative, we treat their ordering for the purpose of generating c_j' coefficients as a rank ordering. We want each RR_j value (for a resistor) to be assigned a higher rank than that assigned to any RR_j^0 value (for a potential resistor). An easy way to do this is to define a value RR_j for each potential resistor given by

$$RR_j = RR_j^0 - RR_1^0 + 1 - \min_{j \in NR} RR_j, \quad j \in N'(0,1) - NR.$$

The set of RR_j values over $j \in N'(0,1)$ then satisfies the desired ordering for both resistors ($j \in NR$) and potential resistors ($j \in N'(0,1) - NR$). (Recall that $NR = NF(0,1)$ by having previously disposed of blatant resistors.)

For the subset $N(k)$ of k non-tabu elements of $N'(0,1)$ (hence $N'(0,1) - T$) that we seek to generate, the ordering over the subset $NR - T$ thus comes ahead of the ordering over the subset $(N'(0,1) - NR) - T$. This allows both resistors and potential resistors to be included among those elements to be assigned new coefficients c_j' and new target values x_j', where the new c_j' coefficients for resistors always have larger absolute values than the c_j' coefficients for potential resistors. If the set of non-tabu resistors $NR - T$ already contains at least k elements, then no potential resistors will be assigned new c_j' or x_j' values.

Overview of Phase 2 Procedure

The rule for generating the target objective $c'x$ that lies at the heart of Phase 2 is based on carrying out the following preliminary steps, where the value k_{max} is determined as previously indicated: (a) re-index the variables x_j, $j \in N'(0,1) - T$, so that the values RR_j are in descending order, thus yielding $RR_1 \geq RR_2 \geq \ldots$; (b) identify the subset $N(k) = \{1, 2, \ldots, k\}$ of NR that references the k largest RR_j values, where $k = \min(|N'(0,1) - T|, k_{max})$; (c) create a rank ordering by letting $R_p, p = 1, \ldots, r$ denote the distinct values among the RR_j, $j \in N(k)$, where $R_1 > R_2 > \ldots R_r, (r \geq 1)$.

Then the rule to determine the c'_j and x'_j values for the variables x_j, $j \in N(k)$, is given as follows:

$N'(0,1) - T$ Rule for Generating c'_j and x'_j, $j \in N(k)$ (for $NR = NF(0,1) \neq \emptyset$):

If $x'_j = 1$, and $RR_j = R_p$, set $c'_j = \langle 1 + 20(r + 1 - p)/r \rangle$ and re-set $x'_j = 0$.
If $x'_j = 0$, and $RR_j = R_p$, set $c'_j = -\langle 1 + 20(r + 1 - p)/r \rangle$ and re-set $x'_j = 1$.

We see that this rule assigns c'_j coefficients so that the $|c'_j|$ values are the positive integers $\langle 1 + 20(1/r) \rangle, \langle 1 + 20(2/r) \rangle, \ldots, \langle 1 + 20(r/r) \rangle = 21$.

We are now ready to specify the Phase 2 procedure in overview, which incorporates its main elements except for the creation and updating of the tabu list T.

Algorithm 5 - Phase 2 Procedure in Overview

1. Solve (LP). (Stop if the first instance of (LP) yields an integer feasible solution x'' which therefore is optimal for (MIP:0–1).) (If the solution x'' to the first instance of (LP) is integer feasible, the method stops with an optimal solution for (MIP:0–1).)
2. There exists at least one fractional variable ($NF \neq \emptyset$). Remove blatant resistors if any exist, from NR and transfer them to $N'(\Phi)$ (or to $N'(1 - x'_j)$) so $NR = NF(0,1)$.
    (a) If $NF = NF(\Phi)$ (hence $NR = \emptyset$), apply the $NF(\emptyset)$ Rule for Generating c'_j and x'_j, $j \in N(k)$, to produce the new target objective $c'x$ and associated target vector x'.
    (b) If instead $NR \neq \emptyset$, then apply the $N'(0,1) - T$ Rule for Generating c'_j and x'_j, $j \in N(k)$, to produce the new target objective $c'x$ and associated target vector x'.
3. Solve LP(x', c'), yielding the solution x''. Update the Problem Inequalities.
4. If x'' is integer feasible: update the best solution $(x^*, y^*) = (x'', y'')$, update $U_0 < x_0^*$, and return to Step 1. Otherwise, return to Step 2.

6.3 Creating and Managing the Tabu List T – Phase 2 Completed

We propose an approach for creating the tabu list T that is relatively simple but offers useful features within the present context. As in a variety of constructions for handling a recency-based tabu memory, we update T by adding a new element j to the first position of the list when a variable x_j becomes tabu (as a result of assigning it a new target value x'_j and coefficient c'_j), and by dropping the "oldest" element that lies in the last position of T when its tabu status expires.

Our present construction employs a rule that may add and drop more than one element from T at the same time. The checking of tabu status is facilitated by using a vector $Tabu(j)$ that is updated by setting $Tabu(j) = true$ when j is added to T and by setting $Tabu(j) = false$ when j is dropped from T. (Tabu status is often monitored by using a vector $TabuEnd(j)$ that identifies the last iteration that element j qualifies as tabu, without bothering to explicitly store the list T, but the current method of creating and removing tabu status makes the indicated handling of T preferable.)

We first describe the method for the case where $k = 1$, i.e., only a single variable x_j is assigned a new target value (and thereby becomes tabu) on a given iteration. The modification for handling the case $k > 1$ is straightforward, as subsequently indicated. Two parameters $T_{\min}$ and $T_{\max}$ govern the generation of T, where $T_{\max} > T_{\min} \geq 1$. For simplicity we suggest the default values $T_{\min} = 2$ and $T_{\max} = n^6$. (In general, appropriate values are anticipated to result by selecting $T_{\min}$ from the interval between 1 and 3 and $T_{\max}$ from the interval between n^5 and n^7.)[14]

The target value x_j' and coefficient c_j' do not automatically change when j is dropped from T and x_j becomes non-tabu. Consequently, we employ one other parameter $AssignSpan$ that limits the duration that x_j may be assigned the same x_j' and c_j' values, after which x_j' is released from the restrictions induced by this assignment. To make use of $AssignSpan$, we keep track of when x_j most recently was added to T by setting $TabuAdd(j) = iter$, where $iter$ denotes the current iteration value (in this case, the iteration when the addition occurred). Then, when $TabuAdd(j) + AssignSpan < iter$, x_j is released from the influence of x_j' and c_j' by removing j from the set $N'(0, 1)$ and adding it to the unassigned set $N'(\Phi)$. As long as x_j is actively being assigned new x_j' and c_j' values, $TabuAdd(j)$ is repeatedly being assigned new values of $iter$, and hence the transfer of j to $N'(\Phi)$ is postponed. We suggest a default value for $AssignSpan$ between $1.5 \times T_{\max}$ and $3 \times T_{\max}$; e.g. $AssignSpan = 2 \times T_{\max}$.

To manage the updating of T itself, we maintain an array denoted $TabuRefresh$ (j) that is initialized by setting $TabuRefresh(j) = 0$ for all $j \in N$. Then on any iteration when j is added to T, $TabuRefresh(j)$ is checked to see if $TabuRefresh(j) < iter$ (which automatically holds the first time j is added to T). When the condition is satisfied, a *refreshing operation* is performed, after adding j to the front of T, that consists of two steps: (a) the list T is reduced in size to yield $|T| = T_{\min}$ (more precisely, $|T| \leq T_{\min}$) by dropping all but $T_{\min}$ the first elements of T; (b) $TabuRefresh(j)$ is updated by setting $TabuRefresh(j) = iter + v$, where v is a number randomly chosen from the interval $[AssignSpan, 2 \times AssignSpan]$. These operations assure that future steps of adding this particular element j to T will not again shrink T to contain $T_{\min}$ elements until $iter$ reaches a value that exceeds $TabuRefresh(j)$. Barring the occurrence of such a refreshing operation, T is allowed to grow without dropping any of its elements until it reaches a size of $T_{\max}$. Once $|T| = T_{\max}$, the oldest j is removed from the end of T each time a new element j is added to the front of T, and hence T is stabilized at the size $T_{\max}$ until a new refreshing operation occurs.

This approach for updating T is motivated by the following observation. The first time j is added to T (when $TabuRefresh(j) = 0$) T may acceptably be reduced in size to contain not just $T_{\min}$ elements, but in fact to contain only 1 element, and no

---

[14] The small value of $T_{\min}$ accords with an intensification focus, and larger values may be selected for diversification. A procedure that modifies $T_{\max}$ dynamically is indicated in Sect. 8.1.

matter what element is added on the next iteration the composition of $N'(0, 1)$ cannot duplicate any previous composition. Moreover, following such a step, the composition of $N'(0, 1)$ will likewise not be duplicated as long as T continues to grow without dropping any elements. Thus, by relying on intervening refreshing operations with $TabuRefresh(j) = 0$ and $T_{min} = 1$, we could conceivably allow T to grow even until reaching a size $T_{max} = n$. (Typically, a considerable number of iterations would pass before reaching such a state.) In general, however, by allowing T to reach a size $T_{max} = n$ the restrictiveness of preventing targets from being reassigned for T_{max} iterations would be too severe. Consequently we employ the two mechanisms to avoid such an overly restrictive state consisting of choosing $T_{max} < n$ and performing a refreshing operation that allows each j to shrink T more than once (whenever *iter* grows to exceed the updated value of $TabuRefresh(j)$) The combination of these two mechanisms provides a flexible tabu list that is self-calibrating in the sense of automatically adjusting its size in response to varying patterns of assigning target values to elements.

The addition of multiple elements to the front of T follows essentially the same design, subject to the restriction of adding only up to T_{min} new indexes $j \in N(k)$ to T on any iteration, should k be greater than T_{min}. We slightly extend the earlier suggestion $T_{min} = 2$ to propose $T_{min} = 3$ for $k_{max} \geq 3$.

One further comment is warranted concerning the composition of T. The organization of the method assures $T \subset N'(0, 1)$ and typically a good portion of $N'(0, 1)$ lies outside T. If exceptional circumstances result in $T = N'(0, 1)$, the method drops the last element of T so that $N'(0, 1)$ contains at least one non-tabu element.

Drawing on these observations, the detailed form of Phase 2 that includes instructions for managing the tabu list is specified below, employing the stopping criterion indicated earlier of limiting the computation to a specified maximum number of iterations. (These iterations differ from those counted by *iter*, which is re-set to 0 each time a new solution is found and the method returns to solve the updated (LP).)

The inequalities introduced in Sections 3 and 4 provide a useful component of this method, but the method is organized to operate even in the absence of such inequalities. The intensification and diversification strategies proposed in Section 5 can be incorporated for solving more difficult problems.

The next section gives another way to increase the power of the foregoing procedure when faced with solving harder problems, by providing a class of additional inequalities that are useful in the context of an intensification strategy.

7 Additional Inequalities for Intensification from an Elite Reference Set

We apply a somewhat different process than the type introduced in Section 4 to produce new inequalities for the purpose of intensification, based on a strategy of extracting (or "mining") useful inequalities from a reference set R of elite solutions. The goal in this case is to generate inequalities that reinforce the characteristics of solutions found within the reference set. The resulting inequalities can be exploited in conjunction with inequalities such as (9) and the systems (12)–(15). Such a com-

---

Algorithm 6 – Complete Phase 2 Procedure

---

0. Choose the values T_{min} and T_{max} and $AssignSpan$.
1. Solve (LP). (Stop if the first instance of (LP) yields an integer feasible solution x'', which therefore is optimal for (MIP:0–1).) Set $TabuRefresh(j) = 0$ for all $j \in N$ and $iter = 0$.
2. There exists at least one fractional variable ($NF \neq \varnothing$). Remove each blatant resistor x_j, if any exists, from NR and transfer it to $N'(\Phi)$ (or to $N'(1 - x_j')$), yielding $NR = NF(0, 1)$. If j is transferred to $N'(\Phi)$ and $j \in T$, drop j from T. Also, if $T = N'(0, 1)$, then drop the last element from T.
   (a) If $NF = NF(\Phi)$ (hence $NR = \varnothing$), apply the $NF(\Phi)$ Rule for Generating c_j' and x_j', $j \in N(k)$.
   (b) If instead $NR = \varnothing$, then apply the $N'(0, 1) - T$ Rule for Generating c_j' and x_j', $j \in N(k)$.
   (c) Set $iter = iter + 1$. Using the indexing that produces $N(k)$ in (a) or (b), add the elements $j = 1, 2, \ldots, \min(T_{min}, k)$ to the front of T (so that $T = (1, 2, \ldots)$ after the addition). If $TabuRefresh(j) < iter$ for any added element j, set $TabuRefresh(j) = iter + v$, for v randomly chosen between $AssignLength$ and $2 \times AssignSpan$ (for each such j) and then reduce T to at most T_{min} elements by dropping all elements in positions $> T_{min}$.
3. Solve $LP(x', c')$, yielding the solution x''. Update the Problem Inequalities.
4. If x'' is integer feasible: update the best solution $(x^*, y^*) = (x'', y'')$, update $U_0 < x_0^*$, and return to Step 1. Otherwise, return to Step 2.

---

bined approach gives an enhanced means for achieving the previous intensification and diversification goals.

The basis for this inequality mining procedure may be sketched as follows. Let $Count_j(v)$, for $v \in \{0, 1\}$, denote the number of solutions in R (or more precisely in an updated instance R' of R), such that $x_j = v$. We make use of sets $J(0)$ and $J(1)$ that record the indexes j for the variables x_j that most frequently receive the values 0 and 1, respectively, over the solutions in R. In particular, at each iteration either $J(0)$ or $J(1)$ receives a new index j^* for the variable x_{j^*} that receives either the value 0 or the value 1 in more solutions of R' than any other variable; i.e., x_{j^*} is the variable having the maximum $Count_j(v)$ value over solutions in R.

Associated with x_{j^*}, we let v^* (= 0 or 1) denote the value v that achieves this maximum $Count_j(v)$ value. The identity of j^* and v^* are recorded by adding j^* to $J(v^*)$. Then $J(v^*)$ is removed from future consideration by dropping it from the current N', whereupon R' is updated by removing all of its solutions x that contain the assignment $x_{j^*} = v^*$. The process repeats until no more solutions remain in R'. At this point we have a minimal, though not necessarily minimum, collection of variables such that every solution in R satisfies the inequality (16) indicated in Step 4 below.

7.1 Generating Multiple Inequalities

The foregoing Inequality Mining Method can be modified to generate multiple inequalities by the following simple design. Let $n(j)$ be the number of times the variable

---

Algorithm 7 – Inequality Mining Method (for the Elite Reference Set R)

---

0. Begin with $R' = R$, $N' = N$ and $J(0) = J(1) = \varnothing$.
1. Identify the variable x_{j*}, $j^* \in N'$, and the value $v^* = 0$ or 1 such that

$$Count_{j*}(v^*) = \max_{j \in N', v \in \{0,1\}} Count_j(v)$$

   Add j^* to the set $J(v^*)$.
2. Set $R' = R' - \{x \in R' : x_{j*} = v^*\}$ and $N' = N' - \{j^*\}$.
3. If $R' = \varnothing$ or $N' = \varnothing$ proceed to Step 4. Otherwise, determine the updated values of $Count_j(v)$, $j \in N'$, $v \in \{0,1\}$ (relative to the current R' and N') and return to Step 1.
4. Complete the process by generating the inequality

$$\sum_{j \in J(1)} x_j + \sum_{j \in J(0)} (1 - x_j) \geq 1 \tag{16}$$

---

x_j appears in one of the instances of (16). To initialize these values we set $n(j) = 0$ for all $j \in N$ in an initialization step that precedes Step 0. At the conclusion of Step 4, the $n(j)$ values are updated by setting $n(j) = n(j) + 1$ for each $j \in J(0) \cup J(1)$.

In the simplest version of the approach, we stipulate that each instance of (16) must contain at least one x_j such that $n(j) = 0$, thus automatically assuring every instance will be different. Let L denote a limit on the number of inequalities we seek to generate. Then, only two simple modifications of the preceding method are required to generate multiple inequalities.

(A) The method returns to Step 0 after each execution of Step 4., as long as $n(j) = 0$ for at least one $j \in N$, and as long as fewer than L inequalities have been generated;

(B) Each time Step 1 is visited immediately after Step 0 (to select the first variable x_{j*} for the new inequality), we additionally require $n(j^*) = 0$, and the method terminates once this condition cannot be met when choosing the first x_{j*} to compose a given instance of (16). Hence on each such "first execution" of Step 1, j^* is selected by the rule

$$Count_{j*}(v^*) = \max_{j \in N', n(j)=0, v \in \{0,1\}} Count_j(v).$$

As a special case, if there exists a variable x_j such that $x_j = 1$ (respectively, $x_j = 0$) in all solutions $x \in R$, we observe that the foregoing method will generate the inequality $x_j \geq 1$ (respectively, $x_j \leq 0$) for all such variables.

The foregoing approach can be given still greater flexibility by subdividing the value $n(j)$ into two parts, $n(j : 0)$ and $n(j : 1)$, to identify the number of times x_j appears in (16) for $j \in J(0)$ and for $j \in J(1)$, respectively. In this variant, the values $n(j : 0)$ and $n(j : 1)$ are initialized and updated in a manner exactly analogous to the initialization and updating of $n(j)$. The restriction of the choice of j^* on Step 1, immediately after executing Step 0, is simply to require $n(j^* : v^*) = 0$, by means of the rule

$$Count_{j*}(v^*) = \max_{j \in N', n(j:v)=0, v \in \{0,1\}} Count_j(v).$$

For additional control, the $n(j:0)$ and $n(j:1)$ values (or the $n(j)$ values) can also be constrained not to exceed some specified limit in subsequent iterations of Step 1, in order to assure that particular variables do not appear a disproportionate number of times in the inequalities generated.

7.2 Additional Ways for Exploiting R

It is entirely possible that elite solutions can lie in "clumps" in different regions. In such situations, a more effective form of intensification can result by subdividing an elite reference set R into different components by a clustering process, and then treating each of the individual components as a separate reference set.

One indication that R should be subdivided is the case where the Inequality Mining Method passes from Step 3 to Step 4 as a result of the condition $N' = \varnothing$. If this occurs when $R' \neq \varnothing$, the inequality (16) is valid only for the subset of R given by $R - R'$, which suggests that R is larger than it should be (or that too many inequalities have been generated). Clustering is also valuable in the context of using the inequalities of (12) for intensification, by dividing the target solutions x' underlying these inequalities into different clusters.

There is, however, a reverse consideration. If clustering (or some other construction) produces an R that is relatively small, there may be some risk that the inequalities (16) derived from R may be overly restrictive, creating a form of intensification that is too limiting (and hence that has a diminished ability to find other good solutions). To counter this risk, the inequalities of (16) can be expressed in the form of goal programming constraints, which are permitted to be violated upon incurring a penalty.

Finally, to achieve a greater degree of intensification, the Inequality Mining Method can employ more advanced types of memory to generate a larger number inequalities (or even use lexicographic enumeration to generate all inequalities of the indicated form). Additional variation can be achieved by introducing additional binary variables as products of other variables, e.g., representing a product such as $x_1 x_2 (1 - x_3)$ as an additional binary variable using standard rules (that add additional inequalities to those composing the system (16)). These and other advanced considerations are addressed in the approach of *satisfiability data mining* (Glover [8]).

8 Supplemental Strategic Considerations

This section identifies a number of supplemental strategic considerations to enhance the performance of the approaches described in preceding sections.

8.1 Dynamic Tabu Condition for $T_{\max}$

The value of $T_{\max}$ can be translated into a tabu tenure that varies within a specified range, and that can take a different value each time a variable x_j (i.e., its index j) is added to the tabu list T. This can be done by employing an array $TabuEnd(j)$ initialized at 0 and updated as follows. Whenever j is added to T, a value v is selected randomly from an interval $[T_a, T_b]$ roughly centered around T_{max}. (For example, if $T_{max} = n^6$ the interval might be chosen by setting $T_a = n^5$ and $T_b = n^7$). $TabuEnd(j)$ is assigned the new value $TabuEnd(j) = v + iter$. Subsequently, whenever j is examined to see if it belongs to T (signaled by $Tabu(j) = true$), if $iter > TabuEnd(j)$ then j is dropped from T.

In place of the rule that removes an element from T by selecting the element at the end of the list as the one to be dropped, we instead identify the element to be dropped as the one having the smallest $TabuEnd(j)$ value. When j is thus removed from T, we re-set $TabuEnd(j) = 0$.

The condition $TabuEnd(j) \geq iter$ could be treated as equivalent to $Tabu(j) = true$, and it would be possible to reference the $TabuEnd(j)$ array in place of maintaining the list T, except for the operation that drops all but the T_{min} first elements of T. Because of this operation, we continue to maintain T as an ordered list, and manage it as specified. (Whenever an element is dropped from T, it is dropped as if from a linked list, so that the relative ordering of elements remaining on T is not disturbed.) Alternatively, T can be discarded if $TabuEnd(j)$ is accompanied by an array $TabuStart(j)$, where $TabuStart(j) = iter$ at the iteration where j becomes tabu, thus making it possible to track the longevity of an element on T.

8.2 Using Model Embedded Memory to Aid in Generating New Target Objectives

We may modify the specification of the c'_j values in Phase 2 by using model embedded memory, as proposed in parametric tabu search. For this, we replace the value 20 in the c'_j generation rules of Sect. 6 by a value $BaseCost$ which is increased on each successive iteration, thus causing the new $|c'_j|$ values to grow as the number of iterations increases. The influence of these values in driving variables to reach their targets will thus become successively greater, and targets that have been created more recently will be less likely to be violated than those created earlier. (The larger the absolute value of c'_j the more likely it will be that x_j will not resist its target value x'_j by becoming fractional.)

Consequently, as the values $|c'_j|$ grow from one iteration to the next, the variables that were given new targets farther in the past will tend to be the ones that become resistors and candidates to receive new target values. As a result, the c'_j coefficients produced by progressively increasing $BaseCost$ emulate a tabu search recency memory that seeks more strongly to prevent assignments from changing the more recently that they have been made.

The determination of the c'_j values can be accomplished by the same rules specified in Sect. 6 upon replacing the constant value 20 by $BaseCost$. Starting with $BaseCost = 20$ in Step 1 of Phase 2, the value of $BaseCost$ is updated each time $iter$ is incremented by 1 in Step 3 to give $BaseCost = \lambda \times BaseCost$ where the parameter λ is chosen from the interval $\lambda \in [1.1, 1.3]$. (This value of λ can be made the same for all iterations, or can be selected randomly from such an interval at each iteration.)

To prevent the $|c'_j|$ values from becoming excessively large, the current $|c'_j|$ values can be reduced once $BaseCost$ reaches a specified limit by applying the following rule.

Reset $BaseCost = 20$ and index the variables $x_j, j \in N'(0,1)$ so that
$\quad |c'_1| \geq |c'_2| \geq \dots \geq |c'_p|$ where $p = |N'(0,1)|$.
Define $\Delta_j = |c'_j| - |c'_{j+1}|$ for $j = 1, \dots, p-1$.
Select $\lambda \in [1.1, 1.3]$.
Set $|c'_p| = BaseCost$ and $|c'_j| = \min(|c'_{j+1}| + \Delta_j, \lambda|c'_{j+1}|)$ for $j = p-1, \dots, 1$.
Let $sign(c'_j) = $"+" if $x'_j = 0$ and $sign(c'_j) = $"−" if $x'_j = 1, j \in N'(0,1)$.
Finally, reset $BaseCost = |c'_1|$ $(=\max_{j \in N'(0,1)} |c'_j|)$.

The new $|c'_j|$ values produced by this rule will retain the same ordering as the original ones and the signs of the c'_j coefficients will be preserved to be consistent with the target values x'_j.

In a departure for diversification purposes, the foregoing rule can be changed by modifying the next to last step to become

$$\text{Set} \quad |c'_1| = BaseCost \quad \text{and} \quad |c'_{j+1}| = \min(|c'_j| + \Delta_{j+1}, \lambda|c'_j|) \quad \text{for} \quad j = 1, \dots, p-1$$

and concluding by resetting $BaseCost = |c'_p|$.

8.3 Multiple Choice Problems

The Phase 2 procedure can be specialized to provide an improved method for handling (MIP:0–1) problems that contain multiple choice constraints which take the form

$$\sum_{j \in N_q} x_j = 1, \; q \in Q$$

where the sets $N_q, q \in Q$, are disjoint subsets of N.

Starting with all $j \in N_q$ unassigned (hence $N_q \subset N'(\Phi)$), the specialization is accomplished by only allowing a single $j \in N_q$ to be transferred from $N'(\Phi)$ to $N'(1)$. Once this transfer has occurred, let $j(q)$ denote the unique index $j \in N'(1) \cap N_q$ and let $x_{j(q)}$ denote a resisting variable, hence $j(q) \in NR$. After disposing of blatant resistors, we are assured that such a resisting variable satisfies $j(q) \in NF$ (i.e., $x_{j(q)}$ is fractional).

We seek a variable x_{j^*} to replace $x_{j(q)}$ by selecting

$$j^* = \arg \max_{j \in NR \cap (N_q - \{j(q)\})} RR_j - T$$

(note $j \in NR \cap (N_q - \{j(q)\})$ implies $j \in N'(0)$), or if no such index j^* exists, selecting

$$j^* = \arg \max_{j \in N'(\Phi) \cap N_q} RR_j - T.$$

Then j^* is transferred from its present set, $N'(0)$ or $N'(\Phi)$, to $N'(1)$, and correspondingly $j(q)$ is transferred from $N'(1)$ to either $N'(1)$ or $N'(\Phi)$. After the transfer, $j(q)$ is re-defined to be given by $j(q) = j^*$.

8.4 Generating the Targets x'_j and Coefficients c'_j Post-optimization

More advanced evaluations for generating new target assignments and objectives for the Phase 2 procedure can be created by using linear programming post-optimization. The approach operates as follows.

The procedures described in Sect. 6 are used to generate a candidate set $N(k)$ of some number k of "most promising options" for further consideration. The resulting variables $x_j, j \in N(k)$, are then subjected to a post-optimization process to evaluate them more thoroughly. We denote the value k for the present approach by $k^\#$, where $k^\#$ can differ from the k used in the component rules indicated in Sect. 6. ($k^\#$ may reasonably be chosen to lie between 3 and 8, though the maximum value of $k^\#$ can be adapted from iteration to iteration based on the amount of effort required to evaluate the current variables that may be associated with $N(k^\#)$.) By the nature of the process described below, the value of k in Phase 2 will be limited to satisfy $k \leq k^\#$.

As in the Phase 2 procedure, there are two cases, one where no resistors exist and $N(k^\#)$ is composed of fractional variables from the set $NF(\Phi)$, and the other where resistors exist and $N(k^\#)$ is composed of resistors and potential resistors from the set $N'(0,1)$. To handle both of these cases, let c_{max} denote the absolute value of the maximum c'_j coefficient normally assigned on the current iteration, i.e., $c_{max} = |c'_1|$ where c'_1 is identified by the indexing used to create the candidate set $N(k^\#)$. Also, let $v''_0 (= c'x'')$ denote the objective function value for the current solutions x'' of $LP(c', x')$, where we include reference to the associated value x''_0 by including x_0 as a secondary objective, as discussed in Sect. 1. (Implicitly, $v''_0 = c'x'' + \varepsilon x''_0$ for some small value ε. The reference to x''_0 is particularly relevant to the case where no resistors exist, since then $c'x'' = 0$.)

We then evaluate the assignment that consists of setting $x'_j = 0$ and $c'_j = c_{max}$ or setting $x'_j = 1$ and $c'_j = -c_{max}$ for each $j \in N(k^\#)$, to determine the effect of this assignment in changing the value of v''_0 upon solving the new $LP(c', x')$ (i.e., the form of $LP(c', x')$ that results for the indicated new value of x'_j and c'_j). To avoid undue computational expense, we limit the number of iterations devoted to the post-optimization performed by the primal simplex method to solve the new $LP(c', x')$. (The post-optimization effort is unlikely to be excessive in any event since the new objective changes only the single coefficient c'_j.)

Denote the new x'_j and c'_j values for the variable x_j currently being evaluated by $x^\#_j$ and $c^\#_j$ and denote the new v''_0 that results from the post-optimization process by $v^\#_0$. Then we employ $v^\#_0$ to evaluate the merit of the option of assigning $x'_j = x^\#_j$ and $c'_j = c^\#_j$.

Case 1. $N(k^\#) \subset NF(\Phi)$ (and there are no resistors).
Both of the options consisting of setting $x^\#_j = 0$ and $c^\#_j = c_{max}$ and of setting

$x_j^\# = 1$ and $c_j^\# = -c_{\max}$ exist for each $j \in N(k^*)$. Denote the quantity $v_0^\#$ for each of these two options respectively by $v_0^\#(j : 0)$ and $v_0^\#(j : 1)$. Then we prefer to make the assignment $x_j^\# = v^\#$ where

$$v^\# = \arg\min_{v \in \{0,1\}} v_0^\#(j : v),$$

and we select the index $j^\# \in N(k^*)$ for this assignment by

$$j^\# = \arg\max_{j \in N(k^*)} EV_j,$$

where EV_j is the evaluation given by $EV_j = |v_0^\#(j : 0) - v_0^\#(j : 1)|$. (Greater refinement results by stipulating that

$$\min_{v_0^\#(j:1)} v_0^\#(j : 0)$$

equals or exceeds a specified threshold value, such as the average of the min $(v_0^\#(j : 0), v_0^\#(j : 1))$ values over $j \in N(k^*)$.)

Case 2. $N(k^*) \subset N'(0, 1) - T$ (and resistors exist)

In this case only a single option exists, which consists of setting $x_j^\# = 1 - x_j'$ and setting $c_j^\# = c_{\max}$ or $-c_{\max}$ according to whether the resulting $x_j^\#$ is 0 or 1. The evaluation rule is therefore simpler than in Case 1: the preferred $j^\#$ for implementing the single indicated option is given simply by

$$j^\# = \arg\min_{j \in N(k^*)} v_0^\#.$$

In both Case 1 and Case 2, if more than one x_j is to be assigned a new target, the elements of $N(k^*)$ can be ordered by the indicated evaluation to yield a subset that constitutes the particular $N(k)$ used in Phase 2 (where possibly $k < k^*$).

The foregoing approach can be the foundation of an aspiration criterion for determining when a tabu element should be relieved of its tabu status. Specifically, for Case 2 the set $N(k^*)$ can be permitted to include some small number of tabu elements if they would qualify to belong to $N(k^*)$ if removed from T. (k^* might be correspondingly increased from its usual value to allow this eventuality.) Then, should the evaluation $v_0^\#(j : x_j^\#)$ be 0 for some such $j \in T$, indicating that all variables achieve their target values in the solution to the associated problem LP(c', x'), and if no $j \in N(k^*) - T$ achieves the same result, then the identified $j \in T$ may shed its tabu status and be designated as the preferred element $j^\#$ of Case 2.

8.5 Target Analysis

Target analysis is a strategy for creating a supervised learning environment to determine effective parameters and choice rules (Glover and Greenberg [10]; Glover and Laguna [11]). We refer to a target solution in the context of target analysis as an *ultimate* target solution to avoid confusion with the target solutions discussed in preceding sections of this chapter. Such an ultimate target solution, which we denote by x^t, is selected to be an optimal (or best known) solution to (MIP:0–1). The supervised learning process of target analysis is applied to identify decision rules for a particular method to enable it to efficiently obtain solutions x^t to a collection of problems from a given domain. The approach is permitted to expend greater effort than normally would be considered reasonable to identify the solutions x^t used to guide the learning process (unless by good fortune such solutions are provided by independent means).

Target analysis can be applied in the context of both the Phase 1 and Phase 2 procedures. We indicate the way this can be done for Phase 2, since the Phase 1 approach is simpler and can be handled by a simplified variant. The target analysis operates by examining the problems from the collection under consideration one at a time to generate information that will then be subjected to a classification method to determine effective decision rules. At any given iteration of the Phase 2 procedure, we have available a variety of types of information that may be used to compose a decision rule for determining whether particular variables x_j should be assigned a target value of $x'_j = 0$ or $x'_j = 1$. The goal is to identify a classification rule, applied to this available information, so that we can make correct decisions; i.e., so that we can choose x_j to receive its value in the solution x^t, given by $x'_j = x^t_j$.

Denote the information associated with a given variable x_j as a vector I_j ("I" for "information"). For example, in the present setting, I_j can consist of components such as $x''_j, x^0_j, V_j, |c'_j|, \Delta_j, RM_j, \lambda_j, RR_j, R\lambda_j$, and so forth, depending on whether x_j falls in the category identified in Step 2(a) or Step 2(b) of the Phase 2 procedure. The advanced information given by the values $v^\#_0(j : v)$ discussed in Sect. 8.4 is likewise relevant to include. For those items of information that can have alternative definitions, different components of I_j can be created for each alternative. This allows the classification method to base its rules on multiple definitions, and to identify those that are preferred. In the case of parameterized evaluators such as RR_j, different instances of the evaluator can be included for different parameter settings (values of λ), thus allowing preferred values of these parameters likewise to be identified. On the other hand, some types of classification procedures, such as separating hyperplane methods, automatically determine weights for different components of I_j and for such methods the identification of preferred parameter values occurs implicitly by reference to the weights of the basic components, without the need to generate multiple additional components of I_j.

From a general perspective, a classification procedure for Phase 2 may be viewed as a method that generates two regions $R(0)$ and $R(1)$, accompanied by a rule (or collection of rules) for assigning each vector of information I_j to exactly one of these regions. The goal is to compose $R(0)$ and $R(1)$ and to define their associated assignment rule in such a fashion that I_j will be assigned to $R(0)$ if the correct value for x_j is given by $x^t_j = 0$, and will be assigned to $R(1)$ if the correct value for x_j is given by $x^t_j = 1$. Recognizing that the classification procedure may not be perfect, and that the information available to the procedure may not be ideal, the goal more precisely is to make "correct assignments" for as many points as possible. We will not discuss here the relative merits of different types of classification methods, but simply keep in mind that the outcome of their application is to map points I_j into regions $R(0)$ and $R(1)$. Any reasonable procedure is likely to do a very much better job of creating such a mapping, and hence of identifying whether a given variable x_j should be assigned the value 0 or 1, than can be accomplished by a trial and error process to combine and calibrate a set of provisional decision rules. (An effort to simply "look at" a range of different data points I_j and figure out an overall rule for matching them with the decisions $x_j = 0$ and $x_j = 1$ can be a dauntingly difficult task.)

In one sense, the classification task is easier than in many classification settings. It is not necessary to identify a correct mapping of I_j into $R(0)$ or $R(1)$ for all variables x_j at any given iteration of Phase 2. If we can identify a mapping that is successful for any one of the variables x_j in the relevant category of Step 2(a) or Step 2(b) of the Phase 2 procedure, then the procedure will be able to quickly discover the solution x^t. Of course, the mapping must be able to detect the fact that assigning a particular I_j to $R(0)$ or $R(1)$ is more likely to be a correct assignment than one specified for another I_j. (For example, if a particular point I_j is mapped to lie "deeply within" a region $R(0)$ (or $R(1)$), then the classification of I_j as implying $x_j^t = 0$ (or $x_j^t = 1$) is presumably more likely to be correct. In the case of a separating hyperplane procedure, for instance, such a situation arises where a point lies far from the hyperplane, hence deeply within one of the two half-spaces defined by the hyperplane.)

The points I_j to be classified, and that are used to generate the regions $R(0)$ and $R(1)$ (via rules that map the points into these regions), are drawn from multiple iterations and from applications of Phase 2 on multiple problems. Consequently, the number of such data points can potentially be large and discretion may be required to limit them to a manageable number. One way to reduce the number of points considered is to restrict the points I_j evaluated to those associated with variables x_j for $j \in N(k)$. The smaller the value of k_{max}, the fewer the number of points generated at each iteration to become inputs for the classification procedure. Another significant way to reduce the number of points considered derives from the fact that we may appropriately create a different set of rules, and hence different regions $R(0)$ and $R(1)$, for different conditions. A prominent example concerns the conditions that differentiate Step 2(a) from Step 2(b).

Still more particularly, the condition $NF = NF(\Phi)$ of Step 2(a), which occurs when there are no fractional resistors, can receive its own special treatment. In this case we are concerned with determining target values for fractional variables that are not currently assigned such x_j' values. In an ideal situation, we would identify an optimal target value x_j' for some such variable at each step (considering the case for $k_{max} = 1$, to avoid the difficulty of simultaneously identifying optimal x_j' values for multiple variables simultaneously), and Phase 2 would then discover the solution x^t almost immediately. In addition, no resistors would ever arise, and the condition $NF = NF(\Phi)$ would be the only one relevant to consider at any step.

Consequently, to create a mapping and associated sets $R(0)$ and $R(1)$ for the condition $NF = NF(\Phi)$, it is appropriate to control each iteration of Phase 2 for the purpose of target analysis so that only "correct decisions" are implemented at each iteration. Then the data points I_j for $j \in NF(\Phi)$ are based on information that is compatible with reaching the ultimate target x^t at each step. (If an incorrect target x'_j were produced on some step, so that x_j is induced to receive the wrong value, then it could be that the "correct rule" for a new (different) variable x_h would not be to assign it the target $x'_h = x^t_h$, because the target $x'_h = 1 - x^t_h$ might lead to the best solution compatible with the previous assignment $x_j = x'_j$.)

Once such a controlled version of target analysis produces rules for classifying vectors I_j for $j \in N(\Phi)$, then these decision rules can be "locked into" the Phase 2 procedure, and the next step is to determine rules to handle the condition $NR \neq \emptyset$ of Step 2(b). Thus the target analysis will execute Phase 2 for its "best current version" of the rules for determining assignments $x_j = x'_j$ (which for Step 2(b) amounts to determining which variables should reverse their assignments to set $x'_j = 1 - x'_j$). This procedure can also be controlled to an extent to prevent the current collection of x'_j targets from diverging too widely from the x^t_j values. The resulting new rules generated by the classification method can then be embedded within a new version of Phase 2, and this new version can be implemented to repeat the target analysis and thereby uncover still more refined rules.

This process can also be used to identify aspiration criteria for tabu search. Specifically, an additional round of target analysis can be performed that focuses strictly on the tabu variables $x_j, j \in T$ on iterations where Step 2(b) applies. Then the classification procedure identifies a mapping of the vectors I_j for these tabu variables into the regions $R(0)$ and $R(1)$. A version of Phase 2 can then use this mapping to identify vectors I_j for $j \in T$ that lie deeply within $R(0)$ and $R(1)$, and to override the tabu status and drop j from T if I_j lies in $R(v)$ but $x'_j = 1 - v$. This rule can be applied with additional safety by keeping track of how often a tabu variable is evaluated as preferably being assigned a target value that differs from its current assignment. If such an evaluation occurs sufficiently often, then the decision to remove j from T can be reinforced.

8.6 Incorporating Frequency Memory

Tabu search methods typically incorporate frequency memory to improve their efficacy, where the form of such memory depends on whether it is intended to support intensification or diversification strategies.

Frequency memory already implicitly plays a role in the method for handling the tabu list T described in Sect. 6.3, since a variable that is frequently added to T is automatically prevented from initiating a refreshing operation, and hence T will continue to grow up to its limit of $T_{\max}$ elements until a variable that is less frequently added (or more specifically, that has not been added for a sufficient duration) becomes a member of T and launches an operation that causes T to shrink.

We consider two additional ways frequency memory can be employed within the Phase 2 procedure. The first supports a simple diversification approach by employing an array $Target(j : v)$ to record how many iterations x_j have been assigned the target value $v \in \{0, 1\}$ throughout previous search, or throughout search that has occurred since x^*_0 was last updated. (The type of frequency memory is called residence frequency memory.) The diversification process then penalizes the choice of an assignment $x_j = v$ for variables x_j and associated values v for which $Target(j:v)$ lies within a chosen distance from

$$\max_{j \in N, v \in \{0,1\}} Target(j:v),$$

motivated by the fact that this maximum identifies a variable x_j and value v such that the assignment $x_j = v$ has been in force over a larger span of previous iterations than any other target assignment.

A more advanced form of frequency memory that supports an intensification process derives from parametric tabu search. This approach creates an intensification score $InScore(x_j = x_j^\#)$ associated with assigning x_j the new target value $x_j^\#$, and takes into account the target assignments $x_h = x_h'$ currently active for other variables x_h for $h \in N'(0,1) - \{j\}$. The score is specifically given by

$$InScore(x_j = x_j^\#) = \sum_{h \in N'(0,1) - \{j\}} Freq(x_j = x_j^\#, x_h = x_h')$$

where $Freq(x_j = x_j^\#, x_h = x_h')$ denotes the number of times that the assignments $x_j = x_j^\#$ and $x_h = x_h'$ have occurred together in previously identified high quality solutions, and more particularly in the elite solutions stored in the reference set R as described in Sect. 7. Abstractly, $Freq(x_j = x_j^\#, x_h = x_h')$ constitutes a matrix with $4n^2$ entries (disregarding symmetry), one for each pair (j,h) and the 4 possible assignments of 0–1 values to the pair x_j and x_h. However, in practice this frequency value can be generated as needed from R, without having to account for all assignments (all combinations of x_j and $x_h = 0$ and 1) since only a small subset of the full matrix entries will be relevant to the set R. The portion of the full matrix relevant to identifying the values $Freq(x_j = x_j^\#, x_h = x_h')$ is also further limited by the fact that the only variables x_j considered on any given iteration are those for which $j \in N(k)$ where k is a relatively small number. (In the case treated in Step 2(b) of the Phase 2 Procedure, a further limitation occurs since only the single $x_j^\#$ value given by $x_j^\# = 1 - x_j'$ is relevant.)

By the intensification perspective that suggests the assignments that occur frequently over the elite solutions in R are also likely to occur in other high quality solutions, the value $InScore(x_j, x_j^\#)$ is used to select a variable x_j to assign a new target value $x_j^\#$ by favoring those variables that produce higher scores. (In particular, the assignments $x_j = x_j^\#$ for such variables occur more often in conjunction with the assignments $x_h = x_h', h \in N'(0,1) - \{j\}$ over the solutions stored in R.)

9 Conclusions

Branch-and-bound (B&B) and branch-and-cut (B&C) methods have long been considered the methods of choice for solving mixed integer programming problems. This orientation has resulted in attracting contributions to these classical methods from many researchers, and has led to successive improvements in these methods extending over a period of several decades. In recent years, these efforts to create improved B&B and B&C solution approaches have intensified and have produced significant benefits, as evidenced by the existence of MIP procedures that are appreciably more effective than their predecessors.

It remains true, however, that many MIP problems resist solution by the best current B&B and B&C methods. It is not uncommon to encounter problems that confound the leading commercial solvers, resulting in situations where these solvers are unable to find even moderately good feasible solutions after hours, days, or weeks of computational effort. As a consequence, metaheuristic methods have attracted attention as possible alternatives or supplements to the more classical approaches. Yet to date, the amount of effort devoted to developing good metaheuristics for MIP problems is almost negligible compared to the effort being devoted to developing refined versions of the classical methods.

The view adopted in this chapter is that metaheuristic approaches can benefit from a change of perspective in order to perform at their best in the MIP setting. Drawing on lessons learned from applying classical methods, we anticipate that metaheuristics can likewise profit from generating inequalities to supplement their basic functions. However, we propose that these inequalities be used in ways not employed in classical MIP methods, and indicate two principal avenues for doing this: the first by generating the inequalities with reference to strategically created target solutions and target objectives, as in Sects. 3 and 4, and the second by embedding these inequalities in special intensification and diversification processes, as in Sect. 5 and 7 (which also benefit by association with the targeting strategies).

The use of such strategies raises the issue of how to compose the target solutions and objectives themselves. Classical MIP methods such as B&B and B&C again provide a clue to be heeded, by demonstrating that memory is relevant to effective solution procedures. However, we suggest that gains can be made by going beyond the rigidly structured memory employed in B&B and B&C procedures. Thus we make use of the type of adaptive memory framework introduced in tabu search, which offers a range of recency and frequency memory structures for achieving goals associated with short term and long term solution strategies. Section 6 examines ways this framework can be exploited in generating target objectives, employing both older adaptive memory ideas and newer ones proposed here for the first time. Additional opportunities to enhance these procedures described in Sect. 8 provide a basis for future research.

Acknowledgement. I am grateful to Said Hanafi for a preliminary critique of this chapter and for useful observations about connections to other work. I am also indebted to César Rego for helpful suggestions that have led to several improvements.

References

1. Balas E Jeroslow R (1972) Canonical Cuts on the Unit Hypercube. SIAM Journal of Applied Mathematics, 23(1):60–69
2. Dantzig G (1963) Linear Programming and Extensions. Princeton University Press, Princeton, NJ.

3. Davoine T Hammer PL Vizviári B (2003). A Heuristic for Boolean Optimization Problems. Journal of Heuristics 9:229–247
4. Fischetti M Glover F Lodi A (2005) Feasibility Pump. Mathematical Programming – Series A 104:91–104
5. Glover F (1978) Parametric Branch and Bound. OMEGA, The International Journal of Management Science 6(2):145–152
6. Glover F (2005) Adaptive Memory Projection Methods for Integer Programming. In: Rego C Alidaee B (eds) Metaheuristic Optimization Via Memory and Evolution: Tabu Search and Scatter Search. Kluwer Academic Publishers.
7. Glover F (2006) Parametric Tabu Search for Mixed Integer Programs. Computers and Operations Research 33(9):2449–2494
8. Glover F (2006a) Satisfiability Data Mining for Binary Data Classification Problems. Research Report, University of Colorado, Boulder
9. Glover F (2007) Infeasible/Feasible Search Trajectories and Directional Rounding in Integer Programming. Journal of Heuristics, Kluwer Publishing (to appear)
10. Glover F Greenberg H (1989) New Approaches for Heuristic Search: A Bilateral Linkage with Artificial Intelligence. European Journal of Operational Research 39(2):119–130
11. Glover F Laguna M (1997) Tabu Search. Kluwer Academic Publishers
12. Glover F Sherali HD (2003) Foundation-Penalty Cuts for Mixed-Integer Programs. Operations Research Letters 31:245–253
13. Guignard M Spielberg K (2003) Double Contraction, Double Probing, Short Starts and BB-Probing Cuts for Mixed (0,1) Programming. Wharton School Report
14. Hanafi S Wilbaut C (2006) Improved Convergent Heuristic for 0–1 Mixed Integer Programming. Research Report, University of Valenciennes
15. Hvattum LM Løkketangen A Glover F (2004) Adaptive Memory Search for Boolean Optimization Problems. Discrete Applied Mathematics 142:99–109
16. Nediak M Eckstein J (2007) Pivot, Cut, and Dive: A Heuristic for Mixed 0–1 Integer Programming. Journal of Heuristics, Kluwer Publishing (to appear)
17. Pardalos PS Shylo OV (2006) An Algorithm for Job Shop Scheduling based on Global Equilibrium Search Techniques. Computational Management Science (Published online), DOI: 10.1007/s10287-006-0023-y
18. Patel J Chinneck JW (2007) Active-Constraint Variable Ordering for Faster Feasibility of Mixed Integer Linear Programs. Mathematical Programming Series A, 110:445–474
19. Pedroso JP (2005) Tabu Search for Mixed Integer Programming. In: Rego C Alidaee B (eds) Metaheuristic Optimization via Memory and Evolution: Tabu Search and Scatter Search. Kluwer Academic Publishers
20. Shylo OV (1999) A Global Equilibrium Search Method. (Russian) Kybernetika I Systemniy Analys 1:74–80
21. Soyster AL Lev B Slivka W (1978) Zero–One Programming with Many Variables and Few Constraints. European Journal of Operational Research 2(3):195–201.
22. Spielberg K Guignard M (2000) A Sequential (Quasi) Hot Start Method for BB (0,1) Mixed Integer Programming. Mathematical Programming Symposium, Atlanta
23. Ursulenko A (2006) Notes on the Global Equilibrium Search. Working paper, Texas A&M University
24. Wilbaut C Hanafi S (2006) New Convergent Heuristics for 0–1 Mixed Integer Programming. Research Report, University of Valenciennes

Index

Natural Computing Series

Printing: Krips bv, Meppel, The Netherlands
Binding: Stürtz, Würzburg, Germany